Lecture Notes in Computer Science 6502

Commenced Publication in 1973
Founding and Former Series Editors:
Gerhard Goos, Juris Hartmanis, and Jan van Leeuwen

Ulrik Brandes   Sabine Cornelsen (Eds.)

Graph Drawing

18th International Symposium, GD 2010
Konstanz, Germany, September 21-24, 2010
Revised Selected Papers

 Springer

Volume Editors

Ulrik Brandes
Department of Computer and Information Science
University of Konstanz, Germany
E-mail: ulrik.brandes@uni-konstanz.de

Sabine Cornelsen
Department of Computer and Information Science
University of Konstanz, Germany
E-mail: sabine.cornelsen@uni-konstanz.de

ISSN 0302-9743   e-ISSN 1611-3349
ISBN 978-3-642-18468-0   e-ISBN 978-3-642-18469-7
DOI 10.1007/978-3-642-18469-7
Springer Heidelberg Dordrecht London New York

Library of Congress Control Number: 2010942929

CR Subject Classification (1998): E.1, G.2.2, G.1.6, I.5.3, H.2.8

LNCS Sublibrary: SL 1 – Theoretical Computer Science and General Issues

Typesetting: Camera-ready by author, data conversion by Scientific Publishing Services, Chennai, India

Printed on acid-free paper

Springer is part of Springer Science+Business Media (www.springer.com)

Preface

The 18th International Symposium on Graph Drawing (GD 2010) was held in Konstanz, Germany, September 21–24, 2010, and was attended by 108 participants from 20 countries.

In response to the call for papers, the Program Commitee received 77 submissions. Each submission was reviewed by at least three Program Committee members. Following substantial discussions, the committee accepted 30 long papers, 5 short papers, as well as 8 posters.

GD 2010 invited two keynote speakers. Carsten Thomassen from the Technical University of Denmark gave a talk on graph decomposition, and Peter Eades from the University of Sydney outlined the future of graph drawing. Both talks were recorded and can be accessed via the conference website, `http://www.graphdrawing.org/gd2010/`.

Keeping with tradition, the symposium was accompanied by the 17th Annual Graph Drawing Contest, including an onsite Graph Drawing Challenge for conference attendees. A detailed report about the event is contained in this volume.

The conference was preceded by *pro*GD, a free training event with tutorials on data mining, biological networks, and social networks. *pro*GD was organized in cooperation with the Research Training Group *Explorative Analysis and Visualization of Large Information Spaces* and the *Konstanz Research School on Chemical Biology.*

For the first time, child care services were offered during the conference, and we do hope that the four little ones whose parents requested it enjoyed GD 2010 as much as we did.

Many people contributed to the success of the conference. First of all, the authors of the submitted contributions deserve special thanks, as well as the members of the Program Committee and all external referees for their careful work and extensive discussions. In addition to the members of the Organizing Committee, we would like to thank especially Christine Agorastos, Gabriela Kruse-Niermann, and our excellent crew of student helpers who took care of all things big and small that have to be considered during such a meeting.

The conference received considerable support from the hosting organization, University of Konstanz, and Deutsche Forschungsgemeinschaft (DFG). Furthermore, we are grateful to the various sponsors listed herein.

The 19th International Symposium on Graph Drawing (GD 2011) will be held September 21–23, 2011, in Eindhoven, The Netherlands, and hosted by Bettina Speckmann and Marc van Kreveld.

October 2010

Ulrik Brandes
Sabine Cornelsen

Organization

Steering Committee

Franz J. Brandenburg	University of Passau, Germany
Ulrik Brandes	University of Konstanz, Germany
Giuseppe Di Battista	University of Rome III, Italy
Peter Eades	University of Sydney, Australia
David Eppstein	University of California, Irvine, USA
Hubert de Fraysseix	EHESS Paris, France
Emden R. Gansner	AT&T Research Labs, USA
Marc van Kreveld	Utrecht University, The Netherlands
Giuseppe Liotta	University of Perugia, Italy
Takao Nishizeki	Tohoku University, Japan
Pierre Rosenstiehl	EHESS Paris, France
Bettina Speckmann	TU Eindhoven, The Netherlands
Roberto Tamassia (Chair)	Brown University, USA
Ioannis G. Tollis	ICS-FORTH and University of Crete, Greece

Organizing Committee

Melanie Badent	University of Konstanz, Germany
Ulrik Brandes (Co-chair)	University of Konstanz, Germany
Sabine Cornelsen (Co-chair)	University of Konstanz, Germany
Martin Mader	University of Konstanz, Germany
Barbara Pampel	University of Konstanz, Germany

Contest Committee

Christian A. Duncan	Louisiana Tech, USA
Carsten Gutwenger	TU Dortmund, Germany
Lev Nachmanson	Microsoft Research, USA
Georg Sander (Chair)	IBM, Germany

Program Committee

David Auber	University of Bordeaux I, France
Christian Bachmaier	University of Passau, Germany
Ulrik Brandes (Chair)	University of Konstanz, Germany
Sabine Cornelsen	University of Konstanz, Germany
Giuseppe Di Battista	University of Rome III, Italy
Emilio Di Giacomo	University of Perugia, Italy
David Eppstein	University of California, Irvine, USA
Emden R. Gansner	AT&T Research Labs, USA
Michael T. Goodrich	University of California, Irvine, USA
Patrick Healy	University of Limerick, Ireland
Seok-Hee Hong	University of Sydney, Australia
Michael Kaufmann	University of Tübingen, Germany
Stephen G. Kobourov	University of Arizona, USA
Jan Kratochvíl	University of Prague, Czech Republic
Giuseppe Liotta	University of Perugia, Italy
Henk Meijer	Roosevelt Academy, The Netherlands
Petra Mutzel	TU Dortmund, Germany
Patrice Ossona de Mendez	EHESS Paris, France
Maurizio Patrignani	University of Rome III, Italy
Marcus Schaefer	DePaul University, Chicago, USA
Bettina Speckmann	TU Eindhoven, The Netherlands
Antonios Symvonis	NTU Athens, Greece
Stephen Wismath	University of Lethbridge, Canada
Xiao Zhou	Tohoku University, Japan

External Reviewers

Patrizio Angelini	Walter Didimo	Kazuyuki Miura
Christopher Auer	Stefan Felsner	Pietro Palladino
Melanie Badent	Fabrizio Frati	Barbara Pampel
Michael Bekos	Andreas Gleißner	Christian Pich
Carla Binucci	Luca Grilli	Bruno Pinaud
Romain Bourqui	Karsten Klein	Salvatore Romeo
Wolfgang Brunner	Nils Kriege	Hoi-Ming Wong
Markus Chimani	Martin Mader	

Sponsors

Universität
Konstanz
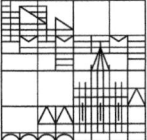

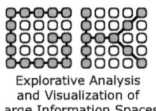
Explorative Analysis
and Visualization of
Large Information Spaces

ChemicalBiology
University of Konstanz

Gold Sponsors

Silver Sponsors

Contributors

SIEMENS

Table of Contents

Papers

Posters

Graph Drawing Contest

On the Size of Graphs That Admit Polyline Drawings with Few Bends and Crossing Angles

Eyal Ackerman[1], Radoslav Fulek[2], and Csaba D. Tóth[3,*]

[1] University of Haifa at Oranim
ackerman@sci.haifa.ac.il
[2] Ecole Polytechnique Fédérale de Lausanne
radoslav.fulek@epfl.ch
[3] University of Calgary
cdtoth@math.ucalgary.ca

Abstract. We consider graphs that admit polyline drawings where all crossings occur at the same angle $\alpha \in (0, \frac{\pi}{2}]$. We prove that every graph on n vertices that admits such a polyline drawing with at most two bends per edge has $O(n)$ edges. This result remains true when each crossing occurs at an angle from a small set of angles. We also provide several extensions that might be of independent interest.

1 Introduction

Graphs that admit polyline drawings with few bends per edge and such that every crossing occurs at a large angle have received some attention lately, since cognitive experiments [7,8] indicate that such drawings are almost as readable as planar drawings. That is, even though they may contain crossings, one can still easily track the edges of such drawings.

A *topological graph* is a graph drawn in the plane with vertices represented by distinct points and edges as arcs connecting its vertices, but not passing through any other vertex. A *polyline drawing* of a graph G is a topological graph where each edge is drawn as a simple polygonal arc between the incident vertices, but not passing through any bend point of other arcs. In a polyline drawing, every crossing occurs in the relative interior of two segments of the two polygonal arcs, and so they have a well-defined crossing angle in $(0, \frac{\pi}{2}]$.

Didimo *et al.* [5] introduced *right angle crossing (RAC)* drawings, which are polyline drawing where all crossings occur at a right angle. They prove that a graph with $n \geq 3$ vertices that admits a straight line RAC drawing has at most $4n - 10$ edges, and this bound is best possible. A simpler proof of this bound was later found by Dujmović et al. [6]. It is not hard to show that any graph admits an RAC drawing with three bends per edge (see Figure 1 for an example). Arikushi *et al.* [3] have recently proved, improving previous results by Didimo *et al.* [5], that if a graph with n vertices admits an RAC drawing with at most two bends per edge, then it has $O(n)$ edges.

* Partially supported by NSERC grant RGPIN 35586.

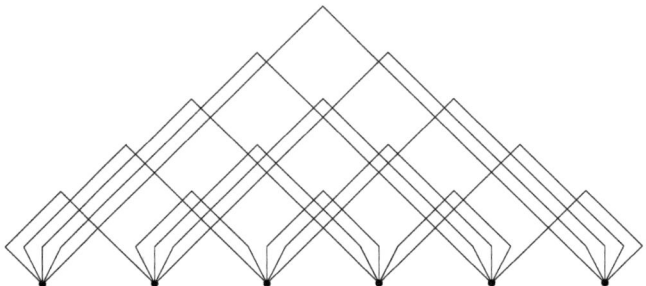

Fig. 1. An RAC drawing of K_6 with 3 bends per edge

Dujmović *et al.* [6] generalized RAC drawings, allowing crossings at a range of angles rather than at right angle. They considered αAC drawings, which are polyline drawings where every crossing occurs at angle *at least* α. They showed that a graph with n vertices and a straight line αAC drawing has at most $\frac{\pi}{\alpha}(3n - 6)$ edges, by partitioning the graph into $\frac{\pi}{\alpha}$ planar graphs. They also proved that their bounds are essentially optimal for $\alpha = \frac{\pi}{k} - \varepsilon$, with $k = 2, 3, 4, 6$ and sufficiently small $\varepsilon > 0$.

Results for polyline drawings. We consider polyline drawings where every crossing occurs at the same angle $\alpha \in (0, \frac{\pi}{2}]$. An $\alpha AC_b^=$ drawing of a graph is a polyline drawing where every edge is a polygonal arc with at most b bends and every crossing occurs at angle *exactly* α. It is easy to see that every graph with n vertices that admits an $\alpha AC_0^=$ drawing has at most $3(3n - 6)$ edges (see Lemma 1 below). Every graph admits an $\alpha AC_3^=$ drawing for every $\alpha \in (0, \frac{\pi}{2}]$: Didimo *et al.* [5] constructed an RAC drawing of the complete graph with three bends per edge (see also Figure 1), where every crossing occurs between a pair of orthogonal segments of the same orientation, so an affine transformation deforms all crossing angles uniformly. It remained to consider graphs that admit $\alpha AC_1^=$ or $\alpha AC_2^=$ drawings. Our main result is:

Theorem 1. *For every* $\alpha \in (0, \frac{\pi}{2}]$, *a graph on n vertices that admits an $\alpha AC_2^=$ drawing has $O(n)$ edges.*

For $\alpha = \frac{\pi}{2}$, this has recently been proved by Arikushi *et al.* [3]. Their proof techniques, however, do not generalize to all $\alpha \in (0, \frac{\pi}{2}]$. Our techniques use some ideas from [3], but are in fact somewhat simpler. The constant hidden in the big-O notation in Theorem 1 is quite moderate. For graphs admitting $\alpha AC_1^=$ drawings (with at most one bend per edge), we give a better upper bound than Theorem 1, using a simpler proof.

Theorem 2. *For any angle* $\alpha \in (0, \frac{\pi}{2}]$, *a graph on n vertices that admits an $\alpha AC_1^=$ drawing has at most $27n$ edges.*

A straightforward generalization of $\alpha AC_2^=$ drawings are polyline drawings where each crossing occurs at an angle from a list of k distinct angles.

Theorem 3. *Let $A \subset (0, \frac{\pi}{2}]$ be a set of k angles. If a graph G on n vertices admits a drawing with at most two bends per edge such that every crossing occurs at some angle from A, then G has $O(k^3 n)$ edges.*

Generalizations to topological graphs. Suppose that every edge in a topological graph is partitioned into edge-segments, such that all crossings occur in the relative interior of the segments. The bends in polyline drawings, for example, naturally define such edge partitions. An *end segment* is an edge-segment incident to a vertex of the edge, while a *middle* segment is an edge-segment not incident to any vertex. The key idea in proving Theorems 1 and 2 is to consider the crossings that involve either two end segments, or an end segment and a middle segment. This idea extends to topological graphs whose edge segments satisfy a few properties (which automatically hold for polyline drawings with same angle crossings). We obtain the following results, which might be of independent interest.

Theorem 4. *Let $G = (V, E)$ be a topological graph on n vertices, in which every edge can be partitioned into two end segments, one colored red and the other colored blue, such that:*

(1) *no two end segments of the same color cross;*
(2) *every pair of end segments intersects at most once; and*
(3) *no end segment is crossed by more than k end segments that share a vertex.*

Then G has $O(kn)$ edges.

Theorem 5. *Let $G = (V, E)$ be a topological graph on n vertices. Suppose that every edge of G can be partitioned into two end segments and one middle segment such that:*

(1) *each crossing involves one end segment and one middle segment;*
(2) *each middle segment and end segment intersect at most once; and*
(3) *each middle segment crosses at most k end segments that share a vertex.*

Then G has $O(k^2 n)$ edges.

Organization. We begin with a few preliminary observations. Our main results appear in Section 3 where we prove Theorems 1, 2, 3, and 5. Theorem 4 is proved in Section 4. It is actually not needed for the proof of Theorem 2, however, it also implies that graphs with $\alpha AC_1^=$ drawings have a linear number of edges.

2   Preliminaries

In a polyline drawing of a graph, the edges are simple polygonal paths, consisting of line segments. We start with a few initial observations about line segments and polygonal paths. We say that two line segments *cross* if their relative interiors intersect in a single point. (Note that intersecting segments that share an endpoint or are collinear do not cross.)

Lemma 1. *Let $\alpha \in (0, \frac{\pi}{2}]$, and let S be a finite set of line segments in the plane such that any two segments may cross only at angle α. Then S can be partitioned into at most three subsets of pairwise noncrossing segments. Moreover, if $\frac{\pi}{\alpha}$ is irrational or if $\frac{\pi}{\alpha} = \frac{p}{q}$, where $\frac{p}{q}$ is irreducible and q is even, then S can be partitioned into at most two subsets of pairwise noncrossing segments.*

Proof. Partition S into maximal subsets of pairwise parallel line segments. Let $\mathcal{S}$ denote the subsets of S. We define a graph $G_{\mathcal{S}} = (\mathcal{S}, E_{\mathcal{S}})$, in which two subsets $S_1, S_2 \in \mathcal{S}$ are joined by an edge if and only if their respective directions differ by angle α. Clearly, the maximum degree of a vertex in $G_{\mathcal{S}}$ is two. Hence, $G_{\mathcal{S}}$ is 3-colorable. In any proper 3-coloring of $G_{\mathcal{S}}$, the union of each color class is a set of pairwise noncrossing segments in S, since they do not meet at angle α.

If $\frac{\pi}{\alpha}$ is irrational, then $G_{\mathcal{S}}$ is cycle-free. If $\frac{\pi}{\alpha} = \frac{p}{q}$, where $\frac{p}{q}$ is irreducible and q is even, then $G_{\mathcal{S}}$ can only have even cycles. In both cases, $G_{\mathcal{S}}$ is 2-colorable, and S has a partition into two subsets of pairwise noncrossing segments.

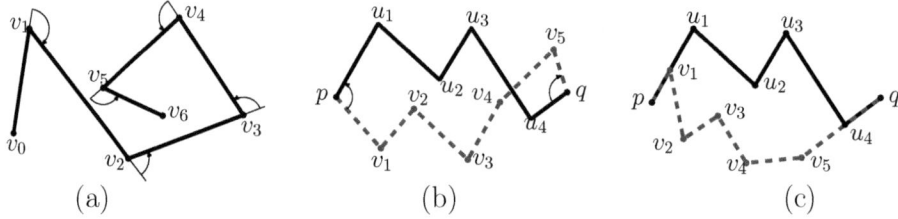

Fig. 2. (a) The turning angles of a polygonal path. (b) Two crossing polygonal paths with the same turning angle between p and q. (c) Two noncrossing polygonal paths with the same turning angle between p and q.

Consider a simple open polygonal path $\gamma = (v_0, v_1, \ldots, v_n)$ in the plane. At every interior vertex v_i, $1 \leq i \leq n-1$, the *turning angle* $\angle(\gamma, v_i)$ is the directed angle in $(-\pi, \pi)$ (the counterclockwise direction is positive) from ray $\overrightarrow{v_{i-1}v_i}$ to $\overrightarrow{v_i v_{i+1}}$; see Figure 2(a). The *turning angle* of the polygonal path γ is the sum of turning angles $\sum_{i=1}^{n-1} \angle(\gamma, v_i)$. We say that two line segments *overlap* if one of them is contained in the other.

Lemma 2. *Let p and q be two points in the plane. Let γ_1 and γ_2 be two directed simple polygonal paths from p to q. If γ_1 and γ_2 have the same turning angle and they do not cross, then the first segment of γ_1 overlaps with the first segment of γ_2 and the last segment of γ_2 overlaps with the last segment of γ_2.*

Proof. Let $\gamma_1 = (u_0, u_1, \ldots, u_m)$ and $\gamma_2 = (v_0, v_1, \ldots, v_n)$, with $p = u_0 = v_0$ and $q = u_m = v_n$. Let β be their common turning angle. Since γ_1 and γ_2 do not cross, they enclose a weakly simple polygon P with $m + n$ vertices (Figure 2(c)). Suppose w.l.o.g. that the vertices of P in clockwise order are $v_0 = u_0, u_1, \ldots, u_m = v_n, v_{n-1} \ldots, v_1$. Every interior angle of P is in $[0, 2\pi]$, and the sum of interior angles is $(m + n - 2)\pi$. The sum of interior angles at the vertices $u_1, \ldots, u_{m-1}$ is $(m-1) \cdot \pi + \beta$; and the sum of interior angles at $v_1, \ldots, v_{n-1}$ is $(n-1) \cdot \pi - \beta$. Hence the interior angles at p and q are both 0.

Let G be a topological multi-graph. We say that two edges *overlap* if their intersection contains a connected set of more than one point. Such a maximal set is called an *overlap* of the two edges. A *common tail* is an overlap of two edges that contains a vertex adjacent to both edges.

Lemma 3. *Let G be a topological multi-graph in which some edges may overlap, but only in common tails. Then the edges of G can be slightly perturbed such that all overlaps are removed and no new crossings are introduced.*

Proof. We successively perturb G and decrease the number of edge pairs that have a common tail. Let $e = (u, v)$ be an edge in G, and let $e_1, e_2, \ldots, e_k$ be edges in G that have a common tail with e, such that their overlaps with e contain the vertex u. Direct all these edges away from u. Then every edge e_i follows an initial portion of e, and then turns either right or left at some *turning point* p_i. Suppose that there is at least one right turning point, and let p_j be the last such point. (Observe that the part of e between u and p_j does not contain turning points of common tails adjacent to vertex v, for otherwise there would be overlaps that are not common tails.)

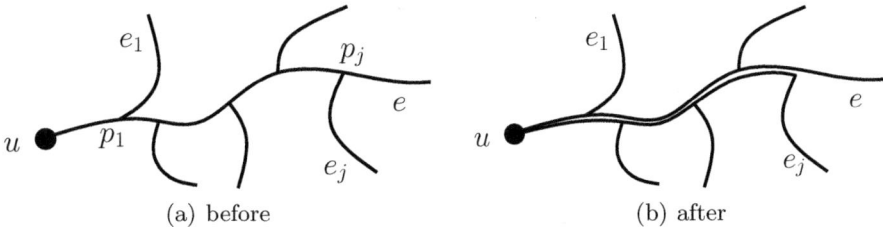

(a) before   (b) after

Fig. 3. Removing overlaps

Redraw all the edges e_i with a right turning point such that they closely follow e on the right. See Figure 3. We have removed the overlap between e and e_j, and decreased the number of edge pairs that have a common tail. If there are no right turning points, then we redraw all the edges with a left turning point such that they closely follow e on the left.

We say that a topological graph is *simple* if any two of its edges meet at most once, either at a common endpoint or at a crossing. By a *quasi-planar* graph we understand a topological graph with no three pairwise crossing edges. In the sequel we will use the linear upper bound on the maximum number of edges in a simple quasi-planar graph by Ackerman and Tardos [1].

3   Polyline Drawing with the Same Crossing Angle

In this section, we prove Theorems 1 and 2. The key new technique can be summarized as follows. Fix an $\alpha AC_2^=$ drawing of a graph $G = (V, E)$. For a constant fraction of the edges $(u, v) \in E$, we draw a new "red" edge that connects u to another vertex in V (which is not necessarily v). The red edges closely follows edges in the $\alpha AC_2^=$ drawing of G, and they only turn at bends and some crossings of G. We show that all crossings among red edges occur on their end segments. This combined with Lemma 1 will complete the proof. We continue with the details.

An $\alpha AC_\infty^=$ drawing of a graph G is polyline drawing where every crossing occurs at angle α. Every edge is a polygonal arc that consists of line segments. The first and last segments of each edge are called *end segments*, all other segments are called *middle segments*. Note that each end segment is incident to a vertex of G. Let $G = (V, E)$ be a graph with an $\alpha AC_\infty^=$ drawing. It is clear that G has at most $3n - 6$ crossing-free edges, since they form a plane graph. All other edges have some crossings. We distinguish several cases below depending on whether the edges have crossings along their end segments.

3.1   Crossings between End Segments

Lemma 4. *Let $0 < \alpha \leq \frac{\pi}{2}$. Let $G = (V, E)$ be a graph on $n \geq 4$ vertices that admits an $\alpha AC_\infty^=$ drawing such that an end segment of every edge $e \in E$ crosses an end segment of some other edge in E. Then $|E| \leq 36n$. Moreover, the number of edges in E whose both end segments cross some end segments is at most $18n$.*

Proof. In the $\alpha AC_\infty^=$ drawing of G, let S be the set of end segments that cross some other end segments. We have $|E| \leq |S| \leq 2|E|$. Direct each segment $s \in S$ from an incident vertex in V to the other endpoint. For a straight line edge, choose the direction arbitrarily.

We construct a directed multi-graph $G' = (V, R)$. We call the edges in R *red*, to distinguish them from the edges of E. For every end segment $s \in S$, we construct a red edge $\gamma(s)$, which is a polygonal path with one bend between two vertices in V. For a segment $s \in S$, the path $\gamma(s)$ is constructed as follows.

Let $u_s \in V$ denote the starting point of s (along its direction). Let c_s be the first crossing of s with an end segment, which we denote by t_s. Let $v_s \in V$ be a vertex incident to the end segment t_s. Now let $\gamma(s) = (u_s, c_s, v_s)$.

Note that for every $s \in S$, the first segment of $\gamma(s)$ is part of the segment s and does not cross any segment in S. Hence the first segments of the red edges $\gamma(s)$ are distinct and do not cross other red edges. However, the second segment of $\gamma(s)$ may cross other red edges or overlap with other red edges. Since the edges of G cross at angle α and c_s is a crossing, the turning angle of $\gamma(s)$ is $\pm \alpha$ or $\pm(\pi - \alpha)$. Note that overlaps may occur only between red end segments. However, they can be removed using Lemma 3.

We show that for any two vertices $u, v \in V$, there are at most 4 directed red edges from u to v. By Lemma 2, any two noncrossing paths of the same turning angle between u and v must overlap in the first and last segments. The first segments of the red edges are pairwise non-overlapping. Furthermore, the red edges from u to v cannot cross, since the segments incident to v cannot cross. Since red edges may have up to 4 distinct turning angles, there are at most 4 pairwise noncrossing edges between u and v. Hence there are at most 8 red edges between u and v (in either direction).

Distinguish two types of red edges. Let R_1 be the set of red edges whose second segment crosses some other red edge, and let R_2 be the set of red edges where both segments are crossing-free. Note that two edges in R_1 cannot follow the same path γ in opposite directions. If R_2 contains two edges that follow the same path γ in opposite directions, then remove one arbitrarily, and denote the remaining edges by $R_2' \subseteq R_2$, with $|R_2'| \geq \frac{1}{2}|R_2|$.

Hence, there are at most 4 red edges in R_1 between any two vertices in V. Let S_1 be the set of second segments of the red edges in R_1. By Lemma 1, there is a subset $S_1' \subseteq S_1$ of pairwise noncrossing segments of size at least $\frac{1}{3}|R_1|$. Let R_1' be the set of red edges containing the segments S_1', with $|R_1'| \geq \frac{1}{3}|R_1|$. Now $(V, R_1' \cup R_2')$ is a simple multi-graph with maximum multiplicity 4, with at most $4(3n-6)$ edges. Hence, $|R| \leq 3 \cdot 4(3n-6) = 36n - 72$, if $n \geq 3$.

For the last part of the statement observe that in the above argument, an edge in E is counted twice if its both end segments are in S.

Theorem 2 follows from Lemma 4.

Proof of Theorem 2: Let $G = (V, E)$ be a graph with $n \geq 4$ vertices drawn in the plane with an $\alpha AC_1^=$ drawing. Let $E_1 \subseteq E$ denote the set of edges in E that have at least one crossing-free end segment. Let $G_1 = (V, E_1)$ and $G_2 = (V, E \setminus E_1)$.

It is easy to see that if $\alpha \neq \frac{\pi}{3}$, then G_1 is a simple quasi-planar graph and so it has at most $6.5n - 20$ edges by a result of Ackerman and Tardos [1]. If $\alpha = \frac{\pi}{3}$, let S_1 be the set of crossed end segments of edges in E_1. By Lemma 1, there is a subset $S_1' \in S_1$ of pairwise noncrossing segments of size $\frac{1}{3}|E_1|$. The graph G_1' corresponding to these edges is planar, with at most $3n - 6$ edges. Hence E_1 contains at most $3 \cdot (3n-6) = 9n - 18$ edges.

By Lemma 4, G_2 has at most $18n$ edges. Hence, G has at most $24.5n$ edges if $\alpha \neq \frac{\pi}{3}$ and at most $27n$ edges otherwise. $\qquad\square$

Remark. It is easy to generalize the proof of Lemma 4 to the case that every two polyline edges cross at one of k possible angles. The only difference is that the red edges may have up to $2k$ different turning angles.

Lemma 5. *Let $G = (V, E)$ be a graph on $n \geq 4$ vertices that admits a polyline drawing such that an end segment of every edge $e \in E$ crosses an end segment of some other edge in E at one of k possible angles. Then $|E| \leq 36kn$.* $\qquad\square$

Corollary 1. *Let $A \subset (0, \frac{\pi}{2}]$ be a set of k angles. If a graph G on n vertices admits a drawing with at most one bend per edge such that every crossing occurs at some angle from A, then G has at most $(36k + 3)n$ edges.* $\qquad\square$

3.2 Crossings between End Segments and Middle Segments

We prove Theorem 5. Note that in an $\alpha AC_2^=$ drawing, we can take $k = 2$.

Proof of Theorem 5: Without loss of generality we can assume that G is drawn in the plane so that the number of edge crossings is minimized under the constraint that it satisfies conditions (1)–(3). Observe that G has at most $3n-6$ edges whose end segments are crossing-free, since two such edges cannot cross each other. Let $E_1 \subset E$ denote the set of edges with at least one crossed end segment. Let S_1 be the set of crossed end segments. It is clear that $|E_1| \le |S_1| \le 2|E_1|$. We will construct a red edge for at least $\frac{1}{k}|S_1|$ end segments in S_1.

Direct every end segment from the incident vertex in V to its other endpoint, and direct every middle segment arbitrarily. For every end segment $s \in S_1$, let c_s be the first crossing along s (that is, closest to its incident vertex in V).

By property (3), for every middle segment m and for every vertex $v \in V$, there are at most k end segments s adjacent to v with $c_s \in m$. Order these at most k end segments according to the position of their crossings c_s along m; and select the one corresponding to the last crossing c_s. Let $S_2 \subset S_1$ be the set of selected end segments, we have $|S_2| \ge \frac{1}{k}|S_1| \ge \frac{1}{k}|E_1|$.

Drawing red edges. We construct a directed multi-graph $G' = (V, E')$. We call the edges in E' *red*, to distinguish them from the edges of E. For every end segment $s \in S_2$, we construct a red edge $\gamma(s)$, which is a polygonal path with two bends between the vertex u_s incident to s and another vertex in V. The red edge $\gamma(s)$ is constructed as follows. Recall that c_s is the first crossing of s with some middle segment, say m_s. Let d_s be the next crossing along m_s (following the direction of m_s), or if c_s is the last crossing along m_s, then let d_s be the endpoint of m_s. In both cases, d_s lies on a unique end segment, which is incident to a unique vertex $v_s \in V$. Now $\gamma(s)$ is the directed path that follows segment s from u_s to c_s; it follows the middle segment m_s from c_s to d_s; and finally follows another end segment from d_s to v_s. See Figure 4(a) for an example. Since we assume that G is drawn with the minimal number of crossings, we have $u_s \ne v_s$.

Indeed, suppose that $u_s = v_s$. If d_s is the endpoint of the middle segment m_s, then we could redraw the edge e_m containing m_s so that edge e_m ends right before reaching point c_s and then continues to u_s closely following along s without crossings. Observe that by this redrawing we reduce the number of crossings without violating conditions (1)–(3). Otherwise d_s is a crossing of m_s with an end segment, say s', which is incident to $u_s = v_s$. In this case, we could redraw the part of s' between v_s and d_s so that it closely follows s from v_s to c_s and then follows m_s to d_s without crossing any edge. Since $d_s \ne c_{s'}$, the end segment s' crosses some middle segment before d_s, and so this redrawing reduces the number of crossings without violating conditions (1)–(3).

Every red edge $\gamma(s)$ is naturally partitioned into two end segments, (u_s, c_s) and (d_s, v_s), and one middle segment (c_s, d_s). The first segment of a red edge $\gamma(s)$ lies along the end segment $s \in S_2$, and it is crossing-free. The middle segment of $\gamma(s)$ lies along a middle segment m_s, follows the direction of m_s, and it is also crossing-free. The last segment of $\gamma(s)$ lies along an end segment, and possibly

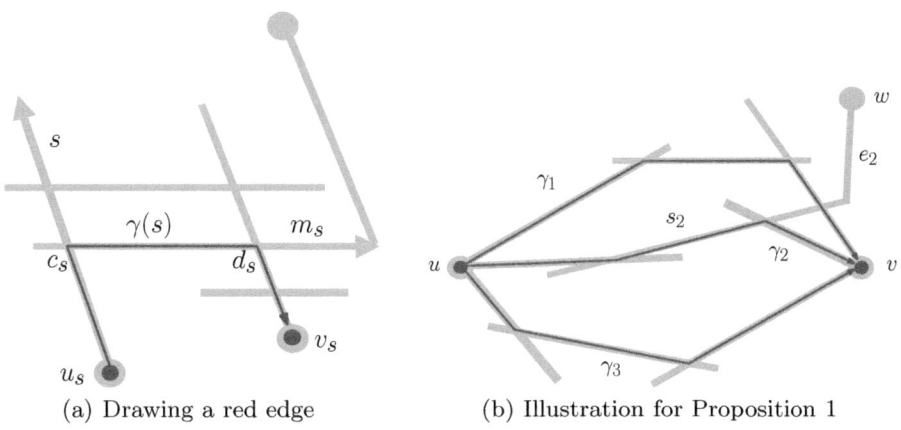

(a) Drawing a red edge   (b) Illustration for Proposition 1

Fig. 4. Illustrations for the proof of Theorem 5

have a common tail with other red edges. Note that the last segment of $\gamma(e)$ does not cross any other red edges, since all crossings in G are between end segments and middle segments, and by construction, the middle segments of red edges do not cross any end segments. Note also that two red edges cannot follow the same polygonal path in opposite directions (e.g., (u, v) and (v, u)), since every red edge follows a prescribed direction along its middle segment. If two red edges overlap, then the overlap lies on their end segments, and they have a common tail. Thus, they can be removed using Lemma 3, and we obtain a planar multi-graph.

Recall that a planar multi-graph on n vertices has at most $3n$ edges, if it has no faces of size 2. We show that by removing at most a $\frac{4k+3}{4k+4}$ fraction of the edges of G', the remaining planar multi-graph will have no 2-faces.

Proposition 1. *Let $\gamma_1, \gamma_2, \gamma_3$ be three parallel red edges from u to v in G'. Let A be the region bounded by γ_1 and γ_3 and containing γ_2. Let $e_2 \in E$ be the edge that contains the middle segment of γ_2, and let s_2 be the middle segment of e_2. If e_2 has an endpoint $w \neq u, v$ (that is, $e_2 \neq (u, v)$), then $w \in A$ or s_2 crosses the last segment of γ_1 or γ_3. (See Figure 4(b).)*

Proof. By construction, only the last segment of a red edge can cross any edge of the original graph G. Hence if e_2 has an endpoint $w \neq u, v$ and $w \notin A$, then s_2 must cross the last segment of γ_1 or γ_3.

For every pair of vertices u, v we consider the two sets of red edges (u, v) and (v, u), and remove the smaller one. Note that at most half of the red edges were deleted. Let $\gamma_1, \gamma_2, \ldots, \gamma_{2k+4}$ be parallel red edges from u to v in G', listed according to their cyclic order around u. Denote by A the region bounded by γ_1 and γ_{2k+4}, and containing the rest of the edges. Let s_i be the middle segment in G that contains the middle segment of γ_i, for $i = 1, 2, \ldots, 2k + 4$, and let e_i be the edge of G that contains s_i. Since G is a simple graph, one of e_{k+2} and e_{k+3} is not an edge between u and v. Assume, w.l.o.g., it is e_{k+2}. Applying Proposition 1 on e_{k+2} and every pair of edges e_i, e_{2k+5-i}, $i = 1, 2, \ldots, k+1$, we conclude that there must be a vertex $w \neq u, v$ of e_{k+2} in A, or s_{k+2} crosses at least $k+1$ end segments that are adjacent to v. However, the latter cannot happen since every middle segment in G crosses by at most k end segments that share a vertex.

Therefore, by deleting all but one of every $2k+3$ consecutive parallel red edges around a vertex, we obtain a planar graph with no 2-faces. In the worst case, we have $4k+4$ consecutive parallel edges between u and v, and keep only one. By removing at most $(4k+3)/(4k+4)$ fraction of the edges in G', we obtain a planar multi-graph with no 2-faces, and thus having at most $3n$ edges. it follows that the number of red edges is bounded by

$$\frac{1}{2}|E'| \leq (4k+4)3n = 12(k+1)n.$$

Since $|E'| \geq \frac{1}{k}|E_1|$, it follows that $|E| < |E_1|+3n \leq k|E'|+3n \leq 24(k+1)kn+3n$. □

3.3   Proofs of Theorems 1 and 3

We are now ready to prove that a graph on n vertices that admits an $\alpha AC_2^=$ drawing has $O(n)$ edges.

Proof of Theorem 1: Let $G = (V, E)$ be a graph with n vertices drawn in the plane with an $\alpha AC_2^=$ drawing. It follows from Lemma 4 that the number of edges in G which have end segments crossing some other end segments, is at most $36n$. Let $G_1 = (V, E_1)$ be the subgraph of G that does not contain two mutually crossing end segments.

Let S_1 be the set of *middle* segments of all edges in E_1. By Lemma 1, there is a subset $S_2 \subset S_1$ of at least $\frac{1}{3}|S_1| = \frac{1}{3}|E_1|$ pairwise noncrossing segments. Let $E_2 \subseteq E_1$ be the set of edges whose middle segments are in S_2, and let $G_2 = (V, E_2)$ be a subgraph of G with at least $\frac{1}{3}|E_1|$ edges.

Note that G_2 has properties (1)–(3) with $k = 2$ in Theorem 5. All crossings are between an end segment and a middle segment. Moreover, no middle segment crosses more than two end segments that share a vertex, since two such end segments form an isosceles triangle with the middle segment.

Therefore, it follows from Theorem 5 that the number of edges in E_2 is at most $147n$. Therefore, G has at most $36n + 3 \cdot 147n = 477n$ edges.   □

With a slightly more effort one can prove Theorem 3.

Proof of Theorem 3: Let $A \subset (0, \frac{\pi}{2}]$ be a set of k angles. Let G be a graph on n vertices that admits a drawing with at most two bends per edge such that every crossing occurs at some angle from A.

Using an easy generalization of Lemma 1 (c.f., [4]), G has a subgraph with at least $|E|/(2k+1)$ edges that does not contain two crossing middle segments. This graph has at most $3n$ edges that have crossing-free end segments. Note that a segment can be crossed by at most $2k$ end segments that share a vertex. Therefore,

The number of edges whose end segments are only crossed by middle segments is at most $O(k^2n)$ by Theorem 5. It remains to count the number of edges that have an end segment which is crossed by another end segment. By Lemma 5, there are at most $O(kn)$ such edges. Therefore, G has $O(k^3n)$ edges.   □

4 Proof of Theorem 4

We prove Theorem 4 using the same ideas as in the proofs of Lemma 4 and Theorem 5.

Proof of Theorem 4: Let G be a topological graph on n vertices, and suppose that every edge of G is partitioned into a red end segment and a blue end segment, such that: (1) no two end segments of the same color cross; (2) every pair of end segments intersects at most once; and (3) no end segment is crossed by more than k end segments that share a vertex. We show that G has $O(kn)$ edges.

By [1] the graph G has at most $6.5n - 20$ edges with a crossing free end segment, since these edges form a simple quasi-planar graph. Denote the rest of the edges by E_1. For every edge $e = (u,v) \in E_1$ we draw a new edge $\gamma(e)$ as follows. Let (u,x) denote the crossed red end segment of e. Direct (u,x) from u to x, and let y be the first crossing point on it. Let e' be the edge that crosses e at y. Then y must lie on the blue end segment of e'. Denote by w the vertex that is adjacent to the blue end segment of e', and observe that $w \neq u$, for otherwise e and e' have end segments that intersect twice. Define $\gamma(e) = (u,y) \cup (y,w)$, orient it from u to w, and call it a *red-blue* edge.

For every pair of vertices u, v, we consider the two sets of parallel edges from u to v and parallel edges from v to u, We remove the smaller set. Denote the resulting multi-graph by $G' = (V, E')$, and observe that $|E'| \geq |E_1|/2$. Note also that G' is a plane graph. Indeed, crossings may occur only between a red end segment and a blue end segment, however, the red end segment of every edge in G' is crossing-free. G' might contain edges with a common tail, however, these overlaps may be removed using Lemma 3.

Proposition 2. *Let $\gamma_1, \gamma_2, \gamma_3$, be three parallel red-blue edges from u to w in G'. Let A be the region bounded by γ_1 and γ_3 and containing γ_2. Let r_2 be the red end segment of G that contains the red end segment of γ_2, and let e_2 be the edge of G that contains r_2. Then if $e_2 = (u,v)$ for $v \neq w$ then either v lies inside A or r_2 crosses the blue end segment of γ_1 or γ_3.*

Proof. Suppose that $e_2 = (u,v)$ such that $v \neq w$. If $v \notin A$ then e_2 must cross γ_1 or γ_3. Since crossings occur only between red and blue end segments, and the red end segment of a red-blue edge is crossing-free, it follows that r_2 must cross the blue end segment of γ_1 or γ_3.

Let $\gamma_1, \gamma_2, \ldots, \gamma_{2k+4}$ be parallel red-blue edges from u to w in G', listed according to their cyclic order around u. Denote by A the region bounded by γ_1 and γ_{2k+4}, and containing the rest of the edges. Let r_i be the red end segment of G that contains the red end segment of γ_i, $i = 1, \ldots, 2k + 4$, and let e_i be the edge of G that contains r_i. One of e_{k+2}, e_{k+3} is not an edge between u and w. Assume, w.l.o.g., it is e_{k+2}. Applying Proposition 2 on γ_{k+2} and every pair of edges $\gamma_i, \gamma_{2k+5-i}$, $i = 1, 2, \ldots, k + 1$, we conclude that there must be a vertex $v \neq u, w$ of e_{k+2} in A, or r_{k+2} crosses at least $k+1$ blue end segment that are adjacent to w. However, the latter cannot happen since every end segment in G is crossed by at most k end segments that share a common vertex.

Therefore, by keeping one out of every $2k + 3$ consecutive parallel red edges around a vertex, we obtain a graph with no 2-faces. In the worst case, we keep only one edge out of $4k + 4$ parallel edges. Therefore, by removing at most $(4k + 3)/(4k + 4)$ of the edges in G' we obtain a planar multi-graph with no 2-faces and therefore with at most $3n$ edges. We conclude that $|E| \leq |E_1| + 6.5n \leq 2|E'| + 6.5n \leq 2(4k + 4)3n + 6.5n = (24k + 30.5)n$.   □

Remark. It follows from Theorem 4 that a graph on n vertices that admits an $\alpha AC_1^=$ drawing has $O(n)$ edges (however, with a worse constant coefficient than in Theorem 2): Let $\alpha \in (0, \frac{\pi}{2}]$ and let $G = (V, E)$ be a graph on n vertices that admits an $\alpha AC_1^=$ drawing. We first partition arbitrarily every straight-line edge of G into two end segments. Next, for every edge of G we assign the color red to one of its end segments, and the color blue to its other end segment. Applying Lemma 1 twice we conclude that there is a set of edges $E' \subseteq E$, such that no two end segments of the same color of two edges in E' cross, and $|E'| \geq |E|/9$. Recall that an end segment in G cannot be crossed by more than two end segments that share a common vertex. Therefore, by Theorem 4 we have that $|E'| \leq 78.5n$ and thus $|E| \leq 706.5n$.

References

1. Ackerman, E., Tardos, G.: On the maximum number of edges in quasi-planar graphs. J.Combinatorial Theory, Ser. A. 114(3), 563–571 (2007)
2. Angelini, P., Cittadini, L., di Battista, G., Didimo, W., Frati, F., Kaufmann, M., Symvonis, A.: On the perspectives opened by right angle crossing drawings. In: Eppstein, D., Gansner, E.R. (eds.) GD 2009. LNCS, vol. 5849, pp. 21–32. Springer, Heidelberg (2010)
3. Arikushi, K., Fulek, R., Keszegh, B., Morić, F., Tóth, C.D.: Drawing graphs with orthogonal crossings. In: Thilikos, D.M. (ed.) WG 2010. LNCS, vol. 6410, Springer, Heidelberg (2010)
4. Arikushi, K., Tóth, C.D.: Crossing angles of geometric graphs (manuscript) (2010)
5. Didimo, W., Eades, P., Liotta, G.: Drawing graphs with right angle crossings. In: Dehne, F., Gavrilova, M., Sack, J.-R., Tóth, C.D. (eds.) WADS 2009. LNCS, vol. 5664, pp. 206–217. Springer, Heidelberg (2009)
6. Dujmović, V., Gudmundsson, J., Morin, P., Wolle, T.: Notes on large angle crossing graphs. In: Computing: The Australian Theory Symp. (CATS 2010). CRPIT (2010)
7. Huang, W.: Using eye tracking to investigate graph layout effects. In: Proc. 6th Asia-Pacific Symp. Visualization (APVIS), pp. 97–100. IEEE, Los Alamitos (2007)
8. Huang, W., Hong, S.-H., Eades, P.: Effects of crossing angles. In: Proc. IEEE Pacific Visualization Symp., pp. 41–46. IEEE, Los Alamitos (2008)

Monotone Drawings of Graphs*

Patrizio Angelini, Enrico Colasante, Giuseppe Di Battista,
Fabrizio Frati, and Maurizio Patrignani

Dipartimento di Informatica e Automazione – Università Roma Tre, Italy
{angelini,colasant,gdb,frati,patrigna}@dia.uniroma3.it

Abstract. We study a new standard for visualizing graphs: A monotone drawing is a straight-line drawing such that, for every pair of vertices, there exists a path that monotonically increases with respect to some direction. We show algorithms for constructing monotone planar drawings of trees and biconnected planar graphs, we study the interplay between monotonicity, planarity, and convexity, and we outline a number of open problems and future research directions.

1 Introduction

A traveler that consults a road map to find a route from a site u to a site v would like to easily spot at least one path connecting u and v. Such a task is harder if each path from u to v on the map has legs moving away from v. Travelers rotate maps to better perceive their content. Hence, even if in the original orientation of the map all the paths from u to v have annoying back and forth legs, the traveler might be happy to find at least one orientation where a path from u to v smoothly flows from left to right.

Leaving the road map metaphora for the Graph Drawing terminology, we say that a path P in a straight-line drawing of a graph is *monotone* if there exists a line l such that the orthogonal projections of the vertices of P on l appear along l in the order induced by P. A straight-line drawing of a graph is *monotone* if it contains at least one monotone path for each pair of vertices. Having at disposal a monotone drawing (map), for each pair of vertices u and v a user (traveler) can find a rotation of the drawing such that there exists a path from u to v always increasing in the x-coordinate.

In a monotone drawing each monotone path is monotone with respect to a different line. *Upward drawings* [7,10] are related to monotone drawings, as in an upward drawing every directed path is monotone. Even more related to monotone drawings are *greedy drawings* [13,12,2]. Namely, in a greedy drawing, between any two vertices a path exists such that the Euclidean distance from an intermediate vertex to the destination decreases at every step, while, in a monotone drawing, between any two vertices a path and a line l exist such that the Euclidean distance from the projection of an intermediate vertex on l to the projection of the destination on l decreases at every step.

Monotone drawings have a strict correlation with an important problem in Computational Geometry: Arkin, Connelly, and Mitchell [3] studied how to find monotone trajectories connecting two given points in the plane avoiding convex obstacles. As a

---

* Supported in part by MIUR (Italy), Projects AlgoDEEP no. 2008TFBWL4 and FIRB "Advanced tracking system in intermodal freight transportation", no. RBIP06BZW8.

U. Brandes and S. Cornelsen (Eds.): GD 2010, LNCS 6502, pp. 13–24, 2011.

corollary, they proved that every planar convex drawing is monotone. Hence, the graphs admitting a convex drawing [6] have a planar monotone drawing. Such a class of graphs is a super-class (sub-class) of the triconnected (biconnected) planar graphs.

In this paper we first deal with trees (Sect. 4). We prove several properties relating the monotonicity of a tree drawing to its planarity and "convexity" [5]. Moreover, we show two algorithms for constructing monotone planar grid drawings of trees. The first one constructs drawings lying on a grid of size $O(n^{1.6}) \times O(n^{1.6})$. The second one has a better area requirement, namely $O(n^3)$, but a worse $\Omega(n)$ aspect ratio.

The existence of monotone drawings of trees allows to construct a monotone drawing of any graph G by drawing any of its spanning trees. However, the obtained monotone drawing could be non-planar even if G is a planar graph. Motivated by this and since every triconnected planar graph admits a planar monotone drawing, we devise an algorithm to construct planar monotone drawings of biconnected planar graphs (Sect. 5). Such an algorithm exploits the SPQR-tree decomposition of a biconnected planar graph. We conclude the paper with several open problems (Sect. 6). Due to space reasons several proofs are omitted and can be found in [1].

2   Definitions and Preliminaries

A *straight-line drawing* of a graph is a mapping of each vertex to a distinct point of the plane and of each edge to a segment connecting its endpoints. A drawing is *planar* if its edges do not cross but, possibly, at common endpoints. A graph is *planar* if it admits a planar drawing. A planar drawing partitions the plane into topologically connected regions, called *faces*. The unbounded face is the *outer face*. A *strictly convex drawing* (resp. a *(non-strictly) convex drawing*) is a straight-line planar drawing in which each face is delimited by a strictly (resp. non-strictly) convex polygon.

We denote by $P(v_1, v_m)$ a path between vertices v_1 and v_m. A graph G is *connected* if every pair of vertices is connected by a path and is *biconnected* (resp. *triconnected*) if removing any vertex (resp. any two vertices) leaves G connected. A *subdivision* of G is obtained by replacing each edge of G with a path. If each path has at most one internal vertex we have a 1-*subdivision*. A *subdivision of a drawing* Γ of G is a drawing Γ' of a subdivision G' of G such that, for every edge (u, v) of G that has been replaced by a path $P(u, v)$ in G', u and v are drawn at the same point in Γ and in Γ', and all the vertices of $P(u, v)$ lie on the segment between u and v.

Let p be a point in the plane and l an half-line starting at p. The *slope* of l, denoted by $slope(l)$, is the angle spanned by a counter-clockwise rotation that brings a horizontal half-line starting at p and directed towards increasing x-coordinates to coincide with l. We consider slopes that are equivalent modulo 2π as the same slope (e.g., $\frac{3}{2}\pi$ is regarded as the same slope as $-\frac{\pi}{2}$). Let Γ be a drawing of a graph G and (u, v) an edge of G. The half-line starting at u and passing through v, denoted by $d(u, v)$, is the *direction* of (u, v). The *slope* of (u, v), denoted by $slope(u, v)$, is the slope of $d(u, v)$. Observe that $slope(u, v) = slope(v, u) - \pi$. When comparing directions and their slopes, we assume that they are applied at the origin of the axes. An edge (u, v) is *monotone* with respect to a half-line l if it has a "positive projection" on l, i.e., if $slope(l) - \frac{\pi}{2} < slope(u, v) < slope(l) + \frac{\pi}{2}$. A path $P(u_1, u_n) = (u_1, \ldots, u_n)$ is *monotone with respect to a half-line* l if (u_i, u_{i+1}) is monotone with respect to l, for $i = 1, \ldots, n - 1$; $P(u_1, u_n)$ is *monotone* if there exists a half-line l such that $P(u_1, u_n)$ is monotone with respect to l. Observe that if a path $P(u_1, u_n) = (u_1, \ldots, u_n)$ is monotone with respect to l, then the orthogonal projections on l of $u_1, \ldots, u_n$ appear in this order along l. A drawing Γ of a graph G is *monotone* if, for each pair of vertices u and v in G, there exists a monotone path $P(u, v)$ in Γ. Observe that monotonicity implies connectivity.

The *Stern-Brocot tree* [14,4] is an infinite tree whose nodes are in bijective mapping with the irreducible positive rational numbers. The Stern-Brocot tree $\mathcal{SB}$ has two nodes $0/1$ and $1/0$ that are connected to the same node $1/1$, where $1/1$ is the right child of $0/1$ and $1/1$ is the left child of $1/0$. An ordered binary tree is then rooted at $1/1$ as follows. Consider a node y/x of the tree. The left child of y/x is the node $(y + y')/(x + x')$, where y'/x' is the ancestor of y/x that is closer to y/x (in terms of graph-theoretic distance in $\mathcal{SB}$) and that has y/x in its right subtree. The right child of y/x is the node $(y + y'')/(x + x'')$, where y''/x'' is the ancestor of y/x that is closer to y/x and that has y/x in its left subtree. The *first level* of $\mathcal{SB}$ is composed of node $1/1$. The *i-th level* of $\mathcal{SB}$ is composed of the children of the nodes of the $(i-1)$-th level of $\mathcal{SB}$. The following property of the Stern-Brocot tree is well-known and easy to observe:

Property 1. The sum of the numerators of the elements of the i-th level of $\mathcal{SB}$ is 3^{i-1} and the sum of the denominators of the elements of the i-th level of $\mathcal{SB}$ is 3^{i-1}.

To decompose a biconnected graph into its triconnected components, we use the *SPQR-tree*, a data structure introduced by Di Battista and Tamassia [8,9]. Definitions about SPQR-trees can be found in [8,9,11] and in [1]. Here we give some notation. Let $\mathcal{T}$ be the SPQR-tree of a graph G. We denote by $pert(\mu)$ the *pertinent* of a node μ of $\mathcal{T}$, that is, the subgraph of G induced by the vertices of G in μ. We denote by $skel(\mu)$ the *skeleton* of a node μ of $\mathcal{T}$, that is, the graph representing the arrangement of the triconnected components composing $pert(\mu)$. The edges of $skel(\mu)$ are called *virtual edges*. The nodes shared by $pert(\mu)$ and the rest of the graph are called *poles* of μ.

3 Properties of Monotone Drawings

Property 2. Any sub-path of a monotone path is monotone.

Property 3. A path $P(u_1, u_n) = (u_1, u_2, \ldots, u_n)$ is monotone if and only if it contains two edges e_1 and e_2 such that the closed wedge centered at the origin of the axes, delimited by the two half-lines $d(e_1)$ and $d(e_2)$, and having an angle smaller than π, contains all the half-lines $d(u_i, u_{i+1})$, for $i = 1, \ldots, n - 1$.

Edges e_1 and e_2 as in Prop. 3 are the *extremal edges* of $P(u_1, u_n)$. The closed wedge delimited by $d(e_1)$ and $d(e_2)$ and containing all the half-lines $d(u_i, u_{i+1})$, for $i = 1, \ldots, n - 1$, is the *range* of $P(u_1, u_n)$ and is denoted by $range(P(u_1, u_n))$, while the closed wedge delimited by $d(e_1) - \pi$ and $d(e_2) - \pi$, and not containing $d(e_1)$ and $d(e_2)$, is the *opposite range* of $P(u_1, u_n)$ and is denoted by $opp(P(u_1, u_n))$.

Property 4. The range of a monotone path $P(u_1, u_n)$ contains the half-line from u_1 through u_n.

Lemma 1. *Let $P(u_1, u_n) = (u_1, u_2, \ldots, u_n)$ be a monotone path and let (u_n, u_{n+1}) be an edge. Then, path $P(u_1, u_{n+1}) = (u_1, u_2, \ldots, u_n, u_{n+1})$ is monotone if and only if $d(u_n, u_{n+1})$ is not contained in $opp(P(u_1, u_n))$. Further, if $P(u_1, u_{n+1})$ is monotone, $range(P(u_1, u_n)) \subseteq range(P(u_1, u_{n+1}))$.*

Corollary 1. *Let $P(u_1, u_n) = (u_1, \ldots, u_n)$ and $P(u_n, u_{n+k}) = (u_n, \ldots, u_{n+k})$ be monotone paths. Then, path $P(u_1, u_{n+k}) = (u_1, \ldots, u_n, u_{n+1}, \ldots, u_{n+k})$ is monotone if and only if $range(P(u_1, u_n)) \cap opp(P(u_n, u_{n+k})) = \emptyset$. Further, if $P(u_1, u_{n+k})$ is monotone, $range(P(u_1, u_n)) \cup range(P(u_n, u_{n+k})) \subseteq range(P(u_1, u_{n+k}))$.*

The following properties relate monotonicity to planarity and convexity.

Property 5. A monotone path is planar.

Lemma 2. *[3] Any strictly convex drawing of a planar graph is monotone.*

In the following we will construct non-strictly convex drawings of graphs. Observe that any graph containing a degree-2 vertex does not admit a strictly convex drawing, but it might admit a non-strictly convex drawing. While not every non-strictly convex drawing is monotone, we can relate non-strict convexity and monotonicity:

Lemma 3. *Any non-strictly convex drawing of a graph such that each set of parallel edges forms a collinear path is monotone.*

On the relationship between convexity and monotonicity, we also have:

Lemma 4. *Consider a strictly convex drawing Γ of a graph G. Let u, v, and w be three consecutive vertices incident to the outer face. Let d be any half-line that splits the angle $\widehat{uvw}$ into two angles smaller than $\frac{\pi}{2}$. Then, for each vertex t of G, there exists a path from v to t in Γ that is monotone with respect to d.*

Next, we provide a powerful tool for "transforming" monotone drawings.

Lemma 5. *An affine transformation of a monotone drawing gives a monotone drawing.*

4   Monotone Drawings of Trees

The first property we present is on the relationship between monotonicity and planarity, and descends from the fact that every monotone path is planar (by Prop. 5) and that in a tree there exists exactly one path between every pair of vertices.

Property 6. Every monotone drawing of a tree is planar.

The second property relates monotonicity and convexity. A *convex drawing* of a tree T [5] is a straight-line planar drawing such that replacing each edge between an internal vertex u and a leaf v with a half-line starting at u through v yields a partition of the plane into convex unbounded polygons. A convex drawing of a tree might not be monotone, because of the presence of two parallel edges. However, if such parallel lines are not used, then a convex drawing is also monotone. Define a *strictly convex drawing* of a tree T as a straight-line planar drawing such that each set of parallel edges forms a collinear path and such that replacing every edge of T between an internal vertex u and a leaf v with a half-line starting at u through v yields a partition of the plane into convex unbounded polygons. We have the following:

Property 7. Every strictly convex drawing of a tree is monotone.

A simple modification of the algorithm presented in [5] constructs strictly convex drawings of trees. Hence, monotone drawings exist for all trees.

We introduce *slope-disjoint drawings* of trees and show that they are monotone. Let T be a tree rooted at a node r. Denote by $T(u)$ the subtree of T rooted at a node u. A *slope-disjoint* drawing of T is such that: (P1) For every node $u \in T$, there exist two angles $\alpha_1(u)$ and $\alpha_2(u)$, with $0 < \alpha_1(u) < \alpha_2(u) < \pi$, such that, for every edge e that is either in $T(u)$ or that connects u with its parent, it holds that $\alpha_1(u) < slope(e) < \alpha_2(u)$; (P2) for every two nodes $u, v \in T$ with v child of u, it holds that $\alpha_1(u) < \alpha_1(v) < \alpha_2(v) < \alpha_2(u)$; (P3) for every two nodes v_1, v_2 with the same parent, it holds that $\alpha_1(v_1) < \alpha_2(v_1) < \alpha_1(v_2) < \alpha_2(v_2)$. We have the following:

Theorem 1. *Every slope-disjoint drawing of a tree is monotone.*

Proof: Let T be a tree and let Γ be a slope-disjoint drawing of T. We show that, for every two vertices $u, v \in T$, a monotone path between u and v exists in Γ. Let w be the lowest common ancestor of u and v in T.

If $w = u$, then, by P1, for every edge e in the path $P(u, v)$, $0 < slope(e) < \pi$. Hence, $P(u, v)$ is monotone with respect to a half-line with slope $\pi/2$. Analogously, if $w = v$ then $P(u, v)$ is monotone with respect to a half-line with slope $-\pi/2$. If $w \neq u, v$, let u' and v' be the children of w in T such that $u \in T(u')$ and $v \in T(v')$. Path $P(u, v)$ is composed of path $P(u, w)$ and of path $P(w, v)$. As before, $P(u, w)$ is monotone with respect to a half-line with slope $-\pi/2$. By P1, for every edge $e \in P(u, w)$, $\alpha_1(u') - \pi < slope(e) < \alpha_2(u') - \pi$. Hence, $\alpha_1(u') < slope(l) < \alpha_2(u')$, for each half-line l contained into the closed wedge $opp(P(u, w))$. Analogously, $P(w, v)$ is monotone with respect to a half-line with slope $\pi/2$ and, by P1, for every edge $e \in P(w, v)$, $\alpha_1(v') < slope(e) < \alpha_2(v')$. Hence, $\alpha_1(v') < slope(l) < \alpha_2(v')$, for each half-line l contained into the closed wedge $range(P(w, v))$. Finally, since u' and v' are children of the same node, by P3 $\alpha_1(u') < \alpha_2(u') < \alpha_1(v') < \alpha_2(v')$ (the case in which $\alpha_1(v') < \alpha_2(v') < \alpha_1(u') < \alpha_2(u')$ being symmetric). Since $\alpha_1(u') < slope(l) < \alpha_2(u')$ for each half-line l contained into the closed wedge $opp(P(u, w))$ and since $\alpha_1(v') < slope(l) < \alpha_2(v')$ for each half-line l contained into the closed wedge $range(P(w, v))$, we have $opp(P(u, w)) \cap range(P(w, v)) = \emptyset$. By Corollary 1, $P(u, v)$ is monotone.   □

By Theorem 1, as long as the slopes of the edges in a drawing of a tree T guarantee the slope-disjoint property, one can *arbitrarily* assign lengths to such edges always obtaining a monotone drawing of T. In the following we present two algorithms for constructing slope-disjoint drawings of any tree T. In both algorithms, we individuate a suitable set of elements of the Stern-Brocot tree $\mathcal{SB}$. Each of such elements, say $s = y/x$, is then used as a slope of an edge of T in the drawing.

Algorithm BFS-based: Consider the first $\lceil \log_2(n) \rceil$ levels of the Stern-Brocot tree $\mathcal{SB}$. Such levels contain a total number of at least $n - 1$ elements y/x of $\mathcal{SB}$. Order such elements by increasing value of the ratio y/x and consider the first $n - 1$ elements in such an order S, say $s_1 = y_1/x_1, s_2 = y_2/x_2, \ldots, s_{n-1} = y_{n-1}/x_{n-1}$. Consider the

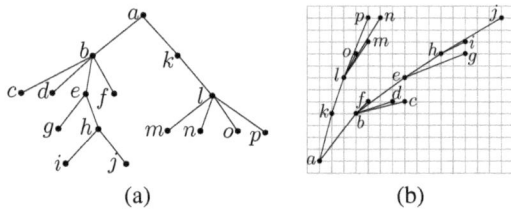

Fig. 1. (a) A tree T. (b) The drawing of T constructed by Algorithm BFS-based.

subtrees of r, say $T_1(r), T_2(r), \ldots, T_{k(r)}(r)$. Assign to $T_i(r)$ the $|T_i(r)|$ elements of S from the $(1 + \sum_{j=1}^{i-1} |T_j(r)|)$-th to the $(\sum_{j=1}^{i} |T_j(r)|)$-th. Consider a node u of T and suppose that a sub-sequence $S(u) = s_a, s_{a+1}, \ldots, s_b$ of S has been assigned to $T(u)$, where $|T(u)| = b - a$. Consider the subtrees $T_1(u), T_2(u), \ldots, T_{k(u)}(u)$ of u and assign to $T_i(u)$ the $|T_i(u)|$ elements of $S(u)$ from the $(1 + \sum_{j=1}^{i-1} |T_j(u)|)$-th to the $(\sum_{j=1}^{i} |T_j(u)|)$-th. Now we construct a grid drawing of T. Place r at $(0,0)$. Consider a node u of T, suppose that a sequence $S(u) = s_a, s_{a+1}, \ldots, s_b$ of S has been assigned to $T(u)$ and suppose that the parent $p(u)$ of u has been already placed at the grid point $(p_x(u), p_y(u))$. Place u at grid point $(p_x(u) + x_b, p_y(u) + y_b)$, where $s_b = y_b/x_b$. See Fig. 1 for an example of application. We have the following:

Theorem 2. *Let T be a tree. Then, Algorithm BFS-based constructs a monotone drawing of T on a grid of area $O(n^{1.6}) \times O(n^{1.6})$.*

Algorithm DFS-based: Consider the sequence S composed of the first $n - 1$ elements $1/1, 2/1, \ldots, n - 1/1$ of the rightmost path of $\mathcal{SB}$. Assign sub-sequences of S to the subtrees of T and construct a grid drawing in the same way as in Algorithm BFS-based. We have the following.

Theorem 3. *Let T be a tree. Then, Algorithm DFS-based constructs a monotone drawing of T on a grid of area $O(n^2) \times O(n)$.*

As a further consequence of Theorem 1, we have the following:

Corollary 2. *Every (even non-planar) graph admits a monotone drawing.*

Namely, for any graph G, construct a monotone drawing of a spanning tree T of G with vertices in general position. Draw the other edges of G as segments, obtaining a straight-line drawing of G in which, for any pair of vertices, there exists a monotone path (the one whose edges belong to T). Such a drawing is a monotone drawing of G.

5   Planar Monotone Drawings of Biconnected Graphs

First, we restate, using the terminology of this paper, the well-known result of [6].

Lemma 6. *[6] Let G be a biconnected planar graph with a given planar embedding such that each split pair u, v is incident to the outer face and each maximal split component of u, v has at least one edge incident to the outer face but, possibly, for edge (u, v). Then, G admits a strictly convex drawing with the given embedding in which the outer face is drawn as an arbitrary strictly convex polygon.*

Let Γ be a monotone drawing, d any direction, and k a positive value. A *directional-scale*, denoted by $\mathcal{DS}(d, k)$, is an affine transformation defined as follows. Rotate Γ by an angle δ until d is orthogonal to the x-axis. Scale Γ by $(1, k)$ (i.e., multiply its y-coordinates by k). Rotate back the obtained drawing by an angle $-\delta$.

Lemma 7. *Let Γ be a monotone drawing and d a direction such that no edge in Γ is parallel to d. For any $\alpha > 0$ a directional-scale $\mathcal{DS}(d - \frac{\pi}{2}, k(\alpha))$ exists that transforms Γ into a monotone drawing in which the slope of any edge is between $d - \alpha$ and $d + \alpha$.*

A path monotone with respect to a direction d is (α, d)-*monotone* if, for each edge e, $d - \alpha < slope(e) < d + \alpha$. A path from a vertex u to a vertex v is an (α, d_1, d_2)-*path* if it is a composition of a (α, d_1)-monotone path from u to a vertex w and of a (α, d_2)-monotone path from w to v. Let p_N, p_S, p_W, and p_E be four points in the plane such that p_W is inside triangle $\triangle(p_N, p_S, p_E)$, $\widehat{p_W p_S p_E} = \widehat{p_W p_N p_E}$, and $\widehat{p_W p_S p_N} + 2\widehat{p_W p_S p_E} < \frac{\pi}{2}$. Quadrilateral (p_N, p_E, p_S, p_W) is a *boomerang* (see Fig. 2(a)).

Let G be a biconnected graph and $\mathcal{T}$ be the SPQR-decomposition of G rooted at an edge e. We prove that G admits a planar monotone drawing by means of an inductive algorithm which, given a component μ of $\mathcal{T}$ with poles u and v, and a boomerang $boom(\mu) = (p_N(\mu), p_E(\mu), p_S(\mu), p_W(\mu))$, constructs a drawing Γ_μ of $pert(\mu)$ satisfying the following properties. Let $d_N(\mu)$ be the half-line starting at $p_E(\mu)$ through $p_N(\mu)$, let $d_S(\mu)$ be the half-line starting at $p_E(\mu)$ through $p_S(\mu)$, let α_μ be $\widehat{p_W(\mu) p_S(\mu) p_E(\mu)} = \widehat{p_W(\mu) p_N(\mu) p_E(\mu)}$, and let $\beta_\mu = \widehat{p_W(\mu) p_S(\mu) p_N(\mu)}$. *(A)* Γ_μ is monotone; *(B)* with the possible exception of edge (u, v), Γ_μ is contained into $boom(\mu)$, with u drawn on $p_N(\mu)$ and v on $p_S(\mu)$; *(C)* each vertex $w \in pert(\mu)$ belongs to a $(\alpha_\mu, -d_N(\mu), d_S(\mu))$-path from u to v. Observe that C implies that in Γ_μ there exists a path between the poles that is monotone with respect to the line through them and that B implies the planarity of Γ_μ.

Lemma 8. *Let μ be a component of $\mathcal{T}$. Every $(\alpha_\mu, -d_N(\mu), d_S(\mu))$-path from u to v is monotone with respect to the half-line from u through v.*

Let $\mu_1, \dots, \mu_k$ be the children of μ in $\mathcal{T}$, with poles $(u_1, v_1), \dots, (u_k, v_k)$. We construct a drawing Γ_μ satisfying A–C by composing drawings $\Gamma_{\mu_1}, \dots, \Gamma_{\mu_k}$, which are constructed inductively, as follows.

If μ is a Q-node, then draw an edge between $p_N(\mu)$ and $p_S(\mu)$.

If μ is an S-node (see Fig. 2(b)), then let p be the intersection point between segment $\overline{p_W(\mu) p_E(\mu)}$ and the bisector line of $\widehat{p_W(\mu) p_N(\mu) p_E(\mu)}$. Consider k equidistant points $p_1, \dots, p_k$ on segment $\overline{p_N(\mu)p}$ such that $p_1 = p_N(\mu)$ and $p_k = p$. For each μ_i, with $i = 1, \dots, k - 1$, consider a boomerang $boom(\mu_i) = (p_N(\mu_i), p_E(\mu_i), p_S(\mu_i), p_W(\mu_i))$ such that $p_N(\mu_i) = p_i$, $p_S(\mu_i) = p_{i+1}$, and $p_E(\mu_i)$ and $p_W(\mu_i)$ determine $\beta_{\mu_i} + 2\alpha_{\mu_i} < \frac{\alpha_\mu}{2}$. Apply the inductive algorithm to μ_i and $boom(\mu_i)$. Also, consider a boomerang $boom(\mu_k) = (p_N(\mu_k), p_E(\mu_k), p_S(\mu_k), p_W(\mu_k))$ such that $p_N(\mu_k) = p$, $p_S(\mu_k) = p_S(\mu)$, and $p_E(\mu_k)$ and $p_W(\mu_k)$ determine $\beta_{\mu_k} + 2\alpha_{\mu_k} < \frac{\alpha_\mu}{2}$. Apply the inductive algorithm to μ_k and $boom(\mu_k)$.

If μ is a P-node (see Fig. 2(c)), then consider $2k$ points $p_1, \dots, p_{2k}$ on segment $\overline{p_W(\mu) p_E(\mu)}$ such that $p_1 = p_W(\mu)$, $p_{2k} = p_E(\mu)$, and $\widehat{p_i p_N(\mu) p_{i+1}} = \frac{\alpha_\mu}{2k-1}$, for each

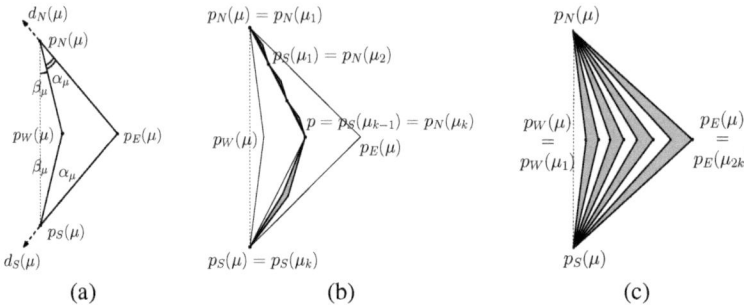

Fig. 2. (a) A boomerang. The construction rules for an S-node (b) and for a P-node (c).

$i = 1, \ldots, 2k - 1$. For each μ_i, with $i = 1, \ldots, k$, consider a boomerang $boom(\mu_i) = (p_N(\mu_i), p_E(\mu_i), p_S(\mu_i), p_W(\mu_i))$ such that $p_N(\mu_i) = p_N(\mu)$, $p_S(\mu_i) = p_S(\mu)$, $p_W(\mu_i) = p_{2i-1}$, and $p_E(\mu_i) = p_{2i}$. Apply the inductive algorithm to μ_i and $boom(\mu_i)$.

If μ is an R-node, then consider the graph G' obtained by removing v and its incident edges from $skel(\mu)$. Since $skel(\mu)$ is triconnected, G' is a biconnected graph whose possible split pairs are on the outer face. Further, each of such split pairs separates at most three maximal split components, and in this case one of them is an edge. By Lemma 6, G' admits a convex drawing whose outer face is any strictly convex polygon. Consider a strictly convex polygon C with one vertex on p_N, one vertex on p_E, and $m - 2$ vertices inside $boom(\mu)$ so that they are visible from $p_S(\mu)$ inside $boom(\mu)$ and the internal angle incident to $p_E(\mu)$ is smaller than $\frac{\pi}{2}$ (see Fig. 3(a)). Construct a convex drawing $\Gamma(G')$ of G' such that the vertices of the outer face of G' are on the vertices of C, with u on $p_N(\mu)$. By Lemma 2, $\Gamma(G')$ is monotone. Slightly perturb the position of the vertices of $\Gamma(G')$ so that no two parallel edges exist and no edge is orthogonal to $d_N(\mu)$. Apply a directional-scale $\mathcal{DS}(d_N(\mu) - \frac{\pi}{2}, k(\frac{\alpha_\mu}{2}))$ to $\Gamma(G')$. By Lemma 7, for every edge $e \in G'$, $slope(-d_N(\mu)) - \frac{\alpha_\mu}{2} < slope(e) < slope(-d_N(\mu)) + \frac{\alpha_\mu}{2})$. Further, by Lemma 4 and by the fact that the internal angle of C incident to $p_N(\mu)$ is smaller than $\frac{\alpha_\mu}{2} < \frac{\pi}{2}$, for every vertex $w \in G'$, a $(\frac{\alpha_\mu}{2}, -d_N(\mu))$-monotone path exists from u to w. Let $\Gamma(skel(\mu))$ be the drawing of $skel(\mu)$ obtained from $\Gamma(G')$ by placing v on $p_S(\mu)$ and drawing its incident edges (see Fig. 3(b)). We have the following:

Claim 1. $\Gamma(skel(\mu))$ *is monotone.*

Consider a drawing $\Gamma'(skel(\mu))$ of a subdivision of $skel(\mu)$ obtained as a subdivision of $\Gamma(skel(\mu))$. We have the following:

Claim 2. $\Gamma'(skel(\mu))$ *is monotone.*

Consider the pair of vertices x, y belonging to the subdivision of $skel(\mu)$ such that the range $range(P(x, y))$ of the monotone path $P(x, y)$ between them in $\Gamma'(skel(\mu))$ creates the largest angle $\angle(x, y)$ among all the pairs of vertices. Let $\gamma = \pi - \angle(x, y)$. Let δ be the smallest angle between two adjacent edges in $\Gamma(skel(\mu))$. For each μ_i, with $i = 1, \ldots, k$, let $p_N(\mu_i)$ and $p_S(\mu_i)$ be the points where u_i and v_i have been drawn in $\Gamma(skel(\mu))$, respectively. Consider a boomerang $boom(\mu_i) = (p_N(\mu_i), p_E(\mu_i),$

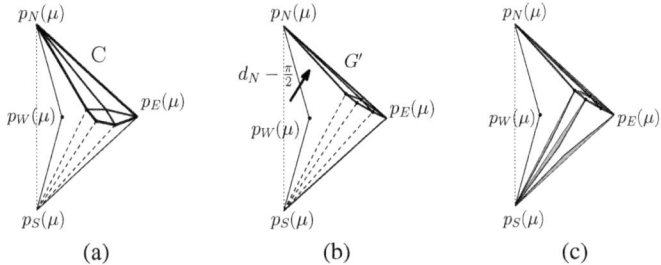

Fig. 3. Two phases of the construction for an R-node: (a) definition of the strictly convex polygon C and (b) the directional-scale applied to G'

$p_S(\mu_i), p_W(\mu_i))$ such that $p_E(\mu_i)$ and $p_W(\mu_i)$ determine $\beta_{\mu_i} + 2\alpha_{\mu_i} < \min\{\frac{\delta}{2}, \frac{\gamma}{2}\}$. For each μ_p such that either $p_N(\mu_p)$ and $p_S(\mu_p)$ lie on the vertices of C or $p_S(\mu_p) = p_S(\mu)$, choose points $p_W(\mu_p)$ and $p_E(\mu_p)$ inside $boom(\mu)$. Then, apply the inductive algorithm to μ_i, with poles u_i and v_i, and $boom(\mu_i)$ (see Fig. 3(c)).

In the following we prove that the above described algorithm constructs a planar monotone drawing of every biconnected planar graph.

Theorem 4. *Every biconnected planar graph admits a planar monotone drawing.*

Proof: Let $\mathcal{T}$ be the SPQR-tree of a biconnected graph G, rooted at any Q-node μ_e corresponding to an edge e. Consider a boomerang $boom(\mu_e) = (p_N(\mu_e), p_E(\mu_e), p_S(\mu_e), p_W(\mu_e))$ such that $x(p_N(\mu_e)) = x(p_S(\mu_e)) < x(p_W(\mu_e)) < x(p_E(\mu_e)), y(p_S(\mu_e)) < y(p_W(\mu_e)) = y(p_E(\mu_e)) < y(p_N(\mu_e))$, and $\beta_{\mu_e} + 2\alpha_{\mu_e} < \frac{\pi}{2}$. Apply the inductive algorithm described above to μ_e and $boom(\mu_e)$. We prove that the resulting drawing is monotone by showing that at each step of the induction the constructed drawing satisfies A–C. This is trivial if μ is a Q-node. Otherwise, μ is an S-node, a P-node, or an R-node and the statement is proved by the following claims:

Claim 3. *If μ is an S-node, Γ_μ satisfies A.*

Proof: Refer to Fig. 4(a). Consider any two vertices $w', w'' \in pert(\mu)$ and the components μ_a and μ_b such that $w' \in pert(\mu_a)$ and $w'' \in pert(\mu_b)$. If $a = b$, then a monotone path between w' and w'' exists by induction. Otherwise, for each μ_i, consider a vertex w_i, where $w_a = w'$ and $w_b = w''$. For each μ_i, with $i = 1, \ldots, k$, consider a $(\alpha_{\mu_i}, -d_N(\mu_i), d_S(\mu_i))$-path $P(u_i, v_i)$ from u_i to v_i containing w_i. Observe that such paths exist since, for each μ_i, Γ_{μ_i} satisfies C. Consider a path $P(u_i, v_i)$ with $1 \le i \le k-1$. Since $\beta_{\mu_i} + 2\alpha_{\mu_i} < \frac{\alpha_\mu}{2}$, and since $p_N(\mu_i)$ and $p_S(\mu_i)$ lie on the bisector line of α_μ, for each edge $e \in P(u_i, v_i)$, it holds $slope(e) < \beta_\mu + \frac{\alpha_\mu}{2} + \beta_{\mu_i} + 2\alpha_{\mu_i} < \beta_\mu + \alpha_\mu < \beta_\mu + 2\alpha_\mu = d_N(\mu) + \alpha$, and $slope(e) > \beta_\mu + \frac{\alpha_\mu}{2} - (\beta_{\mu_i} + 2\alpha_{\mu_i}) > \beta_\mu = d_N(\mu) - \alpha_\mu$. Hence, $P(u_i, v_i)$ is $(\alpha_\mu, -d_N(\mu))$-monotone. Analogously, $P(u_k, v_k)$ is $(\alpha_\mu, d_S(\mu))$-monotone. Therefore, the path $P(u, v)$ composed of all the paths $P(u_i, v_i)$ is an $(\alpha_\mu, -d_N(\mu), d_S(\mu))$-path. By Lemma 8, $P(u, v)$ is monotone. Hence, by Prop. 2, the subpath of $P(u, v)$ between w' and w'' is monotone, as well, and Γ_μ satisfies A. $\square$

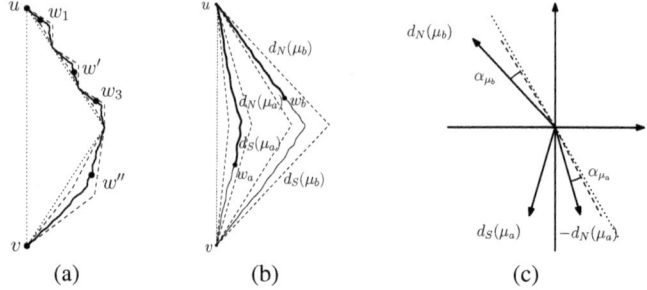

Fig. 4. Γ_μ satisfies A (a) if μ is an S-node and (b)–(c) if μ is a P-node

Claim 4. *If μ is an S-node, Γ_μ satisfies B and C.*

Claim 5. *If μ is a P-node, Γ_μ satisfies A.*

Proof: Consider any two vertices $w_a, w_b \in pert(\mu)$ and the nodes μ_a and μ_b such that $w_a \in pert(\mu_a)$ and $w_b \in pert(\mu_b)$. If $a = b$, then a monotone path from w_a to w_b exists by induction. Otherwise, consider the $(\alpha_{\mu_a}, -d_N(\mu_a), d_S(\mu_a))$-path $P_a(u,v)$ from u to v through w_a and the $(\alpha_{\mu_b}, d_N(\mu_b), -d_S(\mu_b))$-path $P_b(v,u)$ from v to u through w_b, which exist by induction (C). Suppose w_b lies on the $(\alpha_{\mu_b}, d_N(\mu_b))$-monotone path $P(w_b, u)$ from w_b to u that is a subpath of $P_b(v,u)$, the other case being analogous. Consider the $(\alpha_{\mu_a}, -d_N(\mu_a), d_S(\mu_a))$-path $P(u, w_a)$ that is a subpath of $P_a(u,v)$. We show that path $P(w_b, w_a)$ composed of $P(w_b, u)$ and $P(u, w_a)$ is monotone. When translated to the origin of the axes, $d_N(\mu_b)$, $-d_N(\mu_a)$, and $d_S(\mu_a)$ are in the 2nd, 4th, and 3rd quadrant, respectively. By construction, the wedge delimited by $d_N(\mu_b)$ and $-d_N(\mu_a)$ and containing the 3rd quadrant has an angle $\leq \pi - 2\frac{\alpha_\mu}{2k-1}$. Since, by definition, every edge of $P(w_b, u)$ creates an angle with $d_N(\mu_b)$ smaller than $\alpha_{\mu_b} = \frac{\alpha_\mu}{2k-1}$ and every edge of $P(u, w_a)$ creates an angle with $-d_N(\mu_a)$ smaller than $\alpha_{\mu_a} = \frac{\alpha_\mu}{2k-1}$, the slopes of all the edges of $P(w_b, w_a)$ lie inside a wedge having an angle smaller than π. Hence, $P(w_b, w_a)$ is monotone. □

Claim 6. *If μ is a P-node, Γ_μ satisfies B and C.*

Claim 7. *If μ is an R-node, Γ_μ satisfies Prop. A.*

Proof: Consider any two vertices $w_a, w_b \in pert(\mu)$ and the nodes μ_a and μ_b such that $w_a \in pert(\mu_a)$ and $w_b \in pert(\mu_b)$. Let e_a and e_b be the virtual edges of $skel(\mu)$ corresponding to μ_a and μ_b, respectively. If $a = b$ by induction a monotone path from w_a to w_b trivially exists. Otherwise, consider the monotone drawing $\Gamma'(skel(\mu))$ of a 1-subdivision of $skel(\mu)$ and the monotone path $P_{a,b}$ from the subdivision vertex of e_a to the subdivision vertex of e_b. By construction, $\pi - range(P_{a,b}) \geq \gamma$. As $\beta_{\mu_i} + 2\alpha_{\mu_i} < \min\{\frac{\delta}{2}, \frac{\gamma}{2}\} \leq \frac{\gamma}{2}$, for the path $P(w_a, w_b)$ obtained by replacing each edge e_i of $P_{a,b}$ with the corresponding path of $pert(\mu_{e_i})$ it holds that $range(P(w_a, w_b)) < \pi$. □

Claim 8. *If μ is an R-node, Γ_μ satisfies B and C.*

This concludes the proof of Theorem 4. □

6 Conclusions and Open Problems

We initiated the study of monotone graph drawings. Concerning trees, we proved that every monotone drawing is planar, that every strictly convex drawing is monotone, and that monotone drawings exist on polynomial-size grids. We believe that simple modifications of our algorithms allow one to construct strictly convex drawings of trees on polynomial-size grids. Another possible extension of our results is to characterize monotonicity in terms of the angles between adjacent edges. Our definition of slope-disjointness goes in this direction, although it introduces some non-necessary restrictions on the slopes of the edges (like the one that all the slopes are between 0 and π).

We proved that every biconnected planar graph admits a planar monotone drawing. Extending such a result to general simply-connected graphs seems to be non-trivial. There exist planar graphs not having a monotone drawing (see Fig. 5(a)) if the embedding is given. However, we are not aware of any planar graph not admitting a planar monotone drawing for any of its embeddings.

Several area minimization problems concerning monotone drawings are, in our opinion, worth of study. First, determining tight bounds for the area requirements of grid drawings of trees appears to be an interesting challenge. Second, modifying our tree drawing algorithms so that they construct grid drawings in general position would lead to algorithms for constructing monotone drawings of non-planar graphs on a grid of polynomial size. Third, the drawing algorithm we presented for biconnected planar graphs constructs drawings in which the ratio between the lengths of the longest and of the shortest edge is exponential in n. Is it possible to construct planar monotone drawings of biconnected planar graphs in polynomial area?

Finally, we introduce a new drawing standard related to monotone drawings. A path from a vertex u to a vertex v is *strongly monotone* if it is monotone with respect to the half-line from u through v. A drawing of a graph is *strongly monotone* if a strongly monotone path connects each pair of vertices. Strong monotonicity appears to be even more desirable than general monotonicity for the readability of a drawing. However, designing algorithms for constructing strongly monotone drawings seems to be harder than for monotone drawings and only restricted graph classes appear to admit strongly monotone drawings. Note that a subpath of a strongly monotone path is, in general, not strongly monotone; also, while convexity implies monotonicity, it does not imply strong monotonicity, even for planar triangulations (see Fig. 5(b)).

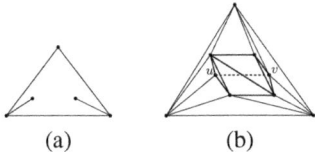

(a)   (b)

Fig. 5. (a) A planar embedding of a graph with no monotone drawing. (b) A drawing of a planar triangulation that is not strongly monotone.

Acknowledgments

We would like to thank Peter Eades for suggesting us to study monotone drawings, triggering all the work that is presented in this paper.

References

1. Angelini, P., Colasante, E., di Battista, G., Frati, F., Patrignani, M.: Monotone drawings of graphs. Tech. Report 178, Dip. di Informatica e Automazione, Università Roma Tre (2010)
2. Angelini, P., Frati, F., Grilli, L.: An algorithm to construct greedy drawings of triangulations. J. Graph Alg. Appl. 14(1), 19–51 (2010)
3. Arkin, E.M., Connelly, R., Mitchell, J.S.: On monotone paths among obstacles with applications to planning assemblies. In: SoCG 1989, pp. 334–343 (1989)
4. Brocot, A.: Calcul des rouages par approximation, nouvelle methode. Revue Chronometrique 6, 186–194 (1860)
5. Carlson, J., Eppstein, D.: Trees with convex faces and optimal angles. In: Kaufmann, M., Wagner, D. (eds.) GD 2006. LNCS, vol. 4372, pp. 77–88. Springer, Heidelberg (2007)
6. Chiba, N., Nishizeki, T.: Planar Graphs: Theory and Algorithms. In: Annals of Discrete Mathematics, vol. 32, North-Holland, Amsterdam (1988)
7. Di Battista, G., Tamassia, R.: Algorithms for plane representations of acyclic digraphs. Theor. Comput. Sci. 61, 175–198 (1988)
8. Di Battista, G., Tamassia, R.: On-line maintenance of triconnected components with SPQR-trees. Algorithmica 15(4), 302–318 (1996)
9. Di Battista, G., Tamassia, R.: On-line planarity testing. SIAM J. Comp. 25(5), 956–997 (1996)
10. Garg, A., Tamassia, R.: On the computational complexity of upward and rectilinear planarity testing. SIAM J. Comp. 31(2), 601–625 (2001)
11. Gutwenger, C., Mutzel, P.: A linear time implementation of SPQR-trees. In: Marks, J. (ed.) GD 2000. LNCS, vol. 1984, pp. 77–90. Springer, Heidelberg (2001)
12. Moitra, A., Leighton, T.: Some results on greedy embeddings in metric spaces. In: Foundations of Computer Science (FOCS 2008), pp. 337–346 (2008)
13. Papadimitriou, C.H., Ratajczak, D.: On a conjecture related to geometric routing. Theoretical Computer Science 344(1), 3–14 (2005)
14. Stern, M.A.: Ueber eine zahlentheoretische funktion. Journal fur die reine und angewandte Mathematik 55, 193–220 (1858)

Upward Geometric Graph Embeddings into Point Sets*

Patrizio Angelini[1], Fabrizio Frati[1], Markus Geyer[2], Michael Kaufmann[2],
Tamara Mchedlidze[3], and Antonios Symvonis[3]

[1] Dipartimento di Informatica e Automazione – Università Roma Tre, Italy
{angelini,frati}@dia.uniroma3.it
[2] Wilhelm-Schickard-Institut für Informatik – Universität Tübingen, Germany
{geyer,mk}@informatik.uni-tuebingen.de
[3] Dept. of Mathematics, National Technical University of Athens, Athens, Greece
{mchet,symvonis}@math.ntua.gr

Abstract. We study the problem of characterizing the directed graphs with an upward straight-line embedding into every point set in general or in convex position. We solve two questions posed by Binucci *et al.* [*Computational Geometry: Theory and Applications, 2010*]. Namely, we prove that the classes of directed graphs with an upward straight-line embedding into every point set in convex position and with an upward straight-line embedding into every point set in general position do not coincide, and we prove that every directed caterpillar admits an upward straight-line embedding into every point set in convex position. Further, we provide new partial positive results on the problem of constructing upward straight-line embeddings of directed paths into point sets in general position.

1 Introduction

Constructing planar straight-line embeddings of graphs into point sets is a well-studied topic of research since more than twenty years. A celebrated result of Gritzmann *et al.* [9] is that the class of graphs that admit a planar straight-line embedding into every point set in general position or in convex position is the one of the *outerplanar graphs*. Efficient algorithms are known to embed outerplanar graphs [4] and trees [5] into any point set in general or in convex position. Further, while testing whether a graph admits a planar straight-line embedding into *every* point set in general or in convex position can be done efficiently, due to the above cited characterization [9] and to the existence of a linear-time algorithm to test whether a graph is outerplanar [11], testing whether a graph admits a planar straight-line embedding into a *given* point set in general position is $\mathcal{NP}$-hard, as proven by Cabello [6]. Planar graph embeddings into point sets have been also studied when edges are allowed to bend (see, e.g., [10,2,7]).

The problem of constructing upward planar straight-line embeddings of directed graphs into point sets has been first suggested by Giordano *et al.* [8] and has been very recently tackled by Binucci *et al.* in [3], who proved the following main results: (a) No biconnected directed graph admits an upward planar straight-line embedding into

---

* This work was partially supported by MIUR (Italy), Projects AlgoDEEP number 2008TF-BWL4 and FIRB "Advanced tracking system in intermodal freight transportation" number RBIP06BZW8, by the German Research Foundation (DFG), project KA 812/15-1 'Graph Drawing for Business Processes', and by the National Technical University of Athens research program ΠEBE 2008.

U. Brandes and S. Cornelsen (Eds.): GD 2010, LNCS 6502, pp. 25–37, 2011.
© Springer-Verlag Berlin Heidelberg 2011

every point set in convex position; (ii) the upward planar straight-line embeddability of a directed graph into every *one-side convex point set* can be characterized and efficiently tested; (iii) there exist directed trees that do not have an upward planar straight-line embedding into every point set in convex position; (iv) every directed path admits an upward planar straight-line embedding into every point set in convex position.

In this paper we continue the study of the straight-line embeddability of directed graphs into planar point sets and show the following results.

In Sect. 3, we study upward planar straight-line embeddings of directed graphs into point sets in general and in convex position. First, we solve an open problem posed in [3], by exhibiting an infinite class of upward planar directed graphs admitting an upward planar straight-line embedding into every point set in convex position, but not into every point set in general position, showing an interesting difference between upward planar straight-line embeddability of directed graphs and planar straight-line embeddability of undirected graphs, as the classes of graphs with a planar straight-line embedding into every point set in convex position and into every point set in general position coincide. Second, we show that every single-source upward planar directed graph with cycles of length at most three admits an upward planar straight-line embedding into every point set in general position. Such a result is the best possible with respect to the number of sources and to the length of the longest cycle.

In Sect. 4, we study upward planar straight-line embeddings of directed trees into point sets in convex position. We solve an open problem posed in [3] by proving that every directed caterpillar admits an upward planar straight-line embedding into every point set in convex position, improving the result in [3] stating that every directed path admits an upward planar straight-line embedding into every point set in convex position.

In Sect. 5, we study upward planar straight-line embeddings of directed paths into point sets in general position. We tackle the problem by considering directed paths with few switches (a switch is either a source or a sink). While the upward planar straight-line embeddability of directed paths with at most two or three switches into point sets in general position can be trivially proven, it is already difficult to deal with directed paths with four or five switches. We prove that directed paths with four (or five) switches admit an upward planar straight-line embedding into every point set in general position, if we suppose that at least one (at least two) of the monotone paths composing the directed paths with four (or five) switches are single edges. Finally, we show that every directed path with at most k switches admits an upward planar straight-line embedding into every point set in general position with $n2^{k-2}$ points.

Omitted proofs can be found in the full version of the paper [1].

2   Preliminaries

A *point set in general position*, or *general point set*, is such that no three points lie on the same line and no two points have the same y-coordinate. The *convex hull* $Ch(S)$ of a point set S is the point set that can be obtained as a convex combination of the points of S. A *point set in convex position*, or *convex point set*, is such that no point is in the convex hull of the others. In a point set S, each point $p \in S$ is given by its coordinates $x(v)$ and $y(v)$ in the plane. We denote by $b(S)$ and by $t(S)$ the lowest and the highest point of S, respectively. A *one-side convex point set* S is a convex point set in which

$b(S)$ and $t(S)$ are adjacent in the border of $Ch(S)$. During the execution of an algorithm which embeds a graph G into a point set S, a *free point* is a point of S to which no vertex of G has been mapped yet. Given a point p in a point set S, a subset S' of S is *clockwise separated around* p if a half-line fixed at p, starting from a horizontal position, directed towards decreasing x-coordinates, and moving clockwise encounters all the points of S' before encountering any other point of S. A *counterclockwise separation* is defined symmetrically.

An *upward planar directed graph* admits a planar drawing where each edge is represented by a curve monotonically increasing in the y-direction. In the following we refer to paths, cycles, caterpillars, and trees meaning upward planar directed graphs whose underlying graphs are paths, cycles, caterpillars, and trees, respectively.

An *upward straight-line embedding* of a graph into a point set is a mapping of each vertex to a distinct point and of each edge to a straight-line segment between its endpoints such that no two edges cross and each edge (u, v) has $y(u) < y(v)$.

A *monotone path* $(v_1, v_2, \ldots, v_k)$ is such that edge (v_i, v_{i+1}) is directed from v_i to v_{i+1}, for $1 \le i \le k - 1$. An upward straight-line embedding of a monotone path into any general point set S can be easily constructed by mapping vertex v_i to the i-th lowest point of S. A monotone path is *trivial* when it consists of a single edge.

3 Embeddings of Directed Graphs into Point Sets

In this section we study the relationship between upward straight-line graph embeddability into convex point sets and upward straight-line graph embeddability into general point sets, and the relationship between upward straight-line graph embeddability, the number of switches, and the length of the longest cycle in the underlying graph.

First, we show an infinite class of graphs that admit an upward straight-line embedding into every convex point set but not into every general point set.

Let G_k be defined as follows, for every $k \ge 3$: G_k has $3k$ vertices, it contains a 3-cycle C^3 composed of edges (u, v), (v, z), and (u, z), and it contains 4-cycles C_i^4, with $i = 1, \ldots, k - 1$, composed of edges (u, v_i), (v_i, w_i), (w_i, z_i), and (u, z_i).

Lemma 1. G_k admits an upward straight-line embedding into every convex point set S with $3k$ points.

Proof: Map u to $b(S)$. Let l and r be the number of points in the subsets L and R of S to the left and to the right, respectively, of the line through $b(S)$ and $t(S)$.

If $l \equiv 0 \bmod 3$ (and $r \equiv 1 \bmod 3$), then iteratively map u_i, v_i, and z_i to the lowest three free points of L, for $i = 1, 2, \ldots, \frac{l}{3}$; further, iteratively map u_i, v_i, and z_i to the lowest three free points of R, for $i = \frac{l+3}{3}, \frac{l+6}{3}, \ldots, k - 1$; finally, map v and z to the highest point of R and to $t(S)$, respectively. If $l \equiv 1 \bmod 3$ (and $r \equiv 0 \bmod 3$), then iteratively map u_i, v_i, and z_i to the lowest three free points of L, for $i = 1, 2, \ldots, \frac{l-1}{3}$; further, map v and z to the highest point of L and to $t(S)$, respectively; finally, iteratively map u_i, v_i, and z_i to the lowest three free points of R, for $i = \frac{l+2}{3}, \frac{l+5}{3}, \ldots, k-1$. If $l \equiv 2 \bmod 3$ (and $r \equiv 2 \bmod 3$), then iteratively map u_i, v_i, and z_i to the lowest three free points of $L \cup \{t(S)\}$, for $i = 1, 2, \ldots, \frac{l+1}{3}$; further, iteratively map u_i, v_i, and z_i to the lowest three free points of R, for $i = \frac{l+4}{3}, \frac{l+7}{3}, \ldots, k - 1$; finally, map v and z to the highest two points of R.

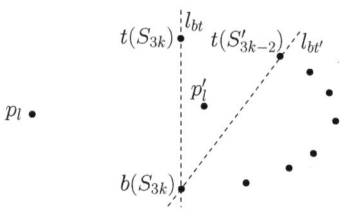

Fig. 1. Point set S_{3k}

In all the cases, the obtained straight-line embedding is upward and planar. ☐

Lemma 2. *There exists a general point set S_{3k} with $3k$ points such that G_k does not admit any upward straight-line embedding into S_{3k}.*

Proof: Point set S_{3k} is any point set that satisfies the following constraints (see Fig. 1). One point p_l is to the left of the line l_{bt} through $b(S_{3k})$ and $t(S_{3k})$. The remaining $3k - 3$ points are to the right of l_{bt} and, together with $b(S)$, they form a convex point set S'_{3k-2} with one point p'_l lying to the left of the line $l_{bt'}$ through $b(S)$ and $t(S'_{3k-2})$.

Observation 1. *Let G be a graph containing a 4-cycle C composed of edges (x_1, x_2), (x_2, x_3), (x_3, x_4), and (x_1, x_4). Let S be a point set such that exactly one point $p_l(S)$ lies to the left of the line through $b(S)$ and $t(S)$. Suppose that an edge of G has been mapped to segment $\overline{b(S)t(S)}$. Then, there exists no upward embedding of G into S in which a vertex of C is mapped to $p_l(S)$.*

Since u is the only source of G_k, such a vertex has to be mapped to $b(S_{3k})$. Further, since a sink of G_k has to be mapped to $t(S_{3k})$ and since every sink of G_k is adjacent to u, segment $\overline{b(S_{3k})t(S_{3k})}$ is part of any embedding. Then, by Observation 1 no vertex of a 4-cycle C_i^4 of G_k is mapped to p_l. Hence, a vertex of C^3 is mapped to p_l. If such a vertex is z, then vertex v is mapped to a point to the right of l_{bt}, hence $\overline{b(S)t(S)}$ crosses the segment between z and v. It follows that v is mapped to p_l. If z is not mapped to $t(S_{3k})$, then $\overline{b(S_{3k})t(S_{3k})}$ crosses the segment between z and v. Hence, z is mapped to $t(S_{3k})$. Then, all the vertices of the 4-cycles of G_k are mapped to the vertices of S'_{3k-2}. A sink of one of the 4-cycles has to be mapped to $t(S'_{3k})$. Since every sink of G_k is adjacent to u, segment $\overline{b(S_{3k})t(S'_{3k})}$ is part of any embedding. Then, by Observation 1 no vertex of a 4-cycle C_i^4 of G_k can be mapped to p'_l, thus proving the lemma. ☐

We get the following:

Theorem 1. *For every $k \geq 3$, there exists a $3k$-vertex upward planar digraph that admits an upward straight-line embedding into every convex point set with $3k$ points but not into every general point set with $3k$ points.*

Next, we show that every single-source graph G whose every simple cycle has length three admits an upward straight-line embedding into every general point set. Such a result is tight both with respect to the maximum length of a cycle in G and with respect to the number of sources in G. Namely, a single-source graph exists whose every simple

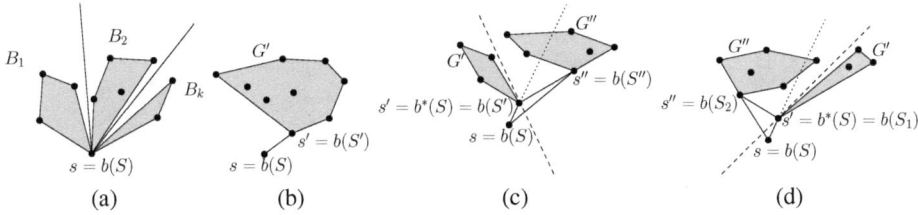

Fig. 2. (a) s is a cut-vertex. (b) B is an edge. (c) B is a 3-cycle (s, s', s'') and l encounters $b(S'')$ after s. (d) B is a 3-cycle (s, s', s'') and l encounters $b(S'')$ before s.

cycle has length at most four not admitting any upward straight-line embedding into some general point set (by Lemma 2). Further, a graph with two sources exists whose every simple cycle has length three not admitting any upward straight-line embedding into some general point set (by the results in [3] on upward straight-line embeddability into one-side convex point sets).

We show a recursive algorithm to construct upward straight-line embeddings of single-source graphs whose every simple cycle has length three into every general point set. The recursion is on the number x of biconnected components of G. If $x = 1$, then the statement is trivially true. If $x > 1$, consider the unique source s of G.

If s is a cutvertex of G, denote by $B_1, \ldots, B_k$ the connected components obtained by removing s from G (see Fig. 2(a)). Clockwise separate the sets $S_1, \ldots, S_k$ with $|B_1|, \ldots, |B_k|$ points, respectively, around $b(S)$. Recursively construct straight-line embeddings of the subgraphs of G induced by the vertices in $B_1 \cup \{b(S)\}$, $\ldots$, in $B_k \cup \{b(S)\}$ into point sets $S_1 \cup \{b(S)\}, \ldots, S_k \cup \{b(S)\}$, respectively.

If s is not a cutvertex of G, consider the biconnected component B incident to s.

If B is an edge (s, s'), denote by S' the point set obtained by removing $b(S)$ from S and by G' the graph obtained by removing s and its incident edge from G (see Fig. 2(b)). Recursively construct an upward straight-line embedding of G' into S'.

If B is a 3-cycle (s, s', s''), denote by $b^*(S)$ the lowest point of S different from $b(S)$. Denote by G' (by G'') the graph composed of s' (resp. of s'') and of every connected component not containing s that is obtained by removing s' (resp. s'') from G. Clockwise separate $|G'| - 1$ points around $b^*(S)$. Such points, together with $b^*(S)$, form a set S'. Let $S'' = S \setminus \{S' \cup \{b(S)\}\}$. Consider a line l fixed at $b^*(S)$ and rotating clockwise starting from a horizontal position. If l encounters $b(S'')$ after s (see Fig. 2(c)), then recursively construct upward straight-line embeddings of G' into S' and of G'' into S''. If l encounters $b(S'')$ before s (see Fig. 2(d)), then counterclockwise separate $|G'| - 1$ points around $b^*(S)$. Such points, together with $b^*(S)$, form a set S_1. Let $S_2 = S \setminus \{S_1 \cup \{b(S)\}\}$. Then, recursively construct upward straight-line embeddings of G' into S_1 and of G'' into S_2.

We get the following:

Theorem 2. *Every single-source upward planar directed graph whose every simple cycle has length three admits an upward straight-line embedding into every point set in general position.*

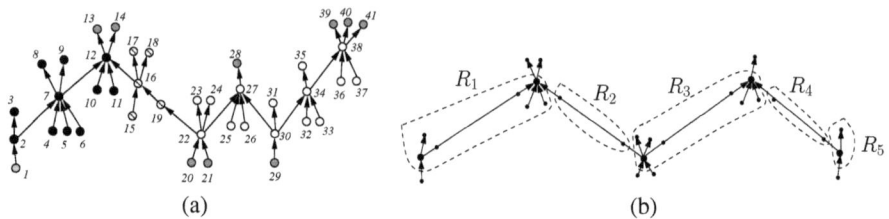

Fig. 3. (a) A caterpillar $G = (V, E)$ and function $\mathcal{E} : V \to \{1, 2, \ldots, n\}$. (b) Decomposition of a caterpillar into increasing caterpillars, decreasing caterpillars, and extremal legs.

4   Embeddings of Directed Caterpillars into Convex Point Sets

In this section we prove that every caterpillar admits an upward straight-line embedding into every point set in convex position.

We introduce some terminology. A *caterpillar* G is a tree such that removing all the degree-1 vertices, called the *legs* of G, yields a path, called the *spine* of G. A caterpillar whose spine is a monotone path is a *monotone caterpillar*. Let v_s and v_t be a source and a sink of the spine of a caterpillar. A vertex w that is connected to v_s by edge (w, v_s) or to v_t by edge (v_t, w) is an *extremal leg* of a caterpillar. In Fig. 3(a) the extremal legs are numbered $1, 13, 14, 20, 21, 28, 29, 39, 40, 41$.

Let G be a caterpillar, let T be its spine, and let u be one of the end-points of T. Let U be the set composed of u and of the extremal legs of G adjacent to u. The following lemma descends from algorithms presented in the literature [8,3].

Lemma 3. *Suppose that u is a source (resp. a sink) of T. Then, G admits an upward straight-line embedding into every one-side convex point set S in which the vertices of G in U are mapped to the $|U|$ lowest (resp. highest) points of S.*

Let G be a monotone caterpillar and let T be its spine. Let s and t be the source and the sink of T, respectively. In addition, suppose that s and t are a source and a sink of G, respectively. We have the following:

Lemma 4. *G admits an upward straight-line embedding into every convex point set S in which s is mapped to the lowest point of S and t is mapped to the highest point of S.*

Next, for a caterpillar $G=(V, E)$, we define a bijective function $\mathcal{E} : V \to \{1, 2, \ldots, n\}$. Let T be the spine of G and let a and b be the end-vertices of T. Function $\mathcal{E}$ is defined according to the following rules (see Fig. 3(a)). ($\mathcal{R}1$): For any two vertices $u, v \in T$ such that u comes before v when traversing T from a to b, the value associated to u and to all the legs adjacent to u is smaller than the value associated to v and to all legs adjacent to v; ($\mathcal{R}2$): For any vertex $u \in T$, the value associated to u is greater than the value associated to all the legs incident to edges entering u; ($\mathcal{R}3$): For any vertex $u \in T$, the value associated to u is smaller than the value associated to all the legs incident to edges exiting u.

We now describe an algorithm to construct an upward straight-line embedding of any caterpillar G into any convex point set S. The idea is to partition G into three

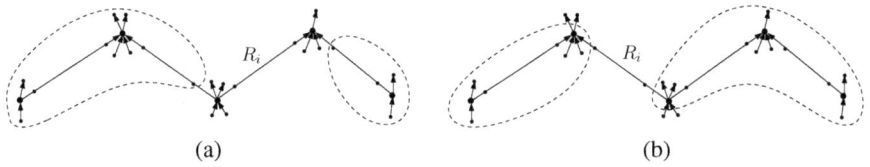

Fig. 4. Sub-caterpillars of (a) an increasing caterpillar R_i and (b) a decreasing caterpillar R_i

smaller caterpillars that can be embedded into suitable subsets of S by means of the two algorithms described above. In the following we formalize this idea.

Let G be a caterpillar with spine T. Let $(T_1, \ldots, T_k)$ be the maximal monotone paths composing T; T_i is an *increasing path* if, for any edge (u, v) in T_i, $\mathcal{E}(u) < \mathcal{E}(v)$ and a *decreasing path* otherwise. Assume the sources and the sinks of T belong to the increasing paths and not to the decreasing paths. Hence, T_1 and T_k are increasing paths, possibly with one vertex. Let R_i be the caterpillar induced by T_i and by the non-extremal legs of G adjacent to T_i; R_i is an *increasing* (resp. *decreasing*) *caterpillar* if T_i is an increasing (resp. decreasing) path. Caterpillar G is partitioned into increasing caterpillars, decreasing caterpillars, and extremal legs (see Fig. 3(b)). If R_i is an increasing caterpillar, let $s(R_i)$ and $t(R_i)$ be the source and the sink of T_i, respectively. If R_i is a decreasing caterpillar, let $t(R_i) = t(R_{i-1})$ and $s(R_i) = s(R_{i+1})$, that is, $s(R_i)$ (resp. $t(R_i)$) is the source (the sink) of T immediately following (preceding) T_i. Observe that, if R_i is a decreasing caterpillar, $s(R_i)$ and $t(R_i)$ do not belong to R_i.

Next, we define the *sub-caterpillars* G_i^1 and G_i^2 of G induced by R_i. If R_i is an increasing caterpillar (see Fig. 4(a)), set G_i^1 (G_i^2) to be the caterpillar induced by the vertices of G preceding $s(R_i)$ (following $t(R_i)$, resp.) in $\mathcal{E}$, except for the extremal legs adjacent to $s(R_i)$ (to $t(R_i)$, resp.). If R_i is a decreasing caterpillar (see Fig. 4(b)), set G_i^1 (G_i^2) to be the caterpillar induced by $t(R_i)$ (by $s(R_i)$, resp.) and by the vertices of G preceding $t(R_i)$ (following $s(R_i)$, resp.) in $\mathcal{E}$.

Consider a line l through $b(S)$ and $t(S)$. Denote by A and B the point sets to the left and to the right of l, resp. (see Fig. 5(a)). Points $b(S)$ and $t(S)$ belong to A. Consider any increasing caterpillar R_i. Denote by i_l (resp. i_h) the number of extremal legs adjacent to $s(R_i)$ (resp. to $t(R_i)$) and by L (resp. H) the set of the $i_l + 1$ lowest (resp. of the $i_h + 1$ highest) points of S. Let $A' = A \setminus (L \cup H)$, $B' = B \setminus (L \cup H)$, $|H \cap A| = h_a$, $|H \cap B| = h_b$, $|L \cap A| = l_a$, and $|L \cap B| = l_b$ (see Fig. 5(a)). We have the following:

Lemma 5. *If* $|G_i^1| \leq |A| - l_a$ *and* $|G_i^2| \leq |B| - h_b$, *then there is an upward straight-line embedding of* G *into* S.

Proof: We distinguish three cases:

Case 1: $|G_i^1| \leq |A'|$ and $|G_i^2| \leq |B'|$. Refer to Fig. 6(a). Map $s(R_i)$ to $t(L)$ and the extremal legs adjacent to $s(R_i)$ to the other points of L. Map $t(R_i)$ to $b(H)$ and the extremal legs adjacent to $t(R_i)$ to the other points of H. Embed G_i^1 into the $|G_i^1|$ lowest points of A' and G_i^2 into the $|G_i^2|$ highest points of B'. Such embeddings can be constructed by Lemma 3, since A' and B' are one-side convex point sets. Embed

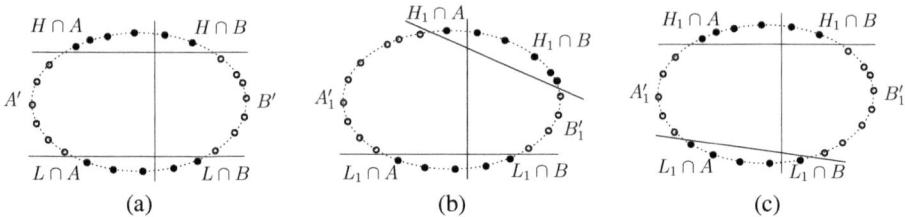

Fig. 5. Point sets for (a) Case 1, (b) Case 2, and (c) Case 3

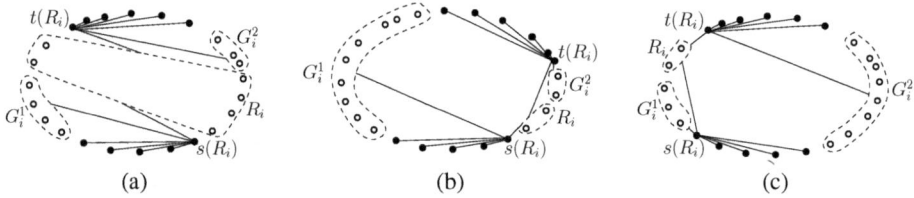

Fig. 6. Embedding G into S if (a) $|G_i^1| \leq |A'|$ and $|G_i^2| \leq |B'|$, (b) $|G_i^1| \leq |A| - l_a$, $|G_i^2| \leq |B| - h_b$, and $|G_i^1| > |A'|$, and (c) $|G_i^2| \leq |B| - h_b$, $|G_i^1| \leq |A'|$, and $|G_i^2| > |B'|$

R_i into the remaining free points of S. This can be done by Lemma 4, since R_i is a monotone caterpillar. We have the following:

Claim 1. *The constructed straight-line embedding of G into S is upward and planar.*

Case 2: $|G_i^1| > |A'|$. We create a partition (A_1', B_1', H_1, L_1) of S such that $|G_i^1| = |A_1'|$, $|G_i^2| + |R_i| = |B_1'|$, $i_l + 1 = |L_1|$, and $i_h + 1 = |H_1|$ (see Fig. 5(b)). Let $d_{G_i^1} = |G_i^1| - |A'|$. By the assumptions of the lemma, $|G_i^1| \leq |A| - l_a$. Since $|A| = |A'| + l_a + h_a$, we have $d_{G_i^1} \leq h_a$. Define A_1' as A' plus the $d_{G_i^1}$ lowest points of $H \cap A$, B_1' as B' minus the $d_{G_i^1}$ highest points of B', $L_1 = L$, and $H_1 = S \setminus (A_1' \cup B_1' \cup L_1)$. Refer to Fig. 6(b). Map $s(R_i)$ to $t(L_1)$ and map the extremal legs adjacent to $s(R_i)$ to the other points of L_1. Map $t(R_i)$ to $b(H_1)$ and the extremal legs adjacent to $t(R_i)$ to the other points of H_1. Embed G_i^1 into A_1' and G_i^2 into the $|G_i^2|$ highest points of B_1'. Such embeddings can be constructed by Lemma 3, since A_1' and B_1' are one-side convex point sets. Embed R_i into the remaining free points of S. This can be done by Lemma 4, since R_i is a monotone caterpillar. We have the following:

Claim 2. *The constructed straight-line embedding of G into S is upward and planar.*

Case 3: $|G_i^1| \leq |A'|$ and $|G_i^2| > |B'|$. We create a partition (A_1', B_1', H_1, L_1) of S such that $|G_i^1| + |R_i| = |A_1'|$, $|G_i^2| = |B_1'|$, $i_l + 1 = |L_1|$, and $i_h + 1 = |H_1|$ (see Fig. 5(c)). Let $d_{G_i^2} = |G_i^2| - |B'|$. By the assumptions of the lemma, $|G_i^2| \leq |B| - h_b$. Since $|B| = |B'| + l_b + h_b$, we have $d_{G_i^2} \leq l_b$. Define B_1' as B' plus the $d_{G_i^2}$ highest points of $L \cap B$, A_1' as A' minus the $d_{G_i^2}$ lowest points of A', $H_1 = H$, and $L_1 = S \setminus (A_1' \cup B_1' \cup H_1)$. Refer to Fig. 6(c). Map $s(R_i)$ to $t(L_1)$ and the extremal legs adjacent to $s(R_i)$ to the other points of L_1. Map $t(R_i)$ to $b(H_1)$ and the extremal legs adjacent to $t(R_i)$ to the other points of H_1. Embed G_i^2 into B_1' and G_i^1 into the $|G_i^1|$ lowest points of A_1'. Such embeddings can be constructed by Lemma 3, since A_1' and B_1' are one-side convex point sets. Embed R_i into the remaining free points of S. This can be done by Lemma 4, since R_i is a monotone caterpillar. We have the following:

Claim 3. *The constructed straight-line embedding of G into S is upward and planar.*

$\square$

Now consider any decreasing caterpillar R_i that is part of G. Define i_l, i_h, L, H, A', B', h_a, h_b, l_a, and l_b as before. We have the following:

Lemma 6. *If $|G_i^1| \leq |A| - h_a + 1$ and $|G_i^2| \leq |B| - l_b + 1$, then there is an upward straight-line embedding of G into S.*

The proof of Lemma 6 is analogous to the one of Lemma 5. The requirements $|G_i^1| \leq |A| - h_a + 1$ and $|G_i^2| \leq |B| - l_b + 1$ of Lemma 6 are weaker than the requirements $|G_i^1| \leq |A| - l_a$ and $|G_i^2| \leq |B| - h_b$ of Lemma 5. This is due to the fact that, if R_i is a decreasing caterpillar, then $t(R_i)$ and $s(R_i)$ belong to G_i^1 and to G_i^2, respectively, and they do not belong to R_i. We are now ready to prove the following:

Theorem 3. *Any n-vertex directed caterpillar G admits an upward straight-line embedding into every convex point set S with n points.*

Proof: Let T be the spine of G and let $\{R_1, \ldots, R_k\}$ be the increasing and decreasing caterpillars of G. Let (A, B) be the partition of S created by a line through $b(S)$ and $t(S)$, where $b(S)$ and $t(S)$ belong to A. Consider the $|A|$-th vertex $v_{|A|}$ in the order $v_1, \ldots, v_n$ of the vertices of G defined by $\mathcal{E}$. We partition G into three smaller caterpillars and we draw each of them on a suitably chosen portion of S. The partition of G is determined by the position of $v_{|A|}$ in G. We distinguish four cases:

Case 1: $v_{|A|}$ *is a vertex of an increasing caterpillar R_i.* Define i_l, i_h, L, H, h_b, l_a, h_a, and l_b as before. Since $v_{|A|} \in R_i$, we have that $|G_i^1| + i_l < |A|$. Since $i_l + 1 = l_a + l_b$, it follows that $l_a \leq i_l + 1$. Thus, $|G_i^1| + l_a \leq |A|$. Analogously, we have that $|G_i^2| + i_h < |B|$. Since $i_h + 1 = h_a + h_b$, it follows that $h_b \leq i_h + 1$. Thus, $|G_i^2| + h_b \leq |B|$. Hence, Lemma 5 applies and the result follows.

Case 2: $v_{|A|}$ *is a vertex of a decreasing caterpillar R_i.* Analogously to Case 1, it can be proven that $|G_i^1| \leq |A| - h_a + 1$ and $|G_i^2| \leq |B| - l_b + 1$. Hence, Lemma 6 applies and the result follows.

Case 3: $v_{|A|}$ *is an extremal leg adjacent to a sink of T.* Let R_i and R_{i+1} be such that $t(R_i) = t(R_{i+1})$ and $v_{|A|}$ is an extremal leg adjacent to $t(R_i)$. Note that R_i is an increasing caterpillar and R_{i+1} is a decreasing caterpillar. Denote by i_h the number of extremal legs adjacent to $t(R_i)$ and by H the set of the $i_h + 1$ highest points of S. Let $|H \cap B| = h_b$ and $|H \cap A| = h_a$. Notice that $h_a + h_b = i_h + 1$. We claim the following.

Claim 4. *Let G_i^1 and G_i^2 (let G_{i+1}^1 and G_{i+1}^2) be the sub-caterpillars of G induced by R_i (resp. by R_{i+1}). At least one of the following inequalities holds: (1) $|G_{i+1}^1| \leq |A| - h_a + 1$; (2) $|G_i^2| \leq |B| - h_b$.*

We further distinguish two cases.

Inequality (1) holds. Consider the decreasing caterpillar R_{i+1}. Let i_l be the number of extremal legs adjacent to $s(R_{i+1})$ and let L be the set of the $i_l + 1$ lowest point of S. Denote $l_a = |L \cap A|$ and $l_b = |L \cap B|$. We have that $|B| > |G^1_{i+1}| + i_l$, hence $|B| \geq |G^1_{i+1}| + l_b$, as $i_l + 1 = l_a + l_b$. Hence, Lemma 6 applies and the result follows.

Inequality (2) holds. Consider the increasing caterpillar R_i. Let i_l be the number of extremal legs adjacent to $s(R_{i+1})$ and let L be the set of the $i_l + 1$ lowest point of S. Denote $l_a = |L \cap A|$ and $l_b = |L \cap B|$. We have that $|A| > |G^1_i| + i_l$, hence $|A| \geq |G^1_i| + l_a$, as $i_l + 1 = l_a + l_b$. Hence, Lemma 5 applies and the result follows.

Case 4: $v_{|A|}$ is an extremal leg adjacent to a source of T. Analogously to Case 3, it can be proven that either $|A| \geq |G^1_{i+1}| + h_a - 1$ and $|B| \geq |G^1_{i+1}| + l_b - 1$ hold simultaneously, thus the result follows from Lemma 6, or $|A| \geq |G^1_i| + l_a$ and $|B| \geq |G^2_i| + h_b$ hold simultaneously, thus the result follows from Lemma 5.   □

5   Embeddings of Directed Paths into General Point Sets

In this section we deal with upward straight-line embeddings of paths with few switches into general point sets. We first deal with paths with four switches.

Theorem 4. *Every path P composed of three monotone paths P_1, P_2, and P_3 admits an upward straight-line embedding into every general point set S if at least one out of P_1, P_2, and P_3 is a trivial path.*

Proof: Let $P_1 = (s_1 = u_1, \ldots, u_U = t_1)$, $P_2 = (t_1 = v_1, \ldots, v_V = s_2)$, and $P_3 = (s_2 = w_1, \ldots, w_W = t_2)$ be the monotone paths composing P, where s_1 and s_2 are sources and t_1 and t_2 are sinks. If P_2 is trivial, a more general result in [3] states that a path P admits an upward straight-line embedding into every general point set if the i-th monotone path composing P is trivial, for every odd i or for every even i. We discuss the case in which P_3 is trivial, the case in which P_1 is trivial being symmetric. Counterclockwise separate a set S_1 of $U - 1$ points around $t(S)$. Construct upward straight-line embeddings of P_1 into $S_1 \cup \{t(S)\}$ and of $P_2 \setminus \{s_2\}$ into the $V - 1$ highest points of $S \setminus S_1$. Map s_2 to $b(S \setminus S_1)$ and t_2 to the only remaining free point of S.

Claim 5. *The constructed straight-line embedding of P into S is upward and planar.*

We now deal with paths with five switches.   □

Theorem 5. *Every path P composed of four monotone paths P_1, P_2, P_3, and P_4 admits an upward straight-line embedding into every general point set S if at least two out of P_1, P_2, P_3, and P_4 are trivial paths.*

Proof: Let $P_1 = (s_1 = u_1, u_2, \ldots, u_U = t_1)$, $P_2 = (t_1 = v_1, v_2, \ldots, v_V = s_2)$, $P_3 = (s_2 = w_1, w_2, \ldots, w_W = t_2)$, and $P_4 = (t_2 = z_1, z_2, \ldots, z_Z = s_3)$ be the monotone paths composing P, where s_1, s_2, and s_3 are sources and t_1 and t_2 are sinks. The case in which P has three sinks and two sources can be discussed analogously. If P_1 and P_3 are trivial or if P_2 and P_4 are trivial, the proof follows from the result in [3] cited in the proof of Theorem 4. We discuss the case in which P_1 and P_4 are trivial. Clockwise separate a set $S_{1,2}$ of V points around $b(S)$. Construct upward straight-line embeddings of $P_2 \setminus \{t_1\}$ into the $V - 1$ lowest points of $S_{1,2} \cup b(S)$ and of $P_3 \setminus \{t_2\}$ into the $W - 1$ lowest points of $S \setminus S_{1,2}$. Map t_1 to $t(S_{1,2})$ and s_1 to the only remaining free point of $S_{1,2}$. Map t_2 to $t(S \setminus S_{1,2})$ and s_3 to the only remaining free point of $S \setminus S_{1,2}$.

Claim 6. *The constructed straight-line embedding of P into S is upward and planar.*

We discuss the case in which P_1 and P_2 are trivial, the case in which P_3 and P_4 are trivial being symmetric. Clockwise separate a set S_4 of $Z - 1$ points around $t(S)$. Construct upward straight-line embeddings of P_4 into $S_4 \cup \{t(S)\}$ and of $P_3 \setminus \{s_2\}$ into the $W - 1$ highest points of $S \setminus S_4$. Consider the line l' through the point where w_2 is drawn and the lowest free point of S. If both the two remaining free points of S are on the same side of l', then map s_2 to $b(S \setminus S_4)$, map t_1 to the highest free point of S, and map s_1 to the other free point of S. Otherwise, one of the two remaining free points of S is to the left l' and the other one is to its right. Then, map s_2 to the lowest of such two points, map t_1 to the highest of such two points, and map s_1 to $b(S \setminus S_4)$.

Claim 7. *The constructed straight-line embedding of P into S is upward and planar.*

We discuss the more involved case in which P_2 and P_3 are trivial. Let p_1 and p_2 be the two highest points of S, with $y(p_1) \geq y(p_2)$. Counterclockwise separate a set S_1 of $U - 1$ points around p_2. Denote by p the point that is added to S_1 when U points are counterclockwise separated around p_2. We consider the following two cases:

Point p is to the left of $l_{1,2}$: Refer to Fig. 7(a). Consider a half-line l_1 fixed at p_1 and passing through p. Rotate l_1 in counterclockwise direction. Let p' be the last point of S_1 encountered by l_1 before encountering p_2. If no point of S_1 is encountered by l_1 before p_2, then let $p' = p$. Map t_1 to p_1, t_2 to p_2, and s_2 to p'; construct upward straight-line embeddings of P_1 into $S_1 \cup \{p_1, p\} \setminus \{p'\}$ and of P_4 into $S \setminus \{S_1\} \cup \{p_2\} \setminus \{p\}$.

Point p is to the right of $l_{1,2}$: Refer to Fig. 7(b). Consider a half-line l_1 fixed at p_1 and passing through p. Rotate l_1 in clockwise direction. Let p' be the last point of S_1 encountered by l_1 before encountering p_2. If no point of S_1 is encountered by l_1 before p_2, then let $p' = p$. Map t_1 to p_2, t_2 to p_1, and s_2 to p'; construct upward straight-line embeddings of P_1 into $S_1 \cup \{p_2\}$ and of P_4 into $S \setminus \{S_1\} \cup \{p_1, p\} \setminus \{p'\}$.

Claim 8. *The constructed straight-line embedding of P into S is upward and planar.*

$\square$

Next, we tackle the problem of embedding paths with at most k switches into general point sets with more than n points. We show the following:

Theorem 6. *Every directed path P with n vertices and k switches admits an upward straight-line embedding into every general point set S with $|S| \geq n2^{k-2}$.*

Proof: We prove the statement by induction on the number of switches; we suppose inductively that one of the end-vertices of P is mapped to $b(S)$ or $t(S)$, depending on whether such a vertex is a source or a sink. The statement is trivial if $k = 2$, as in such a case P is monotone and any general point set with n points suffices.

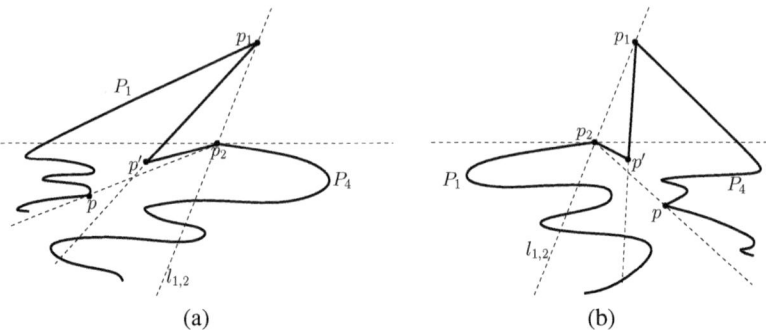

Fig. 7. (a) Point p is to the left of $l_{1,2}$. (b) Point p is to the right of $l_{1,2}$.

Suppose that $k > 2$. Let a_1 be an end-vertex of P. Suppose that a_1 is a source, the case in which it is a sink being analogous. Let $P_a = (a_1, a_2, \ldots, a_l)$ be the maximal monotone path of P containing a_1. Notice that $l \geq 2$. Map $P_a \setminus \{a_l\}$ to the $|P_a| - 1$ points of S with lowest y-coordinate. Denote such a point set by S_a. Map a_l to $t(S)$. Let S_1 and S_2 be the point sets composed of $t(S)$ and of the points of $S \setminus S_a$ to the left and to the right, respectively, of the line through a_{l-1} and a_l. If $|S_1| \geq |S_2|$ (if $|S_2| > |S_1|$), construct an upward straight-line embedding of $P \setminus \{a_1, a_2, \ldots, a_{l-1}\}$ into S_1 (into S_2, resp.) with a_l placed at $t(S_1) = t(S)$ (at $t(S_2) = t(S)$, resp.).

It is easy to see that the constructed straight-line embedding is upward and planar. We show that the cardinality of point sets S_1 and S_2 is sufficient to apply the induction. The number of points in the one of S_1 and S_2 with more points is at least $(|S| - (l-1))/2$. Further, $P \setminus \{a_1, a_2, \ldots, a_{l-1}\}$ has $n - (l-1)$ vertices and $k - 1$ switches. Since $|S| \geq n2^{k-2}$, the one of S_1 and S_2 with more points has at least $(n2^{k-2} - (l-1))/2 = n2^{k-3} - (l-1)/2 > n2^{k-3} - (l-1)2^{k-3}$ and, since $k > 2$, the lemma follows. $\square$

6 Open Problems

In this paper we continued the study of upward straight-line embeddability of directed graphs into point sets initiated in [8,3]. While we solved some of the open questions posed by Binucci *et al.* in [3], the following problems remain open: (i) Is it possible to test in polynomial time whether a directed graph/tree admits an upward straight-line embedding into every point set in general/convex position? (ii) Does every directed path admit an upward straight-line embedding into every point set in general position? (iii) Is there a polynomial function $p(n, k)$ such that every directed path admits an upward straight-line embedding into every point set in general position with at least $p(n, k)$ points? Lemma 6 shows that every directed path admits an upward straight-line embedding into every point set in general position with at least $n2^{k-2}$ points, which is exponential in k.

References

1. Angelini, P., Frati, F., Geyer, M., Kaufmann, M., Mchedlidze, T., Symvonis, A.: Upward geometric graph embeddings into point sets. Tech. Report 177, Dipartimento di Informatica e Automazione, Università Roma Tre (2010)

2. Badent, M., Di Giacomo, E., Liotta, G.: Drawing colored graphs on colored points. Theor. Comput. Sci. 408(2-3), 129–142 (2008)
3. Binucci, C., Di Giacomo, E., Didimo, W., Estrella-Balderrama, A., Frati, F., Kobourov, S., Liotta, G.: Upward straight-line embeddings of directed graphs into point sets. Computat. Geom. Th. Appl. 43, 219–232 (2010)
4. Bose, P.: On embedding an outer-planar graph in a point set. Computat. Geom. Th. Appl. 23(3), 303–312 (2002)
5. Bose, P., McAllister, M., Snoeyink, J.: Optimal algorithms to embed trees in a point set. J. Graph Alg. Appl. 1(2), 1–15 (1997)
6. Cabello, S.: Planar embeddability of the vertices of a graph using a fixed point set is NP-hard. J. Graph Alg. Appl. 10(2), 353–366 (2006)
7. Di Giacomo, E., Didimo, W., Liotta, G., Meijer, H., Trotta, F., Wismath, S.K.: k-colored point-set embeddability of outerplanar graphs. J. Graph Alg. Appl. 12(1), 29–49 (2008)
8. Giordano, F., Liotta, G., Mchedlidze, T., Symvonis, A.: Computing upward topological book embeddings of upward planar digraphs. In: Tokuyama, T. (ed.) ISAAC 2007. LNCS, vol. 4835, pp. 172–183. Springer, Heidelberg (2007)
9. Gritzmann, P., Pach, B.M.J., Pollack, R.: Embedding a planar triangulation with vertices at specified positions. Amer. Math. Mont. 98, 165–166 (1991)
10. Kaufmann, M., Wiese, R.: Embedding vertices at points: Few bends suffice for planar graphs. J. Graph Alg. Appl. 6(1), 115–129 (2002)
11. Wiegers, M.: Recognizing outerplanar graphs in linear time. In: Tinhofer, G., Schmidt, G. (eds.) WG 1986. LNCS, vol. 246, pp. 165–176. Springer, Heidelberg (1987)

On a Tree and a Path with No Geometric Simultaneous Embedding[*]

Patrizio Angelini[1], Markus Geyer[2], Michael Kaufmann[2], and Daniel Neuwirth[2]

[1] Dipartimento di Informatica e Automazione – Università Roma Tre, Italy
angelini@dia.uniroma3.it
[2] Wilhelm-Schickard-Institut für Informatik – Universität Tübingen, Germany
geyer/mk/neuwirth@informatik.uni-tuebingen.de

Abstract. Two graphs $G_1 = (V, E_1)$ and $G_2 = (V, E_2)$ admit a geometric simultaneous embedding if there exists a set of points P and a bijection $M : P \rightarrow V$ that induce planar straight-line embeddings both for G_1 and for G_2. The most prominent problem in this area is the question whether a tree and a path can always be simultaneously embedded. We answer this question in the negative by providing a counterexample. Additionally, since the counterexample uses disjoint edge sets for the two graphs, we also prove that it is not always possible to simultaneously embed two edge-disjoint trees. Finally, we study the same problem when some constraints on the tree are imposed. Namely, we show that a tree of height 2 and a path always admit a geometric simultaneous embedding. In fact, such a strong constraint is not so far from closing the gap with the instances not admitting any solution, as the tree used in our counterexample has height 4.

1 Introduction

Embedding planar graphs is a well-established field in graph theory and algorithms with many applications. Keystones in this field are the works of Thomassen [18], Tutte [19], and Pach and Wenger [17], dealing with planar and convex representations of graphs.

Recently, motivated by the need of concurrently represent different relationships among the same elements, a major focus in the research lies on *simultaneous graph embedding*, in which, given a set of graphs with the same vertex-set, the goal is to place the vertices on the plane so that all the graphs are planar, when drawn separately. Problems of this kind frequently arise in the visualization of evolving networks and in the visualization of huge and complex relationships, as the graph of the Web.

Among the many variants of this problem, the most important and natural one is the *geometric simultaneous embedding* (GSE). Given two graphs $G_1 = (V, E')$ and $G_2 = (V, E'')$, the task is to find a set of points P and a bijection $M : P \rightarrow V$ that induce planar *straight-line* embeddings for both G_1 and G_2.

In the seminal paper on this topic [3], Brass *et al.* proved that GSEs of pairs of paths, of cycles, and of caterpillars always exist. A *caterpillar* is a tree such that deleting all its leaves yields a path. On the other hand, they provided negative results for a pair of

---

[*] This work was supported in part by MIUR of Italy, project AlgoDEEP prot. 2008TFBWL4, and by the German Research Foundation (DFG), project KA 812/15-1 'Graph Drawing for Business Processes'.

U. Brandes and S. Cornelsen (Eds.): GD 2010, LNCS 6502, pp. 38–49, 2011.

outerplanar graphs and for three paths. Erten and Kobourov [6] found a planar graph and a path not allowing any GSE. Geyer *et al.* [14] proved that there exist two edge-disjoint trees not admitting any GSE. Finally, Cabello *et al.* [4] showed a planar graph and a matching not admitting any GSE and gave algorithms to obtain a GSE of a matching and a wheel, an outerpath, or a tree. The most important open problem is the question whether a tree and a path always admit a GSE or not, that is the subject of this paper.

Many variants of the problem, where some constraints are relaxed, have been studied. In the *simultaneous embedding* setting, where the edges do not need to be straight-line segments, any number of planar graphs admit a simultaneous embedding, since any planar graph can be planarly embedded on any given set of points in the plane [16,17]. However, the same does not hold in the *simultaneous embedding with fixed edges* setting [11,13,8], in which the edges shared by the two graphs have to be represented by the same Jordan curve. Finally, the research on this problem opened a new exciting field of problems and techniques, like ULP trees and graphs [7,9,10], colored simultaneous embedding [2], near-simultaneous embedding [12], and matched drawings [5], deeply related to the general fundamental question of point-set embeddability.

In this paper we study the GSE problem of a tree and a path. We answer the question in the negative by providing a counterexample, that is, a tree and a path not admitting any GSE. Moreover, since the tree and the path used in our counterexample do not share any edge, we also negatively answer the question on two edge-disjoint trees.

The main idea is to use the path to enforce part of the tree to be in a non-planar configuration. Namely, we consider level nonplanar trees [7,10], that is, trees not admitting any planar embedding if their vertices must be placed inside certain regions according to a particular leveling. The tree of the counterexample contains many copies of such trees, while the path creates the regions. To prove that at least one copy has the non-planar leveling, we need a huge number of vertices, which is often needed just to ensure the existence of certain structures playing a role in our proof. A much smaller counterexample could likely be constructed with the same techniques, but we decided to prefer the simplicity of the arguments rather than the search for the minimum size.

In Sect. 2 we give define level nonplanar trees. In Sect. 3 we describe a tree T and a path $\mathcal{P}$, and in Sect. 4 we show that T and $\mathcal{P}$ do not admit any GSE. In Sect. 5 we give an algorithm to construct a GSE of a tree of height 2 and a path and in Sect. 6 we make some final remarks. Omitted proofs can be found in the full version of the paper [1].

2 Preliminaries

A (undirected) *k-level tree* $T = (V, E, \phi)$ is a tree $T' = (V, E)$, called the *underlying tree* of T, together with a leveling of its vertices given by a function $\phi : V \mapsto \{1, \ldots, k\}$, such that for every edge $(u, v) \in E$, it holds $\phi(u) \neq \phi(v)$ (see [7,10]). A drawing of $T = (V, E, \phi)$ is a *level drawing* if each vertex $v \in V$ such that $\phi(v) = i$ lies on a horizontal line $l_i = \{(x, i) \mid x \in \mathbb{R}\}$. A level drawing of T is *planar* if no two edges intersect except, possibly, at common end-points. A k-level tree is *level nonplanar* if it does not admit any planar level drawing. We extend this concept to the one of *region-level drawing* by enforcing the vertices of each level to lie inside a region rather than on a horizontal line. Let $l_1, \ldots, l_k$ be k non-crossing straight-line segments and let $r_1, \ldots, r_{k+1}$ be the regions such that any straight-line segment connecting a

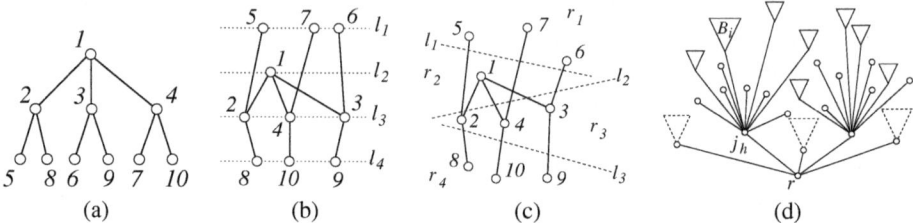

Fig. 1. (a) A tree T_u. (b)-(c) A level nonplanar tree and a region-level nonplanar tree whose underlying tree is T_u. (d) A schematization of $\mathcal{T}$. Joints and stabilizers are small circles. A solid triangle represents a branch, while a dashed triangle represents the subtree connected to a joint.

point in r_i and a point in r_h, with $1 \leq i < h \leq k + 1$, cuts all and only the segments $l_i, l_{i+1}, \ldots, l_{h-1}$, in this order. A drawing of a k-level tree $T = (V, E, \phi)$ is a *region-level drawing* if each vertex $v \in V$ such that $\phi(v) = i$ lies inside region r_i. A k-level tree is *region-level nonplanar* if it does not admit any planar region-level drawing. The 4-level tree T whose underlying tree is shown in Fig. 1(a) is level nonplanar [10] (see Fig. 1(b)). We show that T is also region-level nonplanar (see Fig. 1(c)).

Lemma 1. *The 4-level tree T whose underlying tree is shown in Fig. 1(a) is region-level nonplanar.*

Lemma 1 will be vital for proving that a tree $\mathcal{T}$ and a path $\mathcal{P}$ exist not admitting any GSE. In fact, $\mathcal{T}$ contains many copies of the underlying tree of T, while $\mathcal{P}$ connects vertices in such a way to create the regions satisfying the above conditions and to enforce at least one of such copies to lie inside them according to the nonplanar leveling.

3   The Counterexample

In this section we describe a tree $\mathcal{T}$ and a path $\mathcal{P}$ not admitting any GSE.

The tree $\mathcal{T}$ has a root r and q vertices $j_1, \ldots, j_q$ at distance 1 from r, called *joints*. Each joint is connected to $l := (s-1)^4 \cdot 3^2 \cdot x$ vertices of degree 1, called *stabilizers* and to x subtrees B_i, $i = 1, \ldots, x$, called *branches*, each one consisting of a root r_i, $(s-1) \cdot 3$ vertices of degree $(s-1)$ adjacent to r_i, and $(s-2) \cdot (s-1) \cdot 3$ leaves at distance 2 from r_i. See Fig. 1(d). Vertices of the branches are called *B-vertices* and denoted by 1-, 2-, or 3-vertices, according to their distance from their joint.

For the sake of readability, we use variables q, s, and x as parameters describing the size of certain structures, that will be given a value when the technical details are described. We claim that a number $n \geq \binom{2^7 \cdot 3 \cdot x + 2}{3}$ of vertices suffices for the counterexample. Despite the oversized number of vertices, tree $\mathcal{T}$ has limited *height*, that is, every vertex is at distance from the root at most 4. This leads to the following property.

Property 1. Any simple path of tree edges starting at the root has at most 3 bends.

Path $\mathcal{P}$ is given by describing some basic recurring subpaths on the vertices of $\mathcal{T}$. The idea is to partition the set of branches adjacent to each joint into subsets of s branches each and to connect the vertices of each set with path edges, according to some features of the tree structure, so defining the first building block, called *cell*. Then, cells belonging to the same joint are connected to create *formations*, for which we can ensure some properties on the intersection between tree and path edges. Further, formations are connected to create *extended formations*, which, in their turn, are connected to create *sequences of extended formations*. These structures are constructed in such a way that there exists a set of cells, connected to the same joint and being part of the same formation or extended formation, such that any four of these cells contain a copy of a region-level nonplanar tree, where the level of a vertex is determined by the cell it belongs to. Proving that four of such cells lie in different regions satisfying the properties of separation described above is equivalent to proving the existence of a crossing in T. This allows us not to deal with single copies of the region-level nonplanar tree.

In the following we define such structures more formally and state their properties.

The most basic structure is defined by looking at how P connects the vertices of a set of s branches connected to the same joint of T. For each joint j_h, $h = 1, \ldots, q$, and for each disjoint subset of s branches B_i, $i = 1, \ldots, s$, connected to j_h, we construct a set of s cells as follows. For each $i = 1, \ldots, s$, a *cell* $c_i(h)$ is composed of its *head*, its *tail*, and a number t of stabilizers. The *head* of $c_i(h)$ consists of the unique 1-vertex of B_i, the first three 2-vertices of each branch B_k, with $1 \leq k \leq s$ and $k \neq i$, that are not already used in a cell $c_a(h)$ with $1 \leq a < r$ and, for each 2-vertex not in $c_i(h)$ and not in B_i, the first 3-vertices not already used in a cell $c_a(h)$, with $1 \leq a < i$. The *tail* of $c_i(h)$ is created by considering a set of $3 \cdot s \cdot (s-1)^2$ branches adjacent to j_h, partitioned into $3 \cdot (s-1)^2$ subsets of s subtrees each. The vertices of each subset are distributed between the cells in the same way as for the vertices of the head. Path P visits the vertices of a cell $c_i(h)$ as follows: Starting at the unique 1-vertex of the head, it reaches the 2-vertices of the head, then the 3-vertices of the head, then the 2-vertices of the tail, and finally the 3-vertices of the tail. After each occurrence of a 2- or 3-vertex of the head, P visits a 1-vertex of the tail, and after each occurrence of a 2- or a 3-vertex of the tail, it visits a stabilizer of joint j_h (see Fig. 2(a)). Note that each set of s cells constructed starting from the same set of s branches is such that each subset of size four contains a region-level nonplanar tree, where the levels correspond to the membership of the vertices to a cell. Namely, consider four cells $c_1, \ldots, c_4$ belonging to the same set, leveled in this order. A region level nonplanar tree as in Fig. 1 consists of the 1-vertex v of the head of c_2, the three 2-vertices of c_3 connected to v and, for each of them, the 3-vertex of c_1 and the 3-vertex of c_4 connected to it.

The next structure describes how cells from four different sets are connected each other. A *formation* $F(H)$, where $H = (h_1, h_2, h_3, h_4)$ is a 4-tuple of indices of joints, consists of 592 cells. Namely, for each joint j_{h_i}, $1 \leq i \leq 4$, $F(H)$ contains 148 cells of the same set of s cells connected to j_{h_i}. Path P connects these cells in the order $((h_1 h_2 h_3)^{37} h_4^{37})^4$, that is, P repeats four times the following sequence: It connects $c_1(h_1)$ to $c_1(h_2)$, then to $c_1(h_3)$, then to $c_2(h_1)$, and so on until $c_{37}(h_3)$, from which it connects to $c_1(h_4)$, to $c_2(h_4)$, and so on till $c_{37}(h_4)$ (see Fig. 2(b)). Since cells of $F(H)$ connected to the same joint belong to the same set of s cells, the following holds:

Property 2. For any formation $F(H)$ and any joint j_h, with $h \in H$, if four cells $c_r(h) \in F(H)$ are pairwise separated by straight lines, then there exists a crossing in T.

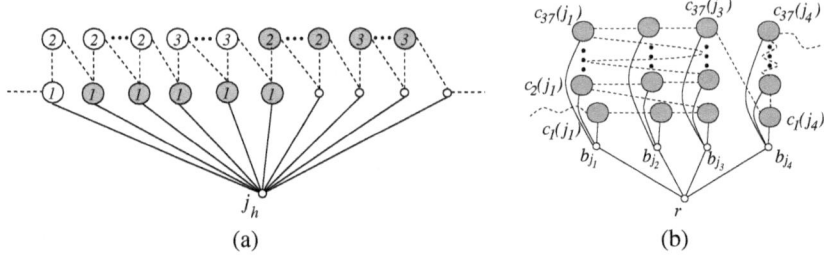

Fig. 2. (a) A cell. Vertices of the head are white and vertices of the tail are grey. B-vertices are large and stabilizers are small circles. (b) A formation.

Formations are connected by $\mathcal{P}$ to create an *extended formation* $EF(H)$, where $H = (H_1 = (h_1, \ldots, h_4), H_2 = (h_5, \ldots, h_8), \ldots, H_x = (h_{4x-3}, \ldots h_{4x}))$ is an x-tuple of 4$-$tuples of disjoint indices of joints. For each 4$-$tuple H_i, $EF(H)$ contains $y - \frac{y}{x}$ formations $F_1(H_i), \ldots, F_{y - \frac{y}{x}}(H_i)$ not belonging to any other extended formation and composed of cells of the same set of s cells connected to the same joint. Formations inside $EF(H)$ are connected in $\mathcal{P}$ in the order $(H_1, H_2, \ldots, H_x)^y$, that is, $\mathcal{P}$ connects $F_1(H_1)$ to $F_1(H_2)$, then to $F_1(H_3)$, and so on until $F_1(H_x)$, then to $F_2(H_1)$, to $F_2(H_2)$, and so on until $F_{y - \frac{y}{x}}(H_x)$. However, in each of these y repetitions one H_i is missing. Namely, in the k-th repetition $\mathcal{P}$ does not reach any formation at H_m, with $m = k \mod x$. We say that the k-th repetition has a *defect* at m and a subsequence $(H_1, H_2, \ldots, H_x)^x$ is a *full repetition*, having exactly one defect at each tuple.

Extended formations are connected by $\mathcal{P}$ in a *sequence of extended formations* $SEF(H)$, where $H = (H_1^*, \ldots, H_{12}^*)$ is a 12$-$tuple of x-tuples of 4$-$tuples of disjoint indices of joints. For each x-tuple H_i^*, with $i = 1, \ldots, 12$, there exist 110 extended formations $EF_j(H_i^*)$, with $j = 1, \ldots, 110$, not belonging to any other sequence of extended formations, that are connected by $\mathcal{P}$ in the order $(H_1^*, \ldots, H_{12}^*)^{(120)}$. We have two types of sequences of extended formations. In the first type, in each repetition $(H_1^*, \ldots, H_{12}^*)$ one extended formation $EF(H_m)$ is missing, creating a *defect* at m. In the second type, in each repetition $(H_1^*, \ldots, H_{12}^*)$ two consecutive extended formations are missing. Namely, in the k-th repetition $\mathcal{P}$ skips the extended formations $EF(H_m^*)$ and $EF(H_{m+1}^*)$, with $m = k \mod 12$, creating a *double defect* at m.

The size of s can now be fixed as the number of formations in an extended formation times the number of cells in a formation, that is, $s := (y - \frac{y}{x}) \cdot 37 \cdot 4$. Further, $q := 48x$, as we need 4 sequences of extended formations (of size 12 each) not sharing any joint. We claim that $x = 7 \cdot 3^2 \cdot 2^{23}$ and $y = 7^2 \cdot 3^3 \cdot 2^{26}$ is sufficient in the proofs.

4 Overview

In this section we present the main arguments leading to the conclusion that $\mathcal{T}$ and $\mathcal{P}$ do not admit any GSE. The main idea is to use the structures given by $\mathcal{P}$ to fix a part of $\mathcal{T}$ in a specific shape creating restrictions for the placement of the further substructures attached to it. Then, we show that such restrictions lead to a crossing in any possible GSE of $\mathcal{P}$ and $\mathcal{T}$. In the following, we will perform an analysis of the geometrical properties of each possible embedding, in order to show that none of them is feasible.

First, we give some definitions and properties of cells enforced by properties of region-level planar drawings and by the order of the cells inside a formation.

Let $c_1(h), c_2(h)$ be two cells connected to a joint j_h that can not be separated by a straight line and a cell $c'(h')$ connected to a joint $j_{h'}$, with $h' \neq h$. A *passage* P between c_1, c_2, and c' exists if the path of c' separates vertices of c_1 from vertices of c_2 (see Fig. 3(a)). Since the separation can not be straight, there is a vertex of c' lying inside the convex hull of the vertices of $c_1 \cup c_2$. Hence, there exist at least two path-edges e_1, e_2 of c' intersected by tree-edges connecting vertices of c_1 to vertices of c_2.

For two passages P_1 between $c_1(h_1)$, $c_2(h_1)$, and $c'(h'_1)$, and P_2 between $c_3(h_2)$, $c_4(h_2)$, and $c'(h'_2)$ (w.l.o.g., assume $h_1 < h'_1$, $h_2 < h'_2$, and $h_1 < h_2$), we distinguish three configurations: (i) If $h'_1 < h_2$, P_1 and P_2 are *independent*; (ii) if $h'_2 < h'_1$, P_2 is *nested* into P_1; and (iii) if $h_2 < h'_1 < h'_2$, P_1 and P_2 are *interconnected* (see Fig. 3(b)).

Let $c_1(h), c_2(h)$, and $c'(h')$ be three cells creating a passage. We call a *door* any triangle given by a vertex v' of c' inside the convex hull of $c_1 \cup c_2$ and by any two vertices of $c_1 \cup c_2$ that encloses neither any other vertex of c_1, c_2 nor any vertex of c' that is closer than v' to $j_{h'}$ in $\mathcal{T}$. A door is *open* if no tree-edge incident to v' crosses the side of the triangle between the vertices of c_1 and c_2, otherwise it is *closed*.

Consider two joints j_a and j_b, with j_h, j_a, j'_h, j_b in this circular order around the root. Any polyline connecting the root to j_a, then to j_b, and again to the root, without crossing tree edges, must traverse each door by crossing both the sides adjacent to v'. If a door is closed, such a polyline has to bend between the two sides adjacent to v'. In the rest of the argument we will exploit this fact to obtain the claimed property that a large part of $\mathcal{T}$ has to follow the same shape. In view of this, we state the following lemmata.

Lemma 2. *For each formation* $F(H)$, *with* $H = (h_1, \ldots, h_4)$, *there exists a passage between some cells* $c_1(h_a), c_2(h_a), c'(h_b) \in F(H)$, *with* $1 \leq a, b \leq 4$.

Lemma 3. *Each passage contains at least one closed door.*

Hence, each formation contains at least one closed door. In the following we prove that the combined effects of closed doors of different formations enforces more restrictions on the shape of the tree. First, we exploit the Ramsey Theorem [15] to state that there exists a set of joints such that any two joints contain cells creating a passage.

Lemma 4. *Given a set of joints* $J = \{j_1, \ldots, j_y\}$, *with* $|J| = y := \binom{2^7 \cdot 3 \cdot x + 2}{3}$, *there exists a subset* $J' = \{j'_1, \ldots, j'_r\}$, *with* $|J'| = r \geq 2^7 \cdot 3 \cdot x$, *such that for each pair of joints* $j'_i, j'_h \in J'$ *there exist two cells* $c_1(i), c_2(i)$ *creating a passage with a cell* $c'(h)$.

Consider two paths $p_1 = \{u_1, v_1, w_1\}$ and $p_2 = \{u_2, v_2, w_2\}$. The bendpoint v_1 of p_1 *encloses* the bendpoint v_2 of p_2 if v_2 is internal to triangle $\triangle(u_1, v_1, w_1)$.

Consider a set of joints $J = \{j_1, \ldots, j_k\}$ in clockwise order around the root. The *channel* c_i of a joint j_i, with $i = 2, \ldots, k-1$, is the region given by the pair of paths, one path of j_{i-1} and one path of j_{i+1}, with the maximum number of enclosing bendpoints with each other. We say that c_i is an *x-channel* if the number of enclosing bendpoints is x. Note that, by Prop. 1, $x \leq 3$. See Fig. 3(c). An *x-channel* c_i is composed of $x + 1$ *channel segments*. The first channel segment cs_1 is the part of c_i visible from the root. The h-th channel segment cs_h is the part of c_i disjoint from cs_{h-1} that is bounded by

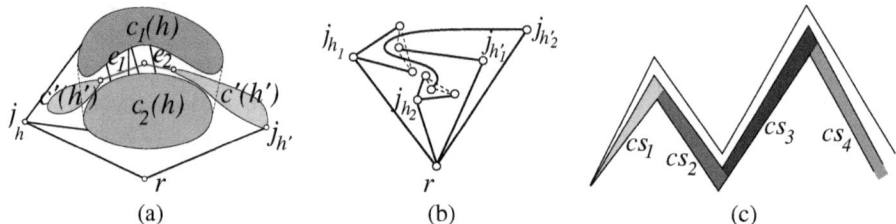

Fig. 3. (a) A passage between cells c_1, c_2, and c'. (b) Two interconnected passages. (c) A 3-channel and its channel segments.

the elongations of the paths of j_{i-1} and j_{i+1} after the h-th bend. The *bending area* $b(a, a + 1)$ of c_i is the region visible from all the points of cs_a and cs_{a+1}. As channels are created by tree-edges, only path-edges can cross the boundaries of the channel.

We study the relationships between path-edges and channels. As every second vertex of $\mathcal{P}$ in a cell is either a 1-vertex or a stabilizer, we have the following property.

Property 3. For any path-edge (a, b), at least one of a and b lies inside either cs_1 or cs_2.

A *blocking cut* is a path edge connecting two consecutive channel segments by cutting some of the other channels twice.

Property 4. If a channel that is cut twice by a blocking cut has vertices in both the channel segments cut by the path-edge, then it has vertices in a different channel segment.

Next, based on Prop. 4, we show that any set of joints as in Lemma 4 contains a subset of joints creating interconnected passages such that each pair of paths of tree-edges starting at the root and containing such joints has at least two common enclosing bend-points, which implies that most of them create 2-channels. From now on, we identify a joint with the channel it belongs to. Then, when dealing with a passage between two cells $c_1(h), c_2(h)$ of a joint j_h and a cell $c'(h')$ of a joint $j_{h'}$, we might also say that there is a passage between joints j_h and j'_h or between the corresponding channels.

Lemma 5. *Let $J = \{j_1, \ldots, j_k\}$ be a set of joints such that there exists a passage between each pair (j_i, j_h), with $1 \le i, h \le k$. Let $\mathcal{P}_1 = \{P \mid P$ connects c_i and $c_{\frac{3k}{4}+1-i}$, for $i = 1, \ldots, \frac{k}{4}\}$ and $\mathcal{P}_2 = \{P \mid P$ connects $c_{\frac{k}{4}+i}$ and c_{k+1-i}, for $i = 1, \ldots, \frac{k}{4}\}$ be two sets of passages between pairs of joints in J (see Fig. 4(a)). Then, for at least $\frac{k}{4}$ of the joints of one set of passages, say $\mathcal{P}_1$, there exist paths in $\mathcal{T}$, starting at the root and containing these joints, which traverse all the doors of $\mathcal{P}_2$ with at least 2 and at most 3 bends. Also, at least half of these joints create an x-channel, with $2 \le x \le 3$.*

By Lemma 5, any formation attached to a certain subset of joints must use at least three different channel segments. In the remainder we focus on this subset and give some properties holding for it. As we need a full sequence of extended formations attached to these joints, k has to be at least eight times the number of channels inside a sequence of extended formations, that is, $k \ge 8 \cdot 48x = 2^7 \cdot 3x$.

A formation F is *nested* in a formation F' if there exist four path-edges $e_1, e_2 \in F$ and $e'_1, e'_2 \in F'$ cutting a boundary cb of a channel c such that all the vertices of the

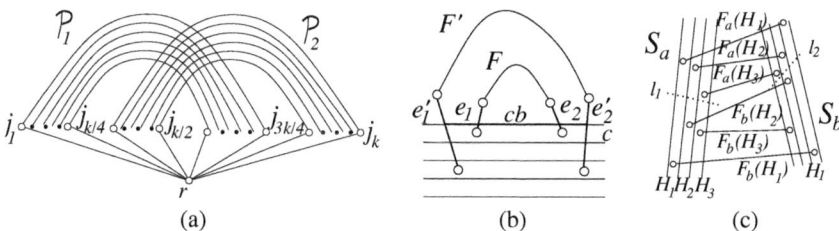

Fig. 4. (a) Two sets of passages $\mathcal{P}_1$ and $\mathcal{P}_2$ as described in Lemma 5. (b) A formation F nested in a formation F'. (c) Two independent sets S_a and S_b.

path in F between e_1 and e_2 lie inside the region delimited by cb and by the path in F' between e_1' and e_2' (see Fig. 4(b)). A series of pairwise nested formations $F_1, \ldots, F_k$ is *r-nested* if there exist r formations $F_{q_1}, \ldots, F_{q_r}$, with $1 \le q_1, \ldots, q_r \le k$, such that the 4-tuples of $F_{q_1}, \ldots, F_{q_r}$ have at least one common joint j, and such that for each pair $F_{q_p}, F_{q_{p+1}}$ there exists at least one formation F_z, with $1 \le z \le k$, such that the 4-tuple of F_z does not contain j, F_{q_p} is nested in F_z and F_z is nested in $F_{q_{p+1}}$.

Let $S_1, \ldots, S_k$ be sets of formations of one extended formation $EF(H)$ such that each set S_i, for $i = 1, \ldots, k$, contains the formations $F_i(H_1), \ldots, F_i(H_r)$, such that $(H_1, \ldots, H_r) \subset H$. Let $F_a(H_c)$ be not nested in $F_b(H_d)$, for each $1 \le a, b \le k$, $a \ne b$, and $1 \le c, d \le r$. If for each two sets S_a, S_b there exists a line l_1 (a line l_2) separating the vertices of S_a (of S_b) inside channel segment cs_1 (channel segment cs_2), then sets $S_1, \ldots, S_k$ are *independent* (see Fig. 4(c)).

In the following lemmata we prove that in any extended formation there exists a nesting of a certain depth (Lemma 8) by first proving that in any extended formation the number of independent sets of formations is limited (Lemma 6) and then by showing that, although some formations might be neither nested nor independent, there exists a certain number that have to be either independent or nested (Lemma 7).

Lemma 6. *No extended formation contains $n \ge 2^{22} \cdot 14$ independent sets of formations $S_1, \ldots, S_n$ such that each set S_i contains formations $F_i(H_1), \ldots, F_i(H_r)$, with $r \ge 22$.*

Lemma 7. *Let EF be an extended formation and let $Q_1, \ldots, Q_4$ be four subsequences of formations, each consisting of a whole repetition $(H_1, H_2, \ldots, H_x)$ of EF. Then, there exists either a pair of nested subsequences or a pair of independent subsequences.*

Lemma 8. *For every extended formation EF, there exists a k-nesting, with $k \ge 6$.*

Then, we study how such a nesting can be performed inside the channels. We will conclude that, in any possible shape of the tree, either it is not possible to draw the nesting formations planar, or that any planar drawing of such formations induces further geometrical constraints not allowing a planar drawing of the rest of the tree.

Let cs_a and cs_b, with $1 \le a, b \le 4$, be two channel segments. If the elongation of cs_a intersects cs_b, then it is possible to connect from cs_b to cs_a by cutting both the sides of cs_a. In this case, cs_a and cs_b have a *2-side connection* (see Fig. 5(b)). On the contrary, if the elongation of cs_a does not intersect cs_b, only one side of cs_a can be used. In this case, cs_a and cs_b have a *1-side connection* (see Fig. 5(a)).

First, we consider the case in which only 1-side connections are possible (Fig. 3(c)). We prove that, in this configuration, the existence of a nesting in one extended formation results in a crossing in either $\mathcal{T}$ or $\mathcal{P}$.

Proposition 1. *If every two channel segments have a* $1-side$ *connection, then* $\mathcal{T}$ *and* $\mathcal{P}$ *do not admit any* GSE.

Next, we study the case in which there exist 2-side connections. We distinguish two types of 2-side connections. If the elongation of channel segment cs_a intersecting channel segment cs_b starts at the bendpoint that is closer to the root, we have a *low Intersection* $I^l_{(a,b)}$ (see Fig. 5(c)). Otherwise, we have a *high Intersection* $I^h_{(a,b)}$ (see Fig. 5(d)). We use notation $I_{(a,b)}$ to describe both $I^h_{(a,b)}$ and $I^l_{(a,b)}$. Two intersections $I_{(a,b)}$ and $I_{(c,d)}$ are *disjoint* if $a, d \in \{1, 2\}$ and $b, c \in \{3, 4\}$. E.g., $I_{(1,3)}$ and $I_{(4,2)}$ are disjoint, while $I_{(1,3)}$ and $I_{(2,4)}$ are not. Since consecutive channel segments can not create 2-side connections, to explore all the possible shapes we consider the combinations of low and high intersections of channel segments cs_1 and cs_2 with cs_3 and cs_4. First, we prove that intersections of adjacent channels have to maintain certain consistencies.

Lemma 9. *Consider two channels* ch_p, ch_q *with the same intersections* $I_{(a,b)}$. *Then, none of channels* ch_i, *where* $p < i < q$, *have an intersection that is disjoint with* $I_{(a,b)}$.

As for Proposition 1, in order to prove that 2-side connections are not sufficient, we exploit the existence of the nesting shown in Lemma 8. Note that every extended formation using a channel segment cs_a to place the nesting must place vertices inside the adjacent bending area. We prove that not many of the nesting formations can use the part of the path that creates the nesting to place vertices in such a bending area.

Lemma 10. *Consider a nesting of formations inside a sequence of extended formations on an intersection* $I_{(a,b)}$, *with* $a \leq 2$. *Then, one of the formations contains a pair of path-edges* (u, v), (v, w), *with* v *lying inside channel segment* cs_a, *separating some formations in* cs_a *from bending area* $b(a, a + 1)$ *or* $b(a - 1, a)$ *(see Fig. 5(e))*.

Let the *inner area* and *outer area* of cs_a be the two parts in which cs_a is split by edges $(u, v), (v, w)$, as in Lemma 10. Since in every extended formation a path exists connecting the inner and the outer area by going around either u or w, the extended formations using such paths create a structure that is analogous to a nesting of formations. We prove that, as every repetition of an extended formation contains a defect, if only 1-side connections are available to host such paths, then a crossing in $\mathcal{T}$ or $\mathcal{P}$ is created.

Lemma 11. *Let* cs_a *be a channel segment that is split into inner area and outer area by two edges such that every extended formation of a sequence of extended formations has vertices in both the areas. If the only possibility to connect vertices from the inner to the outer area is with a 1-side connection, then* $\mathcal{T}$ *and* $\mathcal{P}$ *do not admit any* GSE.

Lemma 11 states that one single 2-side connection is not sufficient to obtain a GSE of $\mathcal{T}$ and $\mathcal{P}$. We prove that a further 2-side connection is not useful if it is not disjoint.

Proposition 2. *If there exists no pair of disjoint 2-side connections, then* $\mathcal{T}$ *and* $\mathcal{P}$ *do not admit any* GSE.

Note that it is sufficient to restrict the analysis to cases $I_{(1,3)}$ (see Figs. 6(a)–(b)) and $I_{(3,1)}$, as the cases involving 2 and 4 can be reduced to them.

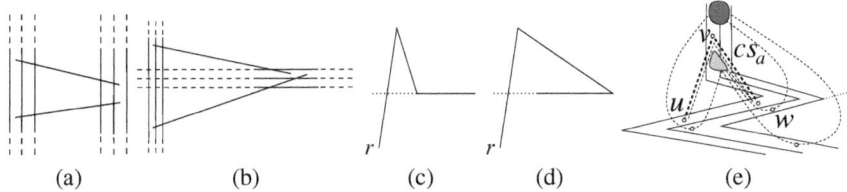

Fig. 5. (a) A 1−side connection. (b) A 2−side connection. (c) A low Intersection. (d) A high Intersection. (e) A situation as in Lemma 10. Inner and the outer areas are represented by a light grey and a dark grey region, respectively.

Lemma 12. *If a shape contains an intersection $I_{(1,3)}$ and does not contain any other intersection that is disjoint with $I_{(1,3)}$, then T and P do not admit any GSE.*

Lemma 13. *If there exists a sequence of extended formations in any shape containing an intersection $I_{(3,1)}$, then T and P do not admit any GSE.*

Finally, we tackle the general case where two disjoint intersections exist.

Proposition 3. *If there exists two disjoint 2-side connections, then T and P do not admit any GSE.*

Note that Lemma 13 stated a property that is stronger than Proposition 2. In fact, a GSE cannot be obtained in any shape containing $I_{(3,1)}$, even if a disjoint intersection is present. Hence, we only consider the eight configurations with $I_{(1,3)}$ and $I_{(4,\{1,2\})}^{h,l}$.

Let cs_i and cs_{i+1} be two consecutive channel segments of a channel ch_k and let e be a path-edge crossing the boundary of one of cs_i and cs_{i+1}, say cs_i. Edge e is a *double cut* at ch_k if the line through e cuts ch_k in cs_{i+1}. A double cut is *simple* if the elongation of e cuts cs_{i+1} (see Fig. 6(c)) and *non-simple* if e itself cuts cs_{i+1} (see Fig. 6(d)). A double cut of an extended formation EF is *extremal* at bending area $b(i, i + 1)$ if no double cut of EF closer to $b(i, i + 1)$ exists. Double cut e blocks visibility to $b(i, i + 1)$ for a part of cs_i in each channel ch_h with $h > k$ or with $h < k$.

We show that a certain ordering of extremal double cuts in two consecutive channel segments leads to a non-planarity, and that, because of the double defect in every repetition of an extended formation, both shapes $I_{(1,3)}^{h}$ $I_{(4,2)}^{h,l}$ induce this order (Lemma 16).

Lemma 14. *Let cs_i and cs_{i+1} be two consecutive channel segments. If there exists an ordered set $S := (1, 2, \ldots, 5)^3$ of extremal double cuts cutting cs_i and cs_{i+1} such that the order of the intersections of the double cuts with cs_i (with cs_{i+1}) is coherent with the order of S, then T and P do not admit any GSE.*

We first show that double cuts exist in $I_{(1,3)}^{h}$ $I_{(4,2)}^{h}$. This is easy to see in $I_{(1,3)}^{h}$ $I_{(4,2)}^{l}$.

Lemma 15. *Shape $I_{(1,3)}^{h}$ $I_{(4,2)}^{h}$ creates double cuts in at least one bending area.*

Lemma 16. *Every sequence of extending formations in shape $I_{(1,3)}^{h}$ $I_{(4,2)}^{h,l}$ contains an ordered set $(1, 2, \ldots, 5)^3$ of extremal double cuts at either $b(2, 3)$ or $b(3, 4)$.*

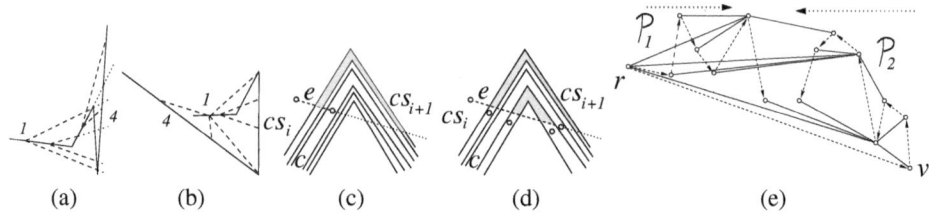

Fig. 6. (a) Case $I_{(1,3)}$ $I^h_{(2,4)}$. As a nesting at $I_{(1,3)}$ must reach $b(2,3)$, it crosses any nesting at $I^h_{(2,4)}$. (b) Case $I_{(1,3)}$ $I^l_{(2,4)}$. A nesting at $I_{(1,3)}$ crosses any nesting at $I^l_{(2,4)}$ and any extended formation nesting at $I_{(1,4)}$ creates a nesting at $I_{(1,3)}$, as it must reach $b(2,3)$ and $b(3,4)$. (c) A simple double cut. (d) A non-simple double cut. (e) Construction of a GSE of a height-2 tree and a path. Arrows indicate that the subpaths before and after r are monotone in opposite directions.

Finally, we consider shapes $I^l_{(1,3)}$ $I^{h,l}_{(4,2)}$. Note that, in both cases, cs_2 lies on the convex hull of the shape. We show that this yields a crossing either in $\mathcal{T}$ or in $\mathcal{P}$.

Lemma 17. *If cs_2 is part of the convex hull, then $\mathcal{T}$ and $\mathcal{P}$ do not admit any GSE.*

Based on the above discussion, we state the following theorem.

Theorem 1. *There exist a tree and a path that do not admit any GSE.*

5 Constructing a GSE of a Tree of Height 2 and a Path

In this section we sketch the algorithm for constructing a GSE of a tree $\mathcal{T}$ of height 2 and a path $\mathcal{P}$. See Fig. 6(e). Draw the root r of $\mathcal{T}$ as the leftmost vertex. Consider the two subpaths $\mathcal{P}_1$ and $\mathcal{P}_2$ of $\mathcal{P}$ starting at r. Assign an orientation to $\mathcal{P}_1$ (to $\mathcal{P}_2$) such that r is the only source of $\mathcal{P}_1$ (of $\mathcal{P}_2$). Draw $\mathcal{P}_1$ to the right of r, placing its vertices from the left to the right following the orientation of $\mathcal{P}_1$. Draw the vertex v following r in $\mathcal{P}_2$ as the rightmost vertex. Finally, draw $\mathcal{P}_2$ to the left of v, placing its vertices from the right to the left following the orientation of $\mathcal{P}_2$, so that the leftmost vertex of $\mathcal{P}_2$ is to the right of the rightmost vertex of $\mathcal{P}_1$.

6 Conclusions

In this paper we have shown that there exist a tree $\mathcal{T}$ and a path $\mathcal{P}$ on the same set of vertices that do not admit any GSE. We first extended the concept of level nonplanar trees [10] to the one of region-level nonplanar trees, and showed that there exist trees not admitting any planar embedding if the vertices are forced to lie inside certain regions according to a prescribed ordering. Then, we constructed $\mathcal{T}$ and $\mathcal{P}$ so that $\mathcal{P}$ creates these regions and enforces at least one of the many region-level nonplanar trees composing $\mathcal{T}$ to lie inside them in the desired order. Our result implies that two edge-disjoint trees exist not admitting any GSE, answering a question posed in [14].

Note that, despite the huge number of vertices, $\mathcal{T}$ can be considered as "simple", as its height is just 4. In this direction, we proved that any tree of height 2 admits a GSE with any path, giving raise to an intriguing question: What about a tree of height 3?

References

1. Angelini, P., Geyer, M., Kaufmann, M., Neuwirth, D.: On a tree and a path with no geometric simultaneous embedding. Tech. Report 176, Dipartimento di Informatica e Automazione, Roma Tre University (2010)
2. Brandes, U., Erten, C., Fowler, J., Fratl, F., Geyer, M., Gutwenger, C., Hong, S H , Kaufmann, M., Kobourov, S., Liotta, G., Mutzel, P., Symvonis, A.: Colored simultaneous geometric embeddings. In: Lin, G. (ed.) COCOON 2007. LNCS, vol. 4598, pp. 254–263. Springer, Heidelberg (2007)
3. Brass, P., Cenek, E., Duncan, C., Efrat, A., Erten, C., Ismailescu, D., Kobourov, S., Lubiw, A., Mitchell, J.: On simultaneous planar graph embeddings. Comp. Geom. 36(2), 117–130 (2007)
4. Cabello, S., van Kreveld, M., Liotta, G., Meijer, H., Speckmann, B., Verbeek, K.: Geometric simultaneous embeddings of a graph and a matching. In: Eppstein, D., Gansner, E.R. (eds.) GD 2009. LNCS, vol. 5849, pp. 183–194. Springer, Heidelberg (2010)
5. Di Giacomo, E., Didimo, W., van Kreveld, M., Liotta, G., Speckmann, B.: Matched drawings of planar graphs. J. Graph Alg. Appl. 13(3), 423–445 (2009)
6. Erten, C., Kobourov, S.G.: Simultaneous embedding of planar graphs with few bends. J. Graph Alg. Appl. 9(3), 347–364 (2005)
7. Estrella-Balderrama, A., Fowler, J., Kobourov, S.G.: Characterization of unlabeled level planar trees. Comp. Geom. 42(6-7), 704–721 (2009)
8. Fowler, J., Jünger, M., Kobourov, S.G., Schulz, M.: Characterizations of restricted pairs of planar graphs allowing simultaneous embedding with fixed edges. In: Broersma, H., Erlebach, T., Friedetzky, T., Paulusma, D. (eds.) WG 2008. LNCS, vol. 5344, pp. 146–158. Springer, Heidelberg (2008)
9. Fowler, J., Kobourov, S.: Characterization of unlabeled level planar graphs. In: Hong, S.H., Nishizeki, T., Quan, W. (eds.) GD 2007. LNCS, vol. 4875, pp. 37–49. Springer, Heidelberg (2008)
10. Fowler, J., Kobourov, S.: Minimum level nonplanar patterns for trees. In: Hong, S.-H., Nishizeki, T., Quan, W. (eds.) GD 2007. LNCS, vol. 4875, pp. 69–75. Springer, Heidelberg (2008)
11. Frati, F.: Embedding graphs simultaneously with fixed edges. In: Kaufmann, M., Wagner, D. (eds.) GD 2006. LNCS, vol. 4372, pp. 108–113. Springer, Heidelberg (2007)
12. Frati, F., Kaufmann, M., Kobourov, S.: Constrained simultaneous and near-simultaneous embeddings. J. Graph Alg. Appl. 13(3), 447–465 (2009)
13. Gassner, E., Jünger, M., Percan, M., Schaefer, M., Schulz, M.: Simultaneous graph embeddings with fixed edges. In: Fomin, F.V. (ed.) WG 2006. LNCS, vol. 4271, pp. 325–335. Springer, Heidelberg (2006)
14. Geyer, M., Kaufmann, M., Vrt'o, I.: Two trees which are self-intersecting when drawn simultaneously. Disc. Math. 309(7), 1909–1916 (2009)
15. Graham, R.L., Rothschild, B.L., Spencer, J.H.: Ramsey Theory. John Wiley & Sons, Chichester (1990)
16. Halton, J.H.: On the thickness of graphs of given degree. Inf. Sc. 54(3), 219–238 (1991)
17. Pach, J., Wenger, R.: Embedding planar graphs at fixed vertex locations. Graphs and Comb. 17(4), 717–728 (2001)
18. Thomassen, C.: Embeddings of graphs. Disc. Math. 124(1-3), 217–228 (1994)
19. Tutte, W.T.: How to draw a graph. London Math. Society 13, 743–768 (1962)

Difference Map Readability for Dynamic Graphs

Daniel Archambault[1], Helen C. Purchase[2], and Bruno Pinaud[3]

[1] Clique Strategic Research Cluster, University College Dublin
`daniel.archambault@ucd.ie`
[2] Department of Computing Science, University of Glasgow
`hcp@dcs.gla.ac.uk`
[3] Université de Bordeaux I, LaBRI UMR CNRS 5800 & INRIA Bordeaux Sud-Ouest
`bruno.pinaud@labri.fr`

Abstract. Difference maps are one way to show changes between times-lices in a dynamic graph. They highlight, using colour, the nodes and edges that were added, removed, or persisted between every pair of adjacent timeslices. Although some work has used difference maps for visualization, no user study has been performed to gauge their performance. In this paper, we present a user study to evaluate the effectiveness of difference maps in comparison with presenting the evolution of the dynamic graph over time on three interfaces. We found evidence that difference maps produced significantly fewer errors when determining the number of edges inserted or removed from a graph as it evolves over time. Also, difference maps were significantly preferred on all tasks.

1   Introduction

Dynamic graph drawing deals with the problem of depicting a graph that evolves over time. Dynamic graph drawing algorithms typically represent the evolving graph as a series of timeslices. A **timeslice** encodes the structure of the graph at a given time. The timeslices, also known as the sequence of graphs, are often placed in chronological order, demonstrating graph evolution.

A few visualization systems [3,6] have exploited difference maps to show the evolution of dynamic series of graphs. A **difference map** does not present the actual timeslices. Rather, for each pair of adjacent timeslices, it presents the union of the nodes and edges in both graphs. The nodes and edges are coloured one of three colours depending on whether they were added, removed, or persisted in the graph over that timeslice. Despite the use of difference maps in visualization systems, the effectiveness of this presentation method has yet to be evaluated.

Many different user interfaces have been used to present dynamic graphs to a user. In an **animation** of the dynamic graph sequence, nodes and edges that are added and removed from the drawing are faded in and out of the display. Node movement is smoothly interpolated so that the user of the system can more easily follow how the data has changed. One could also picture a **slide show** of the data whereby the data is presented like a Powerpoint presentation. No smooth

U. Brandes and S. Cornelsen (Eds.): GD 2010, LNCS 6502, pp. 50–61, 2011.

transitions exist between drawings in the sequence, and the arrow keys are used to cycle through the graphs in the series. In a **small multiples** [21] interface, all timeslices are presented on the screen at once with each timeslice placed inside its own window. The user scans the matrix of windows to see how the graph evolves. All three interfaces could be used to present a series of difference maps, but we currently don't know which is the most effective.

This work presents a user study which investigates two research questions:

1. Do difference maps help improve the readability of dynamic graphs?
2. Under what interface do they help the most: animation, slide show, or small multiples?

We found that difference maps can help answer questions about large scale changes in terms of the number of edges in a graph. Also, difference maps were preferred over simply presenting the dynamic graph series as it evolves over time.

2   Previous and Related Work

Previous and related work is divided into three subsections. First, we present some of the work on difference maps in section 2.1. Section 2.2 presents work in dynamic graph drawing. Finally, section 2.3 presents a few user studies with results on dynamic graph drawing readability.

2.1   Difference Maps

Difference maps were designed to show the differences, in terms of nodes and edges, between a pair of graphs. They do so by taking the union of the nodes and edges in both graphs and colouring them by presence in one graph, the other, or both. Assuming that there is a unique identifier for each node of the graph, a one-to-one correspondence is available and the difference map can be computed in linear time. Fig. 1(c) shows a difference map computed from two graphs. The black nodes are only present in Fig. 1(a). Similarly, the light grey nodes are only present in Fig. 1(b). The grey nodes in Fig. 1(c), however, are present in both graphs.

Archambault [3] used difference maps and graph hierarchies to show where areas of a large graph changed. In this work, graph hierarchies or cluster trees, were used to simplify a large difference map into areas of similar evolution. The work also presented coarsening methods to make the diagrams simpler to read.

Bourqui and Jourdan [6] accentuated areas common to pairs of biological networks using difference maps. In the study of biological networks, structurally similar pathways often have similar function. By emphasizing structural similarities, both structure and context of functionally similar elements can be compared.

Both papers present techniques to visualize difference maps. However, neither of them have a user study to evaluate their effectiveness. These techniques use other graph visualization techniques, such as graph hierarchies and fisheye views, that we do not test in this experiment. In this study, we are interested in evaluating the overall difference map approach in a dynamic graph context.

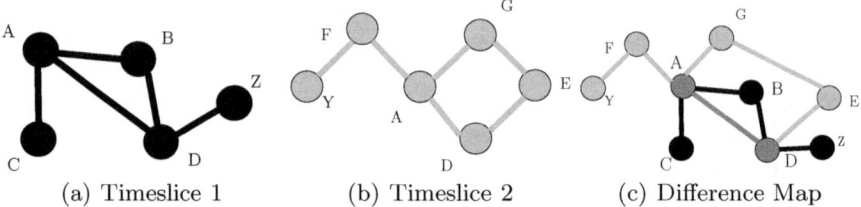

(a) Timeslice 1   (b) Timeslice 2   (c) Difference Map

Fig. 1. Two timeslices and the resulting difference map. **(a)** The graph at time 1. **(b)** The graph at time 2. **(c)** The difference map. Nodes and edges at time 1 are coloured black. Nodes and edges at time 2 are coloured light grey. Nodes that appear in both timeslices are coloured grey. The difference map encodes the nodes and edges that are added, removed, and persist over the dynamic graph series.

2.2 Dynamic Graph Drawing and the Mental Map

A number of dynamic graph drawing algorithms have looked at effective ways of preserving the mental map [16,8,7,9,14,11,13,5], and various experiments have investigated the effect of preserving the mental map [17,20,18,1].

In this study, we use the GraphAEL algorithm [11]. In this approach, inter-timeslice edges exist between nodes that are the same across timeslices. All timeslices are placed into the same plane and laid out using a force directed algorithm. The assigned strength of these inter-timeslice edges controls the amount of mental map preservation between timeslices: the higher the strength of the inter-timeslice edges, the shorter the distance nodes can move, increasing the degree of mental map preservation. Informed by previous experiments [18,1], we use a relatively low level of mental map preservation over all conditions and factors in this experiment.

2.3 Animation vs. Small Multiples for Dynamic Data

Several experiments have evaluated the performance of interfaces on dynamically evolving data. Most of these experiments have compared animation to small multiples on various types of data, and a pair of experiments have looked at this question in the context of dynamic graphs.

Griffen *et al.* [15] found that for clusters of moving hexagons against background noise, animation could be faster and more accurate than small multiples. Robertson *et al.* [19] compared animation, trace line, and small multiples visualization techniques on animated multi-dimensional data. The authors found that animation was the least effective form for analysis. Both small multiples and trace lines were significantly faster than animation, and small multiples was significantly more accurate. These varied results may indicate that the effectiveness of animation or small multiples strongly depends on the data to be visualized.

A pair of experiments have compared animation and small multiples in the context of dynamic graphs. Archambault *et al.* [1] compare small multiples and animation in addition to the effect of the mental map. In this experiment, small

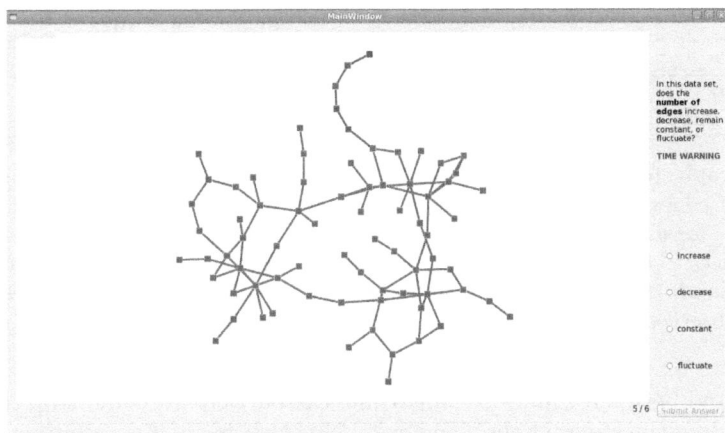

Fig. 2. Experiment interface. The question appears on the right along with four multiple choice answers. The participant selected the appropriate radio button and clicked "submit answer" to respond. The slide show interface under the difference map condition for **threads2** is shown in this figure.

multiples was significantly faster overall and for most tasks. For tasks that involved the simultaneous appearance of nodes or edges in the data, animation was significantly more accurate, but it was not the case that more time led to fewer errors. Farrugia *et al.* [12] compared animation and small multiples on two dynamic graph series. The experiment found that small multiples was significantly faster for most tasks.

In this experiment, we focus on the effectiveness of the difference map and determining the appropriate interface for it. We also compare three interfaces, small multiples, animation, and slide show, in this context.

3   The Experiment

To test the effectiveness of difference maps with respect to interface, we performed a within subject experiment. We employed a 2 condition (no difference map (ND) vs with difference maps (WD)) × 3 factor (animation (Anim) vs slide show (SD) vs small multiples (SM)) × 2 data set (**threads2** and **van de Bunt**) × 4 question design. The following subsections provide the details of this design.

3.1   Interfaces

The animation interface is similar to a movie player. The current view of the graph takes up the entire screen and smooth transitions morph the graph from one timeslice to another. Nodes that are added to the data or removed from it are faded in or out respectively. The positions of nodes and edges are linearly

						avg.			max			min		
Data Set	$|N|$	$|E|$	d	max d	min d	$|N|'$	$|E|'$	d'	$|N|'$	$|E|'$	d'	$|N|'$	$|E|'$	d'
threads2	70.14	83.43	2.38	8	1	2.50	3.83	0.06	4	4	2	0	0	0
van de Bunt	23.14	32.43	2.56	9	1	3.33	6.17	0.78	11	25	6	-6	-24	-6

Fig. 3. Graph statistics where $|N|$, $|E|$, and d are the average number of nodes, average number of edges, and average degree respectively. Primes indicate changes over time. The values for max d and min d are the maximum and minimum degrees observed over all timeslices. The columns labeled avg. correspond to the average change in values over all timeslices. The columns labeled max and min are the maximum and minimum changes observed between any pair of successive timeslices in the data set.

interpolated between frames. At any time, the participant could stop the animation and drag the slider at their own rate. No other form of interaction, including zooming, is allowed. This interface was used previously in Archambault et al. [1].

The slide show interface is very similar to a Powerpoint presentation. In this interface, each timeslice takes up the entire screen as in the animation condition. However, no smooth transition exists between pairs of timeslices. At the bottom right corner of the interface, the current slide number and the total number of slides is indicated. The participant uses the arrow keys to advance to the next timeslice or rewind to the previous one. No other form of interaction, including zooming, is allowed. This interface is shown in Fig. 2.

In the small multiples interface, all timeslices are presented in a matrix ordered left to right and top to bottom. The participant scans the windows to determine the right answer. No other form of interaction, including zooming, is allowed. This interface was used previously in Archambault et al. [1].[1]

3.2   Difference Map Encoding

A difference map, as previously described, is the union of a pair of adjacent timeslices in the dynamic graph sequence. Thus, given a sequence of t graphs, there would be $t - 1$ difference maps, depicting graph evolution. If a node is deleted between a pair of timeslices, it is light blue. If it is added, it is purple. Nodes that persist are brown. The same colour scheme is applied to the edges. In the non difference map condition, all t timeslices are presented using the interface. All nodes, except those pertaining to the question, are coloured grey.

3.3   Data Sets

In this experiment, two graph series of similar size were used to gauge the readability of difference maps. Fig. 3 reports the graph series parameters.

Threads2, used in the work of Frishman and Tal [13], is a graph series representing online newsgroup discussions. Nodes are authors of newsgroup articles,

---

[1] Examples of each interface in operation under each condition × factor pairing for all questions are available at http://www.labri.fr/perso/bpinaud/diffmap/

and an edge exists between two authors if one replied to the posting of another. As authors and postings are never deleted, this data set always grows in size. We selected seven timeslices from this data: timeslices ten through sixteen.

The **van de Bunt** data set [10] is a network that was used previously in the experiment of Farrugia *et al.* [12]. The nodes in the graph are undergraduate students. An edge exists between two undergraduates if they self-reported in a survey that they had a relationship. In the original graph, there were a number of edge types which encoded if the relationship was best friends, friends, friendly neutral, or troubled. In our experiment, we used only the links that were best friends or friends. This data set fluctuates in size over the timeslices.

3.4   Tasks

Our tasks aim to test the readability of both local and global structure in the graph. More importantly, each question should require the participant to look at all timeslices, because the full dynamic evolution of the graph should be taken into account. We chose four questions.

The first question tests the evolution of node degrees in the graph and is a local, topology-based question. It is similar to the types of questions posed in Purchase *et al.* [18]. Four nodes were highlighted different colours, and participants were asked to select the colour of the node as the answer to the question. The remaining nodes and edges were coloured as specified by the condition.

1. Node degree changes. One of the following questions was asked:
   (a) Which vertex **increases** its **degree** over time?
   (b) Which vertex **decreases** its **degree** over time?
   (c) Which vertex **keeps** its **degree constant** over time?

The second question explores when specific edges appear in the graph and gauges if participants can see when a specific pair of edges is added to the data set. The question is local and is one of the most basic questions related to the dynamism of the graph. Four to six nodes were highlighted one of four different colours and participants were asked if a pair of thick edges simultaneously appeared adjacent to a node of a specific colour. The remaining nodes and edges in the graph were coloured as specified by the condition.

2. Which **edges** appear together **exactly once** over all timeslices?

The third question tests the ability of the participant to notice global trends in the graph. Specifically, the question tests if overall trends, in terms of the number of edges in the graph, can be perceived. All nodes and edges, for this question, were coloured as specified by the condition.

3. In this data set, does the **number of edges** increase, decrease, remain constant, or fluctuate?

Finally, we would like to use a question that tests the more global, topology-related readability of a graph. In this case, it involves reading a path through the graph between two nodes. In this question, a black focus node and four coloured nodes appeared in the graph. Participants were asked to select the coloured node that became closer, in a graph connectivity sense, to the black node. The remaining nodes and edges, for this question, were coloured as specified by the condition.

4. When a path exists between the black node and a node of each of the four colours, which path **only decreases** in length?

Nodes pertinent to each question were highlighted with colours to eliminate additional cognitive cost for searching for nodes or reading labels. Keywords appeared in bold, as shown above, so that the participant could easily recognize each question. Multiple choice questions, with four answers, were used. Asking participants to select nodes directly on the screen would put the animation condition at a disadvantage, because in this case, nodes often move on the screen.

3.5   Experimental Design

Each condition × factor pairing (ie: a pairing of difference map and presentation method, for example, difference maps with slide show) was placed in its own block, giving the experiment a total of six blocks. Each block started with a demonstration of the interface, allowing the participant to ask questions, find out about the experiment, and see how the answers could be found.

The blocks each had eight experimental tasks: 2 data sets × 4 questions. These eight tasks were prefixed with a practice block of four questions. During this practice block, each of the four questions was asked exactly once with two on `threads2` and two on `van de Bunt`. Eight versions of each question were found on both data sets. The first six versions were used as experimental data and the last two versions were only used in the practice blocks. Thus, for experimental data, the same version of the same question was never asked twice to the participant. For practice block data, the same version of the same question was never asked under the same condition. For each participant, the order of the questions within each block was randomized with versions of each question randomly selected.

To minimize the cognitive shift incurred by moving from the difference map condition to the non difference map condition, participants answered all questions on one condition followed by the other. However, conditions were counterbalanced by presenting the non difference map condition first to even participants and the difference map condition first to odd participants. The order of the interface blocks within each condition was randomized such that each participant had a unique interface order.

All three interfaces were rendered in real time using the Tulip framework [4]. No time limit was enforced per question or for the experiment overall. However, a warning label appeared on the screen after forty seconds had elapsed for each question, and participants were encouraged to finish their work quickly after that point. The animation started playing automatically after a delay of three seconds had elapsed and took about ten seconds to play in its entirety.

Each experiment was conducted individually with the researcher and took approximately 1 hour and 20 minutes, including the pre-experiment training, practice tasks, experimental tasks for both experimental conditions, and post-experiment questionnaire. Overall, there were twenty-five participants used in the final results. Participants were drawn from members of the Complex and Adaptive Systems Laboratory of University College Dublin (UCD CASL). All but one had computer science experience.

4   Results

In this section, we present the results for our experiment overall, divided by question, and divided by interface. We compare the difference map presentation (WD) to the standard representation (ND). A Shapiro-Wilk test, with a significance level of $\alpha = 0.05$, was used to determine whether or not the data was normally distributed. We found that the error rate data was not normally distributed whereas response time data was. As a consequence, we used an exact Wilcoxon signed rank test on the error rate data and a paired t-test on the response time data. For both tests, a significance level of $\alpha = 0.05$ was used. When we divided the data by question, we applied a Bonferroni correction, thus reducing the significance level to $\alpha = 0.025$.

4.1   WD vs. ND

Overall, we did not find a significant difference, either in terms of error rate or response time, between the difference map condition (WD) and the simple graph timeslice condition without difference maps (ND) independent of interface.

By Question. On questions 1, 2, and, 4, we did not find a significant difference in terms of error rate or response time when comparing WD to ND independent of interface. However, on question 3, we discovered that WD produced significantly fewer errors (WD 0.08, ND 0.25, $p = 0.0035$) as shown in Fig. 4(a). Neither presentation method was significantly faster.

4.2   WD vs. ND Divided by Interface

We subsequently divided the data by interface to determine if, within interface, there were differences between the two presentation methods.

Animation. When considering only the animation interface, we found no significant difference between either condition (WD and ND) overall or on a per question basis.

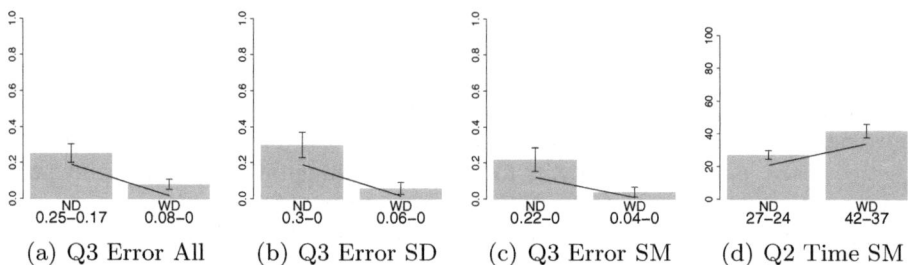

(a) Q3 Error All   (b) Q3 Error SD   (c) Q3 Error SM   (d) Q2 Time SM

Fig. 4. Significant differences found in response time and error rate for this experiment. Error rate is percentage error and time is seconds. The mean and median values, separated by a dash, are written below each bar of the chart. SD and SM are the slide show and small multiples interfaces respectively.

	Q1		Q2		Q3		Q4	
	ND	WD	ND	WD	ND	WD	ND	WD
Mean	12.96	8.08	11.16	9.64	13.40	7.40	12.04	8.96
Std. Deviation	2.61	2.63	2.93	3.15	2.50	2.58	3.27	3.27

Fig. 5. Table of preference data comparing WD to ND aggregated across interface. WD was preferred in all cases and significantly so on questions 1, 3, and 4 ($p = 0.001$, $p = 0.000$, and $p = 0.028$ using a Wilcoxon signed ranks test).

Slide Show. On slide show, we did not find a significant difference between WD and ND overall. However, for question 3, we found that WD produced significantly fewer errors than ND (WD: 0.06, ND 0.30, $p = 0.013$) as shown in Fig 4(b). We did not find a significant difference in terms of response time.

Small Multiples. Under the small multiples interface, we did not find a significant difference overall. However, for question 3, we found that WD produced fewer errors than ND (WD 0.04, ND 0.22, $p = 0.008$). For question 2, we found that ND was significantly faster than WD (ND 27s, WD 42s, $p = 0.007$). These results are shown in Figs. 4(c) and 4(d) respectively. We did not find a significant difference in terms of response time.

4.3 Preference Data

A summary of our findings from the post-experiment questionnaire are shown in Fig. 5. Participants were asked to rank the six condition × interface pairs from 1 to 6, with 1 indicating the most preferred.

When we aggregated the results across interface, WD was preferred to ND for all questions and significantly so for questions 1, 3, and 4. For question 3, *In this data set, does the number of edges increase, decrease, remain constant, or fluctuate?* most participants remarked qualitatively that WD made this question much easier to answer. For question 2, some participants noted that the two colours used in the question made it more difficult to answer.

5   Discussion

5.1   Does the Difference Map Help?

Overall, we did not find that difference maps helped on all tasks. As our tasks are varied, this fact may not be all that surprising as the presentation method may not be suitable for all types of questions.

When we divided the data by question, however, we did find that difference maps were able to significantly reduce the number of errors for the question which asked if the number of edges increased, decreased, remained constant, or fluctuated (question 3). These results were supported by the survey data which found difference maps were preferred significantly on this question. As difference maps highlight, using colour, if the edge has been added or removed, the participant can easily see where and by how much the graph has changed. Thus, difference maps may be helpful when trying to gauge how much the graph has changed between timeslices at a large scale.

5.2   Does It Help for All Interfaces?

There was no significant difference between the two conditions overall for any of the interfaces individually. However, on the slide show interface, WD produced significantly fewer errors than ND for question 3 (change in total number of edges). Using small multiples, ND was significantly faster than WD on question 2 (simultaneous appearance of edges). In terms of error rate for question 3 on this interface, WD produced significantly fewer errors. No other significant differences were found.

The results for question 3 are unsurprising considering that we saw an overall benefit of difference maps on this question globally. It seems that the benefit was achieved mostly using the small multiples and slide show interfaces. As these interfaces do not smoothly fade edges in and out, it may be the case that colour helped gauge when something was inserted or deleted. However, further study is required to confirm this conjecture.

The result on question 2 for the small multiples interface, that ND is significantly faster than WD, may be related to the fact that colour encoded both the answer to the question and graph structure changes. In the WD condition, participants had to contend with two sets of colours for each edge and reason about what the combination meant. Thus, the results suggest that the participants were able to perform the task equally as well, but it took them much longer to find the solution.

The difference map was significantly preferred for questions 1, 3, and 4 according to the survey data. It is surprising to get such strong preference data for one presentation method that does not match the corresponding performance data (which found little benefit in the use of difference maps when performing tasks). This result suggests that even if the use of difference maps may not improve performance, users might feel more comfortable with this presentation method when performing tasks on a dynamic graph sequence.

5.3 Limitations

We have tried to collect data that allows us to make generalizations for each interface, by including more than one data set, and more than one question. However, it would have been impossible to test a wider range of data sets and even more questions. The generalization of these results are therefore limited by these parameters. It was necessary for our participants to have some knowledge of graphs, meaning that our results only hold for this particular population. Running the experiment in a laboratory situation, on context-free graphs (even if based on real data sets) means that these results may not extrapolate to the visualization of graphs within an application context.

6 Conclusions and Future Work

In future work, it would be interesting to see if graph hierarchies can help improve the performance of difference maps. Hierarchies have been used in systems that present difference maps [6,3], and a recent experiment [2] has provided some evidence that hierarchies can improve graph readability for some tasks. It would be interesting to see if difference maps can benefit from these representations.

We performed a user study to gauge the benefit of using difference maps rather than presenting the timeslices directly in a dynamic graph. We tested the readability of difference maps using three interfaces and four questions. In this study, we found that difference maps can help answer questions about large scale changes in a dynamic graph in terms of changes in the number of edges. Also, difference maps were strongly preferred by participants.

Acknowledgments

The first author would like to acknowledge the support of the Clique Strategic Research Cluster funded by Science Foundation Ireland (SFI) Grant No. 08/SRC/I1407. We would also like to thank John Hamer for writing the dynamic graph layout algorithm based on the GraphAEL model. Finally, we would like to thank all of the participants who took part in this experiment.

References

1. Archambault, D., Purchase, H.C., Pinaud, B.: Animation, small multiples, and the effect of mental map preservation in dynamic graphs. IEEE Trans. on Visualization and Computer Graphics (to appear, 2010)
2. Archambault, D., Purchase, H.C., Pinaud, B.: The readability of path-preserving clusterings of graphs. Computer Graphics Forum (Proc. EuroVis) 29(3), 1173–1182 (2010)
3. Archambault, D.: Structural differences between two graphs through hierarchies. In: Proc. of Graphics Interface, pp. 87–94 (2009)

4. Auber, D.: Tulip: A huge graph visualization framework. In: Mutzel, P., Jünger, M. (eds.) Graph Drawing Software. Mathematics and Visualization, pp. 105–126. Springer, Heidelberg (2003)
5. Boitmanis, K., Brandes, U., Pich, C.: Visualizing internet evolution on the autonomous systems level. In: Hong, S.-H., Nishizeki, T., Quan, W. (eds.) GD 2007. LNCS, vol. 4875, pp. 365–376. Springer, Heidelberg (2008)
6. Bourqui, R., Jourdan, F.: Revealing subnetwork roles using contextual visualization: Comparison of metabolic networks. In: Proc. 12th Int. Conf. on Information Visualisation (IV 2008), pp. 638–643 (2008)
7. Brandes, U., Wagner, D.: A bayesian paradigm for dynamic graph layout. In: DiBattista, G. (ed.) GD 1997. LNCS, vol. 1353, pp. 236–247. Springer, Heidelberg (1997)
8. Cohen, R.F., Battista, G.D., Tollis, I.G., Cohen, R.F., Tamassia, R., Tollis, I.G.: A framework for dynamic graph drawing. In: ACM Symp. on Computational Geometry, pp. 261–270 (1992)
9. Diehl, S., Görg, C.: Graphs, they are a changing – dynamic graph drawing for a sequence of graphs. In: Goodrich, M.T., Kobourov, S.G. (eds.) GD 2002. LNCS, vol. 2528, pp. 23–31. Springer, Heidelberg (2002)
10. van Duijn, M., Zeggelink, E., Huisman, M., Stokman, F., Wasseur, F.: Evolution of sociology freshmen into a frendship network. The Journal of Math. Sociology 27(2), 153–191 (2003),
    http://www.stats.ox.ac.uk/~snijders/siena/vdBunt_data.htm
11. Erten, C., Harding, P.J., Kobourov, S., Wampler, K., Yee, G.V.: GraphAEL: Graph animations with evolving layouts. In: Liotta, G. (ed.) GD 2003. LNCS, vol. 2912, pp. 98–110. Springer, Heidelberg (2004)
12. Farrugia, M., Quigley, A.: Effective temporal graph layout: A comparative study of animation versus static display methods. Journal of Information Visualization (to appear, 2010)
13. Frishman, Y., Tal, A.: Online dynamic graph drawing. IEEE Trans. on Visualization and Computer Graphics 14(4), 727–740 (2008)
14. Görg, C., Birke, P., Pohl, M., Diehl, S.: Dynamic graph drawing of sequences of orthogonal and hierarchical graphs. In: Pach, J. (ed.) GD 2004. LNCS, vol. 3383, pp. 228–238. Springer, Heidelberg (2005)
15. Griffen, A.L., MacEachren, A.M., Hardisty, F., Steiner, E., Li, B.: A comparison of animated maps with static small-multiple maps for visually identifying space-time clusters. Annals of the Association of American Geographers 96(4), 740–753 (2006)
16. Moen, S.: Drawing dynamic trees. IEEE Software 7(4), 21–28 (1990)
17. Purchase, H.C., Hoggan, E., Görg, C.: How important is the "mental map"? – an empirical investigation of a dynamic graph layout algorithm. In: Kaufmann, M., Wagner, D. (eds.) GD 2006. LNCS, vol. 4372, pp. 184–195. Springer, Heidelberg (2007)
18. Purchase, H.C., Samra, A.: Extremes are better: Investigating mental map preservation in dynamic graphs. In: Stapleton, G., Howse, J., Lee, J. (eds.) Diagrams 2008. LNCS (LNAI), vol. 5223, pp. 60–73. Springer, Heidelberg (2008)
19. Robertson, G., Fernandez, R., Fisher, D., Lee, B., Stasko, J.: Effectiveness of animation in trend visualization. IEEE Trans. on Visualization and Computer Graphics (Proc. Vis/InfoVis 2008) 14(6), 1325–1332 (2008)
20. Saffrey, P., Purchase, H.C.: The "mental map" versus "static aesthetic" compromise in dynamic graphs: A user study. In: Proc. of the 9th Australasian User Interface Conference, pp. 85–93 (2008)
21. Tufte, E.: Envisioning Information. Graphics Press (1990)

Maximizing the Total Resolution of Graphs

Evmorfia N. Argyriou, Michael A. Bekos, and Antonios Symvonis

School of Applied Mathematical & Physical Sciences,
National Technical University of Athens, Greece
{fargyriou,mikebekos,symvonis}@math.ntua.gr

Abstract. A major factor affecting the readability of a graph drawing is its resolution. In the graph drawing literature, the resolution of a drawing is either measured based on the angles formed by consecutive edges incident to a common node (angular resolution) or by the angles formed at edge crossings (crossing resolution). In this paper, we evaluate both by introducing the notion of "total resolution", that is, the minimum of the angular and crossing resolution. To the best of our knowledge, this is the first time where the problem of maximizing the total resolution of a drawing is studied.

The main contribution of the paper consists of drawings of asymptotically optimal total resolution for complete graphs (circular drawings) and for complete bipartite graphs (2-layered drawings). In addition, we present and experimentally evaluate a force-directed based algorithm that constructs drawings of large total resolution.

1 Introduction

There exist several criteria that have been used to judge the quality of a graph drawing [4,14]. An undesired property that may negatively influence the readability of a graph drawing is the presence of edges that are too close to each other, especially if these edges are adjacent. Thus, maximizing the angles among incident edges becomes an important aesthetic criterion, since there is some correlation between the involved angles and the visual distinctiveness of the edges. On the other hand, recent cognitive experiments by Huang et al. [12,13] indicate that the negative impact of an edge crossing on the human understanding of a graph drawing is eliminated in the case where the crossing angle is greater than 70 degrees. This motivates us to study a new graph drawing scenario in which both *angular* and *crossing resolution* are taken into account in order to produce a straight-line drawing of a given graph. Formally, the term angular resolution denotes the smallest angle formed by two adjacent edges incident to a common node, whereas the term crossing resolution refers to the smallest angle formed by a pair of crossing edges. The angular resolution maximization problem has been extensively studied by the graph drawing community over the last few decades [3,9,10,11,15,16]. On the other hand, the crossing resolution maximization is a relatively new problem [1,6,7]. To the best of our knowledge, this is the first attempt, where both angular and crossing resolution are combined to produce drawings.

U. Brandes and S. Cornelsen (Eds.): GD 2010, LNCS 6502, pp. 62–67, 2011.

2   Drawings with Optimal Total Resolution for Complete and Complete Bipartite Graphs

In this section, we define the total resolution of a drawing and we present drawings of asymptotically optimal total resolution for complete graphs (circular drawings) and complete bipartite graphs (2-layered drawings).

Definition 1. *The* total resolution *of a drawing is defined as the minimum of its angular and crossing resolution.*

We first consider the case of complete graphs K_n, $n \geq 3$. Our aim is to construct a circular drawing of K_n of maximum total resolution. Our approach is constructive and common when dealing with complete graphs. A similar one has been given by Formann et al. [9] for obtaining optimal drawings of complete graphs, in terms of angular resolution. Consider a circle C of radius $r_c > 0$ centered at $(0,0)$ and circumscribe a regular n-polygon Q on C. The nodes of K_n coincide with the vertices of Q. The construction supports the following Theorem. For a detailed proof, refer to [2].

Theorem 1. *A complete graph K_n admits a drawing of total resolution $\Theta(\frac{1}{n})$.*

Obviously, the bound of the total resolution of a complete bipartite graph can be implied by the bound of the complete graph. However, if the nodes of the graph must have integer coordinates, i.e., we restrict ourselves on grid drawings, few results are known regarding the area needed of such a drawing. An upper bound of $O(n^3)$ area can be implied by [3]. This motivates us to separately study the class of complete bipartite graph, since we can drastically improve this bound. Note that tradeoffs between (angular or crossing) resolution and area have been studied by the graph drawing community, in the past [1,5,16].

Let $K_{m,n} = (V_1 \cup V_2, E)$ be a complete bipartite graph, where $V_1 = \{u_1^1, \ldots, u_m^1\}$, $V_2 = \{u_1^2, \ldots, u_n^2\}$ and $E = V_1 \times V_2$. Let $\mathcal{R} = AB\Gamma\Delta$ be a square whose top and bottom sides coincide with $\mathcal{L}_1$ and $\mathcal{L}_2$, respectively. The nodes of V_1 (V_2) reside along side $\Gamma\Delta$ (AB). Let $\ell_1, \ldots, \ell_m$, be a bundle of semi-lines, each of which emanates from vertex B and crosses side $\Gamma\Delta$ of $\mathcal{R}$, so that the angle formed by $B\Gamma$ and semi-line ℓ_i equals to $\frac{(i-1)\cdot\widehat{\Delta B\Gamma}}{m-1}$, for each $i = 1, \ldots, m$. These semi-lines split angle $\widehat{\Delta B\Gamma}$ into $m-1$ angles, each of which is equal to $\frac{\pi}{4\cdot(m-1)}$. Then, we place node u_i^1 at the intersection of semi-line l_i and $\Gamma\Delta$, for each $i = 1, \ldots, m$. We denote by a_i the horizontal distance between two consecutive nodes u_i^1 and u_{i+1}^1, $i = 1, \ldots, m-1$. Symmetrically, we define the position of the nodes of V_2 along side AB of $\mathcal{R}$. We conclude by the following Theorem. For a detailed proof refer to [2].

Theorem 2. *A complete bipartite graph $K_{m,n}$ admits a 2-layered drawing of total resolution $\Theta(\frac{1}{\max\{m,n\}})$.*

Say that the nodes of the graph must have integer coordinates. Then, assuming that $\mathcal{L}_1$ and $\mathcal{L}_2$ coincide with two horizontal grid lines, we can slightly move each node of V_1 and V_2 to the rightmost grid-point to its left. We can prove that there are not two nodes sharing the same grid point, assuming that a_1 is greater than one grid unit. Since no node moves more than one unit of length, the total resolution is not asymptotically affected. Regarding the computation of the area occupied by the drawing, we can further prove that it supports the following Theorem. The reader is referred to [2], for a detailed proof.

Theorem 3. *A complete bipartite graph $K_{m,n}$ admits a 2-layered grid drawing of $\Theta(\frac{1}{\max\{m,n\}})$ total resolution and $O(\max\{m^2, n^2\})$ area.*

3   A Force Directed Algorithm

We present a force-directed algorithm that given a reasonably nice initial drawing, results in a drawing of high total resolution. Our algorithm uses the attractive forces of the classical force-directed algorithm of Eades [8] and some additional forces exerted to the nodes of the graph, that tend to maximize the total resolution of the drawing.

Before we proceed with the description of our algorithm, we present some notation that is heavily used in the remainder. Given a drawing $\Gamma(G)$ of G, we denote by $p_u = (x_u, y_u)$ the position of node $u \in V$ on the plane. The unit length vector from p_u to p_v is denoted, by $\overrightarrow{p_u p_v}$, where $u, v \in V$. The degree of node $u \in V$ is denoted by $d(u)$. Let also $d(G) = \max_{u \in V} d(u)$ be the degree of the graph. Given a pair of points $q_1, q_2 \in \mathbb{R}^2$, with a slight abuse of notation, we denote by $||q_1 - q_2||$ the Euclidean distance between q_1 and q_2. We refer to the line segment defined by q_1 and q_2 as $\overline{q_1 q_2}$. Let $\overrightarrow{\alpha}$ and $\overrightarrow{\gamma}$ be two vectors. The vector which bisects the angle between $\overrightarrow{\alpha}$ and $\overrightarrow{\gamma}$ is $\frac{\overrightarrow{\alpha}}{||\overrightarrow{\alpha}||} + \frac{\overrightarrow{\gamma}}{||\overrightarrow{\gamma}||}$. We denote by $\texttt{Bsc}(\overrightarrow{\alpha}, \overrightarrow{\gamma})$ the corresponding unit length vector. Given a vector $\overrightarrow{\beta}$, we refer to the unit length vector which is perpendicular to $\overrightarrow{\beta}$ and precedes it in the clockwise direction, as $\texttt{Perp}(\overrightarrow{\beta})$.

Let $e = (u, v)$ and $e' = (u', v')$ be two crossing edge and let p_c be their intersection point. Let also $\theta_{vv'}$, $\theta_{v'u}$, $\theta_{uu'}$ and $\theta_{u'v}$ be the angles formed by the intersection of e and e' at p_c, as illustrated in Fig.1a. Then, our preference for right-angle crossings can be captured by spring forces (see Fig.1a), and, by the angles $\theta_{vv'}$ and $\theta_{v'u}$ formed at the crossing (see Fig.1b). The exact formulas of these force are:

$$\mathcal{F}_{\text{spring}}^{\text{cros}}(p_v, p_{v'}) = C_{\text{spring}}^{\text{cros}} \cdot \log \frac{||p_v - p_{v'}||}{\ell_{\text{spring}}^{vv'}} \cdot \overrightarrow{p_v p_{v'}}$$

$$\mathcal{F}_{\text{angle}}^{\text{cros}}(p_v, p_{v'}) = C_{\text{angle}}^{\text{cros}} \cdot sign(\theta_{vv'} - \frac{\pi}{2}) \cdot f(\theta_{vv'}) \cdot \texttt{Perp}(\texttt{Bsc}(\overrightarrow{p_c p_v}, \overrightarrow{p_c p_{v'}}))$$

where the constants $C_{\text{spring}}^{\text{cros}}$ and $C_{\text{angle}}^{\text{cros}}$ are used to control the stiffness of the springs and the strength of the force, respectively, $\ell_{\text{spring}}^{vv'}$, which corresponds to the natural length of the spring, is equal to $\sqrt{||p_c - p_v||^2 + ||p_c - p_v'||^2}$ and

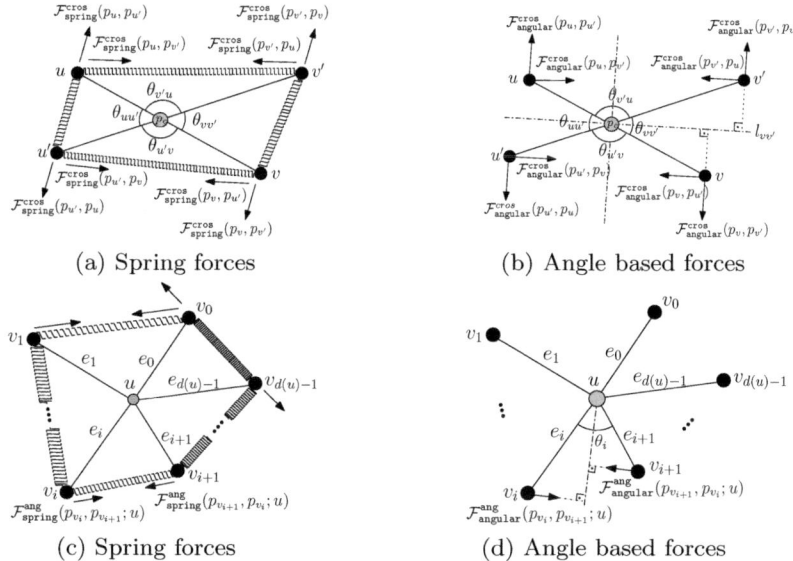

(a) Spring forces   (b) Angle based forces

(c) Spring forces   (d) Angle based forces

Fig. 1. Forces applied on nodes in order to maximize the total resolution

$f : \mathbb{R} \to \mathbb{R}$ is a function so that $f(\theta) = |\frac{\pi}{2} - \theta|/\theta$. The remaining forces of Fig.1a and 1b are defined similarly.

Let u be a node incident to edges $e_0 = (u, v_0), \ldots e_{d(u)-1} = (u, v_{d(u)-1})$. Assume that $e_0, e_1, \ldots, e_{d(u)-1}$ are consecutive in the counter-clockwise order around u in the drawing of the graph. Let θ_i be the angle formed by e_i and $e_{(i+1)mod(d(u))}$, measured in counter-clockwise direction from e_i to $e_{(i+1)mod(d(u))}$. Then, our preference for angles equal to $2\pi/d(u)$ can be captured by spring forces (see Fig.1c), and, by the angles θ_i, $i = 0, 1, \ldots, d(u) - 1$ (see Fig.1d). The exact formulas of these force are:

$$\mathcal{F}_{\text{spring}}^{\text{angular}}(p_{v_i}, p_{v_{(i+1)mod(d(u))}}; u) = C_{\text{spring}}^{\text{angular}} \cdot \log \frac{||p_{v_i} - p_{v_{(i+1)mod(d(u))}}||}{\ell_{\text{spring}}^i}.$$

$$\mathcal{F}_{\text{angle}}^{\text{angular}}(p_{v_i}, p_{v_{(i+1)mod(d(u))}}; u) = C_{\text{angle}}^{\text{angular}} \cdot sign(\theta_i - \frac{2\pi}{d(u)}) \cdot g(\theta_i; u) \cdot$$

$$\text{Perp}(\text{Bsc}(\overrightarrow{p_u p_{v_i}}, \overrightarrow{p_u p_{v_{(i+1)mod(d(u))}}}))$$

where the quantities $C_{\text{spring}}^{\text{angular}}$ and $C_{\text{angle}}^{\text{angular}}$ are constants which captures the stiffness of the spring and the strength of the force, respectively, $g : \mathbb{R} \times V \to \mathbb{R}$ is a function so that $g(\theta; u) = \frac{|\frac{2\pi}{d(u)} - \theta|}{\theta}$ and l_{spring}^i, which corresponds to the natural length of each spring, is equal to:

$$\sqrt{||e_i||^2 + ||e_{(i+1)mod(d(u))}||^2 - 2 \cdot ||e_i|| \cdot ||e_{(i+1)mod(d(u))}|| \cdot \cos(2\pi/d(u))}$$

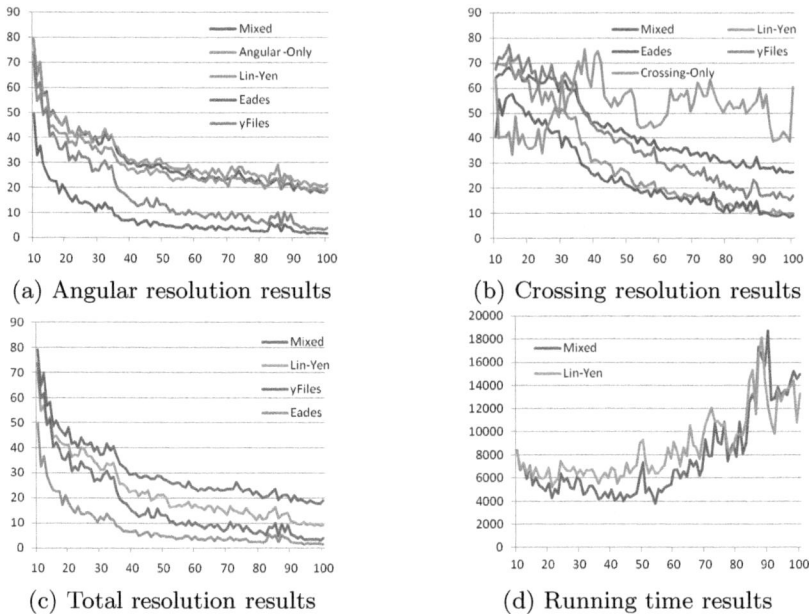

(a) Angular resolution results   (b) Crossing resolution results

(c) Total resolution results   (d) Running time results

Fig. 2. A visual presentation of our experimental results. The X-axis indicates the number of the nodes of the graph. In Fig.(a)-(c) the Y-axis corresponds to the resolution measured in degrees, whereas in Fig.(d) to the running time measured in milliseconds.

Note that by setting zero values to the constants $C_{\mathrm{spring}}^{\mathrm{cros}}$, $C_{\mathrm{angle}}^{\mathrm{cros}}$ or $C_{\mathrm{spring}}^{\mathrm{angular}}$, $C_{\mathrm{angle}}^{\mathrm{angular}}$, our algorithm can be configured to maximize the angular, or the crossing resolution only, respectively. Regarding the time complexity, each iteration of our algorithm takes $O(E^2 + Vd(G)\log d(G))$ time and can be further improved to $O(K + E\log^2 E / \log\log E + Vd(G)\log d(G))$ time per iteration, using standard techniques from computation geometry.

In the following, we present the results of the experimental evaluation of our algorithm. Apart from our algorithm, we have also implemented the algorithms of Eades [8] and Lin and Yen [15]. The implementations are in Java using yFiles library (www.yworks.com). The experiment was performed on a Linux machine with 2.00 GHz CPU and 2GB RAM using the Rome graphs obtained from graphdrawing.org. The experiment was performed as follows. Each Rome graph was laid out using the SmartOrganic layouter of yFiles. This layout was the input of all algorithms. If both the angular and the crossing resolution between two consecutive iterations of each algorithm were not improved more that 0.001 degrees, we assumed that the algorithm has converged. The maximum number of iterations was set to 100.000. Our algorithm is evaluated as (a) Crossing-Only, (b) Angular-Only and (c) Mixed. Fig.2 illustrates the results of the experimental evaluation. For a more detailed analysis refer to [2].

4   Conclusions

We introduced and studied the total resolution maximization problem. Of course, our work leaves several open problems. It would be interesting to try to identify other classes of graphs that admit optimal drawings. Even the case of planar graphs is of interest, as by allowing some edges to cross (say at large angles), we may improve the angular resolution and therefore the total resolution.

References

1. Angelini, P., Cittadini, L., di Battista, G., Didimo, W., Frati, F., Kaufmann, M., Symvonis, A.: On the perspectives opened by right angle crossing drawings. In: Eppstein, D., Gansner, E.R. (eds.) GD 2009. LNCS, vol. 5849, pp. 21–32. Springer, Heidelberg (2010)
2. Argyriou, E.N., Bekos, M.A., Symvonis, A.: Maximizing the total resolution of graphs. Technical Report arXiv:/106871 (2010)
3. Bárány, I., Tokushige, N.: The minimum area of convex lattice n-gons. Combinatorica 24(2), 171–185 (2004)
4. Battista, G.D., Eades, P., Tamassia, R., Tollis, I.G.: Algorithms for drawing graphs: an annotated bibliography. Comput. Geom. 4, 235–282 (1994)
5. Cheng, C.C., Duncanyz, C.A., Goodrichz, M.T., Kobourovz, S.G.: Drawing planar graphs with circular arcs. Discrete and Comp. Geometry 25, 117–126 (1999)
6. Di Giacomo, E., Didimo, W., Liotta, G., Meijer, H.: Area, curve complexity, and crossing resolution of non-planar graph drawings. In: Eppstein, D., Gansner, E.R. (eds.) GD 2009. LNCS, vol. 5849, pp. 15–20. Springer, Heidelberg (2010)
7. Didimo, W., Eades, P., Liotta, G.: Drawing graphs with right angle crossings. In: Dehne, F., Gavrilova, M., Sack, J.-R., Tóth, C.D. (eds.) WADS 2009. LNCS, vol. 5664, pp. 206–217. Springer, Heidelberg (2009)
8. Eades, P.: A heuristic for graph drawing. Congressus Numerantium (42), 149–160 (1984)
9. Formann, M., Hagerup, T., Haralambides, J., Kaufmann, M., Leighton, F., Symvonis, A., Welzl, E., Woeginger, G.: Drawing graphs in the plane with high resolution. SIAM J. Comput. 22(5), 1035–1052 (1993)
10. Garg, A., Tamassia, R.: Planar drawings and angular resolution: Algorithms and bounds. In: Proc. 2nd Annu. Eur. Sympos. Alg., pp. 12–23 (1994)
11. Gutwenger, C., Mutzel, P.: Planar polyline drawings with good angular resolution. In: Whitesides, S.H. (ed.) GD 1998. LNCS, vol. 1547, pp. 167–182. Springer, Heidelberg (1999)
12. Huang, W.: Using eye tracking to investigate graph layout effects. In: Proc. Asia-Pacific Symp. on Visualization, pp. 97–100 (2007)
13. Huang, W., Hong, S.-H., Eades, P.: Effects of crossing angles. In: Proc. of the Pacific Visualization Symp., pp. 41–46 (2008)
14. Kaufmann, M., Wagner, D. (eds.): Drawing Graphs: Methods and Models. LNCS, vol. 2025. Springer, Heidelberg (2001)
15. Lin, C.-C., Yen, H.-C.: A new force-directed graph drawing method based on edge-edge repulsion. In: Proc. of the 9th Int. Conf. on Inf. Vis., pp. 329–334. IEEE, Los Alamitos (2005)
16. Malitz, S.M., Papakostas, A.: On the angular resolution of planar graphs. In: STOC, pp. 527–538 (1992)

Plane Drawings of Queue and Deque Graphs*

Christopher Auer, Christian Bachmaier, Franz Josef Brandenburg,
Wolfgang Brunner, and Andreas Gleißner

University of Passau, Germany
{auerc,bachmaier,brandenb,brunner,gleissner}@fim.uni-passau.de

Abstract. In stack and queue layouts the vertices of a graph are linearly
ordered from left to right, where each edge corresponds to an item and
the left and right end vertex of each edge represents the addition and
removal of the item to the used data structure. A graph admitting a
stack or queue layout is a stack or queue graph, respectively.

Typical stack and queue layouts are rainbows and twists visualizing
the LIFO and FIFO principles, respectively. However, in such visualizations, twists cause many crossings, which make the drawings incompre-
hensible. We introduce linear cylindric layouts as a visualization technique for queue and deque (double-ended queue) graphs. It provides new
insights into the characteristics of these fundamental data structures and
extends to the visualization of mixed layouts with stacks and queues. Our
main result states that a graph is a deque graph if and only if it has a
plane linear cylindric drawing.

1 Introduction

In his pioneering work on the Art of Computer Programming, D. E. Knuth
raises the question "How shall we draw a tree?" [10, p. 306], which can be
seen as the beginning of Graph Drawing. Knuth also studied elementary data
structures, such as stacks, queues and deques, and represented their behavior as
train tracks. In this paper we pose the question "How shall we draw a stack, a
queue, or a deque?". The purpose of such drawings is to visualize the underlying
data structure and the operations applied to it.

Stack and queue layouts have been studied extensively in the past, e.g., in
[1, 2, 4–9, 12–14], and are used for 3D drawings of graphs [12, 13], in VLSI
design [2] and in other application scenarios (see [9] for a short survey). In these
layouts the vertices of a graph are linearly ordered from the left to the right.
The vertices are processed in this order and each edge corresponds to an item
which is inserted into the data structure at its left endpoint and is removed at
its right endpoint. These operations must obey the principles of the underlying
data structure, such as "last-in, first-out" for a stack or "first-in, first-out" for a
queue.

k-stack (k-queue) layouts are a generalization of stack (queue) layouts: Such
layouts consist of a single linear order of the vertices and a partition of the set of

---

* Supported by the Deutsche Forschungsgemeinschaft (DFG), grant BR835/15-1.

U. Brandes and S. Cornelsen (Eds.): GD 2010, LNCS 6502, pp. 68–79, 2011.

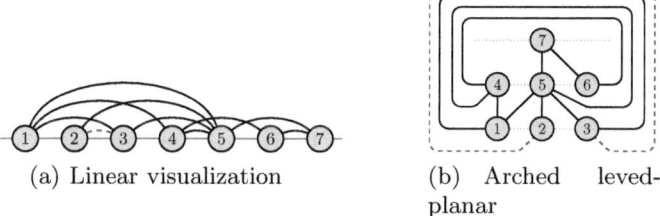

(a) Linear visualization   (b) Arched leved-
                                        planar

Fig. 1. Visualizing queue layouts

edges into k subsets, where each subset permits a stack (queue) layout, see, e.g.,
[5, 9]. k-stack (k-queue) layouts have also been generalized to *mixed* layouts, e.g.,
two stacks and one queue, and have been studied in [5]. The authors motivate
their studies of mixed layouts by their investigations of the deque data structure:
A deque can either emulate two stacks or one queue. Conversely, two stacks and
one queue can emulate one deque.

Stack layouts are also known as book-embeddings of graphs [1] and the number
of pages corresponds to the number of used stacks. These graphs have interesting
graph theoretic properties: A graph G is a stack graph if and only if G is outerplanar [1]. Bernhart and Kainen [1] have characterized the class of 2-stack graphs
as the subgraphs of planar graphs with a Hamiltonian cycle, and every planar
graph has a layout with four stacks [14].

Common visualizations of stack and queue graphs with a given layout place
the vertices from left to right on the x-axis according to the given order. The
edges are drawn as arches, which are x-monotone curves above the x-axis. Stack
layout visualizations show *rainbows* [13] as characteristic structures, which are
properly nested arches, whereas the equivalent structure in queue layouts are
twists [9]. Conversely, in a visualization of queue graphs rainbows are not allowed,
while stack graphs forbid twists.

Fig. 1(a) depicts a queue graph. The edges drawn as solid lines constitute
a valid queue layout since no two arches nest completely (nesting edges with
common end vertices are allowed). A characteristic twist is displayed by edges
$\{1,4\}$ and $\{2,5\}$, i.e., $\{1,4\}$ is added to the queue before $\{2,5\}$ and both are
removed in the same order. However, due to the many crossings (one crossing
per twist), it is hard to follow the routing of the edges and to validate the queue
layout. Introducing the edge $\{2,3\}$ (dashed) destroys the queue layout since it is
completely nested in the arch $\{1,5\}$, i.e., at first $\{1,5\}$ and then $\{2,3\}$ is added
to the queue but $\{2,3\}$ has to be removed before $\{1,5\}$. Nevertheless, it is hard
to recognize immediately that $\{2,3\}$ destroys the queue layout.

Heath et al. [5, 9] characterize the class of queue graphs as the arched leveled-
planar graphs. Such a graph has a planar drawing with vertices placed on levels
and inter-level edges only between two adjacent levels or intra-level edges from
the left-most vertex to vertices on the right side. Fig. 1(b) shows the queue layout
from Fig. 1(a) again, where levels are drawn as dotted lines. Again it is hard to
see immediately that edge $\{2,3\}$ is illegal. Moreover, an invalid drawing of an

arched leveled-planar graph does not necessarily indicate an invalid queue layout since a valid drawing with a different leveling of the vertices is still possible.

In this paper we propose a novel approach for visualizing queue graphs that overcomes the aforementioned drawbacks. We achieve our visualization of a queue graph by winding the arcs at most once around the surface of a 3D cylinder. A 2D representation is achieved by cutting the cylinder and duplicating the vertices. Moreover, our representation is also suitable for deque (double-ended queue) graphs where the deque may even have a restricted set of input or output operations. We can also display graphs with mixed layouts of stacks, queues and deques together in one concise drawing. By applying our drawing technique we immediately arrive at the result that every deque graph is planar.

The remainder of the paper is structured as follows. In Sect. 2 we introduce deque layouts. In Section 3 we define linear cylindric drawings and their equivalent representations. We also characterize the class of graphs permitting a deque layout as the graphs permitting a crossing-free linear cylindric drawing. We describe how linear cylindric drawings help to investigate mixed layouts in Sect. 4. We also revisit queue layouts and show how linear cylindric drawings of queue graphs overcome the drawbacks of the drawings depicted in Fig. 1. Finally, Sect. 5 gives a conclusion and an outlook to future work.

2   Preliminaries

We consider simple undirected graphs $G = (V, E)$ with n vertices and m edges. A linear layout π is a bijective assignment $\pi : V \to \{1, \dots, n\}$ of the vertices to positions $\{1, \dots, n\}$. For each edge $e = \{u, v\}$ we denote by $l(e) = \min\{\pi(u), \pi(v)\}$ and $r(e) = \max\{\pi(u), \pi(v)\}$ the position of its left- and right-hand vertex, respectively. Edge $e \in E$ is said to cover the *range* from $l(e)$ to $r(e)$ (both included).

A deque generalizes a stack and a queue: It has two ends, a *head* **h** and a *tail* **t**, to insert and remove items. If insertions or removals are only allowed at the deque's head, it is called *input* or *output restricted*, respectively. Let α and ω be two functions from E to $\{$**h**, **t**$\}$ that assign to each edge e the side of its addition and its removal, respectively. α/ω are called *input/output assignments (I/O assignments)*. If $\alpha(e) = \omega(e)$, then e is called a *stack edge*, otherwise a *queue edge*, according to the manner the edges are processed by the deque. We denote by $\Delta(G)$ the tuple (π, α, ω) and call it *linear I/O layout*. $\Delta(G)$ is a *deque layout* iff the vertices can be processed from left to right according to π such that all edges can be processed by the deque according to α and ω.

Definition 1. *A graph is a* deque graph *if and only if G has a deque layout. Accordingly, a graph is an* input restricted deque *(an* output restricted deque, *a* stack, *a* queue) *graph if it has a respective layout.*

Note that at each vertex v, before any edge can be inserted to a particular side, e.g., head, all edges that end at v and are accessible from the head need to be removed. Also note that at each vertex v, when inserting edges $e \in E$ into the deque pointing to right-hand neighbors, i.e., $l(e) = \pi(v)$, the queue (stack) edges have to be inserted in (in reverse) order of their respective positions of the right-hand neighbors. This is to ensure that the edges can be removed from the deque when processing the corresponding right-hand neighbor. Furthermore, at first always all queue edges and then all stack edges have to be inserted into the deque since otherwise edges cannot be removed anymore, e.g., consider a stack and a queue edge inserted at the head in that order, then neither of the edges can be removed.

3 Linear Cylindric Drawings of Deque Graphs

In this section we introduce a new type of drawing on the surface of a 3D cylinder (Sect. 3.1) and transform the drawings into equivalent 2D representations. In the case of planar drawings they exactly fit to deque graphs (Sect. 3.2).

3.1 Linear Cylindric Drawings

Definition 2. *In a* linear cylindric drawing $\Gamma(G)$ *of a graph G the vertices are placed disjointly on a straight line L, the* front line, *on the surface of the cylinder parallel to its axis. The edges are drawn as monotone curves in direction of the cylinder's axis and do not cross L.*

For convenience, we consider horizontal cylinders where L is parallel to the x-axis. Moreover, we identify the placement of the vertices on L with the permutation $\pi : V \rightarrow \{1, \ldots, n\}$.

Obviously, every graph has a linear cylindric drawing. The vertices can be arranged arbitrarily on the front line and the edges are drawn as simple curves on the side surface of the cylinder while not crossing L.

For an example, consider graph $G = (\{1, \ldots, 8\}, E)$ as displayed in Fig. 2(a). Fig. 2(b) visualizes a linear cylindric drawing of G: The vertices are drawn on the horizontal front line (dashed) and edges are either drawn as arches, e.g., $\{2, 4\}$, or wrap at most once around the cylinder, e.g., $\{1, 4\}$. Note that the dashed edge $\{3, 8\}$ causes a crossing with edge $\{4, 7\}$.

A linear cylindric drawing $\Gamma(G)$ imposes a direction onto each edge from the lower to the higher π-value of its vertices. We denote an undirected edge $\{u, v\}$

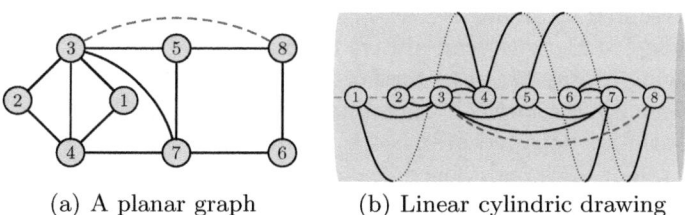

(a) A planar graph   (b) Linear cylindric drawing

Fig. 2. A graph and its linear cylindric drawing

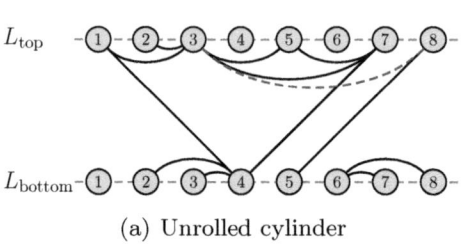 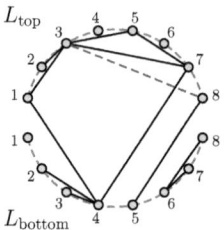

(a) Unrolled cylinder   (b) Unrolled cylinder on the circle

Fig. 3. 2D representations of the linear cylindric drawing of Fig. 2(b)

with $\pi(u) < \pi(v)$ by the directed edge (u, v). Moreover, $\Gamma(G)$ partitions the set of edges into four subsets according to their orientation with respect to the front line. Let $E_\cap$ ($E_\cup$) denote the set of edges leaving and entering their vertices above (below) the front line and let $E_/$ ($E_\backslash$) denote the set of edges leaving the front line above (below) and entering at the opposite side. The subscripts of the sets illustrate the shape of the edges. The edges from $E_/$ and $E_\backslash$ wrap around the cylinder once and the edges from $E_\cap$ ($E_\cup$) can be drawn as arches above (below) the front line.

Definition 3. *The tuple* $\mathcal{C}(G) = (\pi, E_\cup, E_\cap, E_\backslash, E_/)$ *is a* linear cylindric embedding *of* $G = (V, E)$.

We obtain a 2D representation of a linear cylindric drawing, which we call *unrolled cylinder*, by cutting the cylinder along the front line and "bending" the surface of the cylinder until it is plane. The front line with the vertices is duplicated and the two copies constitute the bottom (L_{bottom}) and top (L_{top}) of the so obtained drawing.

For instance, when applied to Fig. 2(b), the result is depicted in Fig. 3(a), where the area that was formerly placed above the front line is now situated at the bottom of the drawing. Then an edge $(u, v) \in E_\backslash$ can be drawn as a straight line from its vertex u on L_{top} to its vertex v on L_{bottom}. Symmetrically, the edges from $E_/$ can be drawn as straight lines L_{bottom} to L_{top}. The edges in $E_\cap$ can be represented as arches above L_{bottom} and, accordingly, the edges from $E_\cup$ can be represented as arches below L_{top}. Note that crossings between each pair of edges $(e_1, e_2) \in E_\cap \times E_\cup$ can always be avoided.

The drawing in Fig. 3(a) can further be continuously transformed to *the unrolled cylinder on the circle* by mapping L_{top} and L_{bottom} to two halves of a circle. The result is depicted in Fig. 3(b). Note that in this drawing all edges are drawn as straight lines and, hence, their routing is uniquely determined.

Without further proof, note that all different types of drawings are topologically equivalent, i.e., all drawings can be transformed into each other by a continuous function without changing the topological structure of the drawing. In particular, planarity is preserved. The equivalence of the drawings has an important implication: Consider a linear cylindric drawing where non-monotonous edges are allowed. Such a drawing can be continuously transformed to a drawing on the unrolled cylinder on the circle with straight-line edges that are monotonous. This justifies our assumption of monotonous edges in Def. 2.

Definition 4. *A graph G is* linear cylindric planar *if G has a linear cylindric embedding that permits a linear cylindric drawing without crossing edges.*

The classes of graphs that can be drawn without crossings on the plane and on the cylinder are equal. We thus obtain:

Corollary 1. *Every linear cylindric planar graph is a planar graph.*

3.2   Characterization of Deque Graphs

In this section we present the main result of this paper:

Theorem 1. *A graph G is a deque graph iff G is linear cylindric planar.*

The idea of the proof is to construct a one-to-one correspondence between a linear I/O layout $\Delta(G) = (\pi', \alpha, \omega)$ and a linear cylindric embedding $\mathcal{C}(G) = (\pi, E_\cup, E_\cap, E_\backslash, E_/)$ as follows: The positions π of the vertices on L are equal to the linear layout π'. The crucial point of the proof concerns the edges: Consider the edges in $E_\cap$ in the linear cylindric embedding that all leave and enter their end vertices above the front line, e.g., $(2,4)$ in Fig. 2(b). Edges in $E_\cap$ are drawn as arches in a linear cylindric layout and, hence, they do not cross iff they form rainbows and, consequently, constitute a stack layout. Hence, an edge $(u,v) \in E_\cap$ is interpreted as an edge that is processed by the deque like by a stack, that is, it is inserted at and removed from the same side, e.g., tail. The same holds true for edges in $E_\cup$ with respect to the deque's head. Edges in $E_\backslash$ or $E_/$ which enter and leave at opposite sides of the front line are interpreted as "moving" from one side of the deque to its opposite side. For instance, $(1,4) \in E_\backslash$, which leaves below and enters above the front line, is interpreted as being inserted at the deque's head and removed from its tail. Consequently, $E_\backslash$ and $E_/$ are queue edges inserted at the head and tail, respectively.

Using this one-to-one correspondence, we are then able to prove that edges cause no crossings in a linear cylindric drawing if and only if they can be processed by the deque. Conversely, any unavoidable crossing in such a drawing can be interpreted as a violation of the deque layout, i.e., crossing edges cannot be processed by the deque by any allowed operation.

First we show that it is sufficient to only consider pairs of edges when investigating a deque layout or a linear cylindric planar embedding. We start with deque layouts:[1]

---

[1] A similar statement is proven in [3], which is concerned with trains entering and leaving a train station from two sides. Such a train station with n tracks can be modelled by n deques and the trains must be assigned to the deques such that they do not block each other.

Lemma 1. $\Delta(G) = (\pi, \alpha, \omega)$ *is a deque layout if and only if for each pair of edges* $e, e' \in E$ *with* $e \neq e'$, $\Delta_{|e,e'}(G) = (\pi, \alpha_{|e,e'}, \omega_{|e,e'})$ *is a deque layout, where* $\alpha_{|e,e'}$ *and* $\omega_{|e,e'}$ *are the restrictions of* α *and* ω *to* $\{e, e'\}$, *respectively.*

Proof. "⇒": If all edges can be processed by the deque, in particular any two edges can be processed.

"⇐": For each pair of distinct edges $e, e' \in E$, $\Delta_{|e,e'}$ is a deque layout. We assume for contradiction that $\Delta(G)$ is not a deque layout.

Since an edge can always be inserted into the deque a problem can only occur when removing an edge.

Let $e \in E$ be the first edge that cannot be removed at some vertex v. W. l. o. g. we assume that $\omega(e) = \mathbf{h}$. Let $e_1, \ldots, e_k$ be the elements between the head and e in the deque. $k \geq 1$ since otherwise e could be removed from the deque. If for all edges e_i with $1 \leq i \leq k$, $r(e_i) = \pi(v)$ and $\omega(e_i) = \mathbf{h}$, then all edges e_i could be removed and e could also be removed. Thus, there exists an edge e_j with $1 \leq j \leq k$ which prevents the removal of e, i.e., $r(e_j) > r(e)$ or $\omega(e_j) = \mathbf{t}$.

In the first case, i.e., $r(e_j) > r(e)$, if $l(e_j) = l(e)$ and $\alpha(e_j) = \alpha(e)$, then at the vertex at position $l(e)$ the edges e and e' would have been inserted into the deque such that e would always be accessible from the head (see also Sec. 2), where four cases have to be distinguished: If both are stack edges added to the head then e_j would be have been inserted before e since $r(e_j) > r(e)$. If both are queue edges then e and e_j would have been inserted at the tail in that order. Similarly if e is a stack and e_j a queue edge, then e_j would have been inserted at the head before e is inserted at the head. If e is a queue edge and e_j a stack edge, then e would have been inserted at the tail before e_j is inserted at the tail as a stack edge. In all cases e is accessible from the head. If $l(e_j) = l(e)$ and $\alpha(e_j) \neq \alpha(e)$, then e_j and e can not be processed in $\Delta_{|e,e_j}(G)$ as well independently of their insertion order. In all other cases the relative order in which e_j and e are inserted into the deque is uniquely determined by $l(e)$ and $l(e_j)$. The same order has to be used in $\Delta_{|e,e_j}(G)$ and causes a problem there as well. The second case $\omega(e_j) = \mathbf{t}$ follows analogously. Hence, in each case $\Delta_{|e,e_j}(G)$ is no deque layout, which is a contradiction. □

Lemma 2 is the corresponding version of Lemma 1 for pairs of edges in linear cylindric planar embeddings:

Lemma 2. $\mathcal{C}(G) = (\pi, E_\cup, E_\cap, E_\setminus, E_/)$ *is a linear cylindric planar embedding of a graph* $G = (V, E)$ *if and only if for each pair of edges* $e, e' \in E$ *with* $e \neq e'$ $\mathcal{C}_{|e,e'}(G) = (\pi, E_\cup \cap \{e, e'\}, E_\cap \cap \{e, e'\}, E_\setminus \cap \{e, e'\}, E_/ \cap \{e, e'\})$ *is a linear cylindric planar embedding.*

Proof. "⇒": Take a linear cylindric drawing without crossings. From this drawing a linear cylindric planar embedding for each pair of edges can be obtained.

"⇐": The drawing of the unrolled cylinder on the circle (e.g., Fig. 3(b)) is a drawing with straight lines and is topologically equivalent to a linear cylindric drawing. Note that the routing of the edges is uniquely determined if the edges are drawn as straight lines. In order to construct a plane drawing with all edges,

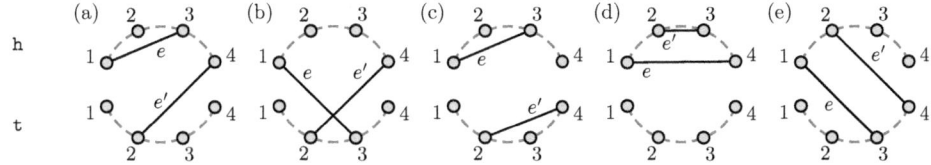

Fig. 4. Different cases in the proof of Lemma 3

simply draw all edges according to their unique straight-line representation. This drawing has no crossings since no pair of edges cross and is conform to embedding $\mathcal{C}(G)$. Hence, $\mathcal{C}(G)$ is linear cylindric planar. $\qquad\qquad\qquad\square$

Lemma 3 is the main step to our theorem. It states that every pair of edges can be processed by a deque if and only if they cause no crossing in the drawing. In order to show this we utilize the aforementioned one-to-one correspondence between a linear cylindric embedding and a linear I/O layout:

Lemma 3. *Given $G = (V, E)$, let $\mathcal{C}(G) = (\pi, E_\cup, E_\cap, E_\backslash, E_/)$ be a linear cylindric embedding and $\Delta(G) = (\pi, \alpha, \omega)$ be a linear I/O layout. If $\forall e \in E$:*

$$\alpha(e) = \mathtt{h} \wedge \omega(e) = \mathtt{h} \Leftrightarrow e \in E_\cup, \qquad \alpha(e) = \mathtt{h} \wedge \omega(e) = \mathtt{t} \Leftrightarrow e \in E_\backslash,$$
$$\alpha(e) = \mathtt{t} \wedge \omega(e) = \mathtt{h} \Leftrightarrow e \in E_/, \qquad \alpha(e) = \mathtt{t} \wedge \omega(e) = \mathtt{t} \Leftrightarrow e \in E_\cap,$$

then, for every pair of distinct edges $e, e' \in E$, $\mathcal{C}_{|e,e'}(G)$ is a linear cylindric planar embedding if and only if $\Delta_{|e,e'}(G)$ is a deque layout.

Proof. Let $e, e' \in E$ be two distinct edges, therefore $l(e) \neq l(e')$ or $r(e) \neq r(e')$. W. l. o. g., we assume $l(e) < l(e')$.

The proof is a complete differentiation between all cases of how two edges are processed by the deque and routed in the embedding: For each case we show that two distinct edges $e, e' \in E$ do not cross if and only if they can be processed by the deque. We assume that e and e' have non-disjoint ranges, i.e., they overlap. Otherwise, they can be drawn without crossings, and, conversely, they can be processed disjointly by the deque according to their I/O assignments. The same holds true if $r(e) = l(e')$. Note that $l(e) = r(e')$ is not possible since we assume $l(e) < l(e')$. The following five cases remain:

Case (a): $\alpha(e) = \omega(e)$, $\alpha(e') \neq \omega(e')$ (e is a stack edge and e' a queue edge): Without loss of generality, we assume $\alpha(e) = \omega(e) = \mathtt{h}$. If $\alpha(e') = \mathtt{h}$, then e and e' do not cross iff $l(e') \leq l(e)$, which is not possible by assumption, or $l(e') \geq r(e)$, which is not possible by assumption since the edges have to overlap.

If $\alpha(e') = \mathtt{t}$, then e and e' do not cross if and only if $r(e') \leq l(e)$, i.e., the edges do not overlap which contradicts our assumption, or $r(e') \geq r(e)$ (Fig. 4 (a)). In the deque first e is inserted at the head, then e' at the tail. Afterwards, e can be removed from the head and then e'.

Conversely, assume that e and e' can be processed by the deque but, for the sake of contradiction, $r(e') < r(e)$, i.e., e and e' cross. This implies that e' must be removed from the head before e. However, at first e is inserted at the head and then e' at the tail and, hence, e' cannot be removed from the head because e is blocking its way.

The case where e is a queue and e' is a stack edge follows analogously.

Case (b): $\alpha(e) \neq \omega(e)$, $\alpha(e') \neq \omega(e')$, $\alpha(e) \neq \alpha(e')$ (two queue edges inserted at different sides): Since e and e' overlap this situation always causes a crossing (Fig. 4(b)). Moreover, since e and e' overlap, there is a time instance where both edges are in the deque. However, since both have to be removed from opposite sides they cannot be removed at all.

Case (c): $\alpha(e) = \omega(e) \neq \alpha(e') = \omega(e')$ (two stack edges inserted at different sides): These edges never cross (Fig. 4(c)) and, in the deque, two stack edges inserted at different sides, can always be processed without any problems.

Case (d): $\alpha(e) = \omega(e) = \alpha(e') = \omega(e')$ (two stack edges inserted at the same side): e and e' do not cross if and only if e nests e' and, hence, $r(e') \leq r(e)$ (Fig. 4(d)). In the deque, at first e and then e' can be inserted at the same side of the deque and both are removed from this side in reverse order.

Conversely, since $l(e) < l(e')$, e is inserted before e' into the deque. Since both are stack edges inserted at the same side, e must not be removed before e' and, hence, $r(e') \leq r(e)$. Thus, e properly nests e' and they cause no crossing.

Case (e): $\alpha(e) = \alpha(e') \neq \omega(e) = \omega(e')$ (two queue edges inserted at the same side): Since e and e' do not cross, we have that $r(e) \leq r(e')$ (Fig. 4(e)). Consequently, e is inserted before e' and both are removed in the same order.

Conversely, since e and e' are both queue edges inserted at the same side and e is inserted before e' it follows that e is removed from the deque before e'. Hence, $r(e) \leq r(e')$ and e and e do not cross in the drawing. □

We are now able to prove our main result of Theorem 1:

Proof. Let $\Delta(G)$ be a linear I/O layout and $\mathcal{C}(G)$ a linear cylindric embedding of G. By Lemma 1, G permits a deque layout iff each pair of edges permits a deque layout. By Lemma 3 this holds true if and only if no pair of edges causes a crossing in the linear cylindric embedding. Finally, by Lemma 2 this is true if and only if $\mathcal{C}(G)$ is a plane embedding. □

By Corollary 1 and Theorem 1 we can conclude:

Corollary 2. *A graph $G = (V, E)$ that permits a deque layout is planar.*

Theorem 1 leads to the following interpretation of a linear cylindric drawing: Consider a vertical line drawn in the middle between vertices 2 and 3 from L_{top} to L_{bottom} in Fig. 3(a). This line intersects the edges $(2,3)$, $(1,3)$, $(1,4)$ and $(2,4)$ in that order from top to bottom. This sequence reflects the content of the deque after vertex 2 and before vertex 3 is processed, where $(2,3)$ is situated at the head and $(2,4)$ at the tail. When moving this vertical line like a scan line further to the right, its crossings with edges always correspond to the content of the deque. If the vertical line passes a crossing between two edges e and e', e.g.,

$(3, 8)$ and $(4, 7)$, then this can be interpreted as swapping the positions of e and e' in the deque, which is an invalid deque operation.

Note that the aforementioned interpretation of a linear cylindric drawing is only true if all edges are monotonously drawn from the left to the right, which is another reason why we assume monotonicity in Definition 2.

4   Linear Cylindric Drawings of Queue and Mixed Layouts

In this section we show how linear cylindric drawings can help to investigate layouts of graphs on data structures that allow only a subset of the operations of a deque and layouts with mixtures of data structures like a deque together with a stack or two stacks. Queue, 1- and 2-stack, and input and output restricted deque layouts are special cases of a deque layout:

Corollary 3

– *A graph is a queue (stack) graph iff it is a deque graph where all edges are queue (stack) edges and are inserted either all at the head or all at the tail.*
– *A graph is a 2-stack graph iff it is a deque graph with stack edges only.*
– *A graph is an input (output) restricted deque graph iff it is a deque graph where $\alpha(e) = \mathbf{h}$ ($\omega(e) = \mathbf{h}$) for all $e \in E$.*

Corollary 4

– *A graph is a queue (stack) graph iff it is linear cylindric planar with the edges either all in $E_\backslash$ or all in $E_/$ (all in $E_\cap$ or all in $E_\cup$).*
– *A graph is a 2-stack graph iff it is linear cylindric planar with all edges in $E_\cup \cup E_\cap$.*
– *A graph is an input (output) restricted deque graph iff it is linear cylindric planar with all edges in $E_\cup \cup E_\backslash$ ($E_\cup \cup E_/$).*

4.1   Queue Graphs

In this section we revisit queue graphs and their drawings that have been discussed in Sect. 1. Consider again the two drawings in Fig. 1. In both drawings it is hard to recognize immediately that the edges drawn as solid lines depict a queue layout and that edge $(2, 3)$ destroys this layout with respect to the depicted linear layout.

The same graph can be depicted with the same linear layout on an unrolled cylinder (Fig. 5) where all edges are in $E_\backslash$. It is immediately visible that none of the solid drawn edges cross and, hence, display a valid queue layout. Edge $(2, 3)$ crosses edges $(1, 4)$ and $(1, 5)$ and, consequently, $(2, 3)$ destroys the queue layout. Moreover, it is immediately visible that exactly these three crossing edges destroy the queue layout. Removing edges until the drawing is crossing-free, e.g., edges $(1, 4)$ and $(1, 5)$, reestablishes a valid queue layout.

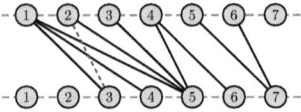

Fig. 5. Unrolled cylinder drawing of a queue graph

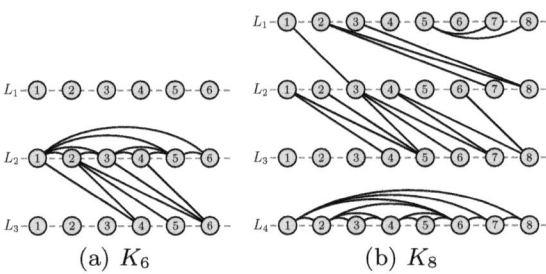

(a) K_6 (b) K_8

Fig. 6. Linear cylindric drawings of mixed layouts

4.2   Linear Cylindric Drawings of Mixed Layouts

Stack, queue, and deque layouts can be extended to *mixed layouts*, where k (possibly different) data structures $\{D_1, \ldots, D_k\}$ are given. Such a mixed layout consists of a single linear layout π of the vertices and a partition of the set of edges E consisting of k subsets $E_1, E_2, \ldots, E_k$, where for each $i \in \{1, \ldots, k\}$, $G = (V, E_i)$ has a layout in data structure D_i with linear layout π. Our technique of representing layouts by an unrolled cylinder straightforwardly extends to mixed layouts: Create $k + 1$ copies $L_1, \ldots, L_{k+1}$ of the front line, place them one upon the other and display the edges E_i of data structure D_i between the i-th and $(i + 1)$-th front line as described in Sect. 3.1.

As an example consider the complete graph K_6 with six vertices. This graph has neither a 2-stack nor a 2-queue layout. Fig. 6(a) shows the K_6 in a linear cylindric drawing with two data structures. Between L_1 and L_2 only stack edges are used. The region between L_2 and L_3 contains only queue edges. Since no edges cross in this drawing we can conclude that the K_6 is a graph with a mixed layout consisting of one stack and one queue.

Figure 6(b) shows a possible mixed layout of the complete graph K_8 with 8 vertices using one input-restricted deque drawn between L_1 and L_2, one queue between L_2 and L_3 and one stack between L_3 and L_4. Note that each edge of the K_8 appears in the representation of exactly one data structure and the same linear layout π is used for all data structures.

5   Conclusion and Future Work

In this paper we introduced the new graph visualization technique by linear cylindric drawings. We proved that the class of graphs that have a plane linear cylindric drawing are exactly those graphs that permit a layout by the deque data structure. We also showed how our new representation gives deeper insights into the fundamental data structures queue, stack and deque and can even be used to investigate graph layouts with mixed data structures.

The decision problem whether or not a graph permits a stack layout can be solved in linear time [1, 11]. In contrast the corresponding decision problem for a queue layout is $\mathcal{NP}$-hard [9]. The question is open whether or not the decision problem in the case of a deque is solvable in polynomial time. Currently we are in the progress of proving that the class of deque graphs coincides with the class of graphs that are a subgraph of a planar graph with a Hamiltonian path. These are new insights we gained by linear cylindric drawings, which also give new characterizations of, e.g., queue graphs and proper level planar graphs.

References

1. Bernhart, F., Kainen, P.: The book thickness of a graph. J. Combin. Theory, Ser. B 27(3), 320–331 (1979)
2. Chung, F.R.K., Leighton, F.T., Rosenberg, A.L.: Embedding graphs in books: A layout problem with applications to VLSI design. SIAM J. Algebra. Discr. Meth. 8(1), 33–58 (1987)
3. Cornelsen, S., Di Stefano, G.: Track assignment. J. Discrete Algorithms 5(2), 250–261 (2007)
4. Dujmović, V., Wood, D.R.: On linear layouts of graphs. Discrete Math. Theor. Comput. Sci. 6(2), 339–358 (2004)
5. Dujmović, V., Wood, D.R.: Stacks, queues and tracks: Layouts of graph subdivisions. Discrete Math. Theor. Comput. Sci. 7, 155–202 (2005)
6. Heath, L.S., Leighton, F.T., Rosenberg, A.L.: Comparing queues and stacks as mechanisms for laying out graphs. SIAM J. Discret. Math. 5(3), 398–412 (1992)
7. Heath, L.S., Pemmaraju, S.V.: Stack and queue layouts of directed acyclic graphs: Part II. SIAM J. Comput. 28(5), 1588–1626 (1999)
8. Heath, L.S., Pemmaraju, S.V., Trenk, A.N.: Stack and queue layouts of directed acyclic graphs: Part I. SIAM J. Comput. 28(4), 1510–1539 (1999)
9. Heath, L.S., Rosenberg, A.L.: Laying out graphs using queues. SIAM J. Comput. 21(5), 927–958 (1992)
10. Knuth, D.E.: The Art of Computer Programming, 1st edn., vol. 1. Addison-Wesley, Reading (1968)
11. Wiegers, M.: Recognizing outerplanar graphs in linear time. In: Tinhofer, G., Schmidt, G. (eds.) WG 1986. LNCS, vol. 246, pp. 165–176. Springer, Heidelberg (1987)
12. Wood, D.R.: Bounded degree book embeddings and three-dimensional orthogonal graph drawing. In: Mutzel, P., Jünger, M., Leipert, S. (eds.) GD 2001. LNCS, vol. 2265, pp. 312–327. Springer, Heidelberg (2002)
13. Wood, D.R.: Queue layouts, tree-width, and three-dimensional graph drawing. In: Agrawal, M., Seth, A. (eds.) FSTTCS 2002. LNCS, vol. 2556, pp. 348–359. Springer, Heidelberg (2002)
14. Yannakais, M.: Embedding planar graphs in four pages. J. Comput. Syst. Sci. 38(1), 36–67 (1989)

An Experimental Evaluation of Multilevel Layout Methods

Gereon Bartel, Carsten Gutwenger, Karsten Klein, and Petra Mutzel

Technische Universität Dortmund, Germany
{carsten.gutwenger,karsten.klein,petra.mutzel}@cs.tu-dortmund.de

Abstract. Applying the multilevel paradigm to energy-based layout algorithms can improve both the quality of the resulting drawings as well as the running time of the layout computation. In order to do this, approaches for the different multilevel phases refinement, placement, layout, and optionally scaling and postprocessing need to be implemented. A number of multilevel layout algorithms have been proposed already, which differ in the way these phases are realized. We present an experimental study that investigates the influence of varying combinations with respect to running time and quality criteria.

1   Introduction

Energy-based layout methods are used in many different application areas such as social sciences and biology to automatically compute straight-line drawings of undirected graphs. They are often preferred over alternative methods because they are reasonably fast, allow straight-forward extensions for a large number of drawing constraints, and are relatively easy to implement. However, in practice they suffer from a number of drawbacks. First of all, they tend to converge to locally optimal solutions far away from the global optimum, e.g., when the graph is not completely unfolded. This is often also a result of poor parameter settings, since it is difficult and time-consuming to optimize the parameters with regard to a large number of input instances, and suitable parameters for a class of graphs may be disadvantageous for other classes. In addition, even though improvements in the implementation and continuous advances in hardware performance allow to compute layouts for medium to large graphs, running times are still too high for many large graphs from practical applications.

The multilevel approach helps to overcome these problems both by avoiding local minima and speeding up the computation due to improved convergence to the final drawing. Due to the step-by-step construction of the final layout starting with the layout of a small graph, there is also no dependency on a good initial drawing to start with. A number of different multilevel approaches have been proposed over the last years, but the influence of the different phases as well as the effects of their combination have not been investigated so far.

There is a wealth of publications concerning energy-based layout methods; see [3, 17] for an overview. An early comparison of such methods was presented

U. Brandes and S. Cornelsen (Eds.): GD 2010, LNCS 6502, pp. 80–91, 2011.

by Brandenburg et al. [4]. Walshaw [19] introduced the multi-level paradigm for graph drawing and Hachul and Jünger [14] presented an experimental study of layout algorithms for large graphs, including energy-based and algebraic approaches. Brandes and Pich [5] presented a study on distance-based graph drawing, and Frishman and Tal [8] and Godiyal et al. [11] investigated the use of the GPU for multi-level layout computation.

We present an experimental study that investigates the performance of a large number of different combinations for the multilevel phases. Our comparison includes a selection of well-established energy-based layout methods, as well as a number of fast methods specifically developed for multilevel approaches. We check if one of the combinations can be recommended as an overall choice or if a number of combinations need to be considered due to the characteristics of the input graphs. Our benchmark set comprises graphs already known from literature and a number of additional generated and real-world graphs.

After a short description of the multilevel paradigm for graph drawing in Sect. 2, we describe the setting and results of our experiments in Sect. 3 and draw a conclusion in Sect. 4.

2   The Multilevel Paradigm

Multilevel layout computation is an iterative process that can be roughly divided in three phases: *coarsening, placement*, and *single level layout*. Starting with the smallest graph, the final layout for the input graph is obtained by successively computing layouts for the graph sequence computed by the coarsening phase. At each level, the additional vertices need to be placed into the layout of the preceding level, optionally after a scaling to provide the necessary space.

Coarsening Phase. Given an input graph G, the coarsening phase builds a multilevel hierarchy by computing a sequence of increasingly smaller graphs $G_0, G_1, \ldots, G_k$ with $G = G_0$. In order to coarsen the graph, sets of vertices in G_i are combined to single vertices in G_{i+1}, where a number of different criteria can be used for deciding which vertices to combine. Subsequent merging or layout steps for the graphs at each level may take into account vertex weights that represent the merging of vertex sets to a single vertex.

The influence of the coarsening method is twofold: On the one hand, the way the graphs are coarsened can have an impact on both the quality of the drawing and the running time on each level. This depends on how well the overall graph structure is represented on the levels, influencing the way the graph is unfolded. On the other hand the number of hierarchy levels may have a significant influence on the total running time.

Placement Phase. The placement phase is responsible for adding vertices to the layout when the algorithm proceeds to the next level in the multilevel hierarchy. Typically, new vertices are placed close to their representative in the previous level. How much the layout computation on each level can profit from the layout given by the previous level, and how a special placement can improve the computation, therefore is influenced both by the way the vertices are merged and the number of vertices that are merged to a single representative. Clever placement may drastically reduce the work needed by the single level layout and also have an impact on the way the graph is unfolded. Instead of using a static position assignment, Frishman and Tal [8], Gajer et al. [10], and Godiyal et al. [11] describe an iterative method with a small number of iterations.

Single Level Layout. On each level i in the multilevel hierarchy, a layout for the corresponding graph G_i has to be computed. The main requirement for the layout method used is that it has to accept the layout resulting from the previous placement phase as initial layout such that the layout is an extended and refined version of the layout of G_{i+1}. In order to allow a running time improvement by the multilevel method, the single level run should either work with a reduced number of iterations or be dependent on a stop criterion, e.g., when the layout energy drops under a predefined threshold.

3   Experimental Study

We use a benchmark set that comprises 43 graphs of varying size and characteristics, including both real world and generated instances[1]. In addition to a selection of graphs used by Hachul and Jünger [14], we use a further real world graphs, comprising a protein interaction network, the RNA network from the InterViewer project [15], a selection of Walshaw's graph archive [18] and the AT&T graph library [2], as well as a number of generated grids. We added a rectangular grid with 400 rows and 20 columns in order to test for distortions, and two smaller squared grids, also in variants where corners of the grids are connected to make the instances more difficult to unfold.

We compare results from multilevel layout computations and also single level runs of the implemented layout methods in order to evaluate the gain from the corresponding multilevel runs. In single level runs, all layout algorithms are called using standard parameters, which are either obtained by empirical evaluation or from the original publications. We adapt these settings for use in the multilevel framework based on experiments on a small "tuning" subset to investigate reasonable ranges for parameter settings. We did this mainly by visual inspection of the resulting layouts.

For our study we ran the full set of method combinations for the three phases on the full benchmark set and mainly evaluated the results of the statistical analysis of the layout characteristics. We still did sample visual inspections, both to check that our analysis is reasonable and to judge the quality when the statistical values are ambiguous. In order to capture the quality of the drawings, we analyze a couple of aesthetic criteria, including standard criteria as number of edge crossings and drawing area, and also typical optimization goals of energy-based layout methods like edge length deviation and node distribution.

[1] More information: `http://ls11-www.cs.tu-dortmund.de/staff/klein/gdmult10`

3.1 Implementations

We implemented a multilevel framework within OGDF [1] that allows to freely combine different strategies for the multilevel phases. Therefore it is not possible to do a fine tuning for combinations to speed up the computation, and the runtime of our approach may be slightly inferior compared to particular implementations of a single multilevel approach.

We used the energy-based layout algorithms provided by OGDF and implemented various. coarsening and placement strategies (see below). In addition to these main phases, the framework provides scaling and local postprocessing at each drawing level. Although these steps may influence running time as well as layout quality, we focus on the three main phases here and use the same minimum scaling and postprocessing options for all combinations.

Coarsening methods. All coarsening methods we use take an approach, where vertices on a level G_i are merged such that a single representative of a group of vertices survives on the next level G_{i+1}.

(NM). The *Null Merger* does not merge any vertices and is used to simply realize a single-level layout within the multilevel framework.

(RM). The *Random Merger* simply selects vertices randomly and merges them with a random neighbor, until the size of the graph decreases by a predefined factor. This method allows a good control over the number of levels in the hierarchy and is used to have a minimal quality threshold that more sophisticated methods should outperform.

(MM). The *Matching Merger* corresponds to the method described in [19], where a matching is computed by visiting the vertices in a random order, matching each unmatched vertex with a random unmatched neighbor if existent. The number of vertices in G_{i+1} is then at least half the number of the vertices in G_i.

(WMM). The *Weighted Matching Merger* is the weighted variant of MM described in [19]. In order to achieve a uniform merging, the matching vertex is chosen to be the neighbor with the smallest weight, where the weight of a vertex v is the number of vertices on lower levels represented by v.

(ECM). The *Edge Cover Merger* is a variant of MM intended to eliminate the problem that for certain graphs containing star-like subgraphs a linear number of levels may be necessary. In addition to the matching, an edge cover is computed such that each contained edge is incident to at least one unmatched vertex. These cover edges are then used to merge their end vertices.

(LBM). The *Local Biconnected Merger* is a variant of ECM that tries to avoid situations where distortions are introduced, since—during the coarsening process—vertices are merged to a cut vertex at which the layout method may twist the orientation. LBM checks if biconnectivity may be lost in the local neighborhood around the potential merging position.

(SM). The *Solar Merger* corresponds to the method described in [13], where the vertices are partitioned into *solar systems*, classifying each vertex as either sun, planet or moon. The resulting solar systems are then collapsed to the sun vertex.

(ISM). The *Independent Set Merger* corresponds to the strategy used within the GRIP [10] approach. It uses a maximal independent set filtration $\mathcal{V} : V = V_0 \supset V_1 \supset \ldots \supset V_k$ with $k = O(\log n)$, such that V_i are the nodes on level i. V_1 is a maximal independent set of G and V_i is a maximal subset of V_{i-1} such that the graph theoretical distance between any pair of its elements is at least $2^{i-1} + 1$. Gajer and Kobourov sped-up the computation using partial BFS-trees, which achieves subquadratic runtime behavior in practice.

Placement methods. We have implemented the following placement methods, most of which were already described in the literature.

(ZP). The *Zero Placer* uses the simplest strategy of all methods under investigation. A vertex is placed at the same position as its representative in the previous level. A small random offset avoids a zero distance between two nodes which may pose problems for some layout methods.

(RP). The *Random Placer* places vertices at random positions within the smallest circle containing all vertices around the barycenter of the current drawing. Clearly, this placement strategy may hinder convergence and lead to distortions in the drawing. It is introduced to give a minimum quality bound that more sophisticated strategies should outperform.

(BP). A well-known method is the *Barycenter Placer*, which places a vertex at the barycenter of its neighbors' positions. The influence of the neighbor position might be weighted, e.g., by the merging weight of the vertices. The barycenter placement should help the energy-based layout algorithm to reach an energy-minimal state because it at least partially considers the influence of the neighbor positions on the movement force for the placed node.

(MP). As an alternative to BP, the *Median Placer* uses the median position of the neighbor nodes for each coordinate axis. This smoothes the effect of neighbors at outlier positions but still takes into account the strong influence of the neighbor positions within the layout methods.

(SP). Developed for use within the FMMM approach [13], the *Solar Placer* is able to use information from the merging phase of the solar merger. A new vertex is placed on the direct line between two suns at the relative position with respect to its position on the intersystem path between these suns. If it lies on more than one intersystem path, the average value of all relevant positions is used. In combination with other mergers, SP resembles ZP.

Layout methods. We investigate the performance of energy-based layout methods that where either implemented as single level methods in OGDF or described as parts of multilevel approaches in the literature, as, e.g., the new multipole method by Hachul. In addition, we implemented two variations which try to exploit the specific multilevel setting. Due to the runtime of the Kamada-Kawai approach, we test a combination of Kamada-Kawai (KK) and the two faster methods Fruchterman-Reingold (FR) and Fast Multipole Embedder (FME), where Kamada-Kawai is used for all levels but the last, where FR or FME is applied instead. For the even slower Davidson-Harel method, we compute the last layout using the very fast New Multipole Method (NMM). Clearly, these combinations will only pay off when the increase in the number of vertices at the last level is relatively large. The following layout methods are used within our study:

(EAD). This method uses the spring model proposed by Eades [7] which represents the classical force directed approach.

(FR). The model proposed by Fruchterman and Reingold [9] is based on the force-directed method, and uses a cooling scheme to limit the total displacement. We used two variations: (FRE) uses the node weighting scheme proposed by Walshaw [19] and (FR) is the original grid variant [9].

(KK). The force-directed method by *Kamada and Kawai* [16] that constitutes an MDS approach, using the graph theoretic distance between vertices as distance.

(NMM). The *New Multipole Method* introduced by Hachul and Jünger [13] that uses a fast multipole approximation of the repulsive forces with $O(N \log N)$ running time leading to an $O(|V| \log(|V| + |E|))$ overall running time.

(FME). The *Fast Multipole Embedder* method by Gronemann [12] uses the same repulsive forces as NMM, but slightly modified attractive forces. The repulsive forces are approximated with the fast multipole method using a reduced quadtree construction and well-separated pair decomposition.

(DH). The layout algorithm introduced by *Davidson and Harel* [6], based on simulated annealing; with a running time of $O(|V|^2|E|)$ by far the slowest algorithm in our comparison.

3.2   Computational Results

We ran the selected combinations and single level layout algorithms on all test graphs, except for Kamada-Kawai and Davidson-Harel, which were only run on graphs with up to 2000 (DH), 6000 (KK), 12000 (KKFR), and 15000 (KKFME) nodes, respectively, due to their space and running time requirements. We used a system with eight quad-core AMD Opteron processors and 16GB memory.

Due to the huge amount of experimental data, we provide detailed results for the most relevant methods and criteria only.

Discussion. We had a number of hypotheses regarding the performance of the methods under investigation when we conducted the experiments. First of all, we expected that combinations from established multilevel approaches should outperform the other combinations at least to a certain extent. Also the methods whose running time includes a significant layout independent computation, like for example KK that comprises an all-pairs shortest paths computation, may not benefit largely by the multilevel approach, whereas the methods FME and NMM especially developed for this scenario should dominate the others.

Our experiments showed that a significant part of our benchmark set did not pose a challenge to the layout algorithms, and that even single level calls were competitive both in speed and quality at least for graphs up to a certain size; cf.

Fig. 2. Especially the NMM method, actually developed for use within a multilevel approach, could often deliver good results in single level runs. However, FME was not able to compute layouts of similar quality with a single level.

From the visual inspection we found that the layouts computed by KK are often the most pleasing, especially for the smaller graphs, but the gain compared to NMM and EAD was not large enough to justify the additional time. E.g., for *add32* KK computes a radial-like layout emphasizing the special structure of the graph better than the best layouts w.r.t. number of crossings and area (see Fig. 3), but also needed 255s compared to 7.7s and 6.1s for FME and NMM, respectively.

(a) SP+ISM+KK   (b) MP+ISM+FME   (c) BP+LBM+KKFME

Fig. 1. Drawings of the *Grid_400_20* and *uk* graphs, which posed a challenge to many combinations regarding the distortions

Grid and mesh-like structures were mostly an easy task for all algorithms, with the exception of the narrow grid, which often suffered from folding in multilevel runs. For some of the layout methods, a specific effect can be observed when the narrow grid is laid out. The graph then is not drawn as a straight line but seems to be compressed; see Fig. 1(b). We found that this is not the result of bad unfolding, but due to the influence of the repulsion forces when vertices from the next hierarchy level are reinserted—the placement close to their parent vertex leads to large repulsion forces that distort the graph layout in the next layout step. The planar graphs *grid_400_20* and *uk* were problematic for nearly all combinations, since they were subject to a large number of distortions; see Fig. 1(a) and 1(c). LBM achieves slightly less crossings on average for these graphs, at the cost of additional running time due to more levels.

Graphs like the flower and spider graphs did a better job in separating single- and multilevel runs. Here it often came to crossings and foldings, and the single level methods typically did not obtain an acceptable layout. LBM only computed a single level for graph *flower_5* and therefore also had inacceptable results.

The interpretation of the results regarding the placement methods did not show big differences. We mainly introduced the random placer to better judge the overall quality of the other methods. Nonetheless it produced good results in most cases, but also some worst-case drawings with a huge number of crossings; therefore it cannot be recommended. Most other placers achieved comparable results without a clear winner; combining SP with SM was not better than any combination with other placers.

For the merger methods, we had a look at the resulting number of levels; see Fig. 4. Whereas RM, ECM, and ISM produced small and comparable numbers (less

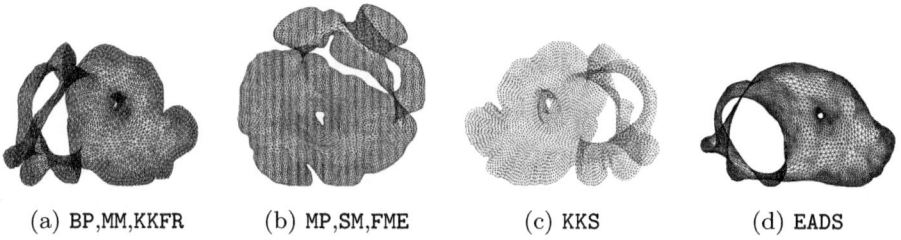

(a) BP,MM,KKFR   (b) MP,SM,FME   (c) KKS   (d) EADS

Fig. 2. Layout of the planar graph *3elt* computed with slowest (BP, MM, KKFR, 15 levels, 91s) and fastest overall combination (MP,SM,FME, 5 levels, 2.27s), compared to results of fastest (EADS, 2.77s) and slowest (KKS, 158s) single level methods

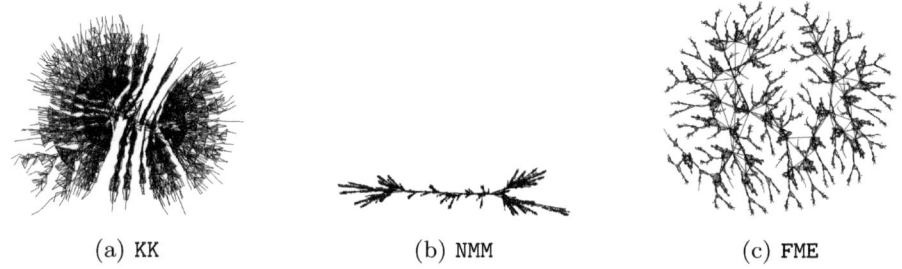

(a) KK   (b) NMM   (c) FME

Fig. 3. Layout of graph *add32* computed with NMM and KK, both with solar placer and merger, and the layout with fewest crossings computed with MP, LBM and FME

than 15), LBM produced slightly more levels, with the exception of *sierpinski_08* and *Grid_400_20* with about 30 levels, graph *flower_005* with only a single level. SM was able to stay below 10 levels in all cases, with a minimum of 3 levels.

The matching mergers MM and WMM produced by far the most levels, especially for the tree graphs, with a maximum of 81 and 76 for *tree_06_05*, and 33 and 31 for the smaller *snowflake_A*. For the graphs with a large star subgraph (*dg_1087* and *snowflake_C*), the number of levels (6565 and 2505) was prohibitively large, making layout computation infeasible.

As expected, there is a clear correlation between the number of levels created by the coarsening phase and the resulting running time. Depending on the graph characteristics, the coarsening may introduce a huge number of levels linear in the number of vertices in the graph. Even though the running time on each of the levels should be much smaller than a single level run (due to the smaller number of iterations and the earlier convergence), this may lead to a tremendous increase in running time without a corresponding gain in layout quality.

Running Time. We first had a look at the overall average running time with respect to the layout methods. Besides DH, which is extremely slow and only

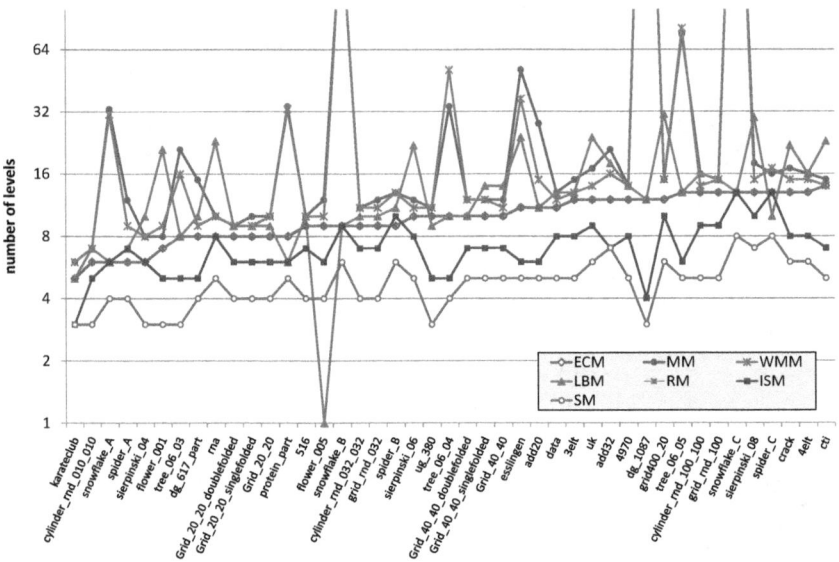

Fig. 4. Number of levels computed by the different merger methods

acceptable for the smallest graphs, we can identify three groups of layout methods: FME, NMM and EAD were always among the fastest layouts, dominating the FR and KK variants, which are an order of magnitude faster than KK alone. Above a graph size of about a few hundred vertices, the single level methods were clearly inferior to the multilevel combinations, where FME was slightly faster than NMM and EAD. We also compared the average running time of the mergers in combination with EAD, relative to SM (see Fig. 5); LBM and the matching mergers show a number of peaks while the other mergers show a balanced behavior and are always slower than SM.

Edge length. Our observation that KK often produced the most pleasing layouts corresponds well with the values of the normalized standard deviations of the edge lengths, where KK is clearly the best with nearly all values below 0.5, and for the fast methods NMM has a slight advantage over FME and EAD; see Fig. 6.

Crossings. We found that the number of crossings was a very good indicator of the layout quality, as layouts with unfolding problems often had a significant larger number of crossings. An exception are graphs which required an extreme number of crossings, e.g. because they contain complete subgraphs as the flower graphs. We observed that the random placer sometimes caused a drastic increase in crossings, especially in combination with the FME. We only report here the number of crossings for a selection of mergers with layout method EAD; see Fig. 7. For the large graphs with a large number of crossings on average, the difference

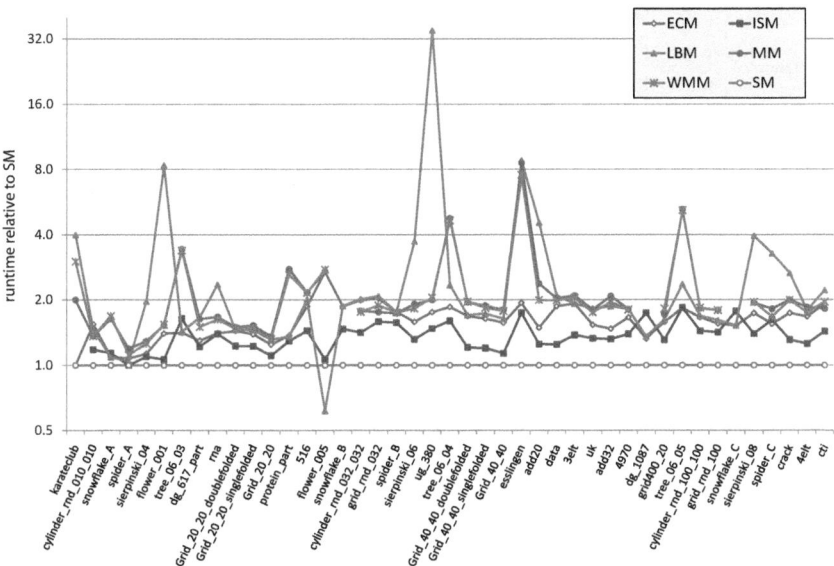

Fig. 5. Average running times of different mergers with EAD relative to SM

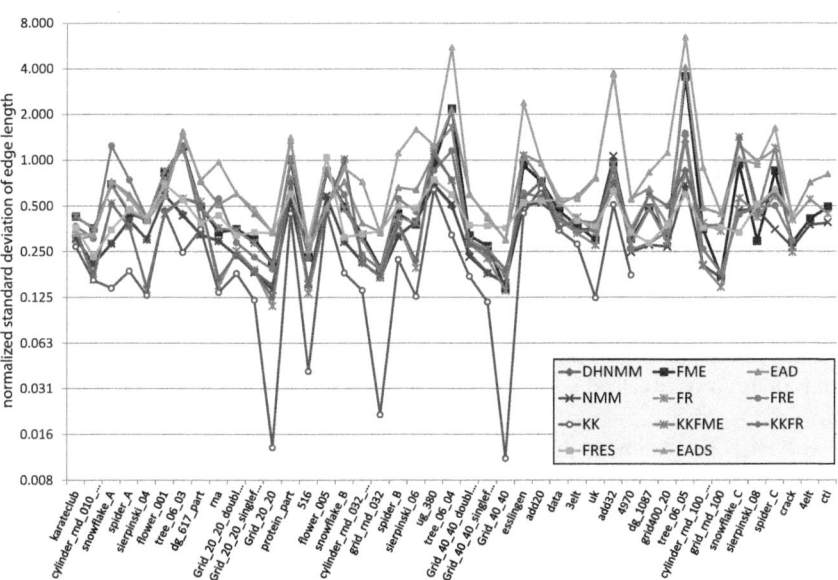

Fig. 6. Normalized standard deviations of the edge lengths for a selection of layout methods

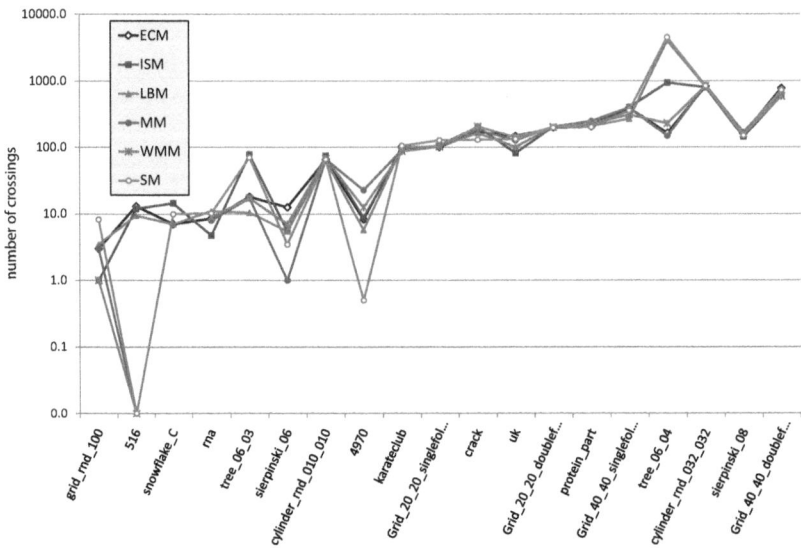

Fig. 7. Details of average number of crossings produced by different mergers with EAD

between the mergers is minimal, whereas for smaller crossing numbers there is no clear winner. SM and also LBM often produced quite good results in comparison.

Drawing Area. The drawing area alone is not a good quality indicator for straight-line drawings, as layouts with a huge number of crossings and overlaps may require a smaller area than drawings that reveal the graph structure well. We therefore do not report on the drawing area here.

4 Conclusions

We presented an experimental study of multilevel layout methods within a framework implemented in OGDF. Though there is no clear winning combination, a number of layout methods, mergers and placers showed a good behavior. The EAD layout algorithm, using the classical force model introduced by Eades worked very well within the multilevel approach and lead to similar results, both in quality and running time, as the models in NMM and FME specifically designed for multilevel computations. Mergers ECM, SM and ISM often produced good results, whereas the others showed bad worst-case behavior. It would be interesting to investigate how the running time is influenced by the different merging due to different graph characteristics, i.e. instead of just looking at the number of levels, to also analyze the effect on the convergence on each level. In addition, instead of looking at single criteria like edge crossings, the combination of different criteria might give a better estimation of the layout quality.

References

[1] The Open Graph Drawing Framework, http://www.ogdf.net
[2] The AT&T graph library, http://www.graphdrawing.org.
[3] Battista, G.D., Eades, P., Tamassia, R., Tollis, I.G.: Graph Drawing. Prentice-Hall, Englewood Cliffs (1999)
[4] Brandenburg, F.J., Himsolt, M., Rohrer, C.: An experimental comparison of force-directed and randomized graph drawing algorithms. In: Brandenburg, F.J. (ed.) GD 1995. LNCS, vol. 1027, pp. 76–87. Springer, Heidelberg (1996)
[5] Brandes, U., Pich, C.: An experimental study on distance-based graph drawing. In: Tollis, I.G., Patrignani, M. (eds.) GD 2008. LNCS, vol. 5417, pp. 218–229. Springer, Heidelberg (2009)
[6] Davidson, R., Harel, D.: Drawing graphs nicely using simulated annealing. ACM Trans. Graph. 15(4), 301–331 (1996)
[7] Eades, P.: A heuristic for graph drawing. Congressus Numerantium 42, 149–160 (1984)
[8] Frishman, Y., Tal, A.: Multi-level graph layout on the GPU. IEEE Transactions on Visualization and Computer Graphics 13(6), 1310–1319 (2007)
[9] Fruchterman, T.M.J., Reingold, E.M.: Graph drawing by force-directed placement. Softw. Pract. Exper. 21(11), 1129–1164 (1991)
[10] Gajer, P., Kobourov, S.G.: GRIP: Graph drawing with intelligent placement. J. Graph Algorithms Appl. 6(3), 203–224 (2002)
[11] Godiyal, A., Hoberock, J., Garland, M., Hart, J.C.: Rapid multipole graph drawing on the GPU. In: Tollis, I.G., Patrignani, M. (eds.) GD 2008. LNCS, vol. 5417, pp. 90–101. Springer, Heidelberg (2009)
[12] Gronemann, M.: Engineering the fast-multipole-multilevel method for multicore and SIMD architectures. Master's thesis, Technische Universität Dortmund (2009)
[13] Hachul, S., Jünger, M.: Drawing large graphs with a potential-field-based multilevel algorithm. In: Pach, J. (ed.) GD 2004. LNCS, vol. 3383, pp. 285–295. Springer, Heidelberg (2005)
[14] Hachul, S., Jünger, M.: Large-graph layout algorithms at work: An experimental study. J. Graph Algorithms Appl. 11(2), 345–369 (2007)
[15] Han, K., Ju, B.-H., Park, J.H.: InterViewer: Dynamic visualization of protein-protein interactions, http://interviewer.inha.ac.kr/
[16] Kamada, T., Kawai, S.: An algorithm for drawing general undirected graphs. Information Processing Letters 31(1), 7–15 (1989)
[17] Kaufmann, M., Wagner, D. (eds.): Drawing Graphs. LNCS, vol. 2025. Springer, Heidelberg (2001)
[18] Walshaw, C.: The graph partitioning archive, http://staffweb.cms.gre.ac.uk/~c.walshaw/partition/
[19] Walshaw, C.: A multilevel algorithm for force-directed graph-drawing. J. Graph Algorithms Appl. 7(3), 253–285 (2003)

Orthogonal Graph Drawing
with Flexibility Constraints

Thomas Bläsius, Marcus Krug, Ignaz Rutter, and Dorothea Wagner

Faculty of Informatics, Karlsruhe Institute of Technology (KIT), Germany
`firstname.lastname@kit.edu`

Abstract. In this work we consider the following problem. Given a planar graph G with maximum degree 4 and a function flex : $E \longrightarrow \mathbb{N}_0$ that gives each edge a *flexibility*. Does G admit a planar embedding on the grid such that each edge e has at most flex(e) bends? Note that in our setting the combinatorial embedding of G is not fixed.

We give a polynomial-time algorithm for this problem when the flexibility of each edge is positive. This includes as a special case the problem of deciding whether G admits a drawing with at most one bend per edge.

1 Introduction

Orthogonal graph drawing is one of the most important techniques for the human-readable visualization of complex data. Its æsthetic appeal derives from its simplicity and straightforwardness. Since edges are required to be straight orthogonal lines—which automatically yields good angular resolution and short links—the human eye may easily adapt to the flow of an edge. The readability of orthogonal drawings can be further enhanced in the absence of crossings, i.e., if the underlying data exhibits planar structure. Unfortunately, not all planar graphs have an orthogonal drawing in which each edge may be represented by a straight horizontal or vertical line. In order to be able to visualize all planar graphs, nonetheless, we allow edges to have bends. Since bends obfuscate the readability of orthogonal drawings, however, we are interested in minimizing the number of bends on the edges. Previous approaches to orthogonal graph drawing in the presence of bends focus on either the minimization of the maximum number of bends per edge or the total number of bends in the drawing.

In typical applications, however, edges have varying importance for the readability depending on their semantic and their importance for the application. Thus, it is convenient to allow some edges to have more bends than others.

We consider the following orthogonal graph drawing problem, which we call FLEXDRAW. Given a *4-planar graph* G, i.e., G is planar and has maximum degree 4, and for each edge e a non-negative integer flex(e), its *flexibility*. Does G admit a planar embedding on the grid such that each edge e has at most flex(e) bends? Such a drawing of G on the grid is called a *flex-drawing*. For a graph with flex(e) > 0 for each edge e in G we shortly say that G has *positive flexibility*.

U. Brandes and S. Cornelsen (Eds.): GD 2010, LNCS 6502, pp. 92–104, 2011.

The problem we consider generalizes a well-studied problem in orthogonal graph drawing, namely the problem of deciding whether a given graph is β-embeddable for some non-negative integer β. A 4-planar graph is *β-embeddable* if it admits an embedding on the grid with at most β bends per edge.

Garg and Tamassia [6] show that it is $\mathcal{NP}$-hard to decide 0-embeddability. The reduction crucially relies on the ability to construct graphs with rigid embeddings. Later, we show that this is impossible if we allow at least one bend per edge. This is a key observation which yields, among others, an efficient algorithm for recognizing 1-embeddable graphs. For special cases, namely planar graphs with maximum degree 3 and series-parallel graphs, Di Battista et al. [1] gave an algorithm that minimizes the total number of bends and hence solves 0-embeddability. On the other hand, Biedl and Kant [2] show that every 4-planar graph admits a drawing with at most two bends per edge with the only exception of the octahedron, which requires an edge with three bends. Similar results are obtained by Liu et al. [9].

Liu et al. [8] claim to have found a characterization of the planar graphs with minimum degree 3 and maximum degree 4 that admit an orthogonal embedding with at most one bend per edge. They also claim that this characterization can be tested in polynomial time. Unfortunately, their paper does not include any proofs and to the best of our knowledge a proof of these results did not appear. Morgana et al. [10] characterize the class of *plane graphs* (i.e., planar graphs with a given embedding) that admit a 1-bend embedding on the grid by forbidden configurations. They also present a quadratic algorithm that either detects a forbidden configuration or computes a 1-bend embedding.

If the combinatorial embedding of a 4-planar graph is given, Tamassia's flow network can be used to minimize the total number of bends [11]. Note that this approach may yield drawings with a linear number of bends for some of the edges. Given a combinatorial embedding that admits a 1-bend embedding, however, the flow network can be modified in a straightforward manner to minimize the total number of bends using at most one bend per edge.

The problem we consider involves considering all embeddings of a planar graph. Many problems of this sort are $\mathcal{NP}$-hard. For instance, 0-embeddability is $\mathcal{NP}$-hard [6], even though it can be decided efficiently if we are given an embedding by minimizing the total number of bends.

Contribution and Outline. In this work we give an efficient algorithm that solves FLEXDRAW for graphs with positive flexibility. Since FLEXDRAW contains the problem of 1-embeddability as a special case this closes the complexity gap between the $\mathcal{NP}$-hardness result for 0-embeddability by Garg and Tamassia [6] and the efficient algorithm for computing 2-embeddings by Biedl and Kant [2].

We present some preliminaries in Section 2. In Section 3 we study orthogonal flex-drawings of graphs with a fixed embedding and introduce the maximum rotation of a graph as a measure of how "flexible" it is. In Section 4 we show that replacing certain subgraphs with graphs that behave similarly does not change the maximum rotation. Based on this fact and the SPQR-tree we give an algorithm that solves FLEXDRAW for biconnected 4-planar graphs with positive flexibility. We extend our algorithm to arbitrary 4-planar graphs with positive flexibility in Section 5. For full proofs we refer the reader to the long version of this article [3].

2 Preliminaries

Orthogonal representation. The *orthogonal representation* introduced by Tamassia [11] describes orthogonal drawings of plane graphs, by listing the faces as sequences of bends. The advantage of the orthogonal representation is, that it neglects the lengths of the segments, thus it is possible to apply different operations on the drawing, without the need to worry about the exact geometry. Our orthogonal representation is always normalized, i.e., each edge has only bends in one direction; this slightly differs from the notion introduced by Tamassia.

The orthogonal representation of a plane graph G is defined as a set of lists $\mathcal{R} = \{\mathcal{R}(f_1), \ldots, \mathcal{R}(f_k)\}$ with a list for each face f_i of G. For each face f_i the list $\mathcal{R}(f_i)$ is a circular list of *edge descriptions* containing the edges on the boundary of f_i in clockwise (counter-clockwise if f_i is the external face) order. Each description $r \in \mathcal{R}(f_i)$ contains the following information: edge(r) denotes the edge represented by r, bends(r) is an integer whose absolute value is the number of 90°-bends of edge(r), where positive numbers represent bends to the right and negative numbers bends to the left. For a given edge description $r \in \mathcal{R}(f_i)$ we denote its successor in $\mathcal{R}(f_i)$ by r' and represent the angle α between edge(r) and edge(r') in f_i by their rotation rot(r, r') = $2 - \alpha/90°$. Every edge has exactly two edge descriptions, if r is one of them, the other is denoted by $\bar{r}$. Since each face forms a rectilinear polygon, every orthogonal representation $\mathcal{R}$ of an orthogonal drawing has the following three properties.

I Each edge description r is consistent with $\bar{r}$, i.e., bends($\bar{r}$) = $-$ bends(r).
II The interior bends of any face f_i sum up to 4 and the exterior bends to -4:

$$\sum_{r\in\mathcal{R}(f_i)} (\text{bends}(r) + \text{rot}(r, r')) = \begin{cases} -4, & \text{if } f \text{ is the external face,} \\ +4, & \text{if } f \text{ is an internal face.} \end{cases}$$

III The angles around every node sum up to 360°.

Given an orthogonal representation $\mathcal{R}$ of a graph, a corresponding orthogonal drawing can be computed efficiently [11]. Hence, it is sufficient to work with orthogonal representations. An orthogonal representation is *valid* for a given flexibility function flex if $|\text{bends}(r)| \leq \text{flex}(\text{edge}(r))$ for each edge description r.

For a planar graph $G = (V, E)$ with orthogonal representation $\mathcal{R}$ and two vertices s and t on the outer face f_1, we denote by $\pi_{\mathcal{R}}(s, t)$ the path in $\mathcal{R}(f_1)$ that connects s and t in counter-clockwise direction. Such a path $\pi = \pi(s, t)$ consists of consecutive edge descriptions $r_1, \ldots, r_k$. We define the *rotation* of π as

$$\text{rot}_{\mathcal{R}}(\pi) = \sum_{i=1}^{k} \text{bends}(r_i) + \sum_{i=1}^{k-1} \text{rot}(r_i, r_{i+1}).$$

Moreover, if v is a vertex of G that has exactly one angle in the outer face, we denote by $\text{rot}_\mathcal{R}(v)$ the rotation of this angle. Note that, for a single edge description r we have $\text{rot}(r) = \text{bends}(r)$. If it is clear from the context which orthogonal representation is meant we omit the indices of π and rot. The concept of rotation is similar to the spirality defined by Di Battista et al. [1].

The value $\text{rot}(\pi(s,t))$ describes the shape of the path $\pi(s,t)$ in the orthogonal representation in terms of the angle between its start- and its endpoint. Fixing the rotation of $\pi(s,t)$, $\pi(t,s)$ and the outer angles at s and t in a sense determines the shape of the outer face. In Section 4, we will exploit this by replacing certain subgraphs of G with simpler graphs whose outer faces have the same shapes.

Connectivity, st-graphs and the SPQR-tree. A graph is *connected* if there exists a path between any pair of vertices. A *separating k-set* is a set of k vertices whose removal disconnects the graph. Separating 1-sets and 2-sets are *cutvertices* and *separation pairs*. A graph is *biconnected* if it does not have a cut vertex and triconnected if it does not have a separation pair. The maximal biconnected components of a graph are called *blocks*.

The *block-cutvertex tree* of a connected graph is a tree whose nodes are the blocks and cutvertices of the graph. In the block-cutvertex tree a block B and a cutvertex v are joined by an edge if v belongs to B.

A *weak st-graph* is a 4-planar graph $G = (V, E)$ with two designated vertices s and t such that the graph $G + st$ is planar and has maximum degree 4. An *st-graph* is a weak st-graph such that $G + st$ is biconnected. An orthogonal representation $\mathcal{R}$ of a (weak) st-graph with positive flexibility is *valid* if each edge e has at most $\text{flex}(e)$ bends and s and t are embedded on the outer face. A valid orthogonal representation of a (weak) st-graph is *tight* if all angles at s and t in inner faces are $90°$.

We distinguish st-graphs with $\deg(s), \deg(t) \leq 2$ by the degrees of s and t. An st-graph is of Type (1,1) if $\deg(s) = \deg(t) = 1$, it is of Type (1,2) if one of them has degree 1 and the other one has degree 2 and it is of Type (2,2) if $\deg(s) = \deg(t) = 2$.

To handle the decomposition of biconnected graphs into triconnected components we use the SPQR-tree, which was introduced by Di Battista and Tamassia [4,5]. A detailed description of the SPQR-tree can be found in the literature [4,5,7]. Here we just give a sketch and some notation.

The SPQR-tree $\mathcal{T}$ of a graph G is a rooted tree that is determined by the *split pairs* of G. A split pair is a pair of vertices that are either connected by an edge or that is a separation pair. In the latter case the corresponding connected components are called the *split components* of the split pair.

The SPQR-tree $\mathcal{T}$ has four different types of nodes, namely S-,P-,Q- and R-nodes. Each node μ of $\mathcal{T}$ has an associated biconnected multigraph, its *skeleton*, denoted by $\text{skel}(\mu)$, which can be seen as a simplified version of the original graph. An edge uv in $\text{skel}(\mu)$ indicates that $\{u, v\}$ is a split pair and the edge uv represents one or more split components of $\{u, v\}$. The *pertinent graph* of a node μ, denoted by $\text{pert}(\mu)$ is the graph that is represented by the subtree of $\mathcal{T}$ with root μ. Note that in particular each pertinent graph is an st-graph.

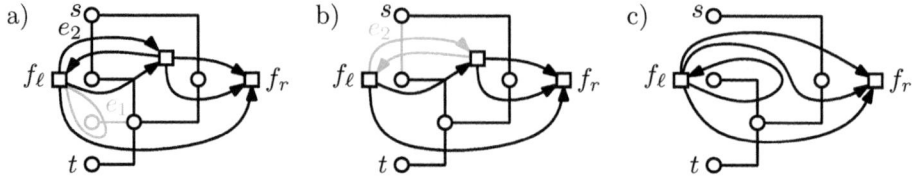

Fig. 1. An st-graph with flexibility 1 for all edges with $\mathrm{rot}(\pi(s,t)) = 1$ and its flex graph $G^{\times}$ (a), after removal of bridge e_1 (b), and removal of edge e_2 (c).

The SPQR-tree of a graph G represents all planar embeddings of G in the sense that choosing planar embeddings for all skeletons of $\mathcal{T}$ corresponds to a choosing a planar embedding of G and vice versa.

Our approach. We start out with an observation. Let G be a 4-planar graph with positive flexibility and let $\{s, t\}$ be a split pair of G that splits G into two subgraphs G_1, G_2 and let e_{ref} be an edge of G_1. Let ρ be the maximum rotation of $\pi(s,t)$ over all embeddings of G_2 where s and t are on the outer face.

If G_2 is of Type (1,1) then obviously the following holds. If G admits a valid orthogonal drawing with the given flexibility such that e_{ref} is embedded on the outer face then also the graph G' where G_2 is replaced by an edge f with flexibility ρ admits such a drawing. Graphs of Type (1,2) and (2,2) allow for similar substitutions.

Thus we can substitute st-graphs of each type with a small gadget graph to obtain a new graph G' such that if G has a valid drawing then also G' has one. We show that the converse is also true, i.e., if the graph G' admits such an embedding then also G does. We then exploit this characterization algorithmically using the SPQR-tree of G to successively replace subgraphs of G by simpler graphs.

3   The Maximum Rotation with a Fixed Embedding

The goal of this section is to derive a description of the valid orthogonal representations of a given (weak) st-graph with positive flexibility and a fixed embedding. Namely, we prove that the values that can be obtained for $\mathrm{rot}(\pi(s,t))$ form an interval for these graphs. We show that if there exists a valid orthogonal representation $\mathcal{R}$ with $\mathrm{rot}_{\mathcal{R}}(\pi(s,t)) \geq 0$ then there exists an orthogonal representation $\mathcal{R}'$ with $\mathrm{rot}_{\mathcal{R}'}(\pi(s,t)) = \mathrm{rot}_{\mathcal{R}}(\pi(s,t)) - 1$, which can be obtained from $\mathcal{R}$ by only altering the number of bends on certain edges.

To model the possible changes of an orthogonal representation $\mathcal{R}$ of a (weak) st-graph G that can be performed by only changing the number of bends on edges we introduce the *flex graph* $G^{\times}$ of G with respect to $\mathcal{R}$, which is based on the bidirected dual graph of G. Thus, the flex graph is a directed multigraph. See Fig. 1a for an illustration. We start out by adding to G the edge st and embed it into the outer face of G thus splitting the outer face into two faces f_{ℓ} and f_r, where f_{ℓ} is bounded by $\pi(s,t)$ and the new edge $\{s,t\}$ and f_r is bounded by $\pi(t,s)$ and $\{s,t\}$. We denote this graph by $\bar{G}$ and its dual graph by $\bar{G}^*$. We set $V^\times = V(\bar{G}^*)$ and we define $E^\times$ as follows. For each edge e of G denote its incident faces in $\bar{G}$ by f_u and f_v and let r_u and r_v be the edge descriptions of e in $\mathcal{R}(f_u)$ and $\mathcal{R}(f_v)$, respectively. We add the edge (f_u, f_v) if $-\operatorname{flex}(e) < \operatorname{bends}(r_u)$ and, analogously, we add (f_v, f_u) if $-\operatorname{flex}(e) < \operatorname{bends}(r_v)$. Consider an edge (f_u, f_v) of $G^\times$ and let r_u and r_v be the edge descriptions of the corresponding edge e in G. The fact that $(f_u, f_v) \in E^\times$ indicates that it is possible to decrease $\operatorname{bends}(r_u)$ (and thus increase $\operatorname{bends}(r_v)$) by at least 1 without violating the flexibility of e.

Assume that there exists a simple directed path from f_ℓ to f_r in $G^\times$. Let $f_\ell = f_1, f_2, \ldots, f_k = f_r$ be this path. We construct a new orthogonal representation $\mathcal{R}'$ from $\mathcal{R}$ as follows. For each edge $f_i f_{i+1}$, $i = 1, \ldots, k-1$, let e_i be the corresponding edge of G and let $r_i \in \mathcal{R}(f_i), \bar{r}_i \in \mathcal{R}(f_{i+1})$ be its edge descriptions. We obtain $\mathcal{R}'$ from $\mathcal{R}$ by decreasing $\operatorname{bends}(r_i)$ by 1 and increasing $\operatorname{bends}(\bar{r}_i)$ by 1 for $i = 1, \ldots, k-1$. First, it is clear that $\mathcal{R}'$ satisfies Properties I and III since we increase and decrease the number of bends consistently and we do not change any angles at vertices. Property II holds since each face of G has either none of its edge descriptions changed or exactly one of them is increased by 1 and exactly one of them is decreased by 1. Moreover, since the path starts at f_ℓ and ends at f_r we have that $\operatorname{rot}_{\mathcal{R}'}(\pi(s,t)) = \operatorname{rot}_{\mathcal{R}}(\pi(s,t)) - 1$. We now show that such a path exists if $\operatorname{rot}(\pi(s,t)) \geq 0$.

Lemma 1. *Let G be a weak st-graph with positive flexibility and let $\mathcal{R}$ be a valid orthogonal representation of G with $\operatorname{rot}_{\mathcal{R}}(\pi(s,t)) \geq 0$. Then the flex graph $G^\times$ contains a directed path from f_ℓ to f_r.*

Proof. Assume that G is a minimal counter example such that $G^\times$ does not contain such a path. First, we show that in $G^\times$ there exists at least one edge starting from f_ℓ. Let $\pi(s,t)$ be composed of the edge descriptions $r_1, \ldots, r_k$ in $\mathcal{R}(f)$, where f is the outer face of G. Then, by assumption we have $\operatorname{rot}(\pi(s,t)) = \sum_{i=1}^{k} \operatorname{bends}(r_i) + \sum_{i=1}^{k-1} \operatorname{rot}(r_i, r_{i+1}) \geq 0$. Since $\operatorname{rot}(r_i, r_{i+1}) \leq 1$ for $i = 1, \ldots, k-1$ we have that $\sum_{i=1}^{k} \operatorname{bends}(r_i) \geq -k+1$ and hence there is at least one r_j with $\operatorname{bends}(r_j) \geq 0$. Hence, $G^\times$ contains an edge corresponding to $\operatorname{edge}(r_j)$ that starts at f_ℓ. This shows that there always exists an edge (f_ℓ, f_u) in $G^\times$. We distinguish three types of edges (f_ℓ, f_u). If $f_u = f_r$ then (f_ℓ, f_u) is the desired path.

If $f_u = f_\ell$ the corresponding edge e of G is a bridge whose removal does not disconnect s and t, see Fig. 1b, then let H be the connected component of $G - e$ containing s and t and let $\mathcal{S}$ be the restriction of $\mathcal{R}$ to H. For the outer face of H we have that $\operatorname{rot}_{\mathcal{S}}(\pi(s,t)) + \operatorname{rot}_{\mathcal{S}}(s) + \operatorname{rot}_{\mathcal{S}}(\pi(t,s)) + \operatorname{rot}_{\mathcal{S}}(t) = -4$. Since $\pi_{\mathcal{R}}(t,s) = \pi_{\mathcal{S}}(t,s)$ we have that $\operatorname{rot}_{\mathcal{S}}(\pi(t,s)) = \operatorname{rot}_{\mathcal{R}}(\pi(t,s))$. Moreover, since we only remove edges the angles at s and t (and thus their rotations) do not decrease, i.e., we have $\operatorname{rot}_{\mathcal{S}}(t) \leq \operatorname{rot}_{\mathcal{R}}(t)$ and $\operatorname{rot}_{\mathcal{S}}(s) \leq \operatorname{rot}_{\mathcal{R}}(s)$. Hence, we have that $\operatorname{rot}_{\mathcal{S}}(\pi(s,t)) \geq -4 - \operatorname{rot}_{\mathcal{R}}(\pi(t,s)) - \operatorname{rot}_{\mathcal{R}}(s) - \operatorname{rot}_{\mathcal{R}}(t) = \operatorname{rot}_{\mathcal{R}}(\pi(s,t)) \geq 0$. Since H has fewer edges than G it is not a counter example and its flex graph $H^\times$ contains a path from f_ℓ to f_r. Since $H^\times$ is a subgraph of $G^\times$ this contradicts the assumption that G is a counter example.

Otherwise, f_u is an internal face of G, see Fig. 1c. Let e be the corresponding edge of G. Let $H := G - e$ and let S be the orthogonal representation $\mathcal{R}$ restricted to H. Note that the flex graph of $H^{\times}$ of H can be obtained from $G^{\times}$ by removing all edges between f_ℓ and f_u and merging f_ℓ and f_u into a single node f'_ℓ. As above we obtain that $\mathrm{rot}_S(\pi(s, t)) \geq 0$ and hence in $H^{\times}$ there exists a path from f'_ℓ to f_r. The corresponding path in $G^{\times}$ (after undoing the contraction of f_ℓ and f_u) either starts at f_ℓ or at f_u and ends at f_r. In the former case we have found our path, in the latter case the path together with the edge (f_ℓ, f_u) forms the desired path. Again this contradicts the assumption that G is a counter example.   □

Recall that a valid orthogonal representation of a (weak) st-graph is tight if the inner angles at s and t are $90°$. We show that a valid orthogonal representation can be made tight without decreasing $\mathrm{rot}(\pi(s, t))$.

Lemma 2. *Let G be a weak st-graph with positive flexibility and let $\mathcal{R}$ be a valid orthogonal representation. Then there exists a valid orthogonal representation $\mathcal{R}'$ of G with the same planar embedding such that $\mathcal{R}'$ is tight, $\mathrm{rot}_{\mathcal{R}'}(\pi(s, t)) \geq \mathrm{rot}_{\mathcal{R}}(\pi(s, t))$ and $\mathrm{rot}_{\mathcal{R}'}(\pi(t, s)) \geq \mathrm{rot}_{\mathcal{R}}(\pi(t, s))$.*

Let G be an st-graph with positive flexibility and a fixed planar embedding $\mathcal{E}$. Lemma 1 shows that the attainable values of $\mathrm{rot}(\pi(s, t))$ for a given st-graph with a fixed embedding form an interval. Hence, the set of possible rotations can be described by the boundaries of this interval and we define the *maximum rotation* of G with respect to $\mathcal{E}$ as $\mathrm{maxrot}_{\mathcal{E}}(G) = \max_{\mathcal{R} \in \Omega} \mathrm{rot}_{\mathcal{R}}(\pi(s, t))$ where Ω contains all valid orthogonal representations of G whose embedding is $\mathcal{E}$.

The following theorem states that indeed the maximum rotation essentially describes the orthogonal representations of st-graphs with fixed embedding and positive flexibility.

Theorem 1. *Let G be an st-graph with positive flexibility and fixed embedding $\mathcal{E}$. Then for each $\rho \in \{-1, \ldots, \mathrm{maxrot}_{\mathcal{E}}(G)\}$ there exists a valid and tight orthogonal representation $\mathcal{R}$ of G with planar embedding $\mathcal{E}$ such that $\mathrm{rot}_{\mathcal{R}}(\pi(s, t)) = \rho$.*

Using a variant of Tamassia's flow network [11] the maximum rotation can be computed efficiently for st-graphs with a fixed embedding.

Theorem 2. *Given an st-graph $G = (V, E)$ with fixed embedding $\mathcal{E}$ with s and t on the outer face we can compute $\mathrm{maxrot}_{\mathcal{E}}(G)$ in $O(n^{3/2})$ time or decide that G does not admit a valid orthogonal representation with this embedding.*

4   Biconnected Graphs

Until now the planar embedding of our input graph was fixed. Now, we assume that this embedding is variable. Following the approach of the previous section we define the maximum rotation of a (weak) st-graph G as $\mathrm{maxrot}(G) = \max_{\mathcal{E} \in \Psi} \mathrm{maxrot}_{\mathcal{E}}(G)$ where Ψ contains all planar embeddings of G such that s and t are embedded on the outer face.

In this section we show that maxrot(G) essentially describes all valid orthogonal representations of G in the sense that substituting a subgraph H of G with a different graph H' with maxrot(H) = maxrot(H') does not change maxrot(G). We further use this substitution to give an algorithm that computes maxrot by successively reducing the size of the graph. To handle the different possible planar embeddings we use the SPQR-tree and we substitute subgraphs with small graphs that have only one embedding. We need the following technical lemma.

Lemma 3. *Let G be an st-graph with $\deg(s), \deg(t) \leq 2$ and let $\mathcal{R}$ be a tight orthogonal representation of G. Then $\mathrm{rot}(\pi(s,t)) + \mathrm{rot}(\pi(t,s)) = -x$ where x is 0,1 and 2 for graphs of Type (1,1), (1,2) and (2,2), respectively.*

The following theorem shows that indeed the maximum rotation describes all possible rotation values of an st-graph.

Theorem 3. *Let G be an st-graph with positive flexibility and let ρ be an integer. Then there exists a tight orthogonal representation $\mathcal{R}$ of G with $\mathrm{rot}(\pi(s,t)) = \rho$ if and only if $-\mathrm{maxrot}(G) - x \leq \rho \leq \mathrm{maxrot}(G)$ where x depends on the Type of G and $x = 0, 1, 2$ for Types (1,1), (1,2) and (2,2), respectively.*

Proof. We first show the only if part. Let $\mathcal{R}$ be any embedding of G. By the definition of maxrot(G) we clearly have that $\mathrm{rot}_{\mathcal{R}}(\pi(s,t)) \leq \mathrm{maxrot}(G)$. By definition we also have that $\mathrm{rot}_{\mathcal{R}}(\pi(t,s)) \leq \mathrm{maxrot}(G)$ (otherwise by mirroring we could obtain an orthogonal representation $\mathcal{R}'$ with $\mathrm{rot}_{\mathcal{R}'}(\pi(s,t)) > \mathrm{maxrot}(G)$) and hence with Lemma 3 we obtain $-\mathrm{rot}(\pi(s,t)) - x \leq \mathrm{maxrot}(G)$.

It remains to show that for any given ρ in the range we can find a valid orthogonal representation. If $-1 \leq \rho \leq \mathrm{maxrot}(G)$ we find an orthogonal representation as follows. Let $\mathcal{R}$ be a valid orthogonal embedding of G with $\mathrm{rot}(\pi(s,t)) = \mathrm{maxrot}(G)$. By Lemma 2 we can reduce the inner angles at s and t to $90°$ without decreasing $\mathrm{rot}(\pi(s,t))$. By Theorem 1 we thus find the desired orthogonal representation.

If $\rho \leq -2$ by Lemma 3 we need to find a valid orthogonal representation $\mathcal{R}$ with $\mathrm{rot}_{\mathcal{R}}(\pi(t,s)) = -\rho - x =: \rho'$. Note that by the definitions of ρ and x we have that $0 \leq \rho' \leq \mathrm{maxrot}(G)$. As above we obtain a valid orthogonal embedding $\mathcal{R}'$ of G with $\mathrm{rot}_{\mathcal{R}'}(\pi(s,t)) = \rho'$. We obtain $\mathcal{R}$ by mirroring $\mathcal{R}'$. □

Note that if s (or t) has degree 1 then its incident edge allows for three different rotations and hence the range of valid rotations contains at least three integers. This observation together with the theorem yields the following.

Corollary 1. *Let G be an st-graph with positive flexibility. If G admits a valid drawing then $\mathrm{maxrot}(G) \geq 1$ if G is of Type (1,1) or (1,2) and $\mathrm{maxrot}(G) \geq -1$ if G is of Type (2,2).*

In particular, Theorem 3 shows that an st-graph G with $\deg(s) = \deg(t) = 1$ essentially behaves like a single edge st with flexibility maxrot(G). The following lemma shows that we can replace any st-graph with $\deg(s), \deg(t) \leq 2$ in a graph G by a different st-graph of the same type and with the same maximum rotation without changing maxrot(G). Fig. 2 illustrates the lemma and its proof.

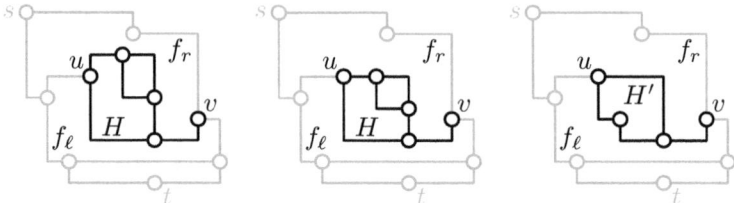

Fig. 2. Illustration of Lemma 4, st-graph G with split pair $\{u, v\}$ splitting off H (left), replacement of H with a tight orthogonal representation (middle) and replacement of H with a graph H' with $\mathrm{maxrot}(H) = \mathrm{maxrot}(H') = 3$ (right)

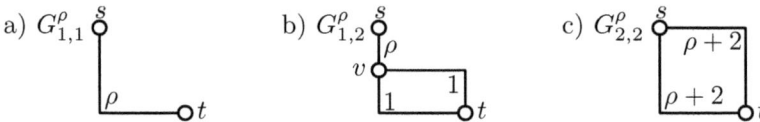

Fig. 3. Gadgets for st-graphs with maximum rotation ρ depending on the Type

Lemma 4. *Let $G = (V, E)$ be an st-graph with positive flexibility and let $\{u, v\}$ be a split pair of G that splits G into two components G^- and H such that G^- contains s and t and H is an st-graph of Type (1,1), Type (1,2) or Type (2,2) (with respect to vertices u and v). Let H' be an st-graph with designated vertices u', v' of the same type as H with $\mathrm{maxrot}(H') = \mathrm{maxrot}(H)$.*

Then G admits a valid orthogonal representation $\mathcal{R}$ with $\mathrm{rot}_{\mathcal{R}}(\pi(s, t)) = \rho$ if and only if the graph G', which is obtained from G by replacing H with H' admits a valid orthogonal representation $\mathcal{R}'$ with $\mathrm{rot}_{\mathcal{R}'}(\pi(s, t)) = \rho$.

We now present three especially simple families of replacement graphs, called *gadgets*, for st-graphs of Types (1,1), (1,2) and (2,2), respectively; see Fig. 3. Let ρ be an integer. The graph $G^{\rho}_{1,1}$ is simply an edge st with $\mathrm{flex}(st) = \rho$. The graph $G^{\rho}_{1,2}$ has three vertices s, v, t and two edges between s and v, both with flexibility 1, and the edge vt with flexibility ρ. The gadget $G^{\rho}_{2,2}$ consists of two parallel edges between s and t, both with flexibility $\rho + 2$. Note that by Corollary 1 all edges of our gadgets have again positive flexibility and that $\mathrm{maxrot}(G^{\rho}_{1,1}) = \mathrm{maxrot}(G^{\rho}_{1,2}) = \mathrm{maxrot}(G^{\rho}_{2,2}) = \rho$. Moreover, each of these graphs has a unique embedding with s and t on the outer face.

We now describe an algorithm that computes $\mathrm{maxrot}(G)$ for a given st-graph G with positive flexibility or decides that G does not admit a valid orthogonal representation. We use the SPQR-tree $\mathcal{T}$ of $G + st$, rooted at the Q-node corresponding to st to represent all planar embeddings of G with s and t on the outer face. Our algorithm processes the nodes of the SPQR-tree in a bottom-up fashion and computes the maximum rotation of each pertinent graph from the maximum rotations of the pertinent graphs of its children. For each node μ we have a variable $\mathrm{maxrot}(\mu)$. We will prove later that after processing a node we have that $\mathrm{maxrot}(\mu) = \mathrm{maxrot}(\mathrm{pert}(\mu))$. For each Q-node μ we initialize maxrot(μ) to be the flexibility of the corresponding edge. We now show how to compute maxrot(μ) from the maximum rotations of its children. We make a case distinction based on the type of μ.

If μ **is an R-node** let $\mu_1, \ldots, \mu_k$ be the children of μ and let $H_1, \ldots, H_k$ be their pertinent graphs. Each virtual edge in skel(μ) represents at least one incidence of an edge of G to its poles. Since skel(μ) is 3-connected each node has at least degree 3 and hence no virtual edge can represent more than two incidences, i.e., the nodes of skel(μ) have degree at most 2 in the subgraphs of G that are represented by the virtual edges of μ. As we already know their maximum rotations we can simply replace each of the graphs by a corresponding gadget; we call the resulting graph G_μ. Since the embeddings of all gadgets are completely symmetric it is enough to compute the maximum rotations of G_μ for the only two embeddings $\mathcal{E}_1$ and $\mathcal{E}_2$ induced by the embeddings of skel(μ). We set maxrot(μ) = max$\{$maxrot$_{\mathcal{E}_1}(G_\mu)$, maxrot$_{\mathcal{E}_2}(G_\mu)\}$ if one of them admits a valid representation. Otherwise we stop and return "infeasible".

If μ **is a P-node** we treat μ similar as in the case where μ is an R-node. Again, we have that each pole has degree at least 3 in skel(μ) and hence no virtual edge can represent more than two edge incidences. We replace each virtual edge with the corresponding gadget and try all possible embeddings of skel(μ), which are at most six and store the maximum rotation or stop if none of the embeddings admits a valid representation.

If μ **is an S-node** let $\mu_1, \ldots, \mu_k$ be the children of μ. We set maxrot(μ) = $\sum_{i=1}^{k}$ maxrot(μ_i) + $k - 1$.

Theorem 4. *Given an st-graph $G = (V, E)$ with positive flexibility it can be checked in $O(n^{3/2})$ time whether G admits a valid orthogonal representation. In the positive case* maxrot(G) *can be computed within the same time complexity.*

Proof. We prove the invariant that after processing node μ we have maxrot(μ) = maxrot(pert(μ)). The proof is by induction on the height h of the SPQR-tree $\mathcal{T}$ of $G + st$. Let μ be the node of $\mathcal{T}$ whose parent corresponds to st.

If $h = 1$ then G is a single edge e and μ its corresponding Q-node. Since maxrot(G) = flex(e) the claim holds. For $h > 1$ let $\mu_1, \ldots, \mu_k$ be the children of μ. By induction we have that maxrot(μ_i) = maxrot(pert(μ_i)) for $i = 1, \ldots, k$. We make a case distinction based on the type of μ.

If μ is an R- or a P-node then by Lemma 4 we have that maxrot(G_μ) = maxrot(pert(μ)) and since the gadgets have a unique embedding we consider all relevant embeddings of G_μ. If none of the embeddings admits a valid orthogonal representation then obviously also pert(μ) and thus G do not admit valid orthogonal representations.

If μ is an S-node and the pertinent graphs of its children admit a valid orthogonal representation then there always exists a valid orthogonal representation of pert(μ). Let $H_1, \ldots, H_k$ be the pertinent graphs of the children of μ and let $v_1, \ldots, v_{k+1}$ be the vertices in skel(μ) such that v_i and v_{i+1} are the poles of H_i. By Theorem 3 there exist tight orthogonal representations $\mathcal{R}_1, \ldots, \mathcal{R}_k$ of $H_1, \ldots, H_k$ with rot($\pi(v_i, v_{i+1})$) = maxrot(μ_i). We put these orthogonal representations together such that the angles at the nodes $v_2, \ldots, v_k$ on

$\pi(v_1, v_{k+1})$ are $90°$. Hence we get an orthogonal representation of $\text{pert}(\mu)$ with $\text{rot}(\pi(v_1, v_{k+1})) = \sum_{i=1}^{k} \text{maxrot}(\mu_i) + k - 1$. On the other hand if we had an orthogonal representation of $\text{pert}(\mu)$ with a higher rotation then at least one of its children μ_i would need to have a rotation that is bigger than $\text{maxrot}(\mu_i)$.

This proves the correctness of the algorithm. For the running time note that the SPQR-tree can be computed in linear time [7]. The time for computing $\text{maxrot}(\mu)$ for a given node μ from the maximum rotations of its children can be done in $O(|\text{skel}(\mu)|^{3/2})$ time by Theorem 4 since $\text{skel}(\mu)$ has only a constant number of embeddings. The total running-time follows from the fact that the total size of all skeletons is in $O(n)$.   □

This theorem can be used to solve FLEXDRAW for biconnected 4-planar graphs with positive flexibility. Such a graph G admits a valid orthogonal representation if and only if one of the graphs $G - e$, $e \in E(G)$ (which is an st-graph with respect to the endpoints of e) admits a valid orthogonal representation such that e can be added to this representation. This can be done if and only if $\text{maxrot}(G - e) + \text{flex}(e) \geq 2$. This can be seen as follows. Let s and t be the endpoints of e. Adding e to $G - e$ creates a new interior face and the total rotation of this new face needs to be 4. We can have at most two $90°$ angles at s and t, hence $\text{maxrot}(G - e) + \text{flex}(e) \geq 2$ is a necessary condition. On the other hand, it is not hard to see that it is possible to add e to a tight orthogonal representation of $G - e$. If $\text{flex}(e) \geq 3$ then we can add e to a tight orthogonal representation of $G - e$ with $\text{rot}(\pi(s, t)) = -1$. Otherwise, we add e to a tight orthogonal representation of $G - e$ with $\text{rot}(\pi(s, t)) = 2 - \text{flex}(e)$, which is possible since $2 - \text{flex}(e) \geq -1$ holds in this case. We obtain the following theorem; the running time is due to $O(n)$ applications of the algorithm for st-graphs.

Theorem 5. FLEXDRAW *can be solved in time* $O(n^{5/2})$ *for biconnected 4-planar graphs with positive flexibility.*

5   Connected Graphs

In this section we generalize our results to connected 4-planar graphs that are not necessarily biconnected. We analyze the conditions under which orthogonal representations sharing a cut vertex can be combined and use the block-cutvertex tree to derive an algorithm that decides whether a connected 4-planar graph with positive flexibility admits a valid orthogonal drawing.

Lemma 5. *Let* G *be a connected 4-planar graph with cutvertex* v *and corresponding cut components* $H_1, \ldots, H_k$. *Then* G *admits a valid orthogonal representation if and only if all cut components* H_i *have valid orthogonal representations such that at most one of them has* v *not on the outer face.*

Now let G be a connected 4-planar graph with positive flexibility and $\mathcal{B}$ its block-cutvertex tree. Let further B be a block of G that is a leaf in $\mathcal{B}$ and let v be the unique cutvertex of B.

If B is the whole graph G we return "true" if and only if G admits any valid orthogonal representation. This can be checked with the algorithm from the previous Section.

If B is not the whole graph G we check whether B admits a valid orthogonal representation having v on its outer face. This can be done with the algorithm from the previous section by rooting the SPQR-tree of B at all edges incident to v. If it does admit such an embedding then by Lemma 5 G admits a valid orthogonal embedding if and only if the graph G', which is obtained from G by removing the block B, admits a valid orthogonal embedding. We check G' recursively. If B does not admit such an embedding we mark B and proceed with another unmarked leaf. If we ever encounter another block B' that has to be marked we return "infeasible". This is correct as in this case B has to be embedded in the interior of B' and vice versa, which is obviously impossible. Checking a single block B can be done in $O(|B|^{5/2})$ time by Theorem 5. Since the total size of all blocks is in $O(n)$ the total running-time is $O(n^{5/2})$. This proves the following theorem.

Theorem 6. FLEXDRAW *can be solved in* $O(n^{5/2})$ *time for 4-planar graphs with positive flexibility.*

Conclusion. We have shown that FLEXDRAW can be solved efficiently for graphs with positive flexibility. Moreover, it is straightforward to generalize our algorithm to positive flexibility functions flex $: E \longrightarrow \mathbb{N} \cup \{\infty\}$, i.e., some edges may be bent arbitrarily often. An interesting open question is whether FLEXDRAW can still be handled if few edges are required to have no bends.

References

1. Battista, G.D., Liotta, G., Vargiu, F.: Spirality and optimal orthogonal drawings. SIAM J. Comput. 27(6), 1764–1811 (1998)
2. Biedl, T., Kant, G.: A better heuristic for orthogonal graph drawings. In: van Leeuwen, J. (ed.) ESA 1994. LNCS, vol. 855, pp. 24–35. Springer, Heidelberg (1994)
3. Bläsius, T., Krug, M., Rutter, I., Wagner, D.: Orthogonal graph drawing with flexibility constraints. Technical Report 2010-18, Faculty of Informatics, Karlsruhe Institute of Technology, KIT (2010), http://digbib.ubka.uni-karlsruhe.de/volltexte/1000019793
4. Di Battista, G., Tamassia, R.: On-line maintenance of triconnected components with SPQR-trees. Algorithmica 15(4), 302–318 (1996)
5. Di Battista, G., Tamassia, R.: On-line planarity testing. SIAM J. Comput. 25(5), 956–997 (1996)
6. Garg, A., Tamassia, R.: On the computational complexity of upward and rectilinear planarity testing. SIAM J. Comput. 31(2), 601–625 (2001)
7. Gutwenger, C., Mutzel, P.: A linear time implementation of SPQR-trees. In: Marks, J. (ed.) GD 2000. LNCS, vol. 1984, pp. 77–90. Springer, Heidelberg (2001)
8. Liu, Y., Marchioro, P., Petreschi, R., Simeone, B.: Theoretical results on at most 1-bend embeddability of graphs. Acta Math. Appl. Sinica (English Ser.) 8(2), 188–192 (1992)

9. Liu, Y., Morgana, A., Simeone, B.: A linear algorithm for 2-bend embeddings of planar graphs in the two-dimensional grid. Discr. Appl. Math. 81(1–3), 69–91 (1998)
10. Morgana, A., de Mello, C.P., Sontacchi, G.: An algorithm for 1-bend embeddings of plane graphs in the two-dimensional grid. Discr. Appl. Math. 141(1-3), 225–241 (2004)
11. Tamassia, R.: On embedding a graph in the grid with the minimum number of bends. SIAM J. Comput. 16(3), 421–444 (1987)

Drawing Ordered $(k-1)$–Ary Trees on k–Grids

Wolfgang Brunner and Marco Matzeder

University of Passau
94030 Passau, Germany
{brunner,matzeder}@fim.uni-passau.de

Abstract. We explore the complexity of drawing ordered $(k-1)$–ary trees on grids with k directions for $k \in \{4, 6, 8\}$ and within a given area. This includes, e. g., ternary trees drawn on the orthogonal grid. For aesthetically pleasing tree drawings on these grids, we additionally present various restrictions similar to the common hierarchical case. First, we generalize the $\mathcal{NP}$–hardness of minimal width in hierarchical drawings of ordered trees to $(k-1)$–ary trees on k–grids and then we generalize the Reingold and Tilford algorithm to k–grids.

1 Introduction

Drawing trees is an important area in graph drawing. There are three main styles: hierarchical, radial and orthogonal [14]. For hierarchical tree drawings the linear time algorithm by Reingold and Tilford [16] is commonly used. These drawings fulfill various aesthetic criteria: vertices on the same level of the tree are placed on a horizontal line, children maintain their order, parents are centered above their children, a certain minimum horizontal distance is guaranteed, edges do not cross, and isomorphic subtrees are identically drawn up to translation. The resulting drawings require $\mathcal{O}(n^2)$ area. However, the problem of determining the minimum width of such drawings of ordered binary trees is $\mathcal{NP}$–hard [1,18]. For unordered trees the $\mathcal{NP}$–hardness was shown by Marriott and Stuckey [15].

Another drawing style are radial drawings. Eades [8] presented an $\mathcal{O}(n)$ time algorithm where the root is placed at the center of the drawing. The subtrees are placed into sectors around the root, whose widths are determined by the number of their leaves. The vertices are placed onto concentric circles and the parents are centered with respect to their children. In contrast to hierarchical drawings, this algorithm positions the vertices on real coordinates.

The third approach are drawings of trees on grids [6,7,12,17]. Vertices are positioned on integer coordinates and the edges are either orthogonal straight lines, arbitrary polylines, or orthogonal polylines. In the following we consider drawings with edges as straight lines. In the case of unordered ternary trees on the orthogonal grid it turns out to be a $\mathcal{NP}$–hard problem to decide whether or not there is a drawing with unit edge length or within given area [5]. The same holds true for 5–ary trees on the hexagonal grid with six directions [3]. Additionally, the logic engine introduced by Eades and Whitesides [10] can be

U. Brandes and S. Cornelsen (Eds.): GD 2010, LNCS 6502, pp. 105–116, 2011.
© Springer-Verlag Berlin Heidelberg 2011

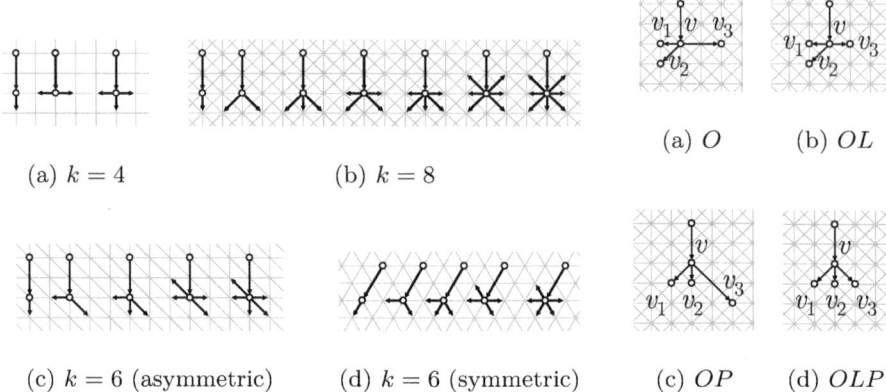

(a) $k = 4$　　　　　　　(b) $k = 8$

(a) O　　　　(b) OL

(c) $k = 6$ (asymmetric)　　(d) $k = 6$ (symmetric)

(c) OP　　　(d) OLP

Fig. 1. Patterns on the k–grids　　　　　**Fig. 2.** Drawing styles

used to prove these $\mathcal{NP}$–hardness results. If only three directions on the orthogonal grid are allowed for straight line edges (west, south, and east), the algorithms use tree folding techniques to achieve tree drawings with a small area [6,7,12,17]. Chan et. al. proved an area bound of $\mathcal{O}(n \log n)$ for upward drawings and of $\mathcal{O}(n \log \log n)$ for non–upward drawings (all four orthogonal directions) of binary trees [6]. However, such drawings are not really pleasing since subtrees may be arbitrarily nested. In h–v layouts using two directions on the orthogonal grid complete binary trees only need $\mathcal{O}(n)$ area and arbitrary binary trees need $\Theta(n \log n)$ area [7]. For binary trees on the orthogonal grid Eades et. al. [9] developed an $\mathcal{O}(n^2)$ time algorithm for non order–preserving h–v drawings on minimal area.

Ordered binary trees can be drawn on the orthogonal grid within $\mathcal{O}(n^{1.5})$ area and $\Theta(n^2)$ area is a tight bound for ternary trees [11]. In this work we draw ordered $(k-1)$–ary trees on k–grids with $k \in \{4,6,8\}$ directions, e. g., ternary trees on the orthogonal grid with four possible directions. We prove that it is a $\mathcal{NP}$–hard problem to decide whether or not an ordered $(k-1)$–ary tree has an order–preserving drawing on the k–grid within a given area. Moreover, we introduce further aesthetics transferred from the hierarchical case [16] leading to more comprehensible and symmetric tree drawings on k–grids. In most cases we show that the decision problem remains $\mathcal{NP}$–hard.

2　Preliminaries

The *orthogonal grid* is the infinite plane graph whose vertices have integer coordinates and whose edges link pairs of vertices at unit distance either vertically or horizontally. For all vertices (x, y) we extend the orthogonal grid with four directions to the *hexa grid* with six directions by adding undirected edges between

(x, y) and $(x+1, y-1)$, see the underlying grid in Fig. 1(c). A further extension is achieved by introducing undirected edges between integer coordinates (x, y) and $(x + 1, y + 1)$ resulting in the *octa grid* with eight directions, see Fig. 1(b). We call these three grids k–*grids* with $k \in \{4, 6, 8\}$. In the literature the hexa grid is called hexagonal grid [2,13] and triangular grid [19] as well. The *distance* between vertices u and v with coordinates (u_x, u_y) and (v_x, v_y) on a k–grid is defined by $\|(u, v)\| = \max(|u_x - v_x|, |u_y - v_y|)$. A *path* $v_1 \rightsquigarrow v_n$ is a sequence of vertices $(v_1, \ldots, v_n)$ in a graph with edges (v_i, v_{i+1}) and $i \in \{1, \ldots, n-1\}$. A *straight path* is a path where all edges have the same direction.

Let $T = (V, E)$ be a *tree* and $v \in V$. We use T_v for the subtree with root v. The *depth* of a vertex v is the number of edges of the path from the root to v. The *height* of T is the depth of the deepest vertex. We call two vertices *siblings* if they have the same parent. Trees with outdegree up to d are called d–*ary trees*. A tree *embedding* $\Gamma_k(T)$ of a tree T on a k–grid is a mapping Γ_k which specifies for each vertex $v \in V$ distinct integer coordinates $\Gamma_k(v) = (x, y)$ on a k–grid. Γ_k maps an edge $e \in E$ on a straight path $\Gamma_k(e)$ of the k–grid whose endpoints are the mappings of vertices linked by e. We use the terms *drawing* and *embedding* synonymously. The *length* of an edge $e \in E$ is the distance between its incident vertices and the *length* of a straight path p is the sum of its edge lengths. We call a vertex v of a $(k-1)$–ary tree on a k–grid *full* if it has $k-1$ children. The relative direction of each outgoing edge of a full vertex is determined, due to the given order of the tree. We call a path of full vertices *full path*.

Similar to the hierarchical drawing style of trees [16,18,20], we produce *planar* drawings preserving a given *minimal distance* and *isomorphic subtrees*. We discuss drawings of ordered trees on k–grids which are *order–preserving* and satisfy the given aesthetics. A tree drawing is *locally uniform* if for each vertex its outgoing edges have the same length, see Fig. 2(b). In a *pattern drawing* of a $(k-1)$–ary tree on the k–grid, the edge directions of the outgoing edges of each vertex have a prescribed angle with respect to the direction of the incoming edge, see Fig. 2(c). All patterns for the various k–grids are shown in Fig. 1. The patterns in Fig. 1(c) do not seem to be symmetric, but changing the underlying grid by rotating two axes yields a more pleasing symmetric drawing, see Fig. 1(d).

In the following we define various restrictions for order–preserving tree drawings on k–grids. Figure 2 presents an ordered tree T with vertices v and its children v_1, v_2 and v_3 in various order–preserving drawing styles. In Fig. 2(a) we show a drawing of T where no additional restrictions are given, called O_k–*drawing*. If a drawing is additionally local uniform on the k–grid, we call it an OL_k–*drawing*, see Fig. 2(b). We call an order–preserving pattern drawing on a k–grid OP_k–*drawing*. For an example see Fig. 2(c). If we combine these two properties, we obtain order–preserving locally uniform pattern drawings, called OLP_k–*drawings*, see Fig. 2(d). In such drawings the children of a vertex v are positioned axially symmetrical with respect to the incoming edge of v, which is analogous to placing the parent centered over its children in the hierarchical case. In Sect. 4 we introduce a drawing algorithm generating OLP_k–drawings.

3 $\mathcal{NP}$–Hardness Results

We show that drawing $(k-1)$–ary trees within a given area is $\mathcal{NP}$–hard for O_k–, OL_4–, OP_k–, and OLP_k–drawings. Clearly all these problems are in $\mathcal{NP}$.

3.1 O–Drawings

In this section we prove the complexity of order–preserving drawings.

Theorem 1. *Let $k \in \{4, 6, 8\}$, T be a rooted $(k-1)$–ary tree, and $A \geq 1$. Determining the existence of an O_k–drawing $\Gamma_k(T)$ with area at most A is $\mathcal{NP}$–hard.*

We reduce *3–SAT* by creating a $(k-1)$–ary tree for a given expression in *3–CNF* similarly to the hierarchical case [1,18] in polynomial time. Let the boolean expression E consist of r clauses $F_1, \ldots, F_r$ with n variables $x_1, \ldots, x_n$. Each clause F_i with $i \in \{1, \ldots, r\}$ has three literals, such that $F_i = (y_{i,1} \vee y_{i,2} \vee y_{i,3})$. A literal $y_{i,j}$ with $j \in \{1, 2, 3\}$ in the i–th clause is either a variable x_l or its negation $\overline{x_l}$. In the following, we treat the case $k = 8$, i.e., we construct a 7–ary tree on the octa grid. To simplify the representation of the tree we replace some full vertices and their adjacent leaves by big vertices, called *box vertices* as can be seen in Fig. 3(a) and 3(b). Any label of a box vertex or near a full vertex and its leaves identifies the center vertex.

Variable tree. For a variable x_l of E with $l \in \{1, \ldots, n\}$, we define a unique 7–ary tree, called *variable tree* $VT(x_l)$ as the shaded part in Fig. 3(a). It is the induced subtree of the root p_l having a full path $p_l \rightsquigarrow u$. The full path $u_1 \rightsquigarrow u_l$ is appended to u having length l depending on the index of x_l. We append a further full vertex h above u and a single vertex z at the top of the full vertex h. Note that (the essential part of) the variable tree $VT(x_l)$ of Fig. 3(a) is shown in Fig. 3(b) as well, but there the vertical outgoing edge of p_l has length four instead of three.

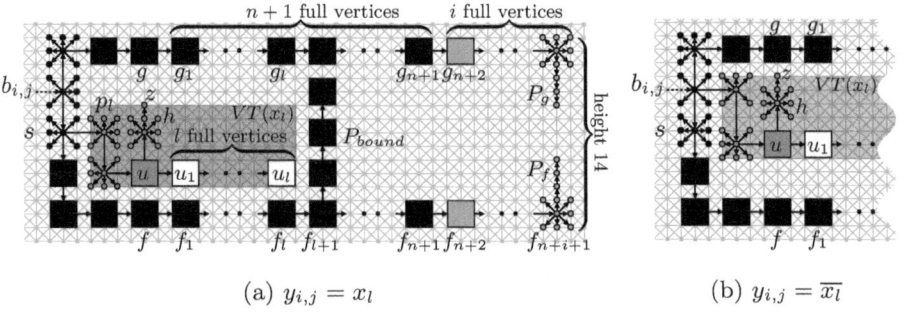

(a) $y_{i,j} = x_l$ (b) $y_{i,j} = \overline{x_l}$

Fig. 3. Literal tree $LT(y_{i,j})$ of $y_{i,j}$ with $j = 3$

Literal tree. The j–th literal $y_{i,j}$ in the i–th clause is either x_l or $\overline{x_l}$. We define a literal tree $LT(y_{i,j})$, which contains the corresponding variable tree $VT(x_l)$. The skeleton of a literal tree (see Fig. 3(a)) is the full path $b_{i,j} \rightsquigarrow f_{l+1}$. At f_{l+1} we append a path of three full vertices P_{bound} in vertical direction and a horizontal full path $f_{l+2} \rightsquigarrow f_{n+i+1}$. A similar symmetrical construction is added to the top from $b_{i,j}$ to g_{n+i+1} without the full path P_{bound}.

The full path P_{bound} ensures that if the skeleton of the literal tree is drawn within minimal width, the variable tree is also drawn within minimal width. At the child above the last vertex f_{n+i+1} we append a path P_f of j single vertices (analogous P_g below g_{n+i+1}) as can be seen in Fig. 3(a). Due to this construction, all these subtrees are non isomorphic.

If $y_{i,j} = x_l$, the root p_l of variable tree $VT(x_l)$ is appended to the vertex s, see Fig. 3(a). Otherwise it is appended to $b_{i,j}$, see Fig. 3(b), where just the essential part of the corresponding literal tree is shown.

Minimal height and width of a literal tree: Let the incoming edge of $b_{i,j}$ face to the east. Fig. 3(a) shows a drawing of $LT(y_{i,j})$ with $y_{i,j} = x_l$ of minimal height 14. Note that the directions of the edges are determined by the incoming edge of the root $b_{i,j}$, except for the edges of P_f and P_g. If all vertical edges have minimal length, the drawing has minimal height 14, otherwise the height may increase.

The minimal height of the drawing of $LT(y_{i,j})$ with $y_{i,j} = \overline{x_l}$ is also 14, as can be seen in Fig. 3(b). Observe that if the vertical outgoing edge of p_l gets length three, the height of the drawing increases due to z. For both cases $y_{i,j} = x_l$ and $y_{i,j} = \overline{x_l}$ the drawing of $LT(y_{i,j})$ has minimal width $8 + 3(n + i + 1)$ due to the paths $f \rightsquigarrow f_{n+i+1}$ and $g \rightsquigarrow g_{n+i+1}$.

Clause tree. For a clause $F_i = (y_{i,1} \lor y_{i,2} \lor y_{i,3})$, we define a 7–ary *clause tree* $CT(F_i)$ with root r_i. Each clause tree $CT(F_i)$ has the paths $r_i \rightsquigarrow c_i$, $c_i \rightsquigarrow e_i$, and $r_i \rightsquigarrow d_i$, see Fig. 4. $CT(F_i)$ contains three literal trees $LT(y_{i,1})$, $LT(y_{i,2})$ and $LT(y_{i,3})$ with roots $b_{i,1}$, $b_{i,2}$ and $b_{i,3}$, appended at the corresponding vertices $a_{i,1}$, $a_{i,2}$ and $a_{i,3}$. For an example in Fig. 4, the edge $(a_{1,3}, b_{1,3})$ connects the first clause tree F_1 with its third literal tree $LT(y_{1,3})$.

Minimal height and width of a clause tree: Let the incoming edge of r_i face to the east. As explained above, all important edge directions are fixed in the drawing then. Each clause tree $CT(F_i)$ has a vertical path $c_i \rightsquigarrow e_i$, which determines the minimal height 49, see Fig. 4. If each literal tree in a clause tree has at least height 15, its height is at least $2 + 1 + 15 + 1 + 15 + 1 + 15 = 50$. If at least one literal tree is drawn with height 14 and the remaining two with height 15, the clause tree can be drawn with height 49. The topmost horizontal path $r_i \rightsquigarrow c_i$ of full vertices determines the minimal width $14 + 3(n + i + 1)$.

Tree $T(E)$. For a boolean expression $E = F_1 \land \ldots \land F_r$ we define a 7–ary tree $T(E)$ with clause trees $CT(F_1), \ldots, CT(F_r)$. The root r_0 of $T(E)$ is the parent of the root r_1 of the first clause tree. Each clause tree $CT(F_i)$ is connected with $CT(F_{i+1})$ $(i \in \{1, \ldots, r-1\})$ by an edge (c_i, r_{i+1}). At the last clause tree $CT(F_r)$ we add a leaf c_{r+1} to the center vertex c_r directing to the east.

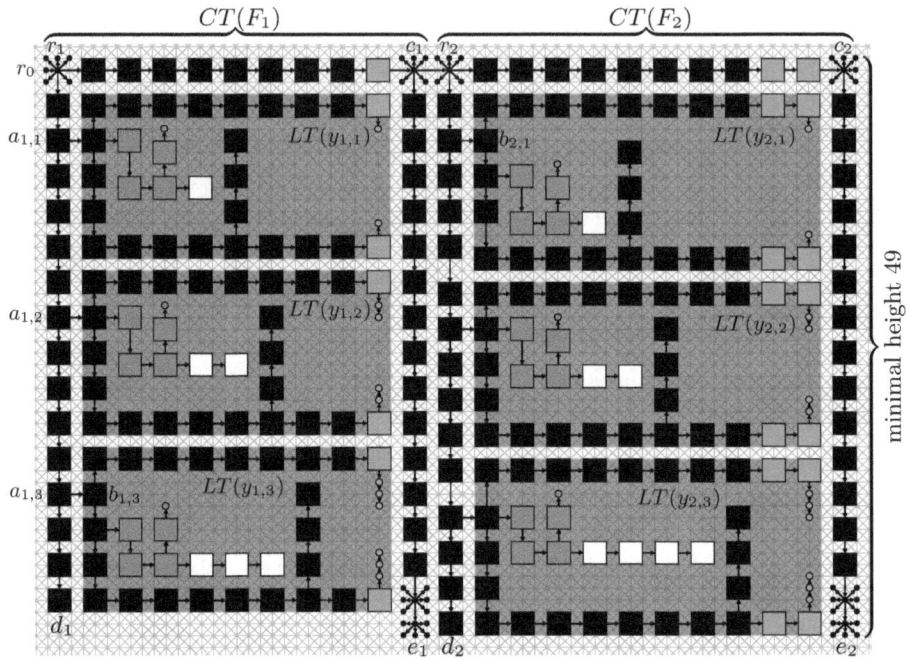

Fig. 4. 7–ary tree $T(E)$ on the octa grid of boolean expression $E = (\overline{x_1} \vee \overline{x_2} \vee x_3) \wedge (x_1 \vee \overline{x_2} \vee \overline{x_4})$ with assignment $\tau(x_1) = \tau(x_2) = false$ and $\tau(x_3) = \tau(x_4) = true$

Minimal height and width of the tree $T(E)$: Let the edge (r_0, r_1) face to the east. The construction generates a horizontal straight path $r_0 \rightsquigarrow c_{r+1}$ at the top ensuring minimal width of $\sum_{i=1}^{r}(14 + 3(n + i + 1)) + (r - 1) = 1.5r^2 + (3n + 19.5)r - 1$. The height of $\Gamma(T(E))$ is the maximal height of the clause trees because the clause trees are placed side by side with aligned top.

Lemma 1. *Let E be a boolean expression in 3–CNF with r clauses and n variables. E is satisfiable if and only if there exists a drawing $\Gamma(T(E))$ of the 7–ary tree $T(E)$ on the octa grid with area at most $A = 49 \cdot W$ with $W = 1.5r^2 + (3n + 19.5)r - 1$.*

Proof. Let E be an arbitrary boolean expression in 3–CNF with n variables $x_1, \ldots, x_n$ and r clauses $F_1, \ldots, F_r$. We construct the 7–ary tree $T(E)$ as described above.

'$\Rightarrow$': Let $\tau : \{x_1, \ldots, x_n\} \to \{true, false\}$ be a satisfying assignment for the boolean expression E. We construct a tree drawing $\Gamma(T(E))$ on the octa grid with area at most A in the following way. If for a variable x_l the assignment $\tau(x_l) = true$ is given, the variable tree $VT(x_l)$ is drawn as in Fig. 3(a) such that all edges are drawn with minimal length. Otherwise for $\tau(x_l) = false$, we draw all edges with minimal length, except the vertical outgoing edge of p_l, which is drawn with length four (see drawing of the variable tree in Fig. 3(b)).

For every satisfied literal in E we construct a drawing with height 14, else 15. Let $y_{i,j}$ be the j–th literal in the i–th clause of E. If $y_{i,j} = x_l$ and $x_l = true$, we draw the literal tree $LT(y_{i,j})$ as shown in Fig. 3(a), where each edge has minimal length and, thus, the literal tree has minimal height 14. If $y_{i,j} = x_l$ and $x_l = false$, we draw the outgoing edge of p_l in $VT(x_l)$ with length four, such that a vertical edge of the skeleton of $LT(y_{i,j})$ has to be drawn with length four. Therefore, the height of the literal tree is 15. If $y_{i,j} = \overline{x_l}$ and $x_l = true$, in the corresponding variable tree $VT(x_l)$ all edges are drawn with minimal length. In this case the drawing of the literal tree needs height 15, due to z. The last of the four cases is $y_{i,j} = \overline{x_l}$ and $x_l = false$, where we draw the literal tree $LT(y_{i,j})$ as in Fig. 3(b) with height 14.

As one of the three literals of each F_i is satisfied, the corresponding literal tree is drawn with height 14 and the other two literal trees have height at most 15, such that the height of the drawing of $CT(F_i)$ is 49. Hence, the described construction produces drawings of clause trees with height 49 and width $14 + 3(n + i + 1)$. The resulting drawing of $T(E)$ is constructed within a rectangle with height at most 49 and width at most W and, hence, area $49 \cdot W$.

'$\Leftarrow$': Let $\Gamma'(T(E))$ be a drawing with area at most $A = 49 \cdot W$. Suppose, the outgoing egde of r_0 of the drawing of $T(E)$ faces to the east determining all essential edge directions. Due to the vertical and horizontal paths in $\Gamma'(T(E))$, the drawing has at least height 49 and at least width W. Thus, the height is exactly 49 and the width exactly W. As a consequence each edge of the path $r_1 \rightsquigarrow c_r$ has minimal length three. For each F_i this determines the horizontal distance between the vertical paths $r_i \rightsquigarrow d_i$ and $c_i \rightsquigarrow e_i$ of $CT(F_i)$. Thus, the width of this clause tree is $14 + 3(n + i + 1)$.

In the following we construct a satisfying assignment $\tau'(E)$ from this drawing $\Gamma'(T(E))$. As explained above in each clause tree $CT(F_i)$ at least one literal tree $LT(y_{i,j})$ has to have height 14. We assign its variable x_l the boolean value satisfying the literal $y_{i,j}$. This is well defined because it is not possible that the literal trees of a non–negated and a negated literal with the same variable x_l both have height 14 due to the subtree isomorphism property.

Consequently if a literal tree $LT(y_{i,j})$ with $y_{i,j} = x_l$ is drawn with height 14, we assign $\tau'(x_l) = true$. For a drawing of a literal tree $LT(y_{i,j})$ with $x_{i,j} = \overline{x_l}$ and height 14, the outgoing edge of p_l has at least length four. We assign $\tau'(x_l) = false$ constructing a satisfiying assignment of the corresponding literal. Since each clause tree $CT(F_i)$ contains a literal tree with height 14, the assignment τ' satisfies each clause F_i. Thus, E is satisfied. If there are remaining variables x_l with no assignment yet, we assign an arbitrary boolean value.   □

For an example, see Fig. 4, where the drawing of the 7–ary tree on the octa grid within height 49 and width $W = 1.5 \cdot 2^2 + (3 \cdot 4 + 19.5) \cdot 2 - 1 = 68$ is given for a boolean expression with two clauses ($r = 2$) and four variables ($n = 4$). The variable $x_1 = false$ appears in the literal $y_{1,1}$ of the first clause F_1 and in $y_{2,1}$ in F_2. The corresponding literal tree of the satisfied literal $y_{1,1} = \overline{x_1}$ has height 14. $LT(y_{2,1})$ has height 15 corresponding to the not satisfied literal $y_{2,1}$. All literals in F_1 are satisfied and all literal trees in $CT(F_1)$ have height 14. Nevertheless, the minimal height of 49 of $CT(F_1)$ is ensured by the vertical path $c_1 \rightsquigarrow e_1$ of box vertices. F_2 has exactly one satisfied literal $y_{2,2}$ and therefore one literal tree $LT(y_{2,2})$ with height 14.

We have shown that for $k = 8$ the problem of determining the existence of a tree drawing on the octa grid on given area is $\mathcal{NP}$–hard. To show the $\mathcal{NP}$–hardness for $k = 6$ and $k = 4$, the construction of the tree $T(E)$ and the values of the heights and widths have to be adjusted. The idea of the constructions remains mainly the same. This completes the proof of Theorem 1.

3.2   *OL*–Drawings

Theorem 2. *Let T be a rooted ternary tree and $A \geq 1$. Determining the existence of an OL_4–drawing $\Gamma_4(T)$ with area at most A is $\mathcal{NP}$–hard.*

Proof. (Sketch). The construction for OL_4–drawings of ternary trees on the orthogonal grid is similar to the construction of the O_4–drawings from above. The local–uniformity property enforces inserting interspaces between vertical paths of the tree leading to other dimensions.   □

For OL_6– and OL_8–drawings this approach does not work because the children of two adjacent locally uniform full vertices would overlap.

3.3   *OP*– and *OLP*–Drawings

Theorem 3. *Let $k \in \{4, 6, 8\}$, T be a rooted $(k-1)$–ary tree, and $A \geq 1$. Determining the existence of an OP_k– or an OLP_k–drawing $\Gamma_k(T)$ with area at most A is $\mathcal{NP}$–hard.*

Proof. The construction for OP_k–drawings for $k \in \{4, 6, 8\}$ is identical to the O_k cases as these drawings already are pattern drawings.

To prove the cases OLP_k for $k \in \{6, 8\}$ we modify the reduction of Sect. 3.1 slightly. The edges connecting full vertices are splitted by one vertex. The outgoing edge of this vertex can have arbitrary length without contradicting the aesthetic local uniformity since it has no siblings. In contrast to Sect. 3.1 and 3.2 the directions of the edges are completely defined by the patterns (see Fig. 1). Thus, for each new vertex the outgoing edge has the same direction as the incoming edge. As a consequence of the construction, the resulting drawing has the same dimensions and the proof is done analogously.

For OLP_4–drawings the proof is identical to the proof of Theorem 2, because an order–preserving local uniform pattern drawing is already given there.   □

4   Heuristic

We present a heuristic in Algorithm 1 which produces drawings of $(k-1)$–ary trees on the k–grid. It is an improvement of our algorithm presented in [3] where enclosing hexagons for distance calculations on the hexa grid were used. Our new algorithm generally creates smaller drawings using more precise contours follow-ing the Reingold and Tilford algorithm [16].

To describe the heuristic we need some definitions. $\mathcal{D}_k$ is a set of *direction vectors* dependent on the current k–grid having $2k$ direction vectors. Let $\mathcal{D}_4$ be the set $\{(0,\pm1),(\pm1,\pm1),(\pm1,0)\}$, $\mathcal{D}_6 = \mathcal{D}_4 \cup \{\pm(\frac{1}{2},1),\pm(1,\frac{1}{2})\}$, and $\mathcal{D}_8 = \mathcal{D}_6 \cup \{\pm(\frac{1}{2},-1),\pm(1,-\frac{1}{2})\}$. For the vector $(d_1,d_2) = D \in \mathcal{D}_k$ we define the *orthogonal vector* $D_o = (-d_2,d_1)$. A *direction contour* $C_{v,D}$ with $D \in \mathcal{D}_k$ is a (non–strictly) monotonic increasing polyline with respect to D_o of $\Gamma(T_v)$. It is the set of vertices and segments of edges visible along the direction vector $D \in \mathcal{D}_k$. If two neighboring segments/points do not have the same end-point, they are connected by a segment parallel to D. For an example see the direction contour $C_{v_4,D_{3,4}}$ in Fig. 5(a). The direction contour $C_{l,D}$ of a leaf l trivially consists of the coordinates of l itself for all $D \in \mathcal{D}_k$. We call the set of all direction contours $C_{v,D}$ with $D \in \mathcal{D}_k$ the *contour* C_v of the subtree T_v.

We use a preprocessing step creating a planar drawing on the k–grid. The outgoing edges of each vertex $v \in V$ get the initial length $3^{height(T)-depth(v)-1}$ as in [3], which guarantees planarity for all k–grids. The recursive Algorithm 1 is called with the root v of T_v as input. The output of each recursion step is the contour of C_v of T_v and the updated edge lengths in the drawing of T_v. If v is a leaf, its trivial contour is returned in line 1. The loop in line 1 realizes the recursive call of $getContour(v_i)$ computing the contours C_{v_i} of the subtrees T_{v_i} for each child v_i. The outgoing edges of v are contracted in line 1, see details in Sect. 4.1. Finally, the contour C_v of T_v is created by merging the contours of the subtrees of the children of v in line 1, see details in Sect. 4.2.

4.1 Contract Edges

To determine the maximal contraction value of the outgoing edges of v, we calculate the distances of all pairs of subtrees and the distances of subtrees to incident edges of v. Let v_i and v_j be children of v with incoming edges e_i and e_j, respectively. First, we calculate the minimal distance between the subtrees T_{v_i} and T_{v_j} using the contours C_{v_i} and C_{v_j}. The difference of the unit vectors $D = \vec{e_i}/\|\vec{e_i}\| - \vec{e_j}/\|\vec{e_j}\|$ determines the necessary direction contours $C_{v_i,D}$ and $C_{v_j,-D}$ facing each other. We calculate the minimal distance $minDist$ between these two direction contours using a scanline moving along D_o. Each bend of

---

Algorithm 1. getContour

Input: A vertex v of a planar drawing of a $(k-1)$–ary tree on a k–grid
Output: Contour C_v of v and the contracted edge lengths of T_v

1 **if** v *is a leaf* **then** return trivial contour of v
2 $\mathcal{C} \leftarrow \emptyset$
3 **foreach** *child* v_i *of* v **do** $\mathcal{C} \leftarrow \mathcal{C} \cup \{getContour(v_i)\}$
4 $contractEdges(\mathcal{C},v)$ // *contract outgoing edges of* v
5 $C_v \leftarrow mergeContours(\mathcal{C},v)$
6 **return** C_v

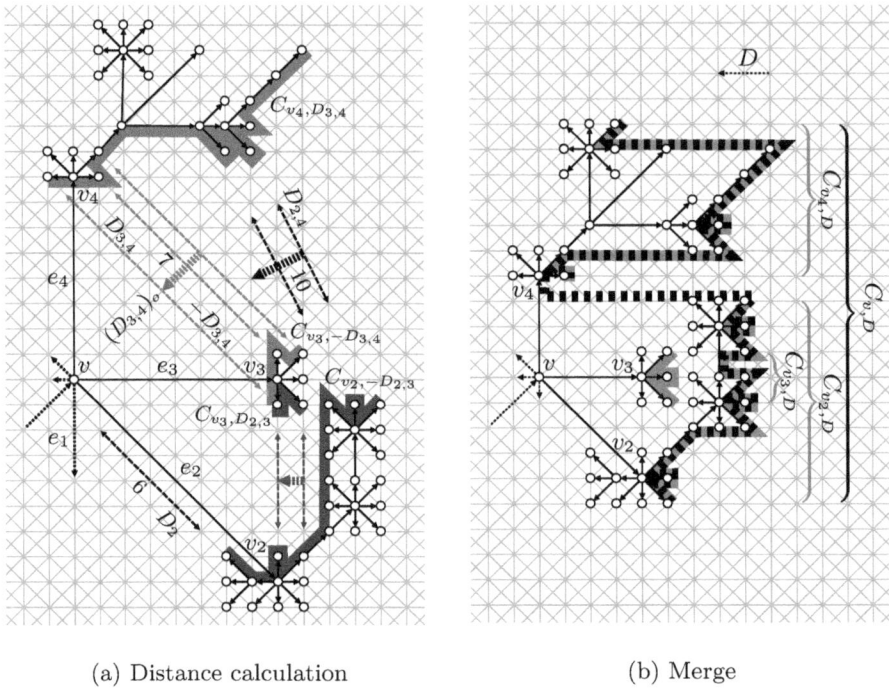

(a) Distance calculation   (b) Merge

Fig. 5. Contraction step of a drawing of a 7–ary tree on the octa grid

both direction contours is visited once to determine the distance to the other
direction contour along the direction D. The complexity of this step is linear in
the number of bends of the direction contours.

Proposition 1. *The number of bends in a direction contour $C_{v,D}$ of a subtree
T_v and direction $D \in \mathcal{D}_k$ is linear in the number of vertices in the subtree T_v.*

The maximal possible contraction value of the two subtrees T_{v_i} and T_{v_j} is determined considering the behavior of the two vertices v_i and v_j while contracting
the edges e_i and e_j simultaneously by one. If the distance between the two vertices is thereby reduced by two, e.g., T_{v_2} and T_{v_4} in Fig. 5(a), we could shorten
e_i and e_j by $\lfloor (minDist - 1)/2 \rfloor$, otherwise the distance is reduced by one and
we could shorten these edges by $minDist - 1$, e.g., T_{v_3} and T_{v_4}. This maximal contraction value is calculated for all pairs of subtrees of the children of
v. For an example see Fig. 5(a) where the distance between the two contours
$C_{v_3,-D_{3,4}}$ and $C_{v_4,D_{3,4}}$ is 7 and the edges e_3 and e_4 could be shortened by 6
without creating an intersection of the two subtrees. To calculate the distance
between the subtrees T_{v_2} and T_{v_4} the direction contours $C_{v_2,-D_{2,4}}$ and $C_{v_4,D_{2,4}}$
are used. The distance between these direction contours is 10. Since the sub-
trees move towards each other, the maximal contraction value considering these
two subtrees is $\lfloor \frac{(10-1)}{2} \rfloor = 4$. To prevent the subtrees of the children of v from

overlapping with its incident edges, we have to determine their distances as well. In this case the distance $minDist$ is calculated between all incident edges of v and each direction contour C_{v_i,D_i} with direction $D_i = \overrightarrow{e_i}/\|\overrightarrow{e_i}\| \in \mathcal{D}_k$. The edge e_i could be shortened by $minDist - 1$. In Fig. 5(a) the distance between C_{v_2,D_2} and e_1 is 6. Thus, e_2 could be shortened by 5. The final contraction value is the minimum of all contraction values and we shorten the outgoing edges of v by this value. In Fig. 5(a) the outgoing edges can be shortened by 4, see Fig. 5(b).

4.2   Merge Contours

After contracting the edges the contour C_v has to be calculated for each direction $D \in \mathcal{D}_k$. Let $v_1, \ldots, v_i$ with $i \in \{1, \ldots, k-1\}$ be the children of v. Again, we use a scanline procedure along D_o. For the new contour $C_{v,D}$ we use the union of all "visible" parts of $C_{v_1,D}, \ldots, C_{v_i,D}$ and of the outgoing edges of v, as can be seen in Fig. 5(b). There, the merge of the contours $C_{v_2,D}$, $C_{v_3,D}$, $C_{v_4,D}$, the edges (v, v_2), (v, v_3), and (v, v_4) is shown. The merge step runs in linear time due to Proposition 1. As can be seen in Fig. 5(b), a part of the edge (v, v_4) becomes part of the dashed direction contour $C_{v,D}$. The direction contour of $C_{v_3,D}$ is not needed anymore because it is no longer visible along direction D.

4.3   Time Complexity

The calculation of the order–preserving locally uniform pattern drawing with Algorithm 1 has time complexity of $\mathcal{O}(n^2)$. In every recursion step of Algorithm 1 the running time is in $\mathcal{O}(n)$ based on the scan line procedures of the contraction step (Sect. 4.1) and the merge step (Sect. 4.2). The recursive method $getContour(v)$ is called for each $v \in V$ exactly once.

Using techniques similar to those used in the Reingold and Tilford algorithm [16] and a more sophisticated merge step, the time complexity of the algorithm can be improved to $\mathcal{O}(n)$. Such an implementation is realized in *Gravisto* [4].

5   Conclusion

We have shown the $\mathcal{NP}$–hardness for several problems of drawing trees on k–grids within a given area for O_{k}–, OP_{k}–, OL_4– and OLP_k–drawings. Furthermore we introduced a heuristic producing OLP_k–drawings of ordered $(k-1)$–ary trees on a k–grid guaranteeing the isomorphic subtree property. For the calculation we use contours similar to *threads* of the algorithm of Reingold and Tilford [16].

For unordered trees we conjecture that the Bhatt and Cosmadakis technique [5] or the logic engine [10] can be used to prove the $\mathcal{NP}$–hardness of drawing unordered trees on the octa grid within a given area or with unit edge length.

References

1. Akkerman, T., Buchheim, C., Jünger, M., Teske, D.: On the complexity of drawing trees nicely: Corrigendum. Acta Inf. 40(8), 603–607 (2004)
2. Aziza, S., Biedl, T.C.: Hexagonal grid drawings: Algorithms and lower bounds. In: Graph Drawing, pp. 18–24 (2004)

3. Bachmaier, C., Brandenburg, F.J., Brunner, W., Hofmeier, A., Matzeder, M., Unfried, T.: Tree drawings on the hexagonal grid. In: Tollis, I.G., Patrignani, M. (eds.) GD 2008. LNCS, vol. 5417, pp. 372–383. Springer, Heidelberg (2009)
4. Bachmaier, C., Brandenburg, F.J., Forster, M., Holleis, P., Raitner, M.: Gravisto: Graph visualization toolkit. In: Graph Drawing, pp. 502–503 (2004)
5. Bhatt, S.N., Cosmadakis, S.S.: The complexity of minimizing wire lengths in VLSI layouts. Inf. Process. Lett. 25(4), 263–267 (1987)
6. Chan, T.M., Goodrich, M.T., Kosaraju, S.R., Tamassia, R.: Optimizing area and aspect ratio in straight-line orthogonal tree drawings. Comput. Geom. Theory Appl. 23(2), 153–162 (2002)
7. Crescenzi, P., Di Battista, G., Piperno, A.: A note on optimal area algorithms for upward drawings of binary trees. Comput. Geom. Theory Appl. 2, 187–200 (1992)
8. Eades, P.: Drawing free trees. Bulletin of the Institute of Combinatorics and its Applications 5, 10–36 (1992)
9. Eades, P., Lin, T., Lin, X.: Minimum size h-v drawings. In: Advanced Visual Interfaces, pp. 386–394 (1992)
10. Eades, P., Whitesides, S.: The logic engine and the realization problem for nearest neighbor graphs. Theor. Comput. Sci. 169(1), 23–37 (1996)
11. Frati, F.: Straight-line orthogonal drawings of binary and ternary trees. In: Hong, S.-H., Nishizeki, T., Quan, W. (eds.) GD 2007. LNCS, vol. 4875, pp. 76–87. Springer, Heidelberg (2008)
12. Garg, A., Rusu, A.: Straight-line drawings of binary trees with linear area and arbitrary aspect ratio. J. Graph Algo. App. 8(2), 135–160 (2004)
13. Kant, G.: Hexagonal grid drawings. In: Mulkers, A. (ed.) Live Data Structures in Logic Programs. LNCS, vol. 675, pp. 263–276. Springer, Heidelberg (1993)
14. Kaufmann, M., Wagner, D. (eds.): Drawing Graphs. LNCS, vol. 2025. Springer, Heidelberg (2001)
15. Marriott, K., Stuckey, P.J.: NP-completeness of minimal width unordered tree layout. J. Graph Algorithms Appl. 8(2), 295–312 (2004)
16. Reingold, E.M., Tilford, J.S.: Tidier drawing of trees. IEEE Trans. Software Eng. 7(2), 223–228 (1981)
17. Shin, C.S., Kim, S.K., Chwa, K.Y.: Area-efficient algorithms for straight-line tree drawings. Comput. Geom. Theory Appl. 15(4), 175–202 (2000)
18. Supowit, K.J., Reingold, E.M.: The complexity of drawing trees nicely. Acta Inf. 18, 377–392 (1982)
19. Tamassia, R.: On embedding a graph in the grid with the minimum number of bends. SIAM J. Comput. 16(3), 421–444 (1987)
20. Walker, J.Q.W.: A node-positioning algorithm for general trees. Softw. Pract. Exper. 20(7), 685–705 (1990)

Optimizing Regular Edge Labelings*

Kevin Buchin[1], Bettina Speckmann[1], and Sander Verdonschot[2]

[1] Dep. of Mathematics and Computer Science, TU Eindhoven, The Netherlands
k.a.buchin@tue.nl, speckman@win.tue.nl
[2] School of Computer Science, Carleton University, Canada
sverdons@connect.carleton.ca

Abstract. A regular edge labeling (REL) of an irreducible triangulation G uniquely defines a rectangular dual of G. Rectangular duals find applications in various areas: as floor plans of electronic chips, in architectural designs, as rectangular cartograms, or as treemaps. An irreducible triangulation can have many RELs and hence many rectangular duals. Depending on the specific application different duals might be desirable. In this paper we consider optimization problems on RELs and show how to find optimal or near-optimal RELs for various quality criteria. Furthermore, we give upper and lower bounds on the number of RELs.

1 Introduction

A *rectangular partition* of a rectangle R is a partition of R into a set $\mathcal{R}$ of non-overlapping rectangles such that no four rectangles in $\mathcal{R}$ meet at one common point. A *rectangular dual* of a plane graph G is a rectangular partition $\mathcal{R}$, such that (i) there is a one-to-one correspondence between the rectangles in $\mathcal{R}$ and the nodes in G; (ii) two rectangles in $\mathcal{R}$ share a common boundary if and only if the corresponding nodes in G are connected. Rectangular duals find applications in various areas: as floor plans of electronic chips or in architectural designs, as rectangular cartograms, or as treemaps.

Not every plane graph has a rectangular dual. A plane graph G has a rectangular dual $\mathcal{R}$ with four rectangles on the boundary of $\mathcal{R}$ if G is an *irreducible triangulation*: (i) G is triangulated and the exterior face is a quadrangle; (ii) G has no separating triangles (a 3-cycle with vertices both inside and outside the cycle) [6,19]. A plane triangulated graph G has a rectangular dual if and only if we can augment G with four external vertices such that the augmented graph is an irreducible triangulation.

The equivalence classes of the rectangular duals of an irreducible triangulation G correspond one-to-one to the *regular edge labelings* (RELs) of G. An REL of an irreducible triangulation G is a partition of the interior edges of G into two subsets of red and blue directed edges such that: (i) around each inner vertex in clockwise order we have four contiguous sets of incoming blue edges, outgoing red

---

* K. Buchin and B. Speckmann are supported by the Netherlands Organisation for Scientific Research (NWO) under project no. 639.022.707.

U. Brandes and S. Cornelsen (Eds.): GD 2010, LNCS 6502, pp. 117–128, 2011.

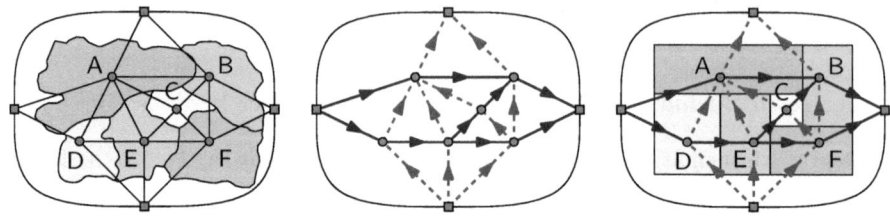

Fig. 1. A subdivision and its augmented dual graph G, a regular edge labeling of G, and a corresponding rectangular dual

edges, outgoing blue edges, and incoming red edges; (*ii*) the left exterior vertex has only blue outgoing edges, the top exterior vertex has only red incoming edges, the right exterior vertex has only blue incoming edges, and the bottom exterior vertex has only red outgoing edges (see Fig. 1, red edges are dashed). Kant and He [17] show how to find a regular edge labeling and construct the corresponding rectangular dual in linear time. Regular edge labelings are also studied by Fusy [15] who calls them *transversal pairs of bipolar orientations*.

An irreducible triangulation can have many RELs and hence many rectangular duals. Depending on the specific application different duals might be desirable. For example, *sliceable* duals—which can be obtained by recursively slicing a rectangle by horizontal and vertical lines—are popular in VLSI design. Not every irreducible triangulation has a sliceable dual. A full characterization of those graphs that do is lacking, but Yeap and Sarrafzadeh [24] prove that irreducible triangulations without separating 4-cycles have a sliceable dual. *Area-universal* duals have the nice property that any assignment of areas to rectangles can be realized by a combinatorially equivalent rectangular dual. Again, not every irreducible triangulation has an area-universal dual, but Eppstein *et al.* [12] show how to find such a dual if it exists.

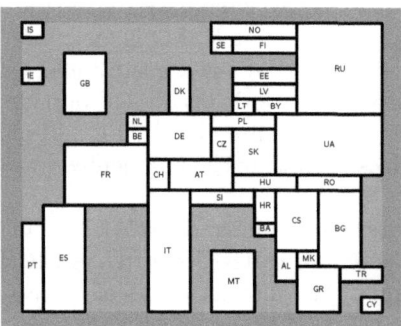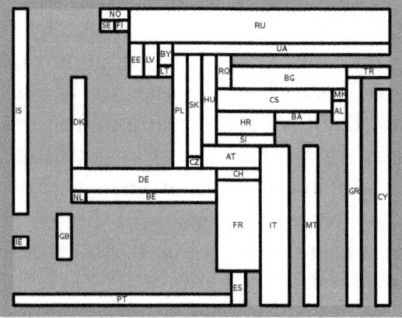

Fig. 2. Two different rectangular duals of the dual graph of a map of Europe. Luxembourg and Moldavia have been removed and "sea regions" have been added to ensure that the dual graph is an irreducible triangulation.

We are particularly interested in the application of rectangular duals to rectangular cartograms. A rectangular cartogram is a thematic map where every region is depicted as a rectangle. The area of the rectangles corresponds to a geographic variable, such as population or GDP. In the context of rectangular cartograms it is desirable that the direction of adjacency between the rectangles of the dual follows the spatial relation of the regions of the underlying map as closely as possible. Consider the two rectangular duals of the dual graph of a map of Europe shown in Fig. 2. The left dual will lead to a recognizable cartogram, whereas the right dual (with France east of Germany and Hungary north of Austria) is useless as basis for a cartogram. Both rectangular duals stem from the same graph G and correspond to two different valid RELs of G.

Previous work on finding RELs that lead to cartograms with geographically suitable adjacency directions has focused on finding RELs that satisfy *user-specified* constraints on a subset of the edges of the input graph. Eppstein and Mumford [11] show how to find RELs that satisfy user-specified orientation constraints, if such labelings exist for the given set of constraints. Van Kreveld and Speckmann [23] search through a user-specified subset of the RELs. Every labeling in this subset is considered acceptable with respect to adjacency directions. In contrast, we consider quality measures that take all edges of G into account and do not concentrate on a fixed, user-specified subset.

Results and organization. In this paper we consider optimization problems on RELs and show how to find optimal or near-optimal RELs for various quality criteria. Furthermore, we give upper and lower bounds on the number of RELs.

Let G be an irreducible triangulation with n vertices. Fusy [15] proves that the RELs of G form a distributive lattice. Hence, one can use reverse search to enumerate all RELs of G and so find optimal RELs for any given quality measure (see Section 2). G can have exponentially many RELs; simple upper and lower bounds are 8^n and $2^{n-O(\sqrt{n})}$. Since the running time and hence the feasibility of any enumeration algorithm depends on the number of RELs, we next give much tighter bounds. In Section 3 we show that G has less than $O(4.6807^n)$ RELs and that there are irreducible triangulations with $\Omega(3.0426^n)$ RELs. Our upper bound relies on Shearer's entropy lemma [10]. Björklund *et al.* [7] recently used this lemma to obtain $(2 - \varepsilon)^n$ algorithms for the TSP problem. In contrast to our application of the lemma, they count vertex sets with certain properties and crucially rely on bounded maximum degree.

In Section 4 we show how to find optimal or near-optimal RELs for rectangular cartogram construction. This step of the construction pipeline has been performed essentially manually in previous work and can now finally be fully automated. We consider two quality criteria: (i) the relative position of the rectangles and (ii) the *cartographic error* of the resulting cartogram. For smaller maps enumeration of all RELs is feasible and we can find optimal solutions. For larger maps enumeration is infeasible. But the diameter of the distributive lattice of RELs is comparatively small and hence simulated annealing performs well. We present experimental results that show that our method can find RELs which result in visually pleasing cartograms with small cartographic error.

2   Useful Facts and Reverse Search

In this section we first collect some useful facts and definitions from previous work and then show how to use reverse search to enumerate RELs.

Regular Edge Labelings. Fusy [15] gives the following facts. A *regular edge coloring* is an REL, with the directions of the edges omitted. A regular edge coloring uniquely determines an REL. An *alternating 4-cycle* is an undirected 4-cycle in which the colors of the edges alternate between red and blue. There are two kinds of alternating 4-cycles, depending on the color of the interior edges incident to the cycle. If these are the same color as the next clockwise cycle edge the cycle is *right alternating*, otherwise it is *left alternating*. The set of RELs of a fixed irreducible triangulation

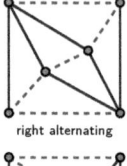

right alternating

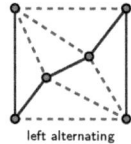

left alternating

form a distributive lattice. The flip operation consists of switching the edge colors inside a right alternating 4-cycle, turning it into a left alternating 4-cycle. An REL with no right alternating 4-cycle is called *minimal*; it is at the bottom of the distributive lattice. An REL induces no monochromatic triangles.

Perron-Frobenius theory. We need the following matrix theory for our lower bound in Section 3.2. For details refer to textbooks on matrices [16,20]. We assume from now on that A is a nonnegative $n \times n$ matrix. A matrix is *nonnegative* if all its elements are nonnegative. The matrix A is *irreducible* if for each (i,j) there is a $k > 0$ such that $(A^k)_{ij} > 0$. Consider the directed graph with adjacency matrix A, where we interpret every non-zero element as an adjacency. The matrix A is irreducible if and only if the associated graph is strongly connected. The matrix A is *primitive* if there is a $k > 0$ such that all elements of A^k are positive. An irreducible matrix with a positive diagonal entry is primitive.

Theorem 1 ([16,20]). *Let A be a primitive non-negative matrix with maximal eigenvalue λ.*

(a) *λ is positive and the unique eigenvalue of largest absolute value. λ has a positive eigenvector and is the only eigenvalue with nonnegative eigenvector.*

(b) *Let $f_A(x) = \min_{x_i \neq 0} \frac{(Ax)_i}{x_i}$ and $g_A(x) = \max_{x_i \neq 0} \frac{(Ax)_i}{x_i}$. Then $f_A(x) \leq \lambda$ for all nonnegative non-zero vectors x, and $g_A(x) \geq \lambda$ for all positive vectors x. If $f_A(x_0) = \lambda$ then x_0 is an eigenvector of A corresponding to λ.*

(c) *Let x be a nonnegative non-zero vector. Then $A^t x / \|A^t x\|$ converges to an eigenvector with eigenvalue λ. Thus, $\lim_{t \to \infty} f_A(A^t x) = \lim_{t \to \infty} g_A(A^t x) = \lambda$.*

Reverse search, proposed by Avis and Fukuda [5], is a general method for enumerating structures that match two criteria: (i) there must be a concept of "neighboring" structures such that the structures form a graph; (ii) there must be a local search operation that moves through this graph in a deterministic way and ends up at a local optimum. The local search defines a forest on the graph, of which each tree is rooted at a local optimum. If the local optima are known and we have a way of enumerating all neighbors of a structure, then we can traverse these trees by starting at a local optimum and testing for each neighbor if the local search ends up at our current structure when applied to that neighbor. If it does, we traverse the edge and recurse.

RELs fit the criteria for reverse search: the distributive lattice is the underlying graph structure and the flip operation is the local search that ends up at the minimal labeling. We need to ensure only that the local search is deterministic. We do so by imposing an ordering on the 4-cycles of the input graph. Then, if an REL has multiple right alternating 4-cycles, we choose the first one according to this ordering. One possible ordering is to sort the vertices lexicographically by their x- and y-coordinates, use this ordering to sort the edges lexicographically by their lower and higher endpoint, and finally use this order on the edges to sort the cycles lexicographically by their lowest edge and the non-adjacent edge.

Avis and Fukuda give an implementation of their algorithm if the graph is given by an *adjacency oracle*. For this we need an upper bound δ on the number of neighbors a labeling can have: δ is the number of 4-cycles in the input graph. The oracle returns the k-th neighbor of a labeling, or $\perp$ if that neighbor does not exist. Using our ordering on the 4-cycles of the input graph, we let the oracle return the resulting labeling after flipping the colors of all edges inside the k-th 4-cycle, or $\perp$ if this 4-cycle is not alternating.

The enumeration takes $O(\delta t(\text{oracle})\Lambda + t(\text{local search})\delta\Lambda)$ time, where Λ is the number of RELs. We have $\delta = O(n^2)$, since a 4-cycle is defined by two edges. The oracle takes linear time, as it might have to switch the color of linearly many edges, and the local search takes quadratic time, as it might have to evaluate all 4-cycles to find the first right-alternating one, for a total of $O(n^4\Lambda)$.

Note that RELs can be represented as lower sets of a directed acyclic graph with polynomial size (see [11,12]). Using this representation one can enumerate RELs more efficiently per REL than with reverse search.

3   Counting Regular Edge Labelings

Here we prove that every irreducible triangulation with n vertices has less than $O(4.6807^n)$ RELs and that there are irreducible triangulations with $\Omega(3.0426^n)$ RELs. Before we present our bounds, we review some additional related work.

Counting all RELs of all n-vertex irreducible triangulations yields the number of combinatorially different rectangular partitions with n rectangles which is in $\Omega(11.56^n)$ [4] and less or equal to 13.5^{n-1} [14]. If we consider partitions to be identical when the incidence structure between rectangles and maximal line segments is the same, then the number of different partitions is in $\Theta(8^n/n^4)$ [2]. RELs are related to *bipolar orientations*—orientations of the edges from a source to a sink—but there is no direct relation between their numbers. Felsner and Zickfeld [13] show that the number of bipolar orientations of a planar graph is in $O(3.97^n)$ and that there are planar graphs with $\Omega(2.91^n)$ bipolar orientations. Many other interesting substructures have been counted in planar graphs (see [3,8,9]), but the upper bounds we obtained by adapting the techniques used for these structures to RELs were far from the bounds that we present.

3.1 Upper Bound

Let $G = (V, E)$ be an irreducible triangulation on n vertices. Since an REL is uniquely determined by a regular edge coloring and G has less than $3n$ edges, we obtain a simple upper bound of 8^n on the number of RELs of G. In the following we refine this bound using Shearer's entropy lemma.

Lemma 1 (Shearer's entropy lemma [10]). *Let S be a finite set and let $A_1, \ldots, A_m$ be subsets of S such that every element of S is contained in at least k of the $A_1, \ldots, A_m$. Let $\mathcal{F}$ be a collection of subsets of S and let $\mathcal{F}_i = \{F \cap A_i : F \in \mathcal{F}\}$ for $1 \leq i \leq m$. Then we have $|\mathcal{F}|^k \leq \prod_{i=1}^{m} |\mathcal{F}_i|$.*

Theorem 2. *The number of regular edge labelings of an irreducible triangulation is in $O(4.6807^n)$.*

Proof (sketch). Let $G = (V, E)$ be an irreducible triangulation on n vertices. Let S be E with the four edges on the exterior face excluded. For a REL $\mathcal{L}$ of G let $E(\mathcal{L})$ be the set of blue edges in $\mathcal{L}$. Let $\mathcal{F} := \{E(\mathcal{L}) \mid \mathcal{L} \text{ is a REL of } G\}$. Since $E(\mathcal{L})$ determines $\mathcal{L}$, the number of RELs is $|\mathcal{F}|$.

For the vertices v_i of G, $1 \leq i \leq n$, let A_i be the set of edges in S of the triangles adjacent to v_i. Every edge $e \in S$ is in four of the sets A_i, namely in the four sets corresponding to the vertices of the two triangles with e as edge. Let F_i be the set of intersections of the set A_i with the sets $E(\mathcal{L})$, i.e., F_i contains all possible ways to choose blue edges around v_i consistent with a REL. By Lemma 1 the number of RELs is bounded by $\prod_{i=1}^{n} |F_i|^{1/4}$.

It is easy to see that $|F_i| \leq 2^5 \binom{d_i}{4}$, where d_i is the degree of v_i. Therefore, the number of RELs of G is bounded by $\left(32^n \prod_{i=1}^{n} \binom{d_i}{4}\right)^{1/4}$, which by convexity (and using a bound of 6 on the average degree) is upper-bounded by $\left(32^n \binom{6}{4}^n\right)^{1/4} = 480^{n/4} < 4.6807^n$. □

3.2 Lower Bound

Our lower bound construction for the number of RELs uses triangulated grids. We refer to the number of rows of a triangulated grid as its *height* h and to the number of columns as its *width* w. We add four vertices to the outside of the grid to turn it into an irreducible triangulation. The total number of vertices of the augmented grid is $n = hw + 4$.

A simple lower bound stems from the following coloring: color all horizontal edges blue and all vertical edges red. Then all diagonals can be colored independently blue or red. This gives a lower bound of $2^{n - O(\sqrt{n})}$ for $h = w$. We prove a stronger bound by coloring only the edges of every h'th row blue (for some choice of h') and by not coloring the edges of columns. We color the parts between the blue rows independently. We assume for now that $h = h' + 1$. Larger values of h do not change the analysis, but do improve the lower bound.

We first describe all steps for $h' = 1$, i.e., the edges of all rows are blue. Then we generalize the method for larger values of h'. We color the triangulated grid from left to right. The edges of the first and last column must be colored red, since a REL has no monochromatic triangles. Assume we have colored the triangulated grid up to the ith column. We call the edges of the ith column and the diagonals connecting to this column from the left the ith *extended column*. How we can color the $(i+1)$st extended column depends on the colors of the ith extended column (assuming we have no restriction from the right).

If $h' = 1$, the previous column can be either red or blue. The color of the previous diagonal does not influence our choices for this column. If the previous column is red, we can make this column red too and choose either color for the diagonal. We can also make this column blue, but then the diagonal needs to be red to 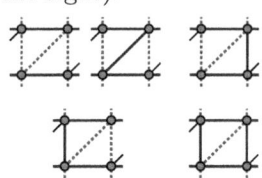 satisfy the constraints around the top vertex of this column. Likewise, if the previous column was blue, our diagonal needs to be red to satisfy the constraints around the bottom vertex of the previous column (see figure).

We represent these coloring options as a transition matrix M, using the column colors as state. With M we can compute the number of colorings up to the ith extended column, by starting in the red state and repeatedly multiplying it with M. The resulting vector gives us the number of colorings ending

From \ To	R	B
R	2	1
B	1	1

in a red or a blue edge. Since M has only positive elements, it is primitive. By Theorem 1(c) the ratio between the number of labelings ending in a red column up to the ith extended column and up to the $(i+1)$th extended column converges towards the largest eigenvalue of M. This eigenvalue is $\phi + 1 > 2.61803$. So for any $\varepsilon > 0$, we obtain more than $(\phi + 1 - \varepsilon)^w$ labelings for sufficiently large w. Since we add two vertices to add a single column, this yields a lower bound of $(\phi + 1 - \varepsilon)^{(n-4)/2}$ for sufficiently large w. If we now increase h, we need to add h vertices to add $h-1$ columns, i.e., we get a lower bound of $(\phi + 1 - \varepsilon)^{(n-4)(h-1)/h}$, which for sufficiently large h and w is larger than 2.61803^n.

Next we consider $h' > 1$. We need to extend the states with information about the vertices, specifically how many color-switches there should be in the next extended column. This information, together with the color of the bottom column edge incident to this vertex, fixes the color of the top edge. So all we need for the state is the color of the bottom edge of the column and the color switches for each vertex, moving upwards. Some states that can be described in this way cannot be part of an REL. We call such states *infeasible*. A state is feasible if it can be reached from the initial all-red state (i.e., all vertical edges of the column are red) and if the all-red state can be reached from it. Thus a state is feasible if and only if it is in the strongly connected component of the all-red state. We remove all infeasible states and consider only the reduced matrix.

The reduced matrix is primitive: it is irreducible by construction, and there is always at least one transition from the all-red state to the all-red state (by coloring all diagonals and horizontal edges between the two columns blue). When constructing a regular edge coloring, we start with the all-red column and color the columns one by one. By Theorem 1(c) the number of regular edge colorings increases with each new column by a factor that converges to the largest eigenvalue of the transition matrix. Therefore, a strict lower bound on this eigenvalue $\lambda_{h'}$ of this matrix gives us a strict lower bound for the growth rate per column (ignoring a constant number of initial columns).

We obtain a strict lower bound on $\lambda_{h'}$ in the following way. We take a non-negative non-zero state vector x, multiply it with the transition matrix, and determine the minimum growth rate (for the non-zero elements). If the vector is not an eigenvector of A (i.e., the growth rate is not the same for all non-zero states) then the minimum growth rate is a strict lower bound on $\lambda_{h'}$ by Theorem 1(b). As vector x we choose $x_0 A^{100}$, where A is the reduced transition matrix and x_0 is the vector with a 1 for the all-red state and 0 otherwise. Since in all the cases that we consider the vector x is positive, we also obtain an upper bound on $\lambda_{h'}$ by the maximum growth rate.

We now again use several copies of h' rows beneath each other to obtain a triangulated grid with $w = h$. The growth rate per vertex in this way approaches $\lambda_{h'}^{1/h'}$. As strict lower bound on $\lambda_{h'}^{1/h'}$ we obtain 2.61803, 2.80921, 2.90453, 2.96067, 2.99746, 3.0233, and 3.04263 for $h' = 1, 2, 3, 4, 5, 6, 7$. Note that the lower bounds are rounded down, and that our upper bounds on $\lambda_{h'}^{1/h'}$ equal the (unrounded) lower bounds up to at least 10 significant digits.

Theorem 3. *The number of regular edge labelings of the triangulated grid is in* $\Omega(3.04263^n)$.

4   Optimizing RELs for Rectangular Cartograms

In this section, we describe how to find good RELs for rectangular cartogram construction and present experimental results. We follow the iterative linear programming method presented in [22] to build a cartogram from an REL.

Quality criteria. We consider two quality criteria: (*i*) the relative position of the rectangles and (*ii*) the cartographic error of the resulting cartogram. Furthermore, we bound the aspect ratio of all rectangles by 20.

To make a rectangular cartogram as recognizable as possible, it is important that the directions of adjacency between the rectangles of the cartogram follow the spatial relation of the regions of the underlying map. An REL specifies the relative directions between adjacent rectangles. We use two quality measures to quantify how well an REL matches the spatial relations between regions in the input map. The first method is based on region centroids [21]. It considers the direction between the centroids of two regions as the "true" direction of adjacency and expresses the quality of a labeling in terms of the deviation from this direction, measured as the smallest angle between the two directions (see Fig. 3 left). The centroid measure tends to perform quite well, although it can lead to counter-intuitive results in some cases. Our second measure is based on the bounding boxes of the regions. The *bounding box separation distance* (bb sep dist) measures the distance these bounding boxes would need to be moved to separate them in the direction indicated by the edge label (see Fig. 3 right). We

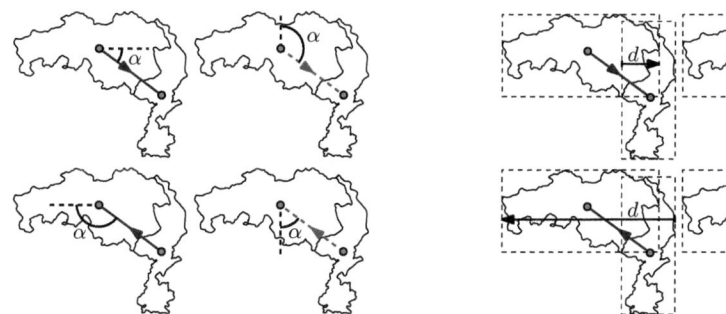

Fig. 3. The angle deviation and bounding box separation distance measures

consider both the average and the maximum error over all edges of a labeling, as well as a *binary* version of the measures: determine the "correct" color and direction for each edge and count the number of edges that are labeled correctly.

Another important quality criterion for cartograms is the *cartographic error* which is defined as $|A_c - A_s|/A_s$, where A_c is the area of the region in the cartogram and A_s is the specified area of that region, given by the geographic variable to be shown. As before, we consider both the maximum and average cartographic error over all regions of the cartogram.

We strive to construct cartograms of low cartographic error and high recognizability, hence we consider various ways to combine the two quality criteria. One possibility is to take weighted averages, another to bound the maximum cartographic error at 5%, while minimizing the maximum angle deviation or bounding box separation distance, which we call a *bounded* measure.

Enumeration. The augmented dual graph of the provinces of the Netherlands has only 408 RELs which can be enumerated in less than a second. Nevertheless the map is large enough to show interesting trends. Fig. 4 shows cartograms produced by enumerating all labelings and taking the best one according to various quality measures. The first data set shows total population on January 1st 2009, the second total livestock in 2009. Both were obtained from the Centraal Bureau voor de Statistiek. The color of a region corresponds to its cartographic error, with red indicating that the region is too small and blue indicating that it is too big. The saturation corresponds to the magnitude of the error, a white region has a cartographic error of at most 5%, while a fully saturated region has a cartographic error of over 30%. The figure clearly shows that combining recognizability measures with cartographic error leads to the best results.

Simulated Annealing. For larger maps—the countries of Europe or the contiguous states of the US—enumeration is infeasible (both have over four billion labelings). Fortunately the diameter of the lattice of RELs is comparatively small (115 for Europe and 278 for the US) and hence simulated annealing [18] performs well. We use a typical static cooling schedule and the original acceptance probability [1] for our experiments. Specifically, given two labelings with qualities q_1 and q_2, the probability that our algorithm moves to the worse labeling is

Fig. 4. Population and livestock cartograms of the provinces of the Netherlands

$e^{|q_1 - q_2|/T}$, where T is the current temperature. We let the temperature decrease exponentially as $T = 0.002^t$, where t is the current time, varying from 0 initially to 1 at the end of the process. The base factor of 0.002 can be increased to produce more random behaviour, or decreased to produce more greedy behavior.

Fig. 5 shows some results of our implementation[1]. Note that we produce only cartograms with correct adjacencies. The top two figures show the total population of the countries of Europe on January 1st 2008, with the populations of Luxembourg and Moldova added to Belgium and Ukraine, respectively. The left cartogram was generated by bounding the maximum cartographic error on 5% and optimizing the average angle deviation, which results in an average

---

[1] Data from Eurostat http://epp.eurostat.ec.europa.eu/portal/page/portal/ eurostat/home, the CIA World Factbook https://www.cia.gov/library/ publications/the-world-factbook/index.html, and the the US Census Bureau http://www.census.gov/

Fig. 5. Top row: two population cartograms of Europe. Bottom row: highway lengths of Europe and population of the US. All with correct adjacencies.

cartographic error of 0.004 and a maximum cartographic error of 0.023. The right cartogram was generated by solely optimizing the average cartographic error. Note that most of the relative positions are suitable while the cartogram still obtains a maximum cartographic error of 0.000.

The two bottom cartograms are produced with the same maps and data sets as the cartograms by Speckmann *et al.* [22] which were based on user-specified RELs. Both cartograms were generated by optimizing a weighted average of the maximum cartographic error with weight 0.7 and the average bb sep distance with weight 0.3. The left cartogram shows the total highway length in Europe. It has an average cartographic error of 0.001 and a maximum error of 0.003. This is a significant improvement over the results of Speckmann *et al.* who achieved 0.022 average and 0.166 maximum cartographic error. The last cartogram shows the total population of the US. It has an average cartographic error of 0.031 and a maximum error of only 0.140. Again a significant improvement over the results of Speckmann *et al.* of 0.086 average and 0.873 maximum cartographic error. We can conclude that our fully automated method to find optimal RELs for cartogram construction performs significantly better than semi-manual methods.

References

1. Aarts, E., Korst, J., Michiels, W.: Simulated annealing. In: Search Methodologies, pp. 187–210. Springer, Heidelberg (2005)
2. Ackerman, E., Barequet, G., Pinter, R.Y.: A bijection between permutations and floorplans, and its applications. Disc. Appl. Math. 154(12), 1674–1684 (2006)
3. Aichholzer, O., Hackl, T., Vogtenhuber, B., Huemer, C., Hurtado, F., Krasser, H.: On the number of plane graphs. In: Proc. 17th SODA, pp. 504–513 (2006)
4. Amano, K., Nakano, S., Yamanaka, K.: On the number of rectangular drawings: Exact counting and lower and upper bounds. TR 2007-AL-115, IPSJ SIG (2007)
5. Avis, D., Fukuda, K.: Reverse search for enumeration. Disc. Appl. Math. 65(1), 21–46 (1996)
6. Bhasker, J., Sahni, S.: A linear algorithm to check for the existence of a rectangular dual of a planar triangulated graph. Networks 7, 307–317 (1987)
7. Björklund, A., Husfeldt, T., Kaski, P., Koivisto, M.: The travelling salesman problem in bounded degree graphs. In: Bugliesi, M., Preneel, B., Sassone, V., Wegener, I. (eds.) ICALP 2006. LNCS, vol. 4051, pp. 198–209. Springer, Heidelberg (2006)
8. Buchin, K., Knauer, C., Kriegel, K., Schulz, A., Seidel, R.: On the number of cycles in planar graphs. In: Lin, G. (ed.) COCOON 2007. LNCS, vol. 4598, pp. 97–107. Springer, Heidelberg (2007)
9. Buchin, K., Schulz, A.: On the number of spanning trees a planar graph can have. In: Proc. 18th ESA (to appear, 2010), arXiv/0912.0712
10. Chung, F.R.K., Graham, R.L., Frankl, P., Shearer, J.B.: Some intersection theorems for ordered sets and graphs. J. Comb. Theory, Ser. A 43(1), 23–37 (1986)
11. Eppstein, D., Mumford, E.: Orientation-constrained rectangular layouts. In: Dehne, F., Gavrilova, M., Sack, J.-R., Tóth, C.D. (eds.) WADS 2009. LNCS, vol. 5664, pp. 49–60. Springer, Heidelberg (2009)
12. Eppstein, D., Mumford, E., Speckmann, B., Verbeek, K.: Area-universal rectangular layouts. In: Proc. 25th ACM Symp. Comp. Geom., pp. 267–276 (2009)
13. Felsner, S., Zickfeld, F.: On the number of planar orientations with prescribed degrees. Electron. J. Comb. 15(1), Research paper R77, 41 (2008)
14. Fujimaki, R., Inoue, Y., Takahashi, T.: An asymptotic estimate of the numbers of rectangular drawings or floorplans. In: Proc. ISCAS, pp. 856–859 (2009)
15. Fusy, É.: Transversal structures on triangulations: A combinatorial study and straight-line drawings. Disc. Math. 309(7), 1870–1894 (2009)
16. Horn, R., Johnson, C.R.: Matrix Analysis. Cambridge University Press, Cambridge (1985)
17. Kant, G., He, X.: Regular edge labeling of 4-connected plane graphs and its applications in graph drawing problems. TCS 172(1-2), 175-193 (1997)
18. Kirkpatrick, S.: Optimization by simulated annealing: Quantitative studies. J. Statistical Physics 34(5), 975–986 (1984)
19. Koźmiński, K., Kinnen, E.: Rectangular dual of planar graphs. Networks 5, 145–157 (1985)
20. Minc, H.: Nonnegative Matrices. Wiley-Interscience, Hoboken (1988)
21. Peuquet, D., Ci-Xiang, Z.: An algorithm to determine the directional relationship between arbitrarily-shaped polygons in the plane. Pattern Rec. 20(1), 65–74 (1987)
22. Speckmann, B., van Kreveld, M., Florisson, S.: A linear programming approach to rectangular cartograms. In: Progress in Spatial Data Handling: Proc. 12th International Symposium on Spatial Data Handling, pp. 529–546 (2006)
23. van Kreveld, M., Speckmann, B.: On rectangular cartograms. Computational Geometry: Theory and Applications 37(3), 175–187 (2007)
24. Yeap, G.K.H., Sarrafzadeh, M.: Sliceable floorplanning by graph dualization. SIAM J. Discrete Mathematics 8(2), 258–280 (1995)

Drawing Graphs in the Plane with a Prescribed Outer Face and Polynomial Area

Erin W. Chambers[1], David Eppstein[2],
Michael T. Goodrich[2], and Maarten Löffler[2]

[1] Dept. of Math and Computer Science, Saint Louis Univ., USA
echambe5@slu.edu
[2] Computer Science Dept., University of California, Irvine, USA
{eppstein,goodrich,loffler}@ics.uci.edu

Abstract. We study the classic graph drawing problem of drawing a planar graph using straight-line edges with a prescribed convex polygon as the outer face. Unlike previous algorithms for this problem, which may produce drawings with exponential area, our method produces drawings with polynomial area. In addition, we allow for collinear points on the boundary, provided such vertices do not create overlapping edges. Thus, we solve an open problem of Duncan *et al.*, which, when combined with their work, implies that we can produce a planar straight-line drawing of a combinatorially-embedded genus-g graph with the graph's canonical polygonal schema drawn as a convex polygonal external face.

1   Introduction

The study of planar graphs has been a driving force for graph theory, graph algorithms, and graph drawing. Our interest in this paper is in drawing planar graphs without edge crossings using straight line segments for edges, in such a way that all faces are convex polygons and the outer face is a given shape. Figure 1 shows an example.

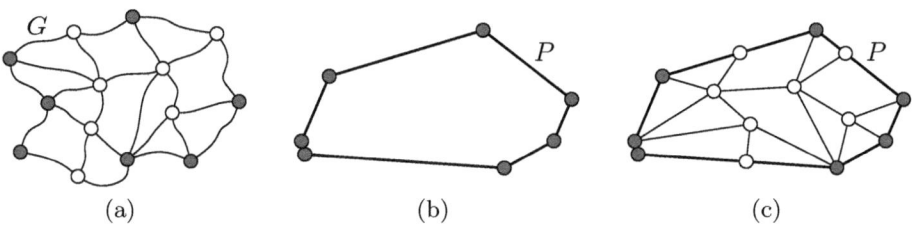

(a)   (b)   (c)

Fig. 1. Our problem: given (a) a combinatorially embedded planar graph G and (b) a polygon P with certain vertices on the outer face of G marked as corresponding to vertices of P, find (c) a straight-line embedding of G that uses P as the shape of its outer face

U. Brandes and S. Cornelsen (Eds.): GD 2010, LNCS 6502, pp. 129–140, 2011.

Related Prior Work. In seminal work that has been highly influential in graph drawing, Tutte [19,20] shows that one can draw any planar graph using non-crossing straight-line edges so that the outer face is drawn as a prescribed convex polygon. This work has influenced a host of subsequent papers, and, according to Google Scholar, Tutte's 1963 paper has been directly cited over 600 times. His work has influenced many methods for drawing maximal planar graphs [10,17,21] and drawing planar graphs using straight-line edges (e.g., see [3,4,5,11,7,14,1,16]). Moreover, not only has Tutte's result itself been highly influential, but because his method is based on a force-directed layout method, it has also influenced a considerable amount of work on force-directed layouts (e.g., see [6,8,12,13,18]).

Unfortunately, one drawback of Tutte's algorithm is that it can result in drawings with exponential area. This area blowup is not inherent in planar straight-line drawing, but known polynomial-area straight-line drawing algorithms (e.g., [4,5,11,14,16]) lose a critical feature of Tutte's drawing algorithm, in that none of them allow the vertices of a planar graph's outer face to be placed on a prescribed convex polygon. Becker and Hotz [2], on the other hand, show how to draw a planar graph with minimum weighted edge length and prescribed outer face but, like Tutte's method, their method may produce drawings with exponential area. Duncan *et al.* [9] pose as an open problem whether an algorithm can produce polynomial-area straight-line drawings with vertices on a given convex polygon.

One motivation for prescribing the outer face of a planar drawing comes from a common way of drawing planar representations of genus-g graphs. Namely, if a graph G is embedded into a genus-g topological surface, the surface may be cut along the edges and vertices of $2g$ fundamental cycles in G to form a topological disk (known as a *canonical polygonal schema*), with a boundary that is made up of $4g$ paths (with multiple copies of the vertices on the fundamental cycles). Moreover, as shown by Duncan *et al.* [9], G can be cut in this way so that each of these $4g$ paths is chord-free, that is, so that there are no non-path edges between two vertices strictly internal to the same path. The standard way of drawing this unfolded version of such an embedding, in the topology literature (e.g., see [15]), is to draw the disk as a convex polygon with each of its $4g$ boundary paths drawn as a straight line segment: the geometric shape is used to make clear the pattern in which the surface was cut to form a disk. Fortunately, given Tutte's seminal result, it is possible to draw any chord-free canonical polygonal schema in this way. The drawback of using Tutte's algorithm for this purpose is that the resulting drawing may have exponential area. Thus, we are interested in drawing the unfolded embedding in polynomial area and in polynomial time.

Our Results. In this paper, we describe an algorithm for drawing a planar graph with a prescribed outer face shape. The input consists of an embedded planar graph G, a partition of the outer face of the embedding into a set $\mathcal{S}$ of k chord-free paths, and a k-sided polygon P; the output of our algorithm is a drawing of G within P with each path in $\mathcal{S}$ drawn along an edge of P. Given the above-mentioned prior result of Duncan *et al.* [9], for finding chord-free canonical polygonal schemas, our result implies that we can solve their open problem: any graph G combinatorially embedded in a genus-g surface has a polynomial-area straight-line planar drawing of a canonical polygonal schema S for G, drawn as a $4g$-sided convex polygon P with the vertices of each path in S drawn along an edge of P.

2 Preliminaries

In this paper, we show how to draw a graph with a given boundary with coordinates of polynomial magnitude. Before treating the main construction, though, we show in this section that we can equivalently state the problem in terms of the *resolution* of the graph. Furthermore, we recall some known results and concepts.

Resolution. Instead of drawing a graph with integer coordinates of small total size, we will make a drawing with real coordinates that stays within a fixed region (inside the input polygon) with a large *resolution*.

Let G be a graph that is embedded in $\mathbb{R}^2$ with straight line segments as edges. We define the *resolution* of G to be the shortest distance between either two vertices of the graph, or between a vertex and a non-incident edge. The *diameter* of G is the largest distance between two vertices of the graph.

We begin by establishing a relation between resolution and size, which basically says that drawing a graph G with small diameter and large resolution also results in another drawing with integer coordinates and small size. Generally it may not be possible to scale a given input polygon such that its coordinates become integers, so we need to do some rounding. We say that two drawings of G are *combinatorially equivalent* if their topology is the same, and any collinear adjacent edges in one drawing are also collinear in the other. We say two drawings are ε-*equivalent* if the distance between the locations of each vertex of G in the two drawings is at most ε. The following lemma is proved in the full version of our paper, which can be found online at arXiv:1009.0088.

Lemma 1. *Let G be a graph, and let Γ be a drawing of G without crossings, with constant resolution, and with diameter D. Then there exists another drawing Γ' of G with integer coordinates and diameter $O(D^2)$, such that a scaled copy of Γ' with diameter D is both combinatorially equivalent and $O(1)$-equivalent to Γ.*

Note that, for a fixed input polygon with non-integer vertex coordinates, this perturbation may slightly modify its shape, since it may not be possible to find a similar copy of the polygon with integer vertex coordinates.

Now, let Q be a set of points in the plane. We define the *potential resolution* of Q to be the resolution of the complete graph on Q. Similarly, for a polygon P, we define its potential resolution to be the potential resolution of its set of vertices. Clearly, the resolution of any drawing we can achieve will depend on the potential resolution of the input polygon, because the drawing could be forced to include any edge of the complete graph.

Next, we make an observation about the nature of the potential resolution of convex polygons.

Observation 1. *If P is a convex polygon, then the potential resolution of P is the minimum over the vertices of P of the distance between that vertex and the line through its two neighboring vertices.*

Thus, for a convex polygon P to allow for a drawing of polynomial area in its interior, we insist that P has a polynomially-bounded aspect ratio. It cannot be arbitrarily thin and still support a polynomial-area drawing in its interior.

Alpha Cuts. We now describe a useful property of the potential resolution of a convex polygon, namely that it can be "distributed" any way we want when cutting the polygon into smaller parts. This will be made more precise later. We first make another observation about convex polygons, which is proven in the full version.

Lemma 2. *Let P be a convex polygon, v a vertex of P, e an edge incident to v, and $\alpha \in (0,1)$ a number. Let P' be a copy of P where v has been replaced by v' by moving v along e over a fraction α of the length of e. Then the potential resolution of P' is at least $1 - \alpha$ times the potential resolution of P.*

Let P be a convex polygon. We will show that we can cut P into two smaller polygons, "distributing" its potential resolution in any way we want.

We define an α-cut of P to be a directed line ℓ that splits P into two smaller polygons, such that if an edge e of P is intersected by ℓ, the length of the piece of e to the left of ℓ is α times the length of e, and the piece of e to the right of ℓ is $(1 - \alpha)$ times the length of e. For a given convex polygon and two features of its boundary (either vertices or edges), there is a unique α-cut that cuts the polygon through those two features in order.

Lemma 3. *Suppose we are given a convex polygon P of resolution d, two features (either vertices or edges) of P, and a fraction $0 < \alpha < 1$. Let ℓ be the α-cut through the two given features that cuts P into a piece P_l to the left of ℓ and a piece P_r to the right of ℓ. Then the potential resolution of P_l is at least αd unless the two features are two adjacent edges that meet to the right of ℓ. Similarly, the potential resolution of P_r is at least $(1 - \alpha)d$ unless the two features are two adjacent edges that meet to the left of ℓ.*

Proof: We will argue about the potential resolution of P_l; the argument for P_r is symmetric. We prove this lemma by applying Lemma 2 to the new vertices of P_l. If both features where ℓ cuts through P are vertices, then all vertices of P_l are also vertices of P and clearly the potential resolution can only become better. However, if one or both of the features are edges, then P_l has one or two new vertices that are not part of P. Figure 2 shows three different cases that can occur. To solve this problem, we first alter P to a different polygon P' that has the new vertices. Let u' be the place where ℓ enters P and u the closest

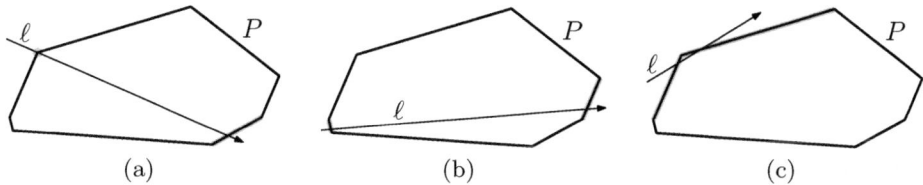

Fig. 2. (a) An α-cut through a vertex and an edge. (b) An α-cut through two non-adjacent edges. (c) An α-cut through two adjacent edges.

vertex below ℓ along the boundary to it (possibly $u = u'$), and similarly let v' be the place where ℓ exits P and v the closest vertex below ℓ. Now, we create P' by moving u to u' and v to v'. Clearly, both will move a fraction $1 - \alpha$ along their edges, so by Lemma 2 P' has a potential resolution of at most α times the potential resolution of P. Therefore, the potential resolution of P_l can only be larger.

The only exception is when $u = v$; in this case we cannot move the vertex to two new places simultaneously, but we have to create two new vertices. Indeed, the result is not true in that case, since the two new vertices can be arbitrarily close to each other as α comes arbitrarily close to 1, so the resolution of P' cannot be expressed in terms of α, as can be seen in Figure 2c. $\qquad\square$

Combinatorial Embeddings. Let $G = (V, E)$ be a plane graph. That is, we consider the combinatorial structure of G's embedding to be fixed, but we are free to move its vertices and edges around. Let F be the set of faces of G, excluding the outer face. We make some definitions about faces. We say that a subset $F' \subset F$ induces a subgraph $G\langle F'\rangle$ of G that consists of all vertices and edges that are incident to the faces in F'. A subset $F' \subset F$ is said to be *vertex-connected* if $G\langle F'\rangle$ is connected; it is said to be *edge-connected* if the dual graph induced by the dual vertices of F' is connected. In other words, faces that share an edge are both edge-connected and vertex-connected, but faces that share only a vertex are only vertex-connected.

We recall a lemma from [9], rephrased in terms of the faces of the graph:

Lemma 4. *Given an embedded plane graph G that is fully triangulated except for the external face and two edges e_1 and e_2 on that external face, it is possible to partition the faces of G into three sets $F_1, F_2, R \subset F$ such that:*

1. *All vertices of G are in either $G\langle F_1\rangle$ or $G\langle F_2\rangle$.*
2. *R is edge-connected and contains the faces incident to e_1 and e_2.*
3. *F_1 and F_2 are both vertex-connected.*
4. *The edge-connected components of F_1 and F_2 all share an edge with the outer face of G.*

Intuitively, R is a path of faces that goes from e_1 to e_2 and that splits the remaining faces into two sets F_1 and F_2.

3   Drawing a Graph with a Given Boundary

We are now ready to formally state the problem and describe the algorithm to solve it.

The Problem. Let G be a triangulated planar graph with a given combinatorial embedding, and let B be the cycle that bounds the outer face of G. Let f be a map from a subset of the vertices of B to points in the plane, such that these points are in convex position and their order along their convex hull is the same as their order along B.

We say that a map g from all vertices of G to points in the plane *respects* f when:

1. The vertices mapped by f are also mapped by g to the same points; these define a convex polygon P.
2. The remaining vertices of B are mapped to the corresponding edges of P.
3. The remaining vertices of G are mapped to the interior of P.
4. If all edges are drawn as straight line segments, they cause no crossings or incidences not present in G.

An example of a respectful embedding was shown in Figure 1.

The input to our problem is a pair (G, f). We will use the notations B and P as above. We will further define F to be the set of faces of G, excluding the outer face. We define $s = |F|$, the number of internal faces, to be the *size* of the problem. We define d to be the *resolution* of the problem, which is the potential resolution of P (recall, that is the resolution of the complete graph on the corners of P). Our goal is to compute a mapping g that respects f and such that the resolution of the embedded graph is bounded by some function of s and d.

Observe that it will not be possible to do so when there are any edges in G between two vertices that have to be on the same edge of P. Therefore, we call a problem *invalid* if this is the case. We will show that for any valid problem, we can find an embedding with a polynomially bounded resolution.

The Main Idea. We want to solve the problem using divide-and-conquer. The idea is to divide P into smaller convex polygons, and F into smaller sets of faces, and map each subset of faces to one of the smaller regions. Then we need to decide which vertices of G are mapped to the new corners of the smaller regions, and solve the subproblems.

A first idea would be to find a path in G between two vertices of B, and lay that out on a straight line, resulting in a split of P into two smaller polygons and solve the two subproblems. There are two issues with this approach though. First, the vertices on the new straight line have to be placed the same way in the two subproblems, which means they are not independent. Second, if there are any chords on this path one of the subproblems will become invalid.

To avoid these issues, we will not split along a single path, but along two paths next to each other. The region between these two paths, which we call a

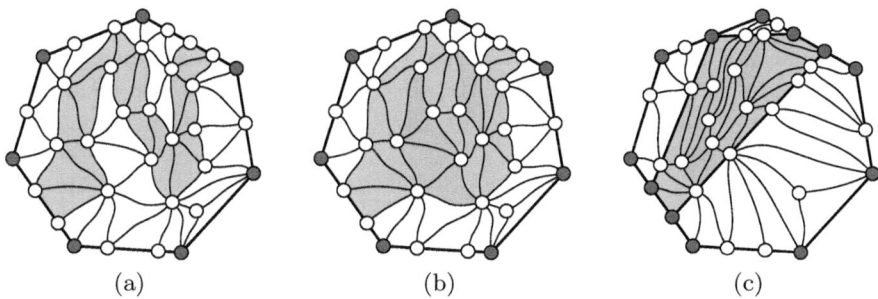

Fig. 3. (a) A "river" (a path in the dual graph that does not reuse any vertices of the primal graph) between two edges on B. (b) The river banks have chords, and so we include the area behind the chords in the river. (c) We fix the vertices on the river boundary that are on B, and draw the rest of the river boundary straight. This results in three smaller problems, plus the area of the river itself.

river, has a controlled structure, which means that we can always complete the interior independently of how the vertices on the edges were placed. Furthermore, if these paths have any chords, we shortcut them along the chords and show how to deal with the added complexity of the river. Because the river may touch the boundary of P in more places, the problem may be decomposed into more than two subproblems. Figure 3(a) shows an example instance, and Figure 3(c) shows a possible decomposition where some vertices on the boundary of P have been fixed, and the paths between them are made straight.

We assume the input is a valid problem with size s and resolution d. We will keep as an invariant the ratio d/s, and show in the next paragraph how to subdivide a problem into smaller valid problems with the same (or better) ratio, plus an extra region (the river). We then recursively solve the independent subproblems, which results in a placement of all vertices that are not in the interior of the river. Finally, we show in the paragraph after that how to we place the vertices inside the river.

Splitting a Problem. Let (G, P) be a valid problem of size s and resolution d, and suppose that P has at least four sides.

Let e_1 and e_2 be two edges of B that lie on two sides of P that are not consecutive. Note that the endpoints of e_1 and e_2 are not necessarily fixed yet. Now, by Lemma 4, there exists a path of faces that connects e_1 to e_2, such that the boundary of this path does not have any repeated vertices. Let R be the union of the faces of F on this path. We call R a *river*; Figure 3(a) shows an example. This river may touch B in other points than e_1 and e_2, so it can subdivide the faces of F into any number of edge-connected subsets (apart from the river itself). We will assign a separate subproblem to each such edge-connected component.

We would like to straighten the banks of the river, but this may lead to invalid subproblems if these banks have any chords. Therefore, we identify any chords that the river has (note that they can only appear on the outside of the river, since the river forms a dual path), and we add the faces of F behind those chords to R. Similarly, if one of the paths touches a side of P more than once, it would create a subproblem that would be flattened. To avoid that, we also incorporate such a region into the river (even though the straight side that lies alongside P is not necessarily a chord).

Next, we count the numbers of faces in the river, as well as those in the parts outside the river. Then we fix the vertices where the river touches P by cutting off the subproblems, using α-cuts where α is the fraction of faces inside the subproblem. Now, by Lemma 3, if we have a problem with parameters s and d, we will construct subproblems with the same (or better) ratio d/s. Finally we straighten the new banks of the river, so that the subproblems have proper convex boundaries. Figure 3 shows an example.

Lemma 5. *Given a valid problem* (F, P) *where* P *has at least four sides, We can subdivide* F *and* P *into disjoint sequences* $F_1, F_2, \ldots, F_h$ *and* $P_1, P_2, \ldots, P_h$ *such that each* $(G\langle F_i \rangle, P_i)$ *is a valid subproblem with ratio* d/s, *and such that the remainders* $F' = F \setminus \bigcup F_i$ *and* $P' = P \setminus \bigcup P_i$ *have the following properties:*

1. F' *and* P' *also have ratio* d/s.
2. *The vertices of* $G\langle F' \rangle$ *that are not vertices of* $G\langle \bigcup F_i \rangle$ *form internally 3-connected components that share at least two vertices of* $G\langle \bigcup F_i \rangle$.

Proof: For the first part of the lemma, we need to show that the subproblems (F_i, P_i) are valid and have a resolution/size ratio at least as good as d/s. First, we define the polygons P_i by applying Lemma 3 to P with $\alpha = s_i/s$ (that is, the fraction of faces in F that is in F_i). The lemma ensures that the new polygons have potential resolution at least $d_i \geq \alpha d = s_i d/s$, so clearly $d_i/s_i \geq d/s$ as required. Second, recall that a subproblem is valid if it does not have any chords between two vertices that have to be drawn on the same side of P_i. For those sides of P_i that are part of P, we already know there are no chords because (G, P) was valid. For the new sides, we explicitly added all faces behind chords to the river $R = F'$.

For the second part of the lemma, we need to show that R has the right ratio and that the internal vertices of the river form 3-connected components that share two or more vertices with the boundary of the river. If we denote $s_R = |R|$ to be the size of the river and d_R to be the potential resolution of the region P' in which it is to be drawn, then by Lemma 3, the potential resolution of R after repeatedly slicing off subpolygons is at least $d_R = d\Pi(1 - s_i/s) \geq d(1 - \Sigma(s_i/s)) = s_R d/s$, so $d_R/s_R \geq d/s$. Finally, since we chose R according to Lemma 4, all edge-connected pieces of F outside R share an edge with the outer face. In particular, this means that the boundary of R does not touch itself and that any subgraphs sliced off by chords are 3-connected and share exactly two vertices with the boundary of R. Furthermore, if the boundary of R touches the same side of P multiple times, then the edge-connected components of F between that side of P and R are also 3-connected and share at least two vertices with the new boundary of R.   □

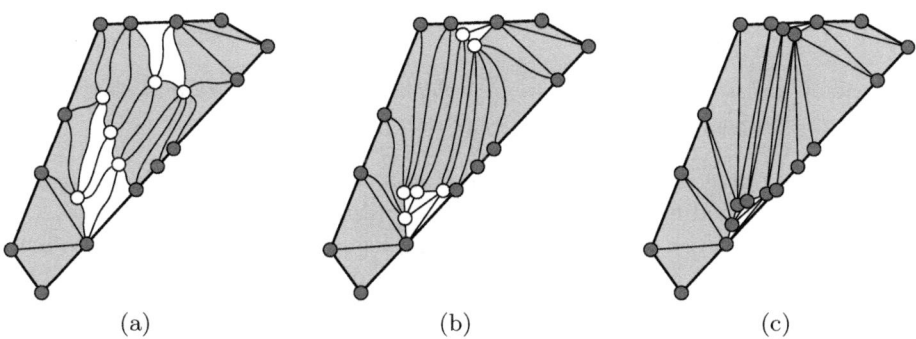

    (a)   (b)   (c)

Fig. 4. (a) The interior of a river, after all vertices on its boundary have already been fixed by recursive calls to the split algorithm. (b) Because of the structure given by the river, we can identify small areas inside the river that we draw using the de Fraysseix-Pach-Pollack algorithm. (c) If we flatten the triangular drawings enough, every vertex is able to see every other one (in fact we need slightly more, namely that no visibility ray comes too close to any vertex). Note that in the figure the drawing is not flat enough for that, but otherwise the structure would become too hard to see. In fact, in this particular case no flattening at all would be required.

When P has only 3 sides, we cannot choose two edges e_1 and e_2 on non-adjacent sides of P. However, we can still use the same basic idea; we just have to be careful because of the special case in Lemma 3. So, let c be a corner of P and let e_1 and e_2 be edges of B on the sides incident to c. The lemma does not give a bound on the resolution of the region on the far side of c. So, let e_1 and e_2 be the edges furthest away from c. Since P is a triangle, the two vertices of B on the far side on e_1 and e_2 are the other two corners of P, and they are joined by a side of P. This means that the region between the river and this side will be included into the river, and there will be no subproblem on the far side of c.

Fixing a River. It remains to show how to place the interior vertices of a river after all vertices on its boundary have been placed recursively. Again, we are given a graph G and a polygon P that it has to be drawn in (P is the boundary of the river, and G is the part of the graph that has to be drawn in it), but there are two important differences with the initial problem: First, now we know that all vertices of B have already been fixed (not only those on the corners but also those on the boundary of P). Second, we know that the remaining vertices of G have a very specific structure, namely, they form internally 2-connected components that share at least two vertices with the vertices fixed along an edge. Furthermore, since all fixed vertices have been placed using the algorithm above, they will never be closer than d/n to each other. This means we can draw these components using the algorithm by de Fraysseix, Pach and Pollack [11], and rotate and scale them to fit inside P. We can then flatten them more such that all remaining edges (between fixed vertices on the boundary of P or vertices on the de Fraysseix-Pach-Pollack drawings) can be drawn with straight line segments. Figure 4 shows an example.

Lemma 6. *Given a river placement problem, of size s and resolution d, we can lay out the graph with a resolution of $\Omega(d/n^3)$.*

Proof: The polygon P that forms a river and the graph G to be drawn in it are formed by subdividing a bigger problem according to Lemma 5. Then, the vertices on the boundary of P are placed during recursive calls to smaller problems. All vertices have to be corners of the polygons of at least one such subproblem, and by our invariant all these problems have ratio d/s, so the distance between any two fixed vertices cannot be smaller than d/n. Furthermore, Lemma 5 tells us that the vertices of G can be grouped into a number of subsets $V_1, V_2, \ldots, V_h$ such that each V_i is internally biconnected, and there is a sequence of at least two vertices in V_i that is in B, so that have been fixed on the boundary of P. Note that there will be exactly two if such a component came from a chord on the river bank, but there can be more if it came from the river touching a side of P multiple times. Then we can place these subgraphs using the de Fraysseix-Pach-Pollack algorithm, starting from the vertices that are already fixed (at distance at least d/n) and adding the remaining $O(|V_i|)$ vertices one by one using 45° edges. This results in a drawing with resolution which can be roughly bounded by d/n^2. Then, it is sufficient to squeeze them by a factor of n to make sure that they do not block any potential edges, and a further factor 2 to make sure that the tips of the de Fraysseix-Pach-Pollack drawings are in fact far enough away from these potential edges, guaranteeing a good resolution. This means the final resolution of the drawing is $\Omega(d/n^3)$. □

Putting It Together. To conclude, Lemmas 5 and 6 together imply:

Theorem 1. *Given a plane graph G with n vertices, a convex polygon P with k corners and potential resolution d, and a map f that maps k vertices on the outer face of G to the k corners of P, we can draw a G in P respecting f using resolution $\Omega(d/n^3)$.*

Note that by Lemma 1, we can rephrase this in terms of the more standard *area* of a drawing when all coordinates are integer.

Corollary 1. *Given a plane graph G with n vertices, a convex polygon P with k corners, at integer coordinates and diameter D, and a map f that maps k vertices on the outer face of G to the k corners of P, we can draw the graph G in a scaled copy P' of P that has diameter $O(D^4 n^6)$, such that the drawing respects f and uses only integer coordinates for the vertices of G.*

Proof: First of all, if P has only integer coordinate vertices and diameter D, then its potential resolution is at least $1/D$. To see this, consider a triangle formed by any three vertices of P: this triangle has area at least $1/2$, and in any direction its base is at most D so its height must be at least $1/D$.

Now, by Theorem 1, we can draw G inside P with resolution $\Omega(d/n^3) = \Omega(1/Dn^3)$. Then we can blow up the drawing by a factor Dn^3, which results in a polygon of diameter D^2n^3 and at least constant resolution. By Lemma 1, there now also exists a drawing of G in a polygon P' of diameter $O(D^4n^6)$ in which all vertices are drawn with integer coordinates. $\qquad\square$

4   Application to Drawing Graphs of Genus g

As mentioned in the introduction, graphs of genus g are often drawn in the plane by drawing their *polygonal schema* in a prescribed convex polygon. Using a canonical polygon schema allows us to draw this outer face as a regular $4g$-gon that has some pairs of edges identified, and vertices on those edges duplicated. Given previous work by Duncan *et al.* [9], which gives us a chord-free polygonal schema derived from a graph G combinatorially-embedded in a genus-g surface, we can complete a straight-line drawing of G using a given regular $4g$-gon as its external face, by applying Theorem 1 and rounding the coordinates. Since a regular $4g$-gon with diameter 1 has potential resolution $\Theta(1/g^2)$, this results in a drawing with resolution $\Omega(1/g^2n^3)$.

Of course, there is a slight issue with using a regular $4g$-gon: not every regular k-gon can be embedded with fractional coordinates. So, such a drawing will not fit exactly on an integer grid no matter how big the integers can be. Thus, we either have to allow for non-integer coordinates or allow for a slight (possibly imperceptible) perturbation of the vertex coordinates.

5   Conclusion and Open Problems

We have given an algorithm to draw any combinatorially-embedded planar graph with a prescribed convex shape as its outer face and polynomial area, with respect to the potential resolution of that shape. That is, if the given convex shape has a polynomially-bounded aspect ratio, then we can draw the graph G in its interior using polynomial area. We have not made a strenuous attempt to optimize the exponent in this area bound. So a natural open problem is to determine the upper and lower bound limits of this function.

With respect to drawings of genus-g graphs using a canonical polygonal schema, although our construction guarantees that copies of corresponding vertices appearing in multiple boundary paths will be drawn in the same relative order, it does not guarantee that they will be drawn with the same inter-path distances. So another open problem is whether one can extend our algorithm to draw such paths with matching inter-path distances for corresponding vertices.

Acknowledgments

This research was supported in part by the National Science Foundation under grant 0830403, and by the Office of Naval Research under MURI grant N00014-08-1-1015.

References

1. Bárány, I., Rote, G.: Strictly convex drawings of planar graphs. Documenta Mathematica 11, 369–391, (2006) arXiv:cs/0507030 ,
   `http://www.math.uiuc.edu/documenta/vol-11/13.html`
2. Becker, B., Hotz, G.: On the optimal layout of planar graphs with fixed boundary. SIAM J. Comput. 16(5), 946–972 (1987), doi:10.1137/0216061
3. Chrobak, M., Goodrich, M.T., Tamassia, R.: Convex drawings of graphs in two and three dimensions. In: Proc. 12th ACM Symp. Comput. Geom., pp. 319–328 (1996), doi:10.1145/237218.237401
4. Chrobak, M., Kant, G.: Convex grid drawings of 3-connected planar graphs. Internat. J. Comput. Geom. Appl. 7(3), 211–223 (1997), doi:10.1142/S0218195997000144
5. Chrobak, M., Payne, T.H.: A linear-time algorithm for drawing a planar graph on a grid. Inf. Proc. Lett. 54(4), 241–246 (1995), doi:10.1016/0020-0190(95)00020-D
6. Davidson, R., Harel, D.: Drawing graphs nicely using simulated annealing. ACM Trans. Graph. 15(4), 301–331 (1996), doi:10.1145/234535.234538
7. Dhandapani, R.: Greedy drawings of triangulations. Discrete Comput. Geom. 43(2), 375–392 (2010), doi:10.1007/s00454-009-9235-6
8. di Battista, G., Eades, P., Tamassia, R., Tollis, I.G.: Graph Drawing. Prentice Hall, Upper Saddle River (1999)
9. Duncan, C.A., Goodrich, M.T., Kobourov, S.G.: Planar drawings of higher-genus graphs. In: Eppstein, D., Gansner, E.R. (eds.) GD 2009. LNCS, vol. 5849, Springer, Heidelberg (2010), doi:10.1007/978-3-642-11805-0_7
10. Fáry, I.: On straight-line representation of planar graphs. Acta Sci. Math. (Szeged) 11, 229–233 (1948)
11. de Fraysseix, H., Pach, J., Pollack, R.: How to draw a planar graph on a grid. Combinatorica 10(1), 41–51 (1990), doi:10.1007/BF02122694
12. Fruchterman, T.M.J., Reingold, E.M.: Graph drawing by force-directed placement. Softw. Pract. Exp. 21(11), 1129–1164 (1991), doi:10.1002/spe.4380211102
13. Gajer, P., Goodrich, M.T., Kobourov, S.G.: A multi-dimensional approach to force-directed layouts of large graphs. Comput. Geom. Theory Appl. 29(1), 3–18 (2004), doi:10.1016/j.comgeo.2004.03.014
14. Kant, G.: Drawing planar graphs using the canonical ordering. Algorithmica 16(1), 4–32 (1996), doi:10.1007/BF02086606
15. Lazarus, F., Pocchiola, M., Vegter, G., Verroust, A.: Computing a canonical polygonal schema of an Orientable Triangulated Surface. In: Proc. 17th ACM Symp. Comput. Geom., pp. 80–89 (2001), doi:10.1145/378583.378630
16. Schnyder, W.: Embedding planar graphs on the grid. In: Proc. 1st ACM-SIAM Symp. Discrete Algorithms, pp. 138–148 (1990),
    `http://portal.acm.org/citation.cfm?id=320191`
17. Stein, S.K.: Convex maps. Proc. Amer. Math. Soc. 2(3), 464–466 (1951), doi:10.1090/S0002-9939-1951-0041425-5
18. Sugiyama, K., Misue, K.: Graph drawing by the magnetic spring model. J. Visual Lang. Comput. 6(3), 217–231 (1995), doi:10.1006/jvlc.1995.1013
19. Tutte, W.T.: Convex representations of graphs. Proc. London Math. Soc. 10(38), 304–320 (1960), doi:10.1112/plms/s3-10.1.304
20. Tutte, W.T.: How to draw a graph. Proc. London Math. Soc. 13(52), 743–768 (1963), doi:10.1112/plms/s3-13.1.743
21. Wagner, K.: Bemerkungen zum Vierfarbenproblem. Jber. Deutsch. Math.-Verein. 46, 26–32 (1936)

Crossing Minimization and Layouts of Directed Hypergraphs with Port Constraints

Markus Chimani[1,*], Carsten Gutwenger[2], Petra Mutzel[2],
Miro Spönemann[3], and Hoi-Ming Wong[2,**]

[1] Algorithm Engineering Group, Friedrich-Schiller-Universität Jena
markus.chimani@uni-jena.de
[2] Chair of Algorithm Engineering, Technische Universität Dortmund
{carsten.gutwenger,petra.mutzel,hoi-ming.wong}@tu-dortmund.de
[3] Real-Time and Embedded Systems Group, Christian-Albrechts-Universität zu Kiel
msp@informatik.uni-kiel.de

Abstract. Many practical applications for drawing graphs are modeled by directed graphs with domain specific constraints. In this paper, we consider the problem of drawing directed hypergraphs with (and without) port constraints, which cover multiple real-world graph drawing applications like data flow diagrams and electric schematics.

Most existing algorithms for drawing hypergraphs with port constraints are adaptions of the framework originally proposed by Sugiyama et al. in 1981 for simple directed graphs. Recently, a practical approach for upward crossing minimization of directed graphs based on the planarization method was proposed [7]. With respect to the number of arc crossings, it clearly outperforms prior (mostly layering-based) approaches. We show how to adopt this idea for hypergraphs with given port constraints, obtaining an upward-planar representation (UPR) of the input hypergraph where crossings are modeled by dummy nodes.

Furthermore, we present the new problem of computing an orthogonal upward drawing with minimal number of crossings from such an UPR, and show that it can be solved efficiently by providing a simple method.

1 Introduction

The visualization of directed graphs in an upward fashion is a central research field in graph drawing. Thereby we ask for a drawing where all arcs are drawn monotonously increasing in one direction, in order to make it easy to follow the overall direction in the process that is modeled by the graph. There is a large number of publications regarding this topic. Most known approaches follow the layer-based scheme originally proposed by Sugiyama et al. [18]. However, many important applications such as data flow diagrams or electric schematics require

---

* Markus Chimani was funded by a Carl-Zeiss-Foundation juniorprofessorship.
** Hoi-Ming Wong was supported by the German Research Foundation (DFG), priority project (SPP) 1307 "Algorithm Engineering", subproject "Planarization Practices in Automatic Graph Drawing".

directed hypergraphs rather than traditional directed graphs; moreover they often come with further specific drawing constraints. Layer-based methods that consider these specialties [17] often suffer from too many edge crossings, and thus planarization-based methods—where minimizing the number of crossings is the main objective—may be preferable. In order to keep a consistent arc direction, the technique of *upward-planarization* is needed. Recent results for traditional graphs have shown that, by using this approach, the number of edge crossings can be reduced by 50% compared to layer-based methods [7, 8].

In this paper we consider the problem of drawing directed hypergraphs with port constraints. Basic definitions on hypergraphs are given in Sect. 1.1, and port constraints are introduced in Sect. 1.2. We describe how to adapt an existing upward-planarization method to handle directed hyperarcs as well as prescribed port positions. Our upward-planarization algorithm produces an upward-planar representation of the hypergraph and is covered in Sect. 2. Using this representation, we show how to construct an orthogonal layout that respects the arc directions using a derived layering in Sect. 3. By exploiting the topological information of the planarization phase, we can efficiently minimize the number of crossings with respect to the computed embedding. We conclude with Sect. 4.

1.1 Hypergraphs

A *directed graph* is a pair $G = (V, A)$, where V is a finite set of nodes and A is a set of ordered node pairs called *directed edges* or *arcs*. A directed acyclic graph (DAG) with exactly one source node is called an *sT-graph*. A node u *dominates* a node v in G if there exists a directed path from u to v in G. A *directed hypergraph* is a pair $H = (V, \mathcal{A})$, where V is a finite set of nodes and $\mathcal{A}$ is a set of pairs (S, T) with non-empty sets $S, T \subseteq V$. The elements of $\mathcal{A}$ are called *directed hyperedges* or *hyperarcs*, S are the *source* nodes, and T are the *target* nodes. While our definition conceptually allows $S \cap T \neq \emptyset$ (i.e., a hyperarc may be or contain a self-loop), we will not consider such a case in this paper.

Now let H be a self-loop free directed hypergraph and $\phi = (S, T)$ a hyperarc of H. A directed tree $\mathcal{T} = (V_\phi, A_\phi)$ with $V_\phi = (S \cup T \cup N)$ is an *underlying tree* of ϕ if: (i) for each source node $s \in S$ there is a node $n \in N$ with $(s, n) \in A_\phi$; (ii) for each target node $t \in T$ there is a node $n' \in N$ with $(n', t) \in A_\phi$; (iii) the degree of each $v \in S \cup T$ is exactly 1 within $\mathcal{T}$; and (iv) each $n \in N$ is only adjacent to vertices of V_ϕ and has degree at least 2. We call N the *hypernodes* of ϕ. Informally, the source and target nodes are the leaves and the hypernodes are the inner nodes of $\mathcal{T}$. $\mathcal{T}$ is called *confluent* if each source node dominates all target nodes. A *directed underlying graph* H' of a directed hypergraph H is obtained by substituting each hyperarc ϕ by an underlying tree $\mathcal{T}_\phi$, i.e., H' consists of the nodes of H together with the hypernodes and arcs of all underlying trees. If each underlying tree contains only one hypernode, we call H' a *star-based* underlying graph of H (cf. *tree-based* and *point-based* drawing style of hyperedges [6]). H' is *confluent* if all underlying trees are confluent.

In this paper we also consider the problem of orthogonal routing of hyperedges. Layer-based approaches for orthogonal routing of plain edges were given by Sander [15] and Baburin [1]. Eschbach et al. showed that the orthogonal hyperedge routing problem using at most one horizontal line segment per hyperedge is NP-hard by revealing its equivalence to the minimal feedback arc set problem [10], and proposed a greedy assignment heuristic and a sifting heuristic. Sander proposed to use standard cycle breaking heuristics [16], still limited to at most one horizontal line segment for each hyperedge (except for hyperedges that span multiple layers).

Since our approach is based on planarization, we use the basic ideas for handling of hyperedges from previous work on hypergraph planarization [6].

1.2   Port Constraints

The *ports* of a hyperarc $\phi = (S, T)$ are the points in the drawing where ϕ touches the nodes in S and T. In many applications these ports have a specific semantic interpretation, such as being *inputs* or *outputs* for data tokens in data flow diagrams, or *pins* of electric components in circuit schematics. Thus in such applications the positioning of ports is not arbitrary, but may be subject to specific constraints [17]. The strictest variant of port constraints is the one where the exact position of each port, relative to the respective node, is prescribed. This implies that each port has an associated side of the node where it is drawn, i.e., the top, bottom, left, or right side.

Given a node v with incident arcs I, the application's model of port constraints may limit the set of admissible clockwise orders of the arcs I. A generic approach to model this set of orders for each node is by using *embedding constraints*, which define a tree structure of constraint nodes [12]. This structure can be considered by an extended planarization algorithm in order to obtain a planar embedding that respects the pre-defined constraints. The embedding constraints approach is compatible with the method described in this paper, but for the sake of brevity we mostly consider only the strict port positions variant in the following, which admits merely one port order for each node.

First approaches to include ports in layer-based drawings were given by Gansner et al. [11] and Sander [14]. A more advanced adaption considers different types of port constraints as well as hyperarcs [17], as they are required for the layout of data flow diagrams. That method leads to quite acceptable results for different types of hypergraphs with port constraints, but is still limited by the fact that a bad layering can lead to an unnecessarily high number of arc crossings. As an alternative that is based on planarization, Eiglsperger et al. [9] proposed a method to include constraints in the orthogonalization phase for the *topology-shape-metrics* approach.

2   Upward-Planarization

The currently strongest upward-planarization approach for sT-graphs is due to [7]. The key ingredients of this algorithm can be summarized as follows. Let $G = (V, A)$ be the graph to draw. Similar to the traditional undirected planarization approach [2], we start with an upward-planar subgraph $U = (V, A')$ of G and then iteratively insert the edges $A \setminus A'$ into U with as few crossings as possible. All arising crossings on an inserted edge are replaced by dummy nodes (*crossing dummies*), such that the iteration always considers the problem of inserting a single arc into an otherwise upward-planar graph.

Yet, unlike for the undirected planarization approach, our upward requirement results in two subtle but central difficulties: not every upward-planar subgraph can be used as a starting point, and a crossing minimal insertion of some arc a may leave a graph in which the remaining, not yet inserted arcs cannot be inserted in an upward fashion anymore, see [7] for details.

In the following, we will investigate how this approach can be extended to hypergraphs and port constraints.

2.1   Preprocessing

We can assume that the given directed hypergraph H contains no hyperacrs with self-loops, as they could be easily reinserted as a postprocessing step without requiring any further crossings. Let H' be the star-based directed underlying graph of hypergraph H. As we require an sT-graph, we may have to insert dummy edges from an artificial super source node to the source nodes of H'. These edges will have weight 0 for the purpose of crossing minimization and can be removed for the final layout computation (Sect. 3).

It remains to make H' acyclic by reversing the direction of the arcs in a *minimal* feedback arc set. Although this problem is NP-hard, any heuristic finding a minimal (in contrast to a *minimum*) arc set suffices for our purpose. Also note that, under the assumption that an upward drawing is actually a suitable drawing paradigm for H, this set will typically be small or even empty. In the final drawing, we re-reverse these arcs again. Yet, for the upward-planarization approach, we still have to take special care of such reversed arcs, as we require each hypergraph to be drawn in a confluent way. Let Rev be the arcs that were reversed in the preprocessing step.

Our port constraints may require arcs $a = (u, v)$, with u being properly drawn below v, (a) to leave u downwards, or (b) to enter v from above. This clearly invalidates the pure upward drawing style. Nevertheless we want to allow such constructs, but have to ensure that such "misdirected" pieces of the arcs are only drawn close to u or v, respectively (see below). Hence any such arc requiring (a) and (b) is substituted by a *chain* of *subarcs* $(d_a^1, u), (d_a^1, d_a^2), (v, d_a^2)$, where d_a^1, d_a^2 are two new dummy nodes. Arcs only requiring either (a) or (b) are analogously replaced by simpler chains of only two subarcs and a single dummy node. For notational simplicity, we will continue to store the original unmodified arcs in the graph, denoted by the set Chn. Whenever we consider an arc $a \in Chn$, we in fact use the complete chain corresponding to a.

So after preprocessing we have a simple sT-graph $G = (V, A)$, with $Rev \subset A$ and $Chn \subset A$. Note that these two subsets may not be disjoint.

2.2   Feasible Upward-Planar Subgraph

Besides ensuring upward-feasibility of the initial upward-planar subgraph (i.e., ensuring that all temporarily removed arcs are re-insertable), our subgraph has to additionally be feasible with respect to the port constraints. As described in Sect. 1.2, embedding constraints can be seen as a specific model of admissible arc orders for planarization. We can handle such constraints by replacing each constrained node by its constraint-tree [12], thus in the following we only consider nodes with no constraints or strict port constraints.

We first compute a spanning tree $U' = (V, A'')$ of G, with $A'' \cap Rev = \emptyset$, which clearly exists due to the minimality of Rev.

We then insert the arcs $a = (x, y) \in A \setminus (A'' \cup Rev)$ one by one into U'. After each insertion step we perform an upward-planarity test, obtaining some feasible embedding, and check whether the graph still allows the insertion of all remaining arcs within this embedding, using the merge-graph paradigm introduced in [7]. By the implicit transformation of the arcs Chn we thereby also ensure that the port constraints can be satisfied. If any of the above two tests fails, we remove the arc again, and store it in a set B instead.

After that, we fix the current embedding of the graph, and try to re-insert the arcs in B a second time, as the previously considered embedding might just have been a bad choice. In the end we have a maximal port-feasible upward-planar subgraph $U = (V, A')$ with a feasible embedding Γ, and the set $B = A \setminus A'$ of arcs not yet in U.

Remark. The runtime of the arc-wise test for insertion-feasibility is still dominated by the acyclicity test in the merge-graph, and we can hence obtain U in $\mathcal{O}(|A|^2)$ time.

2.3   Arc Insertion

Again, we will start our investigation by outlining how arc insertion works for traditional arcs without port constraints, cf. [7] for details. We will then use this algorithm as a building block when considering hypergraphs and port constraints. Sometimes we thereby require seemingly minor internal modifications to this algorithm, which in fact require intricate modifications in the proofs of the algorithm's validity. Due to space constraints, we will only touch upon these issues and focus on the overall idea.

Considering a fixed embedding, the non-upward edge insertion problem can be solved by considering the dual graph of the embedding, and finding a shortest path in this dual graph between two faces adjacent to the edge's start and end node. When considering inserting an arc $a = (u, v)$ in an upward fashion, this dual graph has to be substituted by a somewhat similar routing network that ensures that the identified insertion path is monotonously going upward and is non-self-intersecting.

Furthermore, not every such found insertion path is feasible with respect to the remaining, not yet inserted arcs. Therefore, during the shortest path computation within the routing network, we additionally have to check feasibility of the intermediate merge-graph (simply put, the merge graph is the current graph, augmented with the not-yet inserted arcs in such a way that the remaining arcs are insertable if and only if the merge graph is acyclic).

Hyperarc Insertion. Starting from the subgraph U, we will now iteratively insert full hyperarcs, until we obtain a upward-planarization of G and hence of H. Note that hyperedge insertion in the tree-based paradigm is already NP-hard in the undirected, non-upward setting [6]. We therefore introduce a novel piecewise insertion strategy which realizes a low-crossing number hyperarc insertion, still ensuring confluency of the hyperarc drawing.

For any original hyperarc $\phi \in \mathcal{A}$ in H, G contains a set of arcs $A_\phi \subset A$ and a set of hypernodes V_ϕ. Initially, $|V_\phi| = 1$ since we started with the star-based underlying graph of H. We will see that both sets will grow during the subsequent insertion steps. Also note that, in general, some arcs of A_ϕ may be in A' and some in B.

Let $U^\circ = (V, A^\circ)$ be a port-feasible upward-planar planarization (i.e., crossings are modeled via dummy nodes) of some subgraph of G during the insertion process, B° the not yet inserted arcs, and $\phi \in \mathcal{A}$ the hyperarc to insert. The edge set $A_\phi^\circ := A_\phi \cap A^\circ$ forms a tree, corresponding to the partially tree-based-drawn hyperarc. Our induction hypothesis states that this tree is confluent: In the initial subgraph U, this tree is at most a star (and at least a single edge) and therefore clearly confluent. The arcs $B_\phi := A_\phi \cap B^\circ$ are the arcs corresponding to ϕ that have yet to be inserted in our current iteration step, extending the tree A_ϕ° in a confluent way.

We will insert these arcs one by one. Let $a = (x, y) \in B_\phi$ and assume for now that $a \notin Rev \cup Chn$. We first compute the minimal subtrees T_s, T_t of A_ϕ° that contain all source and target nodes, respectively, that are already connected by A_ϕ°. By the induction hypothesis on confluency, T_s and T_t are disjoint except for at most one hypernode. Let h_s (h_t) be the hypernode of T_s (T_t) closest to T_t (T_s, respectively) on the tree A_ϕ°.

If x is a hypernode, then y is a target vertex of ϕ, and we search for a shortest feasible upward-insertion path from h_s to y. Otherwise, y is a hypernode, x is a source node of ϕ, and we search for a shortest feasible upward-insertion path from x to h_t. Thereby we modify the routing network in three ways:

Port Constraints. If our port constraints require us to leave x or enter y at some special positions with respect to already inserted arcs, we can easily restrict the routing network to use only applicable start or end faces for the routing. Yet, there are additional augmentations necessary, if $a \in Chn$, see below.

Arc Reuse. We want to reuse already drawn (sub)arcs corresponding to ϕ, in order to generate hyperarc drawings with low total number of crossings. Therefore, our routing allows 0-cost crossings over A_ϕ° and next to crossing dummies in A_ϕ°. In other words, the new path may reuse already established paths of ϕ, cf. Fig. 1.

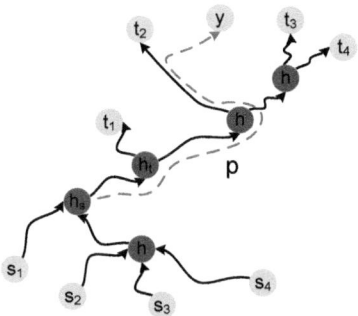

Fig. 1. Arc Reuse for the case when the source node x is one of the red colored hypernodes. The insertion path p for $a = (x, y)$ starts at node h_s.

Splitting other Hyperarcs/-nodes. The knowledge of the properties of hypernodes allows a further improvement, lifted from [6] where it was used in the context of undirected edge insertion for the so-called *minor-monotone* crossing number [3]. We can cross "through" a hypernode h of another hyperedge ψ, as long as we thereby do not separate both source and target nodes from each other, cf. Fig. 2: By our induction hypothesis on confluency, crossing through h means that we split h into two hypernodes h and h' and add the arc (h, h'). Then we change the source (or target) vertex of at least two arcs from h to h'. Let A_{in} and A_{out} be the arcs formerly entering and leaving h, respectively. To ensure confluency, we only allow a split where (a) all A_{in} remain incident to h, (b) all A_{out} become incident to h', or (c) neither of both, but either A_{in} or A_{out} is an empty set. After this split, our routing path can cross over the arc (h, h'). Note that, since we consider a fixed embedding, all valid hypernode-crossings can be modeled via arcs in the routing network directly connecting two faces (e.g., f_6, f_3 in Fig. 2). Such arcs have cost 1, as they induce a hypernode split such that a single crossing suffices.

Inserting chain-transformed arcs, i.e., $a \in Chn$. If $a = (x, y)$ has to leave x downwards or enter y upwards, we have to extend our routing network further. Assume that a only requires the second property (the first one is independent of the second and can be solved analogously). Usually, all arcs dominated by y will be *statically locked*, i.e., we may not cross through them. Now, we unlock the arcs that directly leave y, and search for a path entering y from there; depending on the exact port constraint, only a single face above y might be a valid entrance point into y. We now place the dummy node of the chain-transformation, where (x, y) was split into (x, d), (y, d), into the face from which y is entered. Then all crossings are still drawn only between upward arcs. The small subarc (y, d), which will be reversed in the final drawing, is even drawn without any crossings. By forbidding crossings to happen over (y, d) in the subsequent steps, we therefore guarantee that the overall drawing is still upward, and the

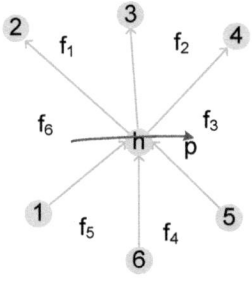
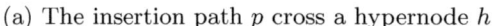

(a) The insertion path p cross a hypernode h.

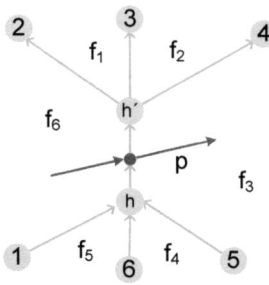

(b) A realization of p.

Fig. 2. Splitting a hypernode by introducing an additional arc can reduce the crossings

small downward pieces of arcs, due to port constraints, are restricted to the direct neighborhood of the corresponding node.

Note that, after the full upward-planarization approach is completed, we can merge multiple such dummy nodes that correspond to the same hyperarc, when they lie in a common face.

Inserting reversed arcs, i.e., $a \in Rev$. To ensure confluency in the hyperarcs, we have to take special care for the arcs that where reversed in the preprocessing step to remove cycles. After the drawing is computed, we will have to reset their original direction. To avoid notational complexity, we will add such arcs only after all other arcs of the hyperarc are already inserted.

Assume the arc $a = (x, y)$ originally connected a source vertex to the hypernode of the star-based underlying graph, but got reversed and hence connects the hypernode to a target vertex. Nonetheless we have to ensure that it connects to the aforementioned subtree T_s, instead of T_t. Let S be the set of original sources in T_s before inserting a. Then a confluency-feasible upward insertion path for a can be found by selecting the minimal insertion path from any node in S to y. Thereby it may cross over $A_\phi^\circ \setminus T_t$ and next to crossing dummies of $A_\phi^\circ \setminus T_t$ at no cost. The analogous holds, if a originally connected a target vertex to the star-based hypernode.

Putting it together. So overall, after first computing a special feasible upward subgraph, our upward-planarization approach inserts one hyperarc after another. Each hyperarc is inserted by incrementally inserting the arcs of the star-based underlying graph, reusing the already established tree-based sub-drawing of the hyperarc as far as possible. By specially considering *original* arc directions, we thereby guarantee that all hyperarcs are drawn as confluent trees. The final result of the planarization is an upward-planar representation (planar, upward-feasible sT-graph) $\mathcal{R}$ of G, and hence of H, together with an embedding Γ of $\mathcal{R}$. Within $\mathcal{R}$, hyperarcs of H are represented as confluent trees, and crossings are represented as dummy nodes.

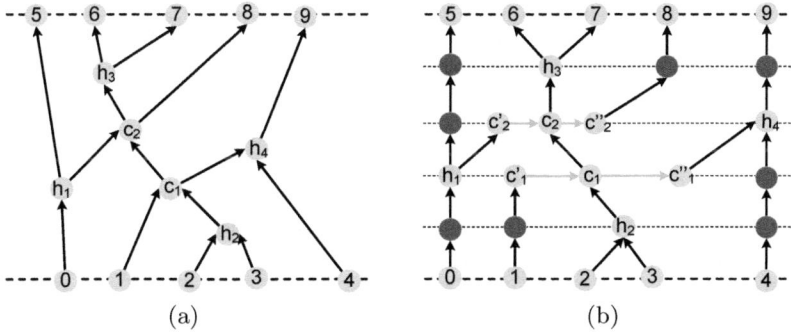

Fig. 3. Steps towards a final layout: (a) the subgraph between consecutive layers of an UPR $\mathcal{R}$, (b) fine-layering of the subgraph with included dummy nodes to split long arcs

3   Layout

Using the embedding Γ of the upward-planar planarization (UPR) $\mathcal{R}$ computed in the previous step, our layout procedure works in three steps:

1. A layering $\mathcal{L}$ of $\mathcal{R}$ is computed.
2. An initial orthogonal drawing of $\mathcal{R}$ is computed.
3. Optional step: Orthogonal compaction is applied to remove unnecessary bend points and improve layout quality.

We finally obtain an upward drawing of $\mathcal{R}$, which induces a drawing of H in a straight-forward way. The number of hyperarc crossings in this drawing equals the number of crossing dummies in $\mathcal{R}$, thus it is minimal with respect to the computed embedding. Our initial method for orthogonal upward drawing may produce an unnecessarily high number of bend points, but it reveals that such a drawing can be computed efficiently. We discuss the individual steps in more detail.

Layering. We compute a layering of $\mathcal{R}$ in two phases using the layering algorithm by Chimani et al. [8], which also induces a node ordering for each layer. In the first phase we compute a layering $\mathcal{L}'$ of the nodes of H, and in the second phase a layering $\mathcal{L}''$ of the subgraph between each two consecutive layers of $\mathcal{L}'$. Notice that the nodes of $\mathcal{L}''$ are either crossing dummies or hypernodes. For each crossing dummy c of $\mathcal{L}''$, we split c by adding new dummy nodes c' and c'' such that c' is the immediate left and c'' is the immediate right neighbor of c. These two nodes will be bend points in the later drawing and ensure the orthogonality of the arcs incident to c. We redirect the left incoming and the right outgoing arc of c such that c' is its new target and c'' is its new source node, respectively. We then merge $\mathcal{L}'$ and $\mathcal{L}''$ into a complete layering $\mathcal{L}$ of $\mathcal{R}$ and create dummy nodes to split long edges that span multiple layers. An example is shown in Fig. 3.

Initial orthogonal drawing. Using the layering $\mathcal{L}$, we apply an arbitrary coordinate assignment algorithm known for the third step of Sugiyama's framework (e.g., [4, 5, 11, 15]) to compute horizontal node coordinates. The orthogonal routing of the edges between each pair of consecutive layers can then be calculated using an existing method for layer-based routing, such as the one proposed by Sander [15]. Since the embedding of $\mathcal{R}$ is planar, such a routing can be constructed without introducing additional arc crossings. As a result, all incoming and outgoing arcs of hypernodes and dummy nodes end in the same point, where incoming arcs reach the nodes from below and outgoing arcs leave the nodes upwards. If there are multiple incoming or multiple outgoing arcs of a node u, the corresponding vertical line segments that touch u overlap each other. These overlapping line segments need to be merged to single line segments.

For the computation of the orthogonal layout, port constraints essentially need to be considered only in the routing algorithm, since they determine where the arcs shall touch the connected nodes. In case of ports that are situated on the left or right side of a node, we can artificially broaden the node prior to the horizontal coordinate calculation and add one bend point per arc such that incoming and outgoing arcs are redirected downwards and upwards, respectively (see [14], Sect. 7).

Applying orthogonal compaction. The resulting initial drawing may contain various unnecessary bends; many such bends can be removed using orthogonal compaction techniques (see [13] for an overview). First, we connect the nodes of each pair of consecutive layers L_{top} and L_{bottom} of $\mathcal{L}'$ by horizontal edges and connect the first nodes and the last nodes on these layers by edges with two bend points, such that all these additional edges form a surrounding rectangular frame (cf. Figure 4). We assign fixed edge lengths to the edges on this frame, so that the nodes on L_{top} and L_{bottom} will remain on their horizontal positions. Then, we compute the orthogonal representation induced by this drawing and apply orthogonal flow-based compaction. This allows us to assign costs to segments. Let $\mathcal{S}_0$ denote the set of line segments adjacent to two bend points with a 90 and a 270 degree angle in a face. The segments in $\mathcal{S}_0$ are assigned maximal cost and zero minimal length. The minimum length of the remaining segments is set to the minimum of their lengths in the initial drawing and a desired spacing, making sure that the initial drawing is a feasible solution for the compaction. Each segment in $\mathcal{S}_0$ for which the compaction achieves zero length is then removed from the orthogonal representation by merging its adjacent segments.

Final layout. The so constructed drawing is an orthogonal drawing and its arc crossings are exactly the crossings modeled by $\mathcal{R}$. In order to obtain a valid drawing of the original hypergraph H, we perform the following post-processing steps:

a) Replace each crossing dummy c and its corresponding neighboring nodes c' and c'' by a horizontal line segment from c' to c''.

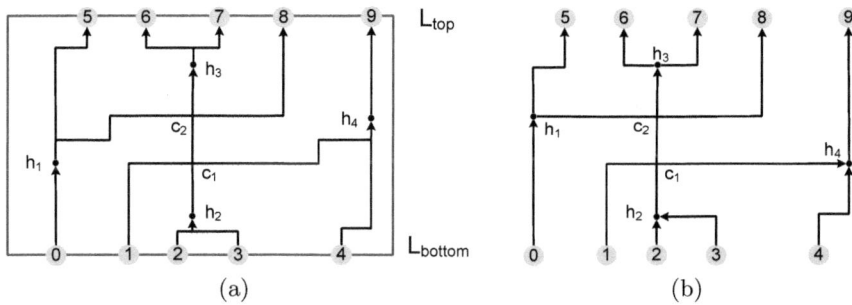

Fig. 4. Orthogonal compaction: (a) An initial drawing of Figure 3(a) with unnecessary bend points. Before starting the compaction, the drawing is framed by a rectangle (red edges) and the new edges are assigned fixed lengths. (b) The final drawing.

b) Eliminate all remaining dummy nodes and hypernodes by directly connecting the line segments of the incoming and outgoing arcs.

c) Reverse all arcs of *Rev*, which were previously reversed to break cycles.

We can add an additional compaction step on the whole layout—similar as before—to further improve the layout. Notice that when considering the whole layout, we do not need to fix the x-coordinates; we only have to ensure that the horizontal distance between ports is fixed.

4   Conclusion

We presented the first planarization approach for hypergraphs in the context of upward drawings. To this ends, we combined the known ideas of upward arc insertion and insertion of a single edge in the undirected minor crossing number setting with a novel heuristic method to assemble multiple adequately chosen insertion paths to a confluently drawn hyperarc. Furthermore, we dealt with the problem of laying out the so obtained upward-planar representation in an orthogonal upward drawing style. As our hyperarcs offer more freedom than prior approaches within the Sugiyama framework, their orthogonalization step is inapplicable for our needs, even after layering. We therefore introduced a new global scheme based on orthogonal compaction.

We also considered port constraints both in the upward-planarization as well as in the layout step, as they are integral to many hypergraph drawing applications such as pins in electrical circuits. We want to stress that our algorithm not only solves the upward drawing problem for hypergraphs with and without port constraints, but also is the first port-constraint-aware upward-planarization approach that is suitable for regular DAGs.

Based on the experience with prior upward-planarization methods, we expect that our methods should work well in practice—the implementation, together with a thorough experimental investigation comparing this approach to the more traditional hypergraph drawing algorithms within the Sugiyama framework, remains as our next research step. Furthermore, we would be interested in more direct orthogonalization methods, along the lines of, e.g., [15, 10]. These methods solve the problem locally on a layer-by-layer basis, but extending them to allow multiple horizontally drawn hypernodes per hyperarc between two layers is non-trivial.

References

[1] Baburin, D.E.: Using graph based representations in reengineering. In: Proc. CSMR 2002, pp. 203–206 (2002)
[2] Batini, C., Talamo, M., Tamassia, R.: Computer aided layout of entity relationship diagrams. J. Syst. Software 4, 163–173 (1984)
[3] Bokal, D., Fijavz, G., Mohar, B.: The minor crossing number. SIAM J. Discrete Math. 20, 344–356 (2006)
[4] Brandes, U., Köpf, B.: Fast and simple horizontal coordinate assignment. In: Proc. Graph Drawing 2001, London, UK, pp. 31–44. Springer, Heidelberg (2002)
[5] Buchheim, C., Jünger, M., Leipert, S.: A fast layout algorithm for k-level graphs. In: Marks, J. (ed.) GD 2000. LNCS, vol. 1984, pp. 229–240. Springer, Heidelberg (2001)
[6] Chimani, M., Gutwenger, C.: Algorithms for the hypergraph and the minor crossing number problems. In: Tokuyama, T. (ed.) ISAAC 2007. LNCS, vol. 4835, pp. 184–195. Springer, Heidelberg (2007)
[7] Chimani, M., Gutwenger, C., Mutzel, P., Wong, H.-M.: Layer-free upward crossing minimization. ACM Journal of Experimental Algorithmics 15 (2010)
[8] Chimani, M., Gutwenger, C., Mutzel, P., Wong, H.-M.: Upward planarization layout. In: Eppstein, D., Gansner, E.R. (eds.) GD 2009. LNCS, vol. 5849, pp. 94–106. Springer, Heidelberg (2010)
[9] Eiglsperger, M., Fößmeier, U., Kaufmann, M.: Orthogonal graph drawing with constraints. In: Proc. SODA 2000, pp. 3–11. SIAM, Philadelphia (2000)
[10] Eschbach, T., Guenther, W., Becker, B.: Orthogonal hypergraph drawing for improved visibility. J. Graph Algorithms Appl. 10(2), 141–157 (2006)
[11] Gansner, E., Koutsofios, E., North, S., Vo, K.-P.: A technique for drawing directed graphs. Software Pract. Exper. 19(3), 214–229 (1993)
[12] Gutwenger, C., Klein, K., Mutzel, P.: Planarity testing and optimal edge insertion with embedding constraints. J. Graph Algorithms Appl. 12(1), 73–95 (2008)
[13] Klau, G.W., Klein, K., Mutzel, P.: An experimental comparison of orthogonal compaction algorithms. In: Proc. Graph Drawing 2000, London, UK, pp. 37–51. Springer, Heidelberg (2001)
[14] Sander, G.: Graph layout through the VCG tool. Technical Report A03/94, Universität des Saarlandes, FB 14 Informatik, 66041 Saarbrücken (October 1994)
[15] Sander, G.: A fast heuristic for hierarchical Manhattan layout. In: Brandenburg, F.J. (ed.) GD 1995. LNCS, vol. 1027, pp. 447–458. Springer, Heidelberg (1996)
[16] Sander, G.: Layout of directed hypergraphs with orthogonal hyperedges. In: Liotta, G. (ed.) GD 2003. LNCS, vol. 2912, pp. 381–386. Springer, Heidelberg (2004)
[17] Spönemann, M., Fuhrmann, H., von Hanxleden, R., Mutzel, P.: Port constraints in hierarchical layout of data flow diagrams. In: Eppstein, D., Gansner, E.R. (eds.) GD 2009. LNCS, vol. 5849, pp. 135–146. Springer, Heidelberg (2010)
[18] Sugiyama, K., Tagawa, S., Toda, M.: Methods for visual understanding of hierarchical system structures. IEEE Trans. Sys. Man. Cyb. 11(2), 109–125 (1981)

Drawing Graphs on a Smartphone[*]

Giordano Da Lozzo, Giuseppe Di Battista, and Francesco Ingrassia

Dipartimento di Informatica e Automazione – Università Roma Tre, Italy
giordano.dalozzo@gmail.com, gdb@dia.uniroma3.it, fra.ingrassia@gmail.com

Abstract. We present a system for the visualization of relational information on the smartphones. It is implemented on the iPhone and on the Google Android platforms and is based on a new visualization paradigm that poses interesting algorithmic challenges. We also show customizations of the system to explore and visualize popular social networks.

1 Introduction

Millions of people in the world have in their pocket a smartphone. Such a widely used device is exploited to quickly access, from almost everywhere, different types of data on different subjects. A large amount of such data is relational information. As an example, smartphones are used to access social networks like Facebook or Twitter, ontologies like the Wikipedia network of concepts, or technical information related to the job of the owner like the connections of a computer network or the delivery routes of a goods distribution framework.

Graph Drawing can play an important role in supporting information visualization on the smartphones, provided that the methodologies and tools that are typical of this research area are recast to meet the needs of such a challenging device. Indeed, different information contexts have already changed their visualization methods in this direction. E.g., on-line newspapers have special visualization formats that are designed for the smartphone.

Dealing with smartphones the challenges that visualization applications have to face are, of course, the small screen and the limited memory size. On the other hand, such strong limitations come together with new technological opportunities that can be exploited to support the interaction. They are the multi-touch screen that is able to capture commonly used gestures like pinch, flick, and slide, sensors like the accelerometer and the compass, and sounds and vibrations. All this comes with an always-on connection to the Internet.

A previous attempt to draw graphs on smartphones has been done in [6]. However, such a tool uses Graph Drawing techniques that are not specifically designed for smartphones.

---

[*] Supported in part by MIUR (Italy), Projects AlgoDEEP no. 2008TFBWL4 and FIRB "Advanced tracking system in intermodal freight transportation", no. RBIP06BZW8.

U. Brandes and S. Cornelsen (Eds.): GD 2010, LNCS 6502, pp. 153–164, 2011.

Since any graph is too large for the little screen of the smartphones, a pivotal reference point for designing interfaces and algorithms for drawing graphs on such a device is the literature on drawing very large graphs. One, for example, could use the fish-eye approach [7] where the details of the drawing decrease according to the distance that separates them from a point chosen by the user. However, using Shneiderman's information visualization mantra [8] (overview first, zoom and filter, then details-on-demand) in this context seems to be unfeasible.

In this paper we present a system for the visualization and interaction with relational information on the smartphones. Section 2 shows the visualization and interaction paradigm we devised, that is based on visualizing a small subgraph defined by a focus vertex and its neighborhood and exploits smartphone-specific interaction primitives. The approach can be considered similar to the navigation approach of [2]. Section 3 shows how even such a simple visualization paradigm can originate interesting algorithmic problems. Section 4 gives details on the adopted algorithms. Section 5 discusses experimental results on such algorithms. Section 6 gives technical details on the implementation and presents case studies.

2   A Visualization and Interaction Paradigm

Our visualization paradigm is based on a navigation approach. The user selects a *focus vertex* v and the drawing contains vertices and edges as follows. Let $N(v)$ be the set of neighbors of v, that is assumed circularly ordered. About vertices, the drawing contains v and a subset ω_i^L, called *lobe*, of $N(v)$, where L (*lobe size*), is the size of the lobe and the elements of ω_i^L have positions $i, \ldots, (i + L - 1)$ mod $|N(v)|$ in $N(v)$. About edges, the drawing contains the *radial* edges from v to the vertices of ω_i^L, the *inner* edges between vertices of ω_i^L, the *outer* edges that have one end-vertex in ω_i^L and one end-vertex in $N(v) \backslash \omega_i^L$, and the *external* edges that have one end-vertex in ω_i^L and one end-vertex that is not in $N(v)$. The focus vertex is in the center of the bottom side of the drawing, while the lobe vertices lie on an half-ellipse centered at the focus vertex. This, together with other graphical features, suggests to the user that ω_i^L is only a subset of $N(v)$ and that the vertices of $N(v) \setminus \omega_i^L$ (*outer vertices*) are under the bottom side of the screen. External edges exit the drawing from the left, top, right borders of the screen, giving the impression that the rest of the graph is outside the screen. Their external end-vertices are not represented. Outer edges are directed downward, leading to vertices of $N(v)$ that are not represented and that are under the bottom side of the screen. Fig. 1 shows an example of visualization.

Different types of edges have different thicknesses. Radial edges, that do not give additional information, are thin, while inner edges, that represent relationships between vertices of the lobe, are emphasized. External edges could be too many to be represented explicitly, hence they are drawn with straight-lines only up to a certain number. Over that number they are shown with a shadow exiting the drawing. Essentially, the important information for the user is if they are 0, a few, or a lot. Similar graphic features are used for outer edges.

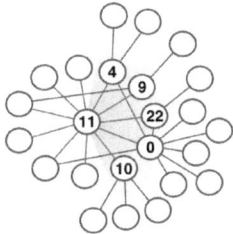 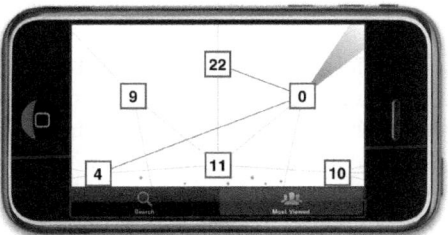

Fig. 1. Visualization of the subgraph induced by the lobe plus the focus vertex (vertex 11)

The user interacts with the graph according to the following primitives. There are *vertex oriented primitives* like focus change, navigation backtrack, and execution of actions associated with vertices (e.g., open a browser or send an sms) and *lobe oriented primitives* like lobe shift, lobe resize, and lobe layout temporary optimization.

The *focus change* primitive substitutes the currently visualized subgraph with the subgraph induced by a new focus vertex and its neighborhood. The gesture used to change the focus consists of dragging the vertex of the lobe that will be the new focus towards the focus position. The *backtrack* primitive shows the previously displayed lobe. The user can navigate the path of the explored focus vertices by performing a double fingers flick towards the bottom or top side of the screen. A side feature of our framework is the possibility of extending the interaction experience by executing an action related to a selected vertex. For example, if the input graph is a social network, possible actions are: showing more information about the individual, sending an e-mail or an sms, deleting him/her from the graph, etc. The gesture associated with this action is a double tapping or a pressure on the vertex.

A primitive that changes the lobe is *lobe shift*. Let ω_i^L be the visualized lobe, the effect of lobe shifting is to substitute ω_i^L with ω_k^L where $k = i \pm 1 \mod |N(v)|$. Sliding a finger on the screen is the simpler way to do this. However, if $|N(v)|$ is very large this interaction can be unsuitable to reach vertices that are far from the current lobe. Hence, the user needs a single gesture to skip a large portion of the subgraph. This is obtained with a flick of finger on the screen. The number of the vertices that are skipped is proportional to the speed of the gesture. The same primitive can be invoked using the accelerometer by changing the slope of the device or using the compass by changing the orientation of the device.

Zooming is a quite common functionality offered by smartphone applications. In our paradigm this function corresponds to a *lobe resize*. Let ω_i^L be the visualized lobe, the effect of this function is to show the lobe ω_k^M where $k = i$ if $(M = L) \vee (M = L + 1)$ or $k = i - 1 \mod |N(v)|$ if $M = L + 2$. The gesture used to resize the current lobe is a multi-touch gesture called pinch, this consists of moving together two fingers on the screen, increasing or decreasing their distance.

The current order of the vertices of a lobe could lead to a drawing that is unpleasant. Hence, we provide a *local optimization* feature that temporary re-orders the vertices of the lobe in order to increase the readability of the drawing. This local optimization is invoked by performing a double tap on the screen.

In order to help the user in maintaining the mental map during the navigation, a mechanism for morphing between successive drawings, supported by a smooth movement of edges and vertices, is necessary. Hence, for all primitives, vertices move slowly towards their final positions. In lobe oriented primitives, vertices entering (exiting) the current lobe come in (out) from the bottom side of the screen. Also, inner edges can become external and vice-versa, and their representation features change coherently. The flick function is accompanied by an inertial rotation of the neighborhood of the focus vertex. The morphing features of the focus change primitive are more complex than those described above. Since they depend on the order of the lobes, they are discussed in Section 4. With the purpose of helping the user interaction experience, sounds and vibration events are associated with the primitives. They are selected in such a way to be consistent with the effects of the corresponding primitives.

3   Choosing the Lobe Order

An important aspect of the visualization paradigm is the left-to-right order of the vertices of the lobe. The specific choice of this order may depend on the specific application. However, the need of preserving the user's mental map implies that the orderings of contiguous lobes of a focus vertex v are consistent. Since this constraint holds for all the lobes of v, this implies the need of choosing a unique order for all the vertices of $N(v)$.

Trivial choices are possible, like using the alphabetical order, that can be suitable for some applications. On the other hand, it is also possible to make different choices, according to aesthetics that are related to the selected visualization paradigm. We deepen two different choices corresponding to two aspects of the paradigm. The first is the one of displaying the relational information as clean as possible, while the second is the one of showing as much relational information as possible. The two choices conflict each other.

The first choice is the one of selecting an order that tries to minimize the visible crossings. Given a focus vertex and a lobe, a *visible crossing* is a crossing between two inner edges. We concentrate only on visible crossings because the only edges that are completely visible for a certain lobe are the inner edges and the radial edges. Also, crossings with radial edges do not compromise readability. More precisely, we try to minimize the average number of visible crossings for all the lobes of a focus vertex.

The problem of minimizing the visible crossings has similarities with the *circular crossings minimization* problem. In that problem, the vertices lie on a circumference, the edges are straight lines, and a circular order is searched that minimizes the total number of crossings. See, e.g. [4,9,1,3]. However, as shown in Fig. 2, the two problems are different.

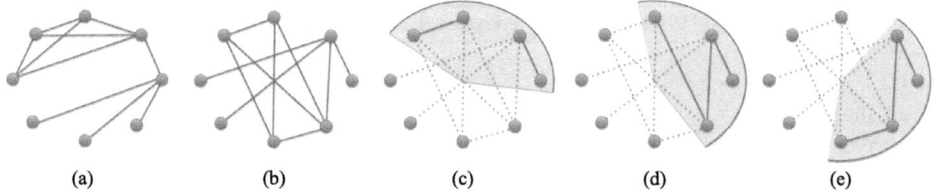

Fig. 2. Circular crossings and visible crossings. (a) Order with minimum total number of crossings. (b) Order with minimum number of visible crossings if the lobe size is equal to 4. (c)–(e) Lobes with 0 visible crossings.

Observe that minimizing the visible crossings with a lobe with size that is equal to $|N(v)|$ is equivalent to solving a circular crossings minimization on the subgraph induced by $N(v)$. But, unfortunately, such a problem has been proved NP-complete [5]. Hence, solving in practice the minimization of the visible crossings requires the usage of heuristics.

One may ask whether heuristics for the circular crossings minimization are effective also for our problem. Our experiments show that this is not the case (see Section 5). Hence, we used in our system a special purpose algorithm that is described in Section 4.

A second alternative for choosing an order is to select one that tries to maximize the *visible edges*, that is to minimize the number of edges that are not inner edges in any lobe. This corresponds to minimizing the information that is lost for a certain focus vertex. Let L be the lobe size and let v be the focus vertex. Observe that it might not exist an order for $N(v)$ in which every edge with end-vertices in $N(v)$ is an inner edge for at least one lobe of size L. As an example, consider the case when the subgraph induced by $N(v)$ contains the clique K_M with $M \geq 2L$. Of course, this information is not lost. In fact, such edges will be visible selecting one of the end-vertices of the missing edges as focus vertex. Observe the similarities with the *Graph Bandwidth Problem*.

4 Algorithmic Framework

The algorithmic framework that we have developed in our system allows to tackle both the visible crossings minimization and the visible edges maximization problems.

The typical parameters of the algorithms for the circular crossings minimization problem presented in [4,1,3] are: 1. the start vertex s of the graph to be drawn; 2. the policy which selects the next vertex to process; and 3. the position that is chosen for the selected vertex. We make use of the same general setting. The algorithmic framework we propose is Algorithm 1.

Notice that Algorithm 1 refers to a vertex processing policy that is not specified. The main processing policies used in the literature for determining the insertion sequence are: *Random,* vertices are processed in random order; *Maximum degree*, at each step, a vertex with the largest number of neighbors is

---

Algorithm 1. Lobe Ordering Algorithmic Framework

$N \leftarrow |N(v)|$
$S \leftarrow N(v)$
while $S \neq \emptyset$ **do**
    $Nexts \leftarrow$ select a start vertex $s \in S$;
    while $Nexts \neq \emptyset$ **do**
        1. extract a vertex $u \in Nexts$ according to the adopted processing policy
        2. $S \leftarrow S \setminus \{u\}$
        3. assign priority $p_u(x)$ to each position $x \in \{0, ..., N-1\}$
        4. place u at position x so that $\min_x p_u(x)$
        5. Nexts $\leftarrow Nexts \cup \{$unplaced vertices of $N(u) \cap N(v)\}$
    end while
end while

---

placed; *Minimum degree,* at each step, a vertex with the least number of neighbors is placed; and *Connectivity,* at each step, a vertex with the largest number of already placed neighbors and (in case of ties) the least number of unplaced neighbors is selected. In our experiment we test the effectiveness of each of them.

Note that the inner cycle of Algorithm 1 involves vertices belonging to the same connected component. Indeed, inside this cycle, a wave-like scanning of the connected component is performed. The outer cycle is used to iterate the inner one on all the connected components, until all vertices are placed.

In order to evaluate the impact (in terms of the qualities of interest) of the placement of the selected vertex in each available position of the current layout, we introduce a real valued *priority function*. The function $p_u(x) : \{0 \ldots N-1\} \rightarrow \mathbb{R}$ estimates the quality of the position x for vertex u considering the *cost* associated with the edges that link u to its already placed neighborhood (i.e. $N_p(u)$). We define $length(s,t) = min(|s-t|, N-|s-t|)$. Let $p_u(x) = \sum_{w \in N_p(u)} c_k(length(x, \pi(w)), L)$ if position x is available and let $p_u(x) = \infty$ otherwise.

With reference to the qualities desired for the representation, we can define different *cost functions* $c_k(d, L) : \{1, \ldots, \lfloor N/2 \rfloor\} \times \{1, \ldots, N\} \rightarrow \mathbb{R}$ that estimate the burden produced by an edge of length d in a layout with lobe size L. Fig. 3 shows two possible cost functions aimed at minimizing the number of crossings between inner edges. The former function $c_L(d, L)$ (*MinLobe*) estimates the cost of an edge of length d in a lobe of size L as the maximum number of visible crossings this edge can produce if there exist (in the lobe) all the edges that are able to cross it. The latter function $c_{LS}(d, L)$ (*MinLobeSum*) computes a weighted average of the previous one with respect to the number of lobes the edge will appear in.

Actually, the current implementation of the algorithm associates with each location a list of priority values that are sequentially evaluated. This feature makes it possible to choose the final position of a vertex using multiple selection criteria (or, equivalently, a *multi-objective function*), whose relative importance is defined by the order they are taken into account.

$$c_L(x, L) = \begin{cases} (x-1)(L-x-1) & x \geq 1 \wedge x < L \\ 0 & otherwise \end{cases}$$

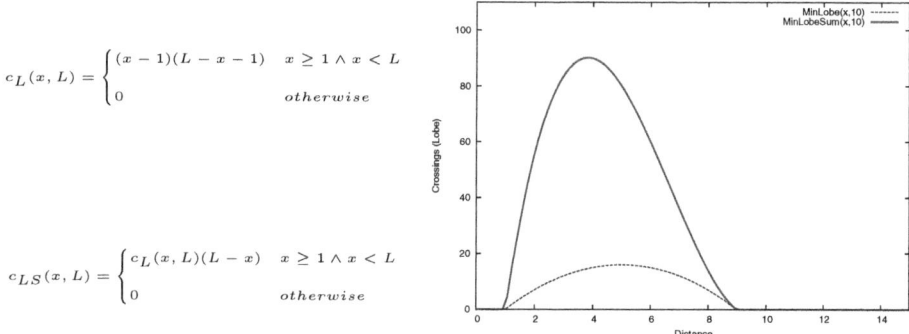

$$c_{LS}(x, L) = \begin{cases} c_L(x, L)(L-x) & x \geq 1 \wedge x < L \\ 0 & otherwise \end{cases}$$

Fig. 3. Cost functions for minimizing visible crossings

If we want to maximize the visible edges, a simple cost function whose purpose is to maximize the number of edges that can become inner edges in some lobe is function $c_I(d, L)$ (*MaxVisibleEdge*), defined as follows: $c_I(d, L) = -1$ if $d < L$ and $c_I(d, L) = 0$ if $d \geq L$.

If we find that two positions have the same priority, we break the tie using a cost function called *MinEdgeLength*, that is $c_D(x, L) = x \ \forall x, L$. It is used to avoid enlarging too much the set of considered positions.

As we said in Section 2, special attention is needed to perform an effective morphing procedure for the focus change primitive from the current focus v to the new focus u. This has important effects also in the algorithmic framework. Let ω_i^L be the visible lobe of v. It is essential, to preserve the user mental map, to order the vertices in $N(u)$ such that the relative order of the vertices in $N(u) \cap \omega_i^L$ remains the same. Also, the external vertices of u that enter the drawing must be placed after the inner vertices at the left of v in the lobe. Hence, the algorithm should be able to impose constraints of the above type on the lobe ordering. Observe that the above algorithmic approach, since the positions are explicitly represented, easily integrates such type of constraints.

5   Experimental Analysis of the Algorithms

In order to asses the effectiveness of our algorithmic techniques we performed several experimental tests using a suite of randomly generated connected graphs, where each graph represents the subgraph induced by $N(v)$ for some choice of focus vertex v. For each test we generated an average of $5,000$ graphs according to two modalities: 1. fixed number of vertices (100) and variable density (from 0.1% to 100%); 2. fixed density and variable number of vertices (from 10 to 100). Both types of graphs are evaluated with a fixed lobe size of 10. This value represents a good compromise between the number of the visible vertices and an effective usage of the available space.

We evaluated the algorithms according to the following requirements: 1. visible crossings, that is the number of crossings between inner edges; 2. visible edges ratio, that is the ratio between the number of visible edges and the number of edges of the graph. A selection of the results is presented below.

In the experiment shown in Fig. 4(a) we compare a simple random order with the BB algorithm [1], chosen as one of the best algorithms for computing a circular layout with few crossings. Each point shows the average number (computed over all lobes) of the visible crossings for a graph whose density is reported on the x-axis. A random order produces fewer visible crossings than the BB algorithm. Such results suggest that the circular approach is not a good choice for the *visible crossings minimization problem*. The tested circular algorithm tries to minimize the total length of the edges but short edges enter more easily in a lobe, increasing the possibility of generating crossings.

In the experiment shown in Fig. 4(c) we compare several variants of our algorithm (see Fig. 6). We changed the processing policy that selects the next vertex to insert (Random or Connectivity) and the cost functions for each level of priority (MinLobe or MinLobeSum). Observe that in all cases we use, for a second priority level, cost function MinEdgeLength. Mixing these features we obtain four different algorithms to evaluate (RLobe, RSum, CLobe, CSum). The graphic shows the comparison between these algorithms with a variable number of vertices and fixed density 15%. It is evident that the choice of a good processing policy has a strong influence on the number of visible crossings. Algorithms that use the connectivity policy perform better than those using the random policy. For a given processing policy, the MinLobeSum cost function is slightly better than the MinLobe.

In the experiment shown in Fig. 4(e) we compare the average number of visible crossings of the CSum algorithm with the minimum number of visible crossings on a lobe of a Random placement. The graphic shows that CSum has a similar trend to Random with slightly better results. Roughly, given an unordered layout and selected the lobe with lower number of visible crossings, the algorithm generates a drawing with the same average number of visible crossings for each lobe.

In the experiment shown in Fig. 4(b) we compare the visible edges ratio between Random order and BB Algorithm. Although BB is not designed for solving this problem, drawing shorter edges decreases the number of edges hidden by the paradigm and therefore increases the visible edges ratio. In the graphic we observe that BB Algorithm has a better behavior than the one of a Random order.

For this problem we follow two different approaches. The first is to reverse the cost function of CSum in order to increase the number of crossings and consequently the number of visible edges. The second approach is a typically greedy approach where a vertex is positioned in the local highest visible edges position. This cost function is associated with a Random and Connectivity processing policy. A Random selection decreases the effectiveness of the results of the MaxVisibleEdge cost function. With an equal processing policy the MaxVisibleEdge function is slightly better than MaxSum. (See Fig. 4(d).)

In the experiment shown in Fig. 4(f) we compare CVis algorithm with BB algorithm. CVis has better results than BB, especially with a small number of vertices (CVis keeps visible 100% of edges for graphs with more vertices than BB). Increasing the number of vertices both algorithms have the same trend and CVis has generally an improvement of $2 - 3\%$.

6   Implementation and Case Studies

The project has been fully driven by the experiments performed on the devices. (See in Fig. 5(a) and 5(b) the usage of the important primitive focus change over the iPhone device.) This led to two SW libraries, one for the iPhone OS 3.1.3 (Objective-C language) and the other for the Google Android 2.1 (Java language) platforms. Both the libraries were designed to have fully customizable graphical and behavioral components and to allow simple usage for the SW developer.

Although the two platforms are very different, we developed the SW so that both prototypes have the following common features: 1. use vector graphics, 2. build a new abstract layer over the platform to manage animations and gestures, and 3. optimize the number of operations and edge drawings needed for each display refresh. Because of the limitations of the platform, the Android release required also to minimize the number of refreshes.

We implemented several case studies. Two of them, that refer to the context of social networks, are especially interesting.

The first case study uses the *Facebook API* to determine the graph of the friendships of a Facebook user (see Fig. 5(c) and 5(d)). These APIs allow to read (and write) objects and social connections of the Facebook Graph. The objects, that are the vertices of the graph, have a unique identifier (ID) and their associated data can be retrieved with a simple fetch of the URL `https://graph.facebook.com/ID`. All objects are linked together through relationships of different types for different objects. We retrieve the connections using URLs of the form: `https://graph.facebook.com/ID/CONNECTION_TYPE`. Obviously, in order for the queries to succeed they must not violate the privacy restrictions set by the users. The user can provide his/her personal credentials (username and password) to our *FacebookView Application* through an input window. The system automatically generates the necessary requests to determine the subgraph of the Facebook Graph induced by the set of vertices consisting of the facebook user and his/her friends. Actually, in the current implementation the set of vertices (and relationships between them) is extended at run-time with queries that refer to resources (events, photos, links, videos, etc.) liked or tagged by other users.

The second case study shows the public relations exposed and available on the Web from the *Google Social Graph API* (see Fig. 5(f)). This information

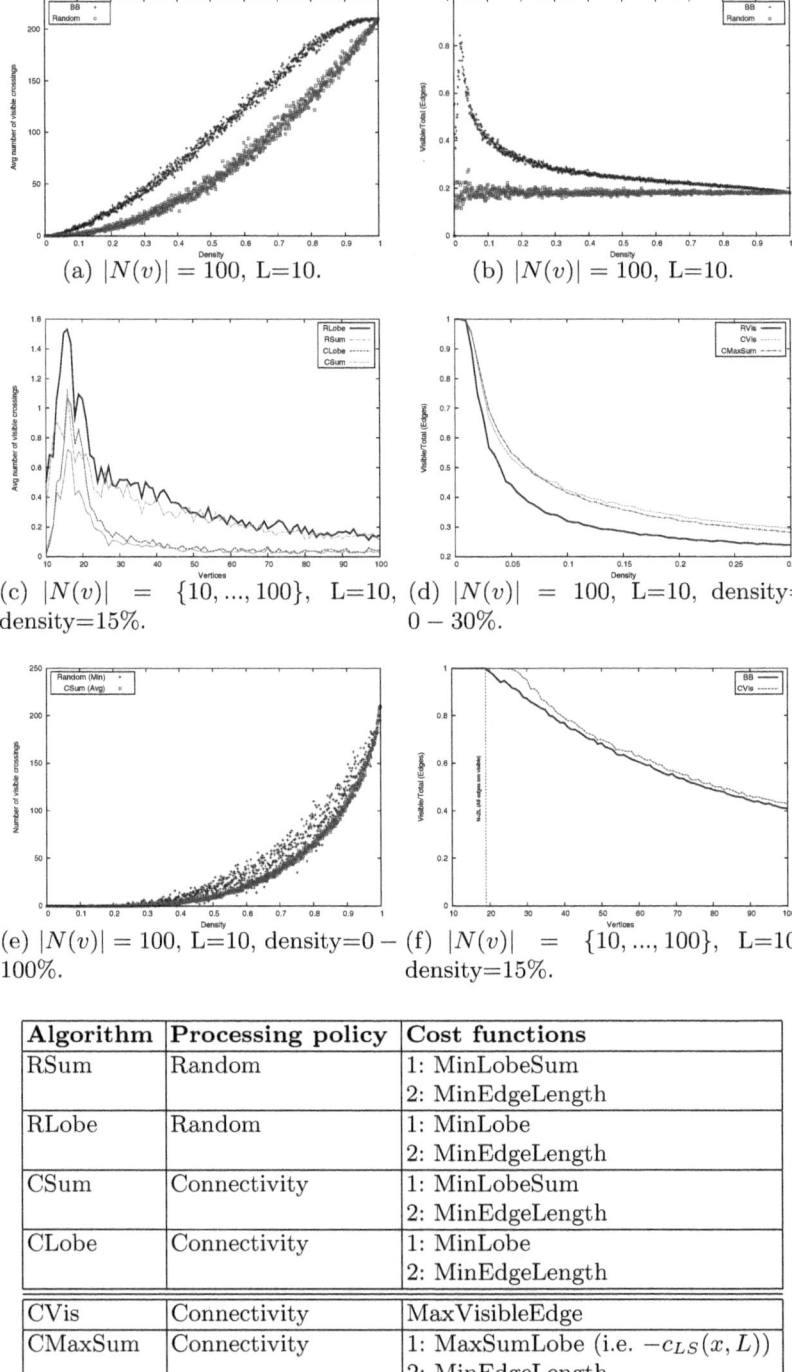

(a) $|N(v)| = 100$, L=10.

(b) $|N(v)| = 100$, L=10.

(c) $|N(v)| = \{10,...,100\}$, L=10, density=15%.

(d) $|N(v)| = 100$, L=10, density= $0-30\%$.

(e) $|N(v)| = 100$, L=10, density=0 – 100%.

(f) $|N(v)| = \{10,...,100\}$, L=10, density=15%.

Algorithm	Processing policy	Cost functions
RSum	Random	1: MinLobeSum 2: MinEdgeLength
RLobe	Random	1: MinLobe 2: MinEdgeLength
CSum	Connectivity	1: MinLobeSum 2: MinEdgeLength
CLobe	Connectivity	1: MinLobe 2: MinEdgeLength
CVis	Connectivity	MaxVisibleEdge
CMaxSum	Connectivity	1: MaxSumLobe (i.e. $-c_{LS}(x, L)$) 2: MinEdgeLength

Fig. 4. Algorithms experimental results

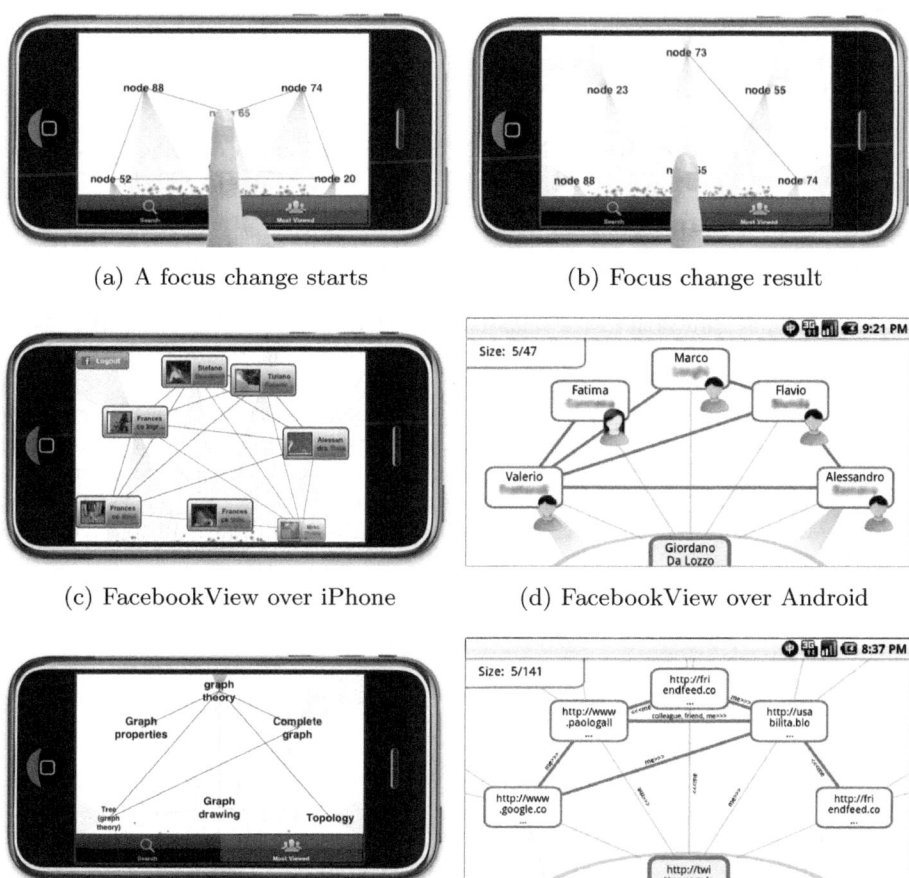

(a) A focus change starts

(b) Focus change result

(c) FacebookView over iPhone

(d) FacebookView over Android

(e) WikipediaView over iPhone

(f) SocialView over Android

Fig. 5. Application samples

is declared within public profiles via XFN (XHTML Friends Network), FOAF (Friend Of A Friend), and other declared public connections. For example, XFN provides a simple way to define human relationships through Web links using the *rel* attribute of the <*a href*> tag (e.g. <a href="http://pino.example.com/" rel="colleague met">Pino</a>). The user can provide to our *SocialView Application* the URL of a public account through an input window. The system automatically generates a query for acquiring as much relational information as possible.

We also implemented a case study to explore Wikipedia (see Fig. 5(e)). A user selects a word and the smartphone shows the related concepts. For example if the user selects Graph Drawing the device shows the Graph Theory, Topology, Geometry, etc. When the user finds an interesting concept, he/she can expand the vertex and visualize the Wikipedia page.

References

1. Baur, M., Brandes, U.: Crossing reduction in circular layouts. In: Hromkovič, J., Nagl, M., Westfechtel, B. (eds.) WG 2004. LNCS, vol. 3353, pp. 332–343. Springer, Heidelberg (2004)
2. Eades, P., Cohen, R.F., Huang, M.L.: Online animated graph drawing for web navigation. In: di Battista, G. (ed.) GD 1997. LNCS, vol. 1353, pp. 330–335. Springer, Heidelberg (1997)
3. He, H., Sykora, O.: New circular drawing algorithms. In: Proc. ITAT 2004 (2004)
4. Makinen, E.: On circular layouts. Internal Journal of Computer Mathematics, 29–37 (1988)
5. Masuda, S., Kashiwabara, T., Nakajima, K., Fujisawa, T.: An NP-hard crossing minimization problem for computer network layout. Technical report (1986)
6. Pixelglow. Instaviz (2008), `http://instaviz.com/`
7. Sarkar, M., Brown, M.H.: Graphical fisheye views of graphs. In: Proceedings of the Conference on Human Factors in Computing Systems CHI 1992 (1992)
8. Shneiderman, B.: The eyes have it: A task by data type taxonomy for information visualizations. In: Proc. of the IEEE Symp. on Visual Lang., pp. 336–343 (1996)
9. Six, J.M., Tollis, I.G.: Circular drawings of biconnected graphs. In: Proc. ALENEX (1999)

Topology-Driven Force-Directed Algorithms*

Walter Didimo, Giuseppe Liotta, and Salvatore A. Romeo

Università di Perugia, Italy
{didimo,liotta,romeo}@diei.unipg.it

Abstract. This paper studies the problem of designing graph drawing algorithms that guarantee good trade-offs in terms of number of edge crossings, crossing angle resolution, and geodesic edge tendency. It describes two heuristics designed within the *topology-driven force-directed* framework that combines two classical graph drawing approaches: the force-directed approach and a planarization-based approach (e.g., the topology-shape-metrics approach). An extensive experimental analysis on two different test suites of graphs shows the effectiveness of the proposed solutions for the optimization of some readability metrics.

1 Introduction and Overview

Several empirical studies compare different aesthetic criteria for drawing graphs and show that the number of edges crossings often has the strongest impact on the readability of a diagram (see, e.g., [21,22]). As a consequence, it has been widely accepted that one of the primary optimization tasks of a good graph drawing algorithm is the minimization of crossings and a large body of literature has been devoted to this topic (see, e.g., [4,20]).

Cognitive experiments by Huang, Huang *et al.*, and Ware *et al.* provide new insights into the classical correlation between edge crossings and human understanding of graph drawings [18,19,24]. As these experiments show, the readability of a diagram not only depends on the edge crossings (which are unavoidable for dense non-planar graphs) but also on the "quality" of these crossings and on the "quality" of the curves that cross with each other. The experiments suggest that the query "Are two vertices adjacent in the drawing?" is easier to answer when the minimum angle formed by the crossing edges is as large as possible and when the curves that represent the edges are as straight as possible. The task of easily following the drawing of an edge from its source to its destination is also listed in the *NetViz Nirvana* of a recent study by Dunne and Shneiderman about HCI design principles for visual analytics [7].

This paper studies the problem of designing graph drawing algorithms that guarantee good trade-offs in terms of number of edge crossings, *crossing angle resolution* and *geodesic edge tendency*. The crossing angle resolution measures the value of the minimum angle formed by two crossing edges. The geodesic edge tendency measures how close a bent edge is to the straight-line segment connecting the end-points of the edge. To this aim, we adopt a graph drawing methodology, which we call *topology-driven force-directed* approach, that combines two classical graph drawing algorithmic frameworks: the *force-directed* approach and *planarization-based* approach.

---

* Work supported in part by the MIUR project AlgoDEEP prot. 2008TFBWL4.

U. Brandes and S. Cornelsen (Eds.): GD 2010, LNCS 6502, pp. 165–176, 2011.

The force-directed approach associates the input graph with a physical system of forces and it tries to place the vertices so that the total energy of the physical system is minimal. Only a few examples of force directed algorithms use edges with bends (see, e.g., [5,10,12]); the vast majority of the force-directed algorithms represent the edges as straight-line segments and thus the computed drawings are optimal in terms of geodesic edge tendency. Unfortunately, force directed algorithms may introduce unnecessarily many crossings. For example, the drawing of Fig. 1(a) computed with a force-directed algorithm of the OGDF Library[1] contains edge crossings even though the input graph is planar.

A planarization-based algorithm first computes a topology (i.e., an embedding) of the graph with a small number of edge crossings and then it applies a drawing algorithm that preserves this topology. Since the crossing minimization problem is NP-hard, a planarization-based algorithm typically relies on heuristics such as first computing a maximal planar subgraph and then inserting an edge per time (see, e.g., [4,15]). Each unavoidable edge crossing is replaced by a dummy vertex and a planar embedding of an augmented planar graph is obtained. The drawing algorithm typically preserves this computed topology by introducing bends along the edges; for example, the well-known *topology-shape-metrics* algorithm by Tamassia [23] is a planarization-based algorithm that computes orthogonal drawings. Planarization-based algorithms are likely to compute drawings with smaller number of crossings than force-directed algorithms and often with a better crossing angle resolution. However, the bends along the edges may strongly affect the geodesic edge tendency. See for example the orthogonal drawing of Fig. 1(b). This drawing, computed with the GDToolkit Library[2], depicts the same graph of Fig. 1(a). Note that, understanding whether the two black vertices are adjacent is more complicated in the orthogonal drawing because their connecting edge is far from being geodesic.

The main results in this paper are as follows:

– We describe ORTHFD and POLYFD, two graph drawing algorithms that were designed within the topology-driven force-directed framework (see Section 3). For example, Fig. 1(c) shows a drawing computed by algorithm ORTHFD. Observe that compared to Fig. 1(a), the drawing does not have crossings; compared to Fig. 1(b) the edge between the black vertices is easier to follow.
– We perform an experimental study that compares our algorithms with respect to some of the most effective force-directed and planarization-based algorithms described in the literature. The results show that the drawings computed with the topology-driven force-directed approach achieve better trade-offs in terms of number of edge crossings, crossing angle resolution, and geodesic edge tendency. They often provide good results with respect to some other important aesthetic criteria as well, such as number of bends and vertex angle resolution (see Section 4).

Throughout this paper we assume that the reader is familiar with the basic concepts of force-directed and planarization-based techniques (see e.g. [4,20]).

---

[1] http://www.ogdf.net/

[2] http://www.dia.uniroma3.it/~gdt

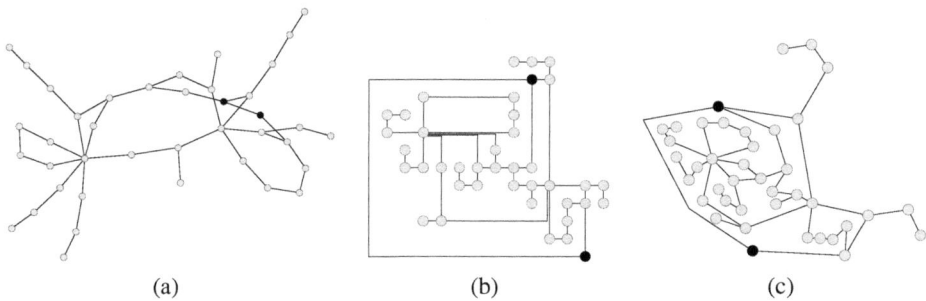

Fig. 1. Drawings of the same graph with different approaches: (a) A force-directed approach; (b) A planarization-based approach; (c) A topology-driven force-directed approach

2 Related Work

There are many examples in the literature describing force-directed algorithms that preserve a given topology (i.e. a given embedding). While we refer the interested reader to the accurate surveys in [8,9,10] for exhaustive lists of references, we mention here only those results that can be more closely related to the approach described in this paper.

Bertault describes a force-directed algorithm, called PRED, that preserves edge crossing properties [2]. Our work and the one of Bertault have some similarities but also substantial differences. Similar to PRED, our algorithmic framework uses force-directed techniques that do not increase the number of crossings with respect to an initial drawing of the graph. Differently from PRED, our work: (i) Combines crossing minimization heuristics and force-directed methods; (ii) Guarantees good crossing angle resolution by allowing bent edges; (iii) Enhances the force-directed model with a set of constraints that guarantees a good geodesic edge tendency in spite of the fact that the edges can bend.

Also, Dwyer *et al.* [9,10] have extensively studied sophisticated stress-optimization techniques that maintain a set of constraints, including a given topology. Similar to the algorithms in those papers, we allow bends along the edges. Differently from those papers, our primary goal is to find good trade-offs between number of edge crossings, crossing angle resolution, and geodesic edge tendency, while we do not focus on the stability problem in dynamic graph layout. This difference impacts on the developed techniques and on the computed drawings. For example, we do not consider the continuous network layout problem and we do not insist on preserving an initial mental map; in fact, our techniques may even change the initial topology if this change reduces the total number of edge crossings.

3 The Topology-Driven Force-Directed Approach

Let G be a graph. The *topology-driven force-directed* approach computes a drawing Γ of G into two main phases. Phase 1: A drawing Γ_0 of G is computed by applying a planarization-based algorithm. Phase 2: The final drawing Γ is obtained by applying a force-directed algorithm to Γ_0, with the constraint that Γ does not have more edge crossings than Γ_0.

Within this general approach, one can design several algorithms by combining different strategies for the two phases. In this work we describe two specific algorithms that we call ORTHFD and POLYFD, respectively. Both of these algorithms adopt a similar planarization step, that computes a planar embedding of G where crossings are replaced with dummy vertices, called *cross vertices*. In order to have a small number of edge crossings the planarization step uses a classical heuristic based on finding a shortest path in the dual graph for each edge insertion (see, e.g., [4]). Algorithms ORTHFD and POLYFD then construct a drawing Γ_0 of G that preserves the computed planar embedding. Algorithm ORTHFD uses the *topology-shape-metrics* approach for orthogonal drawings in the Kandinsky drawing convention [13]. According to this convention, the vertices are represented as squares of the same dimension and there may be parallel edges incident to vertices whose degree is larger than four. The orthogonal drawing is computed by the algorithm described in [3], which minimizes the total number of bends along the edges in $O(n^{\frac{7}{4}} \log n)$ time if the network flow algorithm of Garg and Tamassia is used [14]. We use the $O(n^2 \log n)$ algorithm described in [3].

Algorithm POLYFD computes a polyline drawing Γ_0 by applying the algorithm described in [6]. Γ_0 is constructed by finding a suitable orientation of the edges of the graph and by using this orientation to define a visibility representation of the graph. The polyline drawing is obtained collapsing the vertices of the visibility representation into points (see, e.g., [4]).

Phase 2 of ORTHFD and of POLYFD uses a common force-directed strategy, that works in different steps:

- **Step 1:** Every bend of Γ_0 is replaced with a dummy vertex, called a *bend vertex*. After this step all the edges are straight-line segments. Call Γ_1 the new drawing.
- **Step 2:** A force directed algorithm is applied, starting with Γ_1 as the initial drawing where the general physical model is that described in [11]: Each vertex is modeled as an electrically charged particle and each edge as a spring. The phisical model is augmented by additional constraints as described below:

    - Let $e = (u, v)$ be an edge of G that has k bends in Γ_0 and let $e_1, e_2, \ldots, e_{k+1}$ be the edges of Γ_1 forming e. Note that, if $k = 0$ then e_1 coincides with e; in this case we call e_1 a *straight edge*. With respect to Γ_1, denote by $\ell(e_i)$ the length of e_i, by $\ell(e)$ the sum of all $\ell(e_i)$, and by $d(u, v)$ the Euclidean distance between u and v. The spring corresponding to e_i is given a zero-energy length equal to $d(u, v) \frac{\ell(e_i)}{\ell(e)}$. Also, the stiffness of the springs that model the straight edges is smaller than the stiffness of the springs associated with the other edges. This is done to ensure that each edge approximates the straight line between its end-vertices, in order to obtain a good geodesic edge tendency.
    - Let $\overrightarrow{f(v)}$ be the resultant of all forces acting on v, and let d be the maximum distance by which v can be moved along the direction of $\overrightarrow{f(v)}$ without creating any new edge crossings. Vertex v is moved along the direction of $\overrightarrow{f(v)}$ by a quantity δ such that $\delta \leq \min\{d, |\overrightarrow{f(v)}|\}$. More precisely, if moving v by a quantity $|\overrightarrow{f(v)}|$ does not create crossings, then we set $\delta = |\overrightarrow{f(v)}|$, otherwise we compute δ by applying a binary search along the direction of $\overrightarrow{f(v)}$; we stop

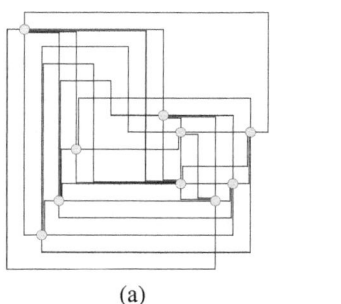 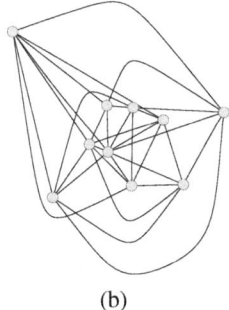

(a)   (b)

Fig. 2. Two drawings of the same graph computed with different algorithms. The drawing computed by ORTHFD, in Fig. 2(b), has six fewer crossings than the one computed by a planarization-based orthogonal drawer of GDToolkit, in Fig. 2(a). The edges in ORTHFD are represented by smoothed curves.

this search as soon as a placement for v is found that does not increase the number of edge crossings. Note that this procedure may lead to a reduction of the number of crossings computed by the planarization step. For example, the drawing computed by ORTHFD in Fig. 2(b) has six fewer crossings than the one in Fig. 2(a), computed by using a planarization-based orthogonal drawer of GDToolkit.

– **Step 3:** This is a post-processing step whose goal is to improve the crossing angle resolution. Let Γ_2 be the drawing at the end of Step 2 and let α be a target value for the minimum angle formed by two crossing edges. Let e_1, e_2 be two crossing edges of Γ_2 that form an angle smaller than α. In order to enlarge the crossing angle formed by $e_1 = (u_1, v_1)$ and $e_2 = (u_2, v_2)$ we adopt a technique similar to that described in [5]. Denote by c the crossing point of e_1, e_2. Define a disk δ centered at c such that δ does not intersect in Γ_2 any edge other than e_1 and e_2. Let p_i, q_i be the intersection points between δ and e_i ($i \in \{1, 2\}$). Edge e_i is split into the path $(u_i, p_i), (p_i, q_i), (q_i, v_i)$. Also, the straight-line edges $(p_1, p_2), (p_2, q_1), (q_1, q_2), (q_2, p_1)$ are added to the drawing. The four-cycle formed by these dummy edges is called the *cage* of crossing c. The zero-energy length of all edges forming the cages is set to be a constant smaller than the zero-energy length of the shortest edge of Γ_2; the stiffness of the edges in the cages is larger than the one of the other edges. Let Γ_2' be the drawing obtained from Γ_2' by inserting all cages. By applying a few iterations of the force-directed method to Γ_2', we obtain a new drawing where the cages are drawn as close as possible to squares, which enforces their diagonals to cross at large angles.

– **Step 4:** Let Γ_3 be the drawing at the end of Step 3. The algorithm iteratively removes unnecessary bends along the edges. A bend is unnecessary if its removal: (i) does not introduce new crossings; (ii) does not make the crossing angle resolution lower than the given threshold α.

4   Experimental Study

We tested algorithms ORTHFD and POLYFD on two different test suites of graphs. The first test suite, called Rome, consists of 300 graphs randomly selected in the popular collection known as the "Rome graphs" [1]; we selected 30 graphs for each fixed number of vertices $n \in \{10, 20, \ldots, 100\}$. The graphs in this test suite are typically sparse (the average density of the Rome graphs is about 1.4) and they reflect the structure of graphs coming from real life applications in the field of databases.

The second test suite, called Rand consists of 300 randomly generated graphs, that are denser than the "Rome graphs". They have number of vertices $n \in \{10, 20, \ldots, 50\}$ and density $d \in [1.5 - 4.0]$. For each pair $\langle n, d \rangle$, a graph with n vertices and $m = d \cdot n$ edges was generated with a uniform probability distribution; we generated 60 graphs for each distinct value of n.

We initially compared the drawings computed by ORTHFD and POLYFD with three effective force-directed and planarization-based algorithms, namely FM^3, ORTH, and POLY. Algorithm FM^3 is the fast force-directed algorithm described by Hachul and Jünger [16]. Among the wide set of force-directed algorithms, we chose FM^3 because an experimental study showed that it typically computes drawings with smaller number of crossings and edge overlaps than other force-directed methods [17]. We used the implementation of FM^3 available in the open source library OGDF. ORTH and POLY are the algorithms that compute an orthogonal drawing and a polyline drawing in Phase 1 of ORTHFD and POLYFD, respectively. We used the implementation of these algorithms available in the GDToolkit Library; the algorithms ORTH and POLY perform well in terms of number of bends and number of edge crossings [4]. However, after a few experimental measures we realized that ORTH outperformed POLY in all aesthetics. Also POLY was often giving rise to drawings with overlapping edge bends, which was making the final drawings rather confusing. For example, Fig. 3 shows five drawings of the same graph, each computed by one of the above algorithms. It is immediate to see that the drawing of Fig. 3(d) is less readable than the others when trying to answer basic queries, such as identifying the neighbors of the black vertex (white vertices). Therefore, we decided to restrict the experimental comparison to algorithms ORTH, FM^3, ORTHFD, and POLYFD. We remark that the problem of overalpping edge bends does not occur with algorithm POLYFD, where the repulsive forces acting between pairs of vertices do not allow two bends to be drawn at the same point.

In the experimental setting of algorithms ORTHFD and POLYFD we set the minimum threshold α for the crossing angle resolution to be $\alpha = 30°$. This is motivated by the work of Ware et al. [24], who observe that crossing angles smaller than $30°$ are more likely to cause visual confusion when reading a drawing of a graph.

In the following we present the charts summarizing our experimental results. In all charts we use bars with the following colors for the different algorithms: (a) FM^3 - White color; (b) ORTH - Light gray color; (c) ORTHFD - Dark gray color; POLYFD - Black color.

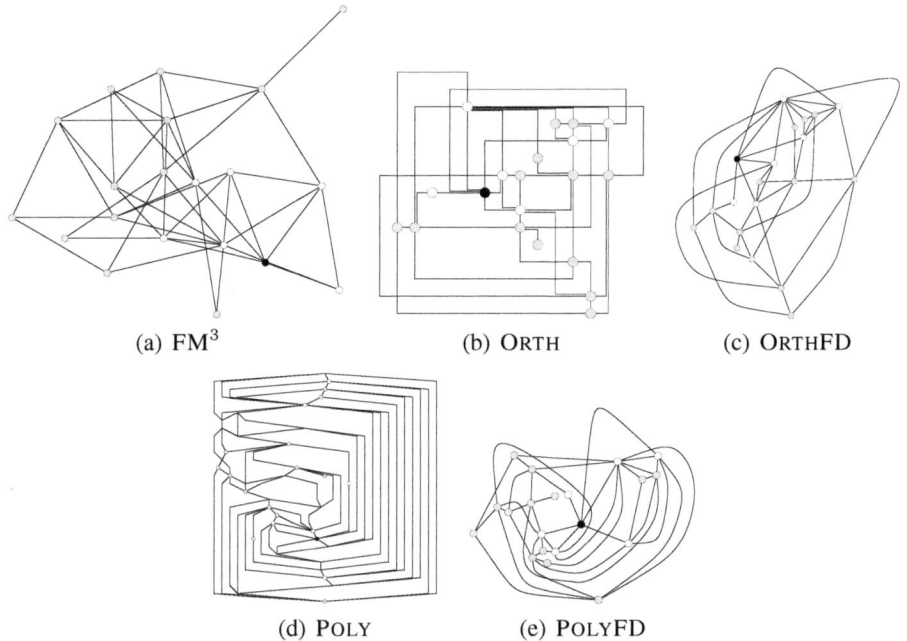

(a) FM^3 (b) ORTH   (c) ORTHFD

(d) POLY   (e) POLYFD

Fig. 3. Five drawings of the same graph computed with different algorithms

Number of Edge Crossings

Fig. 4(a) and Fig. 4(b) show that the number of edge crossings of ORTH, ORTHFD, and POLYFD is, on average, half that computed by FM^3. This behavior is a consequence of the impact of the planarization heuristic used in ORTH, ORTHFD, and POLYFD. The experiments also confirmed the observation made in the previous section that Phase 2 of the topology-driven force-directed framework can further reduce the number of edge crossings of a planarized drawing. In particular, the drawings computed by ORTHFD contain on average 10% less edge crossings than ORTH. We finally remark that some of the drawings computed by FM^3 presented edge overlaps (37 in the Rome test suite and 62 in the Rand test suite). The other algorithms never created edge overlaps.

For example, Fig. 3(a) has 59 edge crossings, Fig. 3(b) and Fig. 3(e) have 17 edge crossings, while Fig. 3(c) has 16 edge crossings.

Crossing Angle Resolution

Fig. 5(a) and Fig. 5(b) show the crossing angle resolution in the drawings computed by FM^3, ORTHFD, and POLYFD, that is, the value of the minimum angle formed by any two crossing edges. Clearly, the crossing angle resolution of the drawings computed by ORTH is always optimal (i.e., 90°) and hence we do not report the data for this algorithm. If a drawing is planar, it was assigned an optimal crossing angle resolution. From the charts it can be seen that both ORTHFD and POLYFD outperform FM^3.

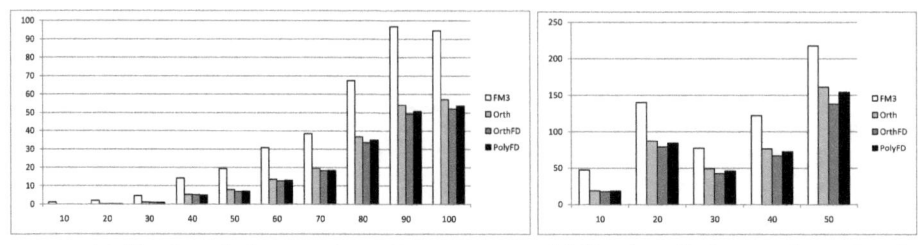

(a) Number of edge crossings - Rome   (b) Number of edge crossings - Rand

Fig. 4. Number of edge crossings for the two test suites of graphs (average values over the number of vertices)

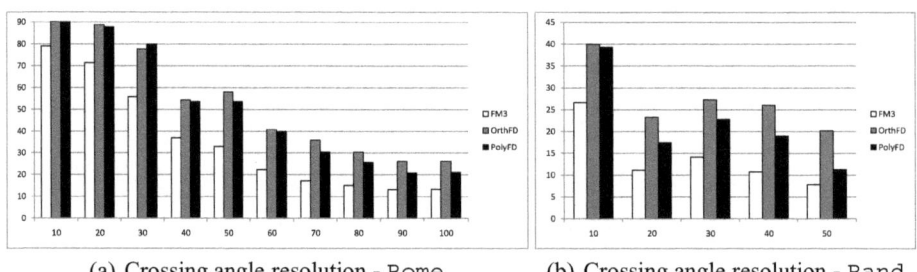

(a) Crossing angle resolution - Rome   (b) Crossing angle resolution - Rand

Fig. 5. Crossing angle resolution for the two test suites (average values over the number of vertices)

In particular, the average crossing angle resolution of the drawings computed by OR-THFD is always above $30°$ for number of vertices up to 80 in the Rome test suite. Also, the crossing angle resolution of the drawings computed by FM^3 on the Rand graphs is often very poor (around $10°$), while it is maintained always above $20°$ by ORTHFD.

Also, we observed that the percentage of edge crossings with an angle smaller than $30°$ is about 6% on average for the drawings computed by FM^3, while it is about 2% for the drawings computed by algorithms ORTHFD and POLYFD. This data is particularly relevant if considered together with the observation that the number of edge crossings created by FM^3 is typically much higher than the one created by ORTHFD and POLYFD.

For example, Fig. 3(a) has two crossings that form an angle less than $30°$, all edge crossings in Fig. 3(c) form angles larger than $30°$, and Fig. 3(e) has one edge crossing that forms an angle less than $30°$.

Geodesic Edge Tendency

Clearly, the drawings computed by FM^3 are optimal in terms of geodesic edge tendency, because they have straight-line edges only. To measure the geodesic edge tendency of the drawings computed by algorithms ORTH, ORTHFD, and POLYFD we considered two different parameters: (i) The percentage of edges (u, v) that are not monotone in

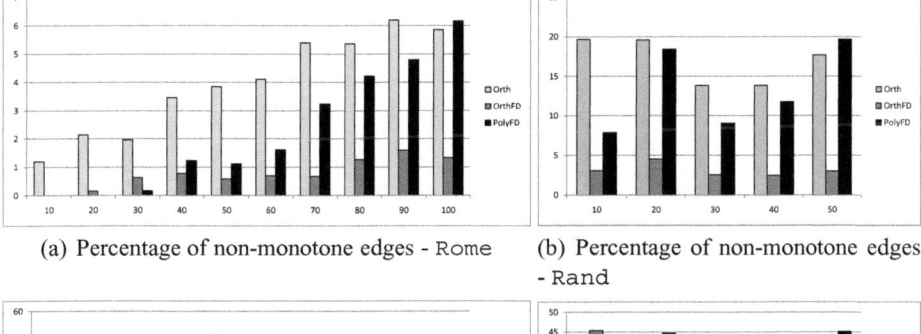

(a) Percentage of non-monotone edges - Rome   (b) Percentage of non-monotone edges - Rand

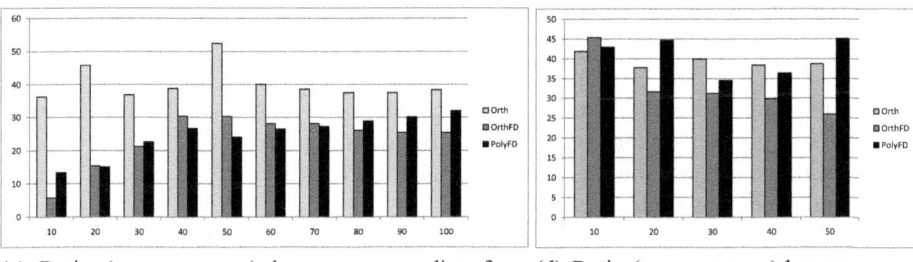

(c) Ratio (as percentage) between max. dist. from (d) Ratio (as percentage) between max.
(u, v) to $\overline{uv}$ and $|\overline{uv}|$ - Rome   dist. from (u, v) to $\overline{uv}$ and $|\overline{uv}|$ - Rand

Fig. 6. Geodesic edge tendency (average values over the number of vertices)

the direction of the straight segment $\overline{uv}$; small percentages of such edges indicate good geodesic edge tendency. (ii) The ratio between the maximum distance from an edge (u, v) to the segment $\overline{uv}$ and the length of $\overline{uv}$. Small values indicate good geodesic edge tendency.

Fig. 6(a) and Fig. 6(b) show that the number of non-monotone edges in the drawings computed by ORTHFD is between than $2 - 4\%$, and dramatically improves the values for the drawings computed by ORTH, namely by 82% on average. Also POLYFD gives some improvement with respect to ORTH for almost all instances, but this improvement is not so relevant as for ORTHFD. For example, Fig. 3(b) has six non-monotone edges, all edge in Fig. 3(c) are monotone, and Fig. 3(e) has four non-monotone edges.

Concerning the second parameter used to measure the geodesic edge tendency, Fig. 6(c) and Fig. 6(d) show the ratio (expressed as percentage) between the maximum distance from an edge (u, v) to the segment $\overline{uv}$ and the length of $\overline{uv}$. It can be observed that for the Rome graphs both the two topology-driven force-directed algorithms significantly improve the results of algorithm ORTH. More precisely, this percentage is on average 40% for ORTH, 23% for ORTHFD, and 24% for POLYFD. For the Rand graphs Orth and PolyFD perform similarly (about 40% on average), while OrthFD perform better for almost all instances (about 32% on average).

Bends and Vertex Angles

From the experiments we also observed that ORTHFD works well also in terms of number of bends and vertex angle resolution. Namely, Fig. 7 shows that the ORTHFD

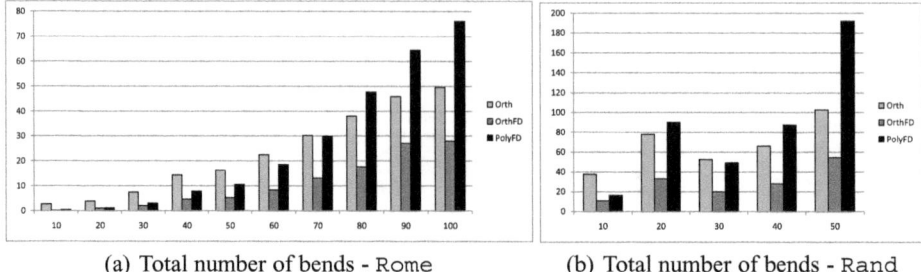

(a) Total number of bends - Rome   (b) Total number of bends - Rand

Fig. 7. Total number of bends for the two test suites (average values over the number of vertices)

reduces the total number of bends of ORTH by 60% on average, both for the Rome and for the Rand graphs. On the contrary, POLYFD usually performs worse than the other two algorithms, because the drawings computed in Phase 1 by POLYFD contain many more bends than those computed by ORTH.

About the vertex angle resolution, we excluded from the comparison the drawings computed by Orth, where two edges incident to the same vertex may form an angle of zero degree (if they are parallel). Fig. 8 shows that ORTHFD and POLYFD perform better than FM^3 except for the sample of 10 vertices in the Rome graphs. In particular, the vertex angle resolution of the drawings computed by ORTHFD is on average 2.5 times higher than that of the drawings computed by ORTH on the Rand graphs.

For example, Fig. 3(b) has 38 edge bends, Fig Fig. 3(c) has 10 edge bends, and Fig. 3(e) has 34 edge bends. Also, Fig. 3(a) has vertex angle resolution equal to 0.64°, Fig. 3(c) has vertex angle resolution equal to 11.2°, and Fig. 3(e) has vertex angle resolution equal to 5.68°.

Running Time

We designed our algorithms to take into account several readability metrics, which typically conflict with each other. The good performances in terms of aesthetic criteria are

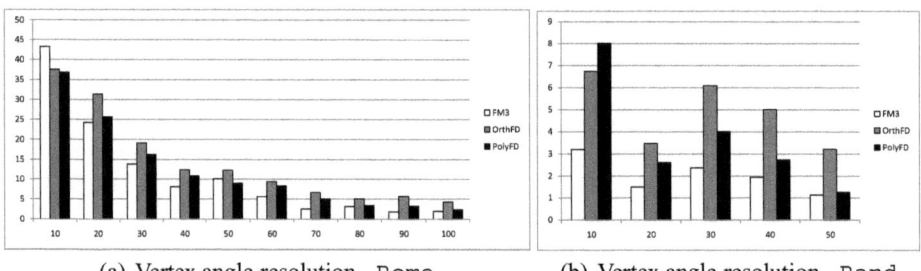

(a) Vertex angle resolution - Rome   (b) Vertex angle resolution - Rand

Fig. 8. Vertex angle resolution (average values over the number of vertices)

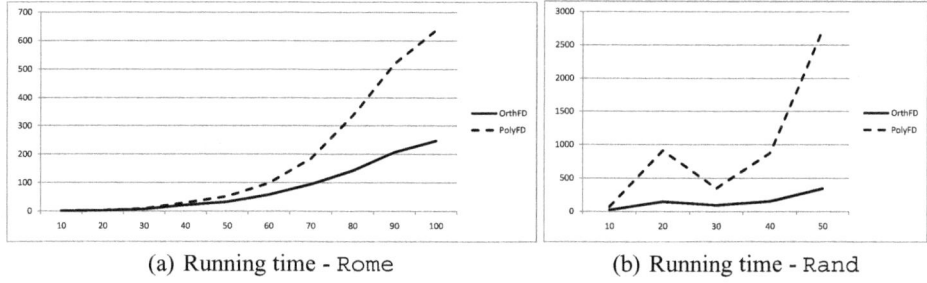

(a) Running time - Rome   (b) Running time - Rand

Fig. 9. Running time (in seconds) of algorithms ORTHFD and POLYFD (average values over the number of vertices)

however paid in terms of computational efficiency. A theoretical analysis shows that the algorithms require $O((n + b)(m \log m))$ time, where n, m, and b are the number of vertices, edges, and bends in the drawing. The time performances of ORTHFD and POLYFD are reported in Fig. 9. The algorithms were executed on a PC Intel Core Duo 2.66 GHz and 2GB RAM. From the charts, it is possible to see that the Rand graphs are computationally much harder than the Rome graphs, due to their higher density. However, the time required by ORTHFD is much smaller than that required by POLYFD.

5   Conclusions and Open Problems

In this paper we concentrated on the design of graph drawing algorithms that guarantee a good trade-off between number of edge crossings, crossing angle resolution, and geodesic edge tendency. We adopted the topology-driven force-directed framework and experimentally studied the performances of two algorithms, ORTHFD and POLYFD designed within this framework. The experimental analysis shows that algorithm OR-THFD has a better trade-off between number of edge crossings, crossing angle resolution, and geodesic edge tendency than existing force-directed and planarization-based algorithms. Also, it behaves rather well in terms of number of bends and in the vertex angle resolution. Furthermore ORTHFD generally outperformed POLYFD in all the experiments we have executed.

As our experiments show, algorithm ORTHFD can be reasonably applied to graphs having up to 100 vertices. It would be interesting to design graph drawing algorithms that perform equally well with respect to the aesthetic criteria taken into account in this paper and that can be applied efficiently to larger graphs. More experimental results that compare our technique with other force-directed algorithms are also of interest.

References

1. Di Battista, G., Garg, A., Liotta, G., Tamassia, R., Tassinari, E., Vargiu, F.: An experimental comparison of four graph drawing algorithms. Comput. Geom. 7, 303–325 (1997)
2. Bertault, F.: A force-directed algorithm that preserves edge crossing properties. In: Kratochvíl, J. (ed.) GD 1999. LNCS, vol. 1731, pp. 351–358. Springer, Heidelberg (1999)

3. Bertolazzi, P., Di Battista, G., Didimo, W.: Computing orthogonal drawings with the minimum number of bends. IEEE Trans. on Computers 49(8), 826–840 (2000)
4. Di Battista, G., Eades, P., Tamassia, R., Tollis, I.G.: Graph Drawing. Prentice-Hall, Upper Saddle River (1999)
5. Didimo, W., Liotta, G., Romeo, S.A.: Graph visualization techniques for conceptual web site traffic analysis. In: PacificVis, pp. 193–200. IEEE, Los Alamitos (2010)
6. Didimo, W., Pizzonia, M.: Upward embeddings and orientations of undirected planar graphs. J. Graph Algorithms Appl. 7(2), 221–241 (2003)
7. Dunne, C., Shneiderman, B.: Improving graph drawing readability by incorporating readability metrics: A software tool for network analysts. Technical report (2009)
8. Dwyer, T.: Scalable, versatile and simple constrained graph layout. Comput. Graph. Forum 28(3), 991–998 (2009)
9. Dwyer, T., Marriott, K., Schreiber, F., Stuckey, P.J., Woodward, M., Wybrow, M.: Exploration of networks using overview+detail with constraint-based cooperative layout. IEEE Trans. Vis. Comput. Graph. 14(6), 1293–1300 (2008)
10. Dwyer, T., Marriott, K., Wybrow, M.: Topology preserving constrained graph layout. In: Tollis, I.G., Patrignani, M. (eds.) GD 2008. LNCS, vol. 5417, pp. 230–241. Springer, Heidelberg (2009)
11. Eades, P.: A heuristic for graph drawing. Congressus Numerantium 42, 149–160 (1984)
12. Finkel, B., Tamassia, R.: Curvilinear graph drawing using the force-directed method. In: Pach, J. (ed.) GD 2004. LNCS, vol. 3383, pp. 448–453. Springer, Heidelberg (2005)
13. Fößmeier, U., Kaufmann, M.: Drawing high degree graphs with low bend numbers. In: Brandenburg, F.J. (ed.) GD 1995. LNCS, vol. 1027, pp. 254–266. Springer, Heidelberg (1996)
14. Garg, A., Tamassia, R.: A new minimum cost flow algorithm with applications to graph drawing. In: North, S.C. (ed.) GD 1996. LNCS, vol. 1190, pp. 201–216. Springer, Heidelberg (1997)
15. Gutwenger, C., Mutzel, P., Weiskircher, R.: Inserting an edge into a planar graph. Algorithmica 41(4), 289–308 (2005)
16. Hachul, S., Jünger, M.: Drawing large graphs with a potential-field-based multilevel algorithm. In: Pach, J. (ed.) GD 2004. LNCS, vol. 3383, pp. 285–295. Springer, Heidelberg (2005)
17. Hachul, S., Jünger, M.: Large-graph layout algorithms at work: An experimental study. J. Graph Algorithms Appl. 11(2), 345–369 (2007)
18. Huang, W.: Using eye tracking to investigate graph layout effects. In: APVIS, pp. 97–100 (2007)
19. Huang, W., Hong, S.-H., Eades, P.: Effects of crossing angles. In: PacificVis, pp. 41–46. IEEE, Los Alamitos (2008)
20. Kaufmann, M., Wagner, D. (eds.): Drawing Graphs. Springer, Heidelberg (2001)
21. Purchase, H.C.: Which aesthetic has the greatest effect on human understanding? In: DiBattista, G. (ed.) GD 1997. LNCS, vol. 1353, pp. 248–261. Springer, Heidelberg (1997)
22. Purchase, H.C., Carrington, D.A., Allder, J.-A.: Empirical evaluation of aesthetics-based graph layout. Empirical Software Engineering 7(3), 233–255 (2002)
23. Tamassia, R.: On embedding a graph in the grid with the minimum number of bends. SIAM Journal on Computing 16(3), 421–444 (1987)
24. Ware, C., Purchase, H.C., Colpoys, L., McGill, M.: Cognitive measurements of graph aesthetics. Information Visualization 1(2), 103–110 (2002)

On Graphs Supported by Line Sets[*]

Vida Dujmović[1], William Evans[2], Stephen Kobourov[3], Giuseppe Liotta[4],
Christophe Weibel[5], and Stephen Wismath[6]

[1] School of Computer Science, Carleton University
cgm.cs.mcgill.ca/~vida
[2] Dept. of Computer Science, Univ. of British Columbia
www.cs.ubc.ca/~will
[3] Dept. of Computer Science, Univ. of Arizona
www.cs.arizona.edu/~kobourov
[4] Dept. of Computer Science, Univ. of Perugia
www.diei.unipg.it/~liotta
[5] Dept. of Mathematics, McGill University
www.math.mcgill.ca/~weibel
[6] Dept. of Computer Science, Univ. of Lethbridge
www.cs.uleth.ca/~wismath

Abstract. For a set S of n lines labeled from 1 to n, we say that S supports an n-vertex planar graph G if for every labeling from 1 to n of its vertices, G has a straight-line crossing-free drawing with each vertex drawn as a point on its associated line. It is known from previous work [4] that no set of n parallel lines supports all n-vertex planar graphs. We show that intersecting lines, even if they intersect at a common point, are more "powerful" than a set of parallel lines. In particular, we prove that every such set of lines supports outerpaths, lobsters, and squids, none of which are supported by any set of parallel lines. On the negative side, we prove that no set of n lines that intersect in a common point supports all n-vertex planar graphs. Finally, we show that there exists a set of n lines in general position that does not support all n-vertex planar graphs.

1 Introduction

We consider the effect of restricting the placement of vertices in a planar, straight-line, crossing-free embedding of a planar graph. Every vertex has an associate region of the plane where it can be placed. If each region is the whole plane then the regions support all planar graphs. If the regions are points then they fail to support even such a simple class of graphs as paths. Our interest is in what classes of planar graphs are supported by particular families of vertex regions. Specifically, in this paper we focus on vertex regions that are lines.

A set of segments is *crossing-free* if no two segments intersect in their interiors. A *vertex labeling* of a graph $G = (V, E)$ is a bijection $\pi : V \rightarrow [n]$. A set R of n regions

---

[*] Research supported in part by: NSERC, MIUR under project AlgoDEEP prot. 2008TFBWL4. The research in this paper started during the *McGill/INRIA Workshop at Bellairs*. The authors thank the organizers and the other participants for useful discussions.

U. Brandes and S. Cornelsen (Eds.): GD 2010, LNCS 6502, pp. 177–182, 2011.

(subsets of $\mathbb{R}^2$) labeled from 1 to n *supports* a graph G *with vertex labeling* π if there exists a set of distinct points $p_1, p_2, \ldots, p_n$ such that p_i lies in region i for all i and the segments $\overline{p_{\pi(u)} p_{\pi(v)}}$ for $(u, v) \in E$ are crossing-free. The set R of n labeled regions *supports* a graph G if R supports G with vertex labeling π for every vertex labeling π. As an example of the use of this terminology, we show that every n-pinwheel (set of n labeled lines that share a common point) supports any n-squid (see definition below).

While we focus on embeddings that prescribe a specific region for each vertex, the problem is also interesting if each vertex may be placed in any one of the regions. In this variant, a set of regions R *supports* a graph $G = (V, E)$ *without mapping* if there *exists* a bijection π from V to R such that R supports G with vertex labeling π. Rosenstiehl and Tarjan [8] posed the question of whether there exists a point set of size n that supports without mapping all n-vertex planar graphs: a *universal* point set for all planar graphs. De Fraysseix *et al.* [3] resolved the question in the negative by presenting a set of n-vertex planar graphs that requires a point set of size $\Omega(n + \sqrt{n})$. For some classes of n-vertex planar graphs, universal point sets of size n have been found. In particular, Gritzmann *et al.* [6] showed that any set of n points in general position forms a universal point set for trees and indeed for all outerplanar graphs, for which Bose [2] gave an efficient drawing algorithm.

Embedding with mapping is an even more restricted version of the problem. For example, any set of n points supports without mapping all n-vertex paths. Whereas, for large enough n, no set of n points supports all n-vertex paths. For $n \geq 5$, every set of n points contains a subset of three collinear points or four points in convex position.[1] In both cases, it is easy to devise a vertex mapping of three (respectively four) consecutive vertices of any n-vertex path to that subset of points that forces an edge crossing.

If we remove the straight-line edge condition in this mapped setting, Pach and Wenger [7] showed that any set of n points supports all n-vertex planar graphs, however $\Omega(n)$ bends per edge may be necessary in any crossing-free drawing. Even if the mapping constraint is relaxed to just two colors: the red vertices must be mapped to any red point and the blue vertices to any blue point, Badent *et al.* [1] proved that $\Omega(n)$ bends per edge are sometimes necessary.

Estrella-Balderrama *et al.* [4] show that any set of n parallel lines supports exactly the class of *unlabeled level planar (ULP) graphs*. This class of graphs contains several sub-classes of trees (namely caterpillars, radius-2 stars and degree-3 spiders) [4] and a restricted set of graphs with cycles (such as generalized caterpillars) [5]. The simplest class of trees not supported by parallel lines is the class of lobsters.

We show that any set of n lines that intersect at a common point supports a larger sub-class of n-vertex trees than the ULP graphs. We further show that no set of n lines that intersect at a common point supports all n-vertex planar graphs. Whether such a set of lines supports all trees is a natural open question. We also show that there exists a set of n lines in general position that does not support all n-vertex planar graphs. Here, a set of lines is considered in general position if no two lines are parallel and no three lines intersect in a common point. The main open question remaining is whether there exists a set of lines in general position that supports all planar graphs.

---

[1] Eszter Klein's Happy Ending problem.

2   Preliminaries

Definition 1. A *pinwheel* is an arrangement of n distinct lines that intersect the origin and are labeled from 1 to n in clockwise order. Each line in the pinwheel is called a *track*.

As the next lemma shows, pinwheels are an interesting family of line sets to consider when investigating whether more general families support planar graphs.

Lemma 1. *Any class of graphs supported by every n-line pinwheel is also supported by every arrangement of n lines, no two of which are parallel.*

Proof: Determine a circle that contains all line intersections. By scaling the arrangement down we can make the radius of this circle arbitrarily small, rendering it effectively into a pinwheel.   ∎

3   Graphs Supported by Arrangements of Lines

In this section we describe non-ULP families of planar graphs that are supported by every arrangement of lines, no two of which are parallel. We use pinwheels as supporting sets for the graphs in these families since by Lemma 1 the results will then apply to the more general arrangements. We begin by studying lobsters, then extend the result to squids and finally consider outerpaths.

- A *caterpillar* is a graph in which the removal of all degree one vertices and their incident edges results in a path. This path is called the *spine* of the graph.
- A *lobster* is a graph in which the removal of all degree one vertices and their incident edges results in a caterpillar.
- A *squid* is a subdivision of a lobster.
- An *outerpath* is an outerplanar graph whose weak dual is a path (where the weak dual is obtained from the dual by removing the vertex corresponding to the outer-face and its adjacent edges).

Lemma 2. *Every n-line pinwheel supports any n-vertex lobster.*

Proof: Let L be a lobster with n vertices and spine vertices $v_1, v_2, \ldots, v_k$. We compute a straight-line embedding of L on any labeled n-pinwheel such that no two edges cross for any vertex labeling of L. We place the vertices in order of a preorder traversal of L, where L is considered as a tree rooted at v_1 and such that each spine vertex is the last descendant of its parent. At any step, there is a set of vertices not all of whose children have been drawn – call these the *active* vertices. We maintain the invariant that all active vertices can "see" the origin (*i.e.*, the segment from the embedded active vertex to the origin does not intersect a segment of the drawing). As a consequence, all active vertices can see an empty disk (*i.e.*, a disk that does not intersect a segment of the current drawing) of nonzero radius centered at the origin and intersecting every track twice; see Fig. 1. At each step, the current vertex is placed at one of the two intersection

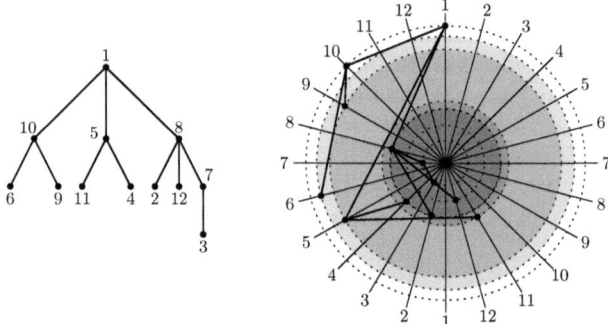

Fig. 1. (a) A labeled lobster with spine vertices $v_1 = 1$, $v_2 = 8$. (b) An embedding using the algorithm. The dotted circles indicate the empty discs at each step.

points of its track and the boundary of the largest empty disk centered at the origin that is seen by all active vertices. The intersection point that is chosen is the one that is encountered first in counter-clockwise radial order from the track of its already-placed parent. (The first vertex, v_1, is initially placed on its corresponding track at an arbitrary point that is not the origin.)

The correctness of the drawing algorithm is proved by induction on the length of the spine. While a spine vertex is active, only vertices at distance at most two from it are drawn. Since the radial distance between a vertex and its parent is less than 180 degrees, we maintain the invariant that each active vertex sees the origin.   ■

Lemma 3. *Every n-line pinwheel supports any n-vertex squid.*

Proof: We extend the algorithm of Lemma 2 to the drawing of squids. A squid G' can be obtained from a lobster G by subdividing edges of G. For each vertex v created by subdividing an edge (u, w) of G, we define v's *lobster parent* as the closer of u or w to the root v_1. We draw the vertices in order of a preorder traversal of the graph. But at each step, the position chosen for a vertex is on the track that is encountered first in counter-clockwise radial order from the track of its lobster parent instead of its parent. As a result, the whole path obtained by subdividing an edge is drawn at the radial distance of at most 180 degrees from its lobster parent. As in the proof of Lemma 2, every active vertex can see the origin.   ■

With similar techniques as those of Lemmas 2 and 3, the following can be proved.

Lemma 4. *Every n-line pinwheel supports any n-vertex outerpath.*

Lemmas 3, 4, and 1 imply the following.

Theorem 1. *Every arrangement of n lines, no two of which are parallel, supports any n-vertex squid and any n-vertex outerpath.*

4   Non-supporting Line Sets

In this section, we show that there is a labeled planar graph that is not supported by any pinwheel. We also show that there exists a family of n-line sets, where each set is in

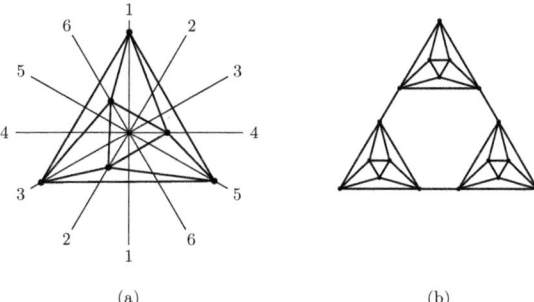

Fig. 2. (a) Graph G_6 can be realized only if the origin of the pinwheel is contained in an internal face. (b) Graph G_6^3 consists of three copies of G_6 connected by three edges.

general position, that does not support all n-vertex planar graphs. Note that this does not rule out the possibility that some family of n-line sets in general position could support all n-vertex planar graphs.

Both arguments rely on graphs that use as a building block the graph G_6 in Fig. 2(a). It is not difficult to show that any straight-line, crossing-free embedding of G_6 with the given labeling requires that the origin of the pinwheel (with tracks labeled in clockwise order) is in an internal face. We prove a slightly stronger statement since we will need it in the proof of Theorem 3. For a set, S, of lines, no two of which are parallel, define the *core*, $C(S)$, of S to be the union of the intersections, finite edges, and bounded cells of the arrangement of S.

Lemma 5. *Let S be any set of lines, no two of which are parallel, labeled so that they intersect some line at infinity in the order 1,2,3,4,5,6. In any straight-line, crossing-free embedding of G_6 (labeled as in Fig. 2(a)) on S, the core $C(S)$ intersects some internal face of the embedding.*

Proof: Suppose for the sake of contradiction that the core $C(S)$ lies in the external face of some embedding of G_6. Thus each vertex of G_6 lies on a half-line (of the arrangement of S) that does not intersect $C(S)$ and these six half-lines intersect a line at infinity in some order. Let us assume initially that this order is $1, 2, 3, 4, 5, 6$.

Consider edge $(1, 5)$ in the embedding of G_6, and the line ℓ that contains edge $(1, 5)$. For any pair of vertices $a, b \in \{2, 3, 4\}$ of G_6, if a and b are in the distinct half-planes bounded by ℓ then a does not see b, that is, the segment between a and b intersects edge $(1, 5)$. Since both 3 and 4 are adjacent to 2 in G_6, all three of these vertices are in the same half-plane determined by ℓ. Moreover, 2 is contained inside of a triangle, T, determined by either $1, 3, 5$ or $1, 4, 5$. Since 6 is adjacent to 2 in G_6, 6 is inside of T as well. However, line 6 does not intersect T, which provides a desired contradiction.

A similar argument holds for the other possible half-line orders. ∎

To construct the graph that is not supported by any pinwheel, we make three copies of G_6 and label them so that the vertex labeled k in the original graph is labeled $6(i-1)+k$ in the ith copy. Finally, the graph G_6^3 is created by connecting the three labeled copies of G_6 with the help of three additional edges, as shown in Fig. 2(b).

Theorem 2. *Planar graph G_6^3 is not supported by any pinwheel.*

Proof: Assume for the sake of contradiction that there is a straight-line, crossing-free drawing of the labeled graph G_6^3 on the pinwheel. By Lemma 5, each of the copies of G_6 can be realized crossing-free and with straight line edges only if the origin of the pinwheel is contained in an internal face. Without loss of generality, that implies that the first copy of G_6 is inside an internal face of the second copy of G_6, and both are inside an internal face of the third copy of G_6. That provides a desired contradiction, since the edge connecting the first copy with the third copy must cross some edge of the second copy.   ■

We now turn our attention to lines in general position. One might hope that lines in general position (*i.e.*, no two lines are parallel and no three lines intersect in a common point) provide enough freedom in the placement of vertices to support any planar graph. We show that the general position assumption alone is not sufficient. Specifically, we can prove that there exists a family of n-line sets such that each set is in general position and not all planar graphs are supported by a line set in the family. Using a parabolic grid of n lines and a graph that contains multiple copies of G_6 we can obtain the following theorem (due to space constraints, we leave the proof out of this abstract):

Theorem 3. *For every $n \geq 24$, the parabolic grid on n lines does not support all n-vertex planar graphs.*

5   Conclusion and Open Problems

Whether there exists some set of n lines that does support all n-vertex planar graphs is a natural question that is still open. It is also not known whether pinwheels support all trees.

References

1. Badent, M., Giacomo, E.D., Liotta, G.: Drawing colored graphs on colored points. Theor. Comput. Sci. 408(2-3), 129–142 (2008)
2. Bose, P.: On embedding an outer-planar graph in a point set. Computational Geometry: Theory and Applications 23(3), 303–312 (2002)
3. de Fraysseix, H., Pach, J., Pollack, R.: How to draw a planar graph on a grid. Combinatorica 10(1), 41–51 (1990)
4. Estrella-Balderrama, A., Fowler, J.J., Kobourov, S.G.: Characterization of unlabeled level planar trees. Computational Geometry: Theory and Applications 42(7), 704–721 (2009)
5. Fowler, J.J., Kobourov, S.G.: Characterization of unlabeled level planar graphs. In: Hong, S.-H., Nishizeki, T., Quan, W. (eds.) GD 2007. LNCS, vol. 4875, Springer, Heidelberg (2008)
6. Gritzmann, P., Mohar, B., Pach, J., Pollack, R.: Embedding a planar triangulation with vertices at specified points. American Math. Monthly 98, 165–166 (1991)
7. Pach, J., Wenger, R.: Embedding planar graphs at fixed vertex locations. Graphs and Combinatorics 17, 717–728 (2001)
8. Rosenstiehl, P., Tarjan, R.E.: Rectilinear planar layouts and bipolar orientations of planar graphs. Discrete Computational Geometry 1(4), 343–353 (1986)

Drawing Trees with Perfect Angular Resolution and Polynomial Area

Christian A. Duncan[1], David Eppstein[2], Michael T. Goodrich[2],
Stephen G. Kobourov[3], and Martin Nöllenburg[2]

[1] Department of Computer Science, Louisiana Tech. Univ., Ruston, Louisiana, USA
[2] Department of Computer Science, University of California, Irvine, California, USA
[3] Department of Computer Science, University of Arizona, Tucson, Arizona, USA

Abstract. We study methods for drawing trees with perfect angular resolution, i.e., with angles at each vertex, v, equal to $2\pi/d(v)$. We show:

1. Any unordered tree has a crossing-free straight-line drawing with perfect angular resolution and polynomial area.
2. There are ordered trees that require exponential area for any crossing-free straight-line drawing having perfect angular resolution.
3. Any ordered tree has a crossing-free ***Lombardi-style*** drawing (where each edge is represented by a circular arc) with perfect angular resolution and polynomial area.

Thus, our results explore what is achievable with straight-line drawings and what more is achievable with Lombardi-style drawings, with respect to drawings of trees with perfect angular resolution.

1 Introduction

Most methods for visualizing trees aim to produce drawings that meet as many of the following aesthetic constraints as possible:

1. straight-line edges,
2. crossing-free edges,
3. polynomial area, and
4. perfect angular resolution around each vertex.

These constraints are all well-motivated, in that we desire edges that are easy to follow, do not confuse viewers with edge crossings, are drawable using limited real estate, and avoid congested incidences at vertices. Nevertheless, previous tree drawing algorithms have made various compromises with respect to this set of constraints; we are not aware of any previous tree-drawing algorithm that can achieve all these goals simultaneously. Our goal in this paper is to show what is actually possible with respect to this set of constraints and to expand it further with a richer notion of edges that are easy to follow. In particular, we desire tree-drawing algorithms that satisfy all of these constraints simultaneously. If this is provably not possible, we desire an augmentation that avoids compromise and instead meets the spirit of all of these goals in a new way, which, in the case of this paper, is inspired by the work of artist Mark Lombardi [18].

U. Brandes and S. Cornelsen (Eds.): GD 2010, LNCS 6502, pp. 183–194, 2011.

Problem Statement. The art of Mark Lombardi involves drawings of social networks, typically using circular arcs and good angular resolution. Figure 1 shows such a work of Lombardi that is crossing-free and almost a tree. Note that it makes use of both circular arcs and straight-line edges. Inspired by this work, let us define a set of problems that explore what is achievable for drawings of trees with respect to the constraints listed above but that, like Lombardi's drawings, also allow curved as well as straight edges.

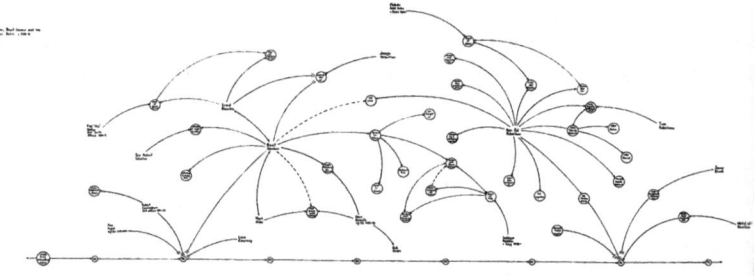

Fig. 1. Mark Lombardi, *Pat Robertson, Beurt Servaas, and the UPI Takeover Battle, ca. 1985-91,* 2000 [18]

Given a graph $G = (V, E)$, let $d(u)$ denote the **degree** of a vertex u, i.e., the number of edges incident to u in G. For any drawing of G, the **angular resolution** at a vertex u is the minimum angle between two edges incident to u. A vertex has **perfect angular resolution** if its minimum angle is $2\pi/d(u)$, and a drawing has perfect angular resolution if *every* vertex does. Drawings with perfect angular resolution cannot be placed on an integer grid unless the degrees of the vertices are constrained, so we do not require vertices to have integer coordinates. We define the **area** of a drawing to be the ratio of the area of a smallest enclosing circle around the drawing to the square of the distance between its two closest vertices.

Suppose that our input graph, G, is a rooted tree T. We say that T is **ordered** if an ordering of the edges incident upon each vertex in T is specified. Otherwise, T is **unordered**. If all the edges of a drawing of T are straight-line segments, then the drawing of T is a **straight-line** drawing. We define a **Lombardi drawing** of a graph G as a drawing of G with perfect angular resolution such that each edge is drawn as a circular arc. When measuring the angle formed by two circular arcs incident to a vertex v, we use the angle formed by the tangents of the two arcs at v. Circular arcs are strictly more general than straight-line segments, since straight-line segments can be viewed as circular arcs with infinite radius. Figure 2 shows an example of a straight-line drawing and a Lombardi drawing for the same tree. Thus, we can define our problems as follows:

1. Is it always possible to produce a straight-line drawing of an unordered tree with perfect angular resolution and polynomial area?
2. Is it always possible to produce a straight-line drawing of an ordered tree with perfect angular resolution and polynomial area?
3. Is it always possible to produce a Lombardi drawing of an ordered tree with perfect angular resolution and polynomial area?

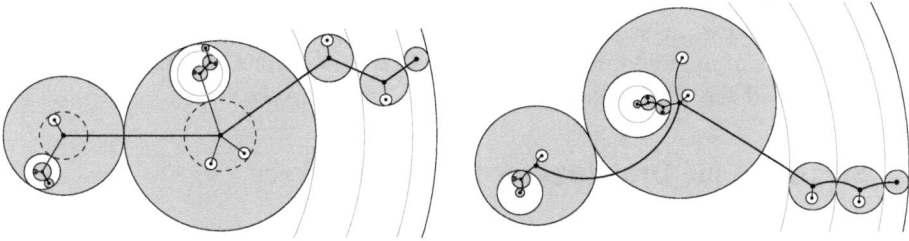

(a) Straight-line drawing for an unordered tree   (b) Lombardi drawing for an ordered tree

Fig. 2. Two drawings of a tree T with perfect angular resolution and polynomial area as produced by our algorithms. Bold edges are heavy edges, gray disks are heavy nodes, and white disks are light children. The root of T is in the center of the leftmost disk.

Related Work. Tree drawings have interested researchers for many decades: e.g., hierarchical drawings of binary trees date to the 1970's [24]. Many improvements have been proposed since this early work, using space efficiently and generalizing to non-binary trees [2, 5, 13, 14, 15, 21, 23, 22]. These drawings do not achieve all the constraints mentioned above, however, especially the constraint on angular resolution.

Alternatively, several methods strive to optimize angular resolution of trees. Radial drawings of trees place nodes at the same distance from the root on a circle around the root node [11]. Circular tree drawings are made of recursive radial-type layouts [20]. Bubble drawings [16] draw trees recursively with each subtree contained within a circle disjoint from its siblings but within the circle of its parent. Balloon drawings [19] take a similar approach and heuristically attempt to optimize space utilization and the ratio between the longest and shortest edges in the tree. Convex drawings [4] partition the plane into unbounded convex polygons with their boundaries formed by tree edges. Although these methods provide several benefits, none of these methods guarantees that they satisfy all of the aforementioned constraints.

The notion of drawing graphs with edges that are circular arcs or other nonlinear curves is certainly not new to graph drawing. For instance, Cheng *et al.* [6] used circle arcs to draw planar graphs in an $O(n) \times O(n)$ grid while maintaining bounded (but not perfect) angular resolution. Similarly, Dickerson *et al.* [7] use circle-arc polylines to produce planar confluent drawings of non-planar graphs, Duncan *et al.* [8] draw graphs with fat edges that include circular arcs, and Cappos *et al.* [3] study simultaneous embeddings of planar graphs using circular arcs. Finkel and Tamassia [12] use a force-directed method for producing curvilinear drawings, and Brandes and Wagner [1] use energy minimization methods to place Bézier splines that represent express connections in a train network. In a separate paper [10] we study Lombardi drawings for classes of graphs other than trees.

Our Contributions. In this paper we present the first algorithm for producing straight-line, crossing-free drawings of unordered trees that ensures perfect angular resolution and polynomial area. In addition we show, in Section 3, that if the tree is ordered (i.e., given with a fixed combinatorial embedding) then it is not always possible to maintain perfect angular resolution and polynomial drawing area when using straight lines for edges. Nevertheless, in Section 4, we show that crossing-free polynomial-area Lombardi drawings of ordered trees are possible. That is, we show that the answers to the questions posed above are "yes," "no," and "yes," respectively.

2   Straight-Line Drawings for Unordered Trees

Let T be an unordered tree with n nodes. We wish to construct a straight-line drawing of T with perfect angular resolution and polynomial area.

The main idea of our algorithm is, similarly to the common bubble and balloon tree constructions [16, 19], to draw the children of each node of the given tree in a disk centered at that node; however, our algorithm differs in several key respects:

- Before drawing the tree, we perform a *heavy path decomposition* [17]: for each node v, the *heavy child* of v is the child with the greatest number of descendants, and the other children are *light children*, denoted $L(v)$. The paths that follow edges from nodes to their heavy children are *heavy paths*, and they form a partition of the input tree with the property that the tree $H(T)$ formed by compressing each heavy path to a node has only logarithmic depth $h(T)$.
- In our drawing, each heavy path P is confined to a disk, whose radius is linear in the number of nodes descending from P and exponential in the level of P in the heavy path decomposition. In this way, at each step downwards in the heavy path decomposition, the total radius of the disks at that level shrinks by a constant factor, allowing room for disks at lower levels to be placed within the higher-level disks.
- For each heavy path P, and each node v on P, we form another disk, contained within the disk for P, that contains v at its center and also contains the disks for the lower-level heavy paths connected to v (the descendants of the light children of v). The disks for the nodes of the heavy path are placed within the disk for the heavy path, with the topmost node of the heavy path at the center of the disk for the heavy path and successive heavy path nodes placed on concentric circles within this disk.
- Because the radii of our disks are exponential in the level of the heavy path decomposition, the radii of the disks for the children of v add up to a constant fraction of the radii of the disk for v itself (Figure 3a). Within the disk centered at node v, we place the smaller disks containing the heavy paths descending from v in two concentric annuli (Figure 3b).
- The outer annulus surrounding v contains light children of v that are the ancestors of many nodes (relative to the total number of descendants of v and the degree of v); the disks for the heavy paths containing these light children are placed using a greedy algorithm so that the edges connecting them to v have angles that are multiples of the proper angular resolution.
- The inner annulus surrounding v contains the remaining light children, each of which is the ancestor of few enough nodes that the disk for its heavy path may be placed in the inner annulus at the perfect angular resolution for v, filling out all the positions incident to v that were not already filled by the two edges of the heavy path for v and the disks in the outer annulus.

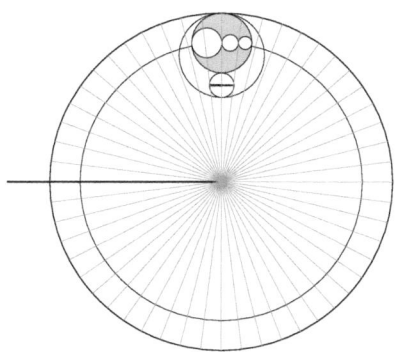 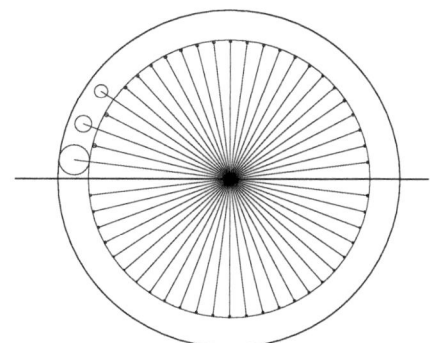

(a) All light children fit into a disk of radius $r_v/4$ and are split into small and large disks.

(b) Large disks are placed in the outer annulus and small disks in the inner disk.

Fig. 3. Drawing a node v and its light children $L(v)$

As we show in the full paper [9], this method draws the given tree with perfect angular resolution and polynomial drawing area. However, our method may reorder the children of each node, so it does not respect a fixed embedding of the given tree. Figure 2a shows a drawing of an unordered tree according to our method.

3   Straight-Line Drawings for Ordered Trees

In many cases, the ordering of the children around each vertex of a tree is given; that is, the tree is ordered (or has a fixed combinatorial embedding). In the previous section we rely on the freedom to order subtrees as needed to achieve a polynomial area bound. Hence that algorithm cannot be applied to ordered trees with fixed embeddings. As we now show, there are ordered trees that have no straight-line crossing-free drawings with polynomial area and perfect angular resolution.

Specifically we present a class of ordered trees for which any straight-line crossing-free drawing of the tree with perfect angular resolution requires exponential area. Figure 4a shows a caterpillar tree, which we call the ***Fibonacci caterpillar*** because of its simple behavior when required to have perfect angular resolution. This tree has as its spine a k-vertex path, each vertex of which has 3 additional leaf nodes embedded on the same side of the spine. When drawn with straight-line edges, no crossings, and with perfect angular resolution, the caterpillar is forced to spiral (a single or a double spiral). The best drawing area, exponential in the number of vertices in the caterpillar, is achieved when the caterpillar forms a symmetric double spiral; see Figure 4c.

The Fibonacci caterpillar shows that we cannot maintain all constraints (straight-line edges, crossing-free, perfect angular resolution, polynomial area) for ordered trees. However, as we show next, using circular arcs instead of straight-line edges allows us to respect the remaining three constraints. See, for example, Figure 4b.

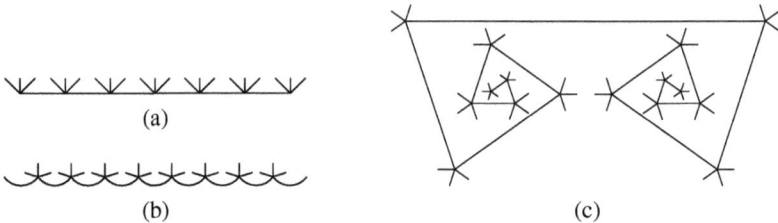

Fig. 4. (a) A Fibonacci caterpillar; (b) Lombardi drawing; (c) Straight-line drawing with perfect angular resolution and exponential area

4   Lombardi Drawings for Ordered Trees

In this section, let T be an ordered tree with n nodes. As we have seen in Section 3, we cannot find polynomial area drawings for all ordered trees using straight-line edges. An augmentation of the straight-line edge requirement is the use of circular arcs as edges. Circular arcs are curves that are not only still easy to follow visually but they also let us achieve all remaining three constraints, i.e., we can find crossing-free circular arc drawings with perfect angular resolution and polynomial area. We call a drawing with circular arcs and perfect angular resolution a Lombardi drawing, so in other words we aim for crossing-free Lombardi drawings with polynomial area.

The flavor of the algorithm for Lombardi tree drawings is similar to our straight-line tree drawing algorithm of Section 2: We first compute a heavy-path decomposition $H(T)$ for T. Then we recursively draw all heavy paths within disks of polynomial area. Unlike before, we need to construct the drawing in a top-down fashion since the placement of the light children of a node v now depends on the curvature of the two heavy edges incident to v.

Our construction in this section uses the invariant that a heavy path P at level j is drawn inside a disk D of radius $2 \cdot 4^{h(T)-j} n(P)$, where $n(P) = |T_v|$ for the root v of P.

4.1   Drawing Heavy Paths

Let $P = (v_1, \ldots, v_k)$ be a heavy path at level j of the heavy-path decomposition that is rooted at the last node v_k. We denote each edge $v_i v_{i+1}$ by e_i. Recall that the angle in an intersection point of two circular arcs is measured as the angle between the tangents to the arcs at that point. We define the angle $\alpha(v_i)$ for $2 \le i \le k-1$ to be the angle between e_{i-1} and e_i in node v_i (measured counter-clockwise). The angle $\alpha(v_k)$ is defined as the angle in v_k between e_{k-1} and the light edge $e = v_k u$ connecting the root v_k of P to its parent u. Due to the perfect angular resolution requirement for each node v_i, the angle $\alpha(v_i)$ is obtained directly from the number of edges between e_{i-1} and e_i and the degree $d(v_i)$.

Lemma 4.1. *Given a heavy path $P = (v_1, \ldots, v_k)$ and a disk D_i of radius r_i for the drawing of each v_i and its light subtrees, we can draw P with each v_i in the center of its disk D_i inside a large disk D such that the following properties hold:*

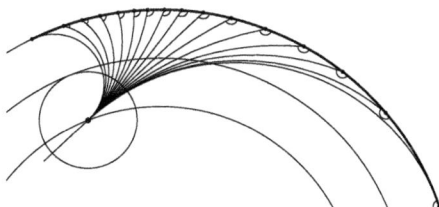

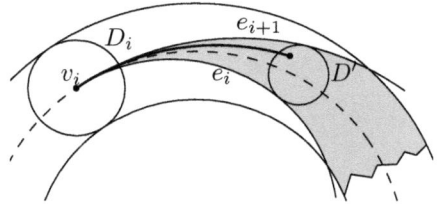

Fig. 5. Any angle $\alpha \in [0, \pi]$ can be realized

Fig. 6. Placing a single disk D' in the extended small zone of D_i (shaded gray)

1. *each heavy edge e_i is a circular arc that does not intersect any disk other than D_i and D_{i+1};*
2. *there is a stub edge incident to v_k that is reserved for the light edge connecting v_k and its parent;*
3. *any two disks D_i and D_j for $i \neq j$ are disjoint;*
4. *the angle between any two consecutive heavy edges e_{i-1} and e_i is $\alpha(v_i)$;*
5. *the radius r of D is $r = 2\sum_{i=1}^{k} r_i$.*

Proof. We draw P incrementally starting from the leaf v_1 by placing D_1 in the center M of the disk D of radius $r = 2\sum_{i=1}^{k} r_i$. We may assume that D_1 is rotated such that the edge e_1 is tangent to a horizontal line at v_1 and that it leaves v_1 to the right. All disks $D_2, \ldots, D_k$ will be placed with their centers $v_2, \ldots, v_k$ on concentric circles $C_2, \ldots, C_k$ around M. The radius of C_i is $r_1 + 2\sum_{j=2}^{i-1} r_j + r_i$ so that D_{i-1} and D_i are placed in disjoint annuli and hence by construction no two disks intersect (property 3). Each disk D_i will be rotated around its center such that the tangent to C_i at v_i is the bisector of the angle $\alpha(v_i)$.

We now describe one step in the iterative drawing procedure that draws edge e_i and disk D_{i+1} given a drawing of $D_1, \ldots, D_i$. Disk D_i is placed such that C_i bisects the angle $\alpha(v_i)$ and hence we know the tangent of e_i at v_i. This defines a family $\mathscr{F}_i$ of circular arcs emitted from v_i that intersect the circle C_{i+1}, see Figure 5. We consider all arcs from v_i until their first intersection point with C_{i+1}. Observe that the intersection angles of $\mathscr{F}_i$ and C_{i+1} bijectively cover the full interval $[0, \pi]$, i.e., for any angle $\alpha \in [0, \pi]$ there is a unique arc in $\mathscr{F}_i$ that has intersection angle α with C_{i+1}. Hence we choose for e_i the unique circular arc that realizes the angle $\alpha(v_{i+1})/2$ and place the center v_{i+1} of D_{i+1} at the endpoint of e_i. We continue this process until the last disk D_k is placed. This drawing of P realizes the angle $\alpha(v_i)$ between any two heavy edges e_{i-1} and e_i (property 4). Note that for the edge from v_k to its parent we can only reserve a stub whose tangent at v_k has a fixed slope (property 2). Figure 7 shows an example.

Note that each edge e_i is contained in the annulus between C_i and C_{i+1} and thus does not intersect any other edge of the heavy path or any disk other than D_i and D_{i+1} (property 1). Furthermore, the disk D with radius $r = 2\sum_{i=1}^{k} r_i$ indeed contains all the disks $D_1, \ldots, D_k$ (property 5). $\qquad\square$

Lemma 4.1 shows how to draw a heavy path P with prescribed angles between the heavy edges and an edge stub to connect it to its parent. Since each heavy path P (except the

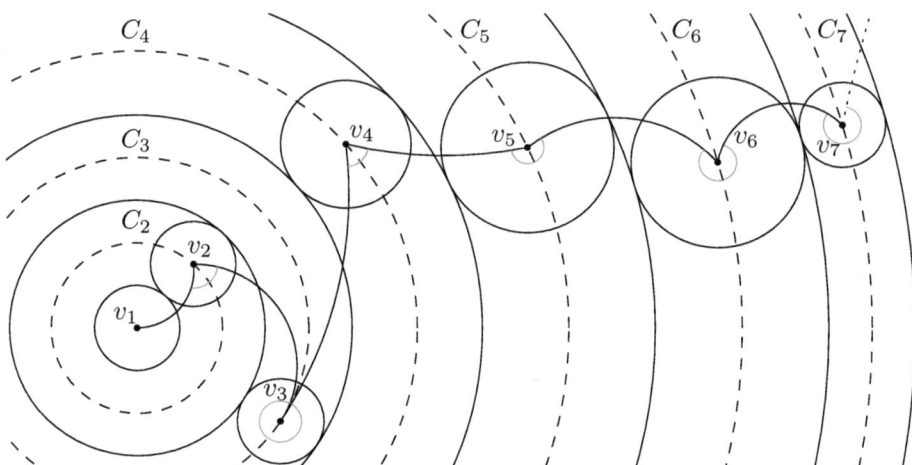

Fig. 7. Drawing a heavy path P on concentric circles with circular-arc edges. The angles $\alpha(v_i)$ are marked in gray; the edge stub to connect v_7 to its parent is dotted.

path at the root of $H(T)$) is the light child of a node on the previous level of $H(T)$ that light edge is actually drawn when placing the light children of a node, which we describe next.

4.2 Drawing Light Children

Once the heavy path P is drawn as described above, it remains to place the light children of each node v_i of P. For each node v_i the two heavy edges incident to it partition the disk D_i into two regions. We call the region that contains the larger conjugate angle the *large zone* of v_i and the region that contains the smaller conjugate angle the *small zone*. If both angles equal π, then we can consider both regions small zones.

For a node v_i at level j of $H(T)$ we define the radius r_i of D_i as $r_i = 4^{h(T)-j}(1 + \sum_{u \in L(v_i)} |T_u|) = 4^{h(T)-j}l(v_i)$. All light children of v_i are at level $j+1$ of $H(T)$ and thus by our invariant every light child u of v_i is drawn in a disk of radius $r_u = 2 \cdot 4^{h(T)-j-1}|T_u|$. Thus we know that $r_u \leq r_i/2$; in fact, we even have $\sum_{u \in L(v_i)} r_u \leq r_i/2$.

Light children in the small zone. Depending on the angle $\alpha(v_i)$, the small zone of a disk D_i might actually be too narrow to directly place the light children in it. Fortunately, we can always place another disk D' of radius at most $r_i/2$ in an extension of the small zone along the annulus of D_i in the drawing of P such that D' touches e_{i-1} and e_i and does not intersect any other previously placed disk, see Figure 6. If there is a single child u in the small zone then $D' = D_u$ and we are done. The next lemma shows how to place more than one child; the proof can be found in the full paper [9].

Lemma 4.2. *If a single disk D' of radius r' can be placed in the possibly extended small zone of the disk D_i, then we can correctly place any sequence of l disks $D'_1, \ldots, D'_l$ with radii $r'_1, \ldots, r'_l$ and $\sum_{i=1}^{l} r'_i = r'$ in the (extended) small zone of D_i.*

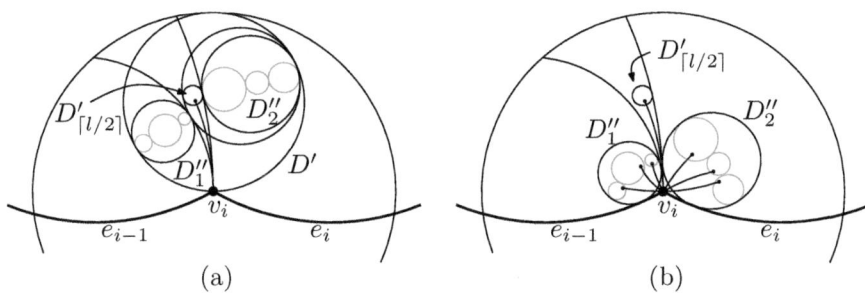

Fig. 8. Placing light children in the large zone by first splitting it into two parts (a) and then applying the algorithm for small zones to each part (b)

Light children in the large zone. Placing the light children of a vertex v_i in the large zone of D_i must be done slightly different from the algorithm for the small zone since Lemma 4.2 holds only for opening angles of at most π. On the other hand, the large zone does not become too narrow and there is no need to extend it beyond D_i. Our approach splits the large zone into two parts that again have an opening angle of at most π so that we can apply Lemma 4.2 and draw all children accordingly.

Let l be the number of light children in the large zone of D_i. We first place a disk D' of radius at most $r_i/2$ such that it touches v_i and such that its center lies on the line bisecting the opening angle of the large zone. The disk D' is large enough to contain the disjoint disks $D'_1, \ldots, D'_l$ for the light children of v_i along its diameter. We need to distinguish whether l is even or odd. For even l we create a container disk D''_1 for disks $D'_1, \ldots, D'_{l/2}$ and a container disk D''_2 for $D'_{l/2+1}, \ldots, D'_l$. Now D''_1 and D''_2 can be tightly packed on the diameter of D'. Using a similar argument as in Lemma 4.2 we separate the two disks by a circular arc through v_i that is tangent to the bisector of $\alpha(v_i)$ in v_i. Since D' is centered on the bisector this is possible even though the actual opening angle of the large zone is larger than π. If l is odd, we create a container disk D''_1 for disks $D'_1, \ldots, D'_{\lfloor l/2 \rfloor}$ and a container disk D''_2 for $D'_{\lceil l/2 \rceil+1}, \ldots, D'_l$. The median disk $D'_{\lceil l/2 \rceil}$ is not included in any container. Then we apply Lemma 4.2 to D' and the three disks $D''_1, D'_{\lceil l/2 \rceil}, D''_2$ along the diameter of D', see Figure 8a. The separating circular arcs in v_i are again tangent to the bisector of $\alpha(v_i)$, which is, since l is odd, also the correct slope for the circular arc connecting v_i to the median disk $D'_{\lceil l/2 \rceil}$.

In both cases we split the large zone and the sequence of light children to be placed into two parts that each have an opening angle at v_i of at most π between a separating circular arc and the edge e_{i-1} or e_i, respectively. Next, we move D''_1 and D''_2 along the separating circular arcs keeping their tangencies until they also touch the edge e_{i-1} or e_i, respectively. Then we can apply Lemma 4.2 to both container disks and thus place all light children in the large zone, see Figure 8b.

Drawing light edges The final missing step is how to actually connect a heavy node v_i to its light children given a position of v_i and positions of all disks containing its light subtrees. Let u be a light child of v_i and let D_u be the disk containing the drawing of T_u. When placing the disk D_u in the small or large zone of v_i we made sure that a circular arc from v_i with the tangent required for perfect angular resolution at v_i can reach any point inside D_u without intersecting any other edge or disk.

On the other side, we know by Lemma 4.1 that u is placed in the outermost annulus of D_u and that it has a stub for the edge $e = uv_i$. This stub is the required tangent for e in order to obtain perfect angular resolution in u. Let C_u be the circle that is the locus of u if we rotate D_u and the drawing of T_u around the center of D_u.

There is again a family $\mathcal{F}$ of circular arcs with the correct tangent in u that lead towards D_u and intersect the circle C_u. As observed in Lemma 4.1 the intersection angles formed between $\mathcal{F}$ and C_u bijectively cover the full interval $[0, \pi]$, i.e., for any angle $\alpha \in [0, \pi]$ there is a unique circular arc in $\mathcal{F}$ that has an intersection angle of α with C_u. In order to correctly attach u to v_i we first choose the arc a in $\mathcal{F}$ that realizes an intersection angle of $\alpha(u)/2$ with C_u, where $\alpha(u)$ is the angle between e and the heavy edge from u to its heavy child that is required for perfect angular resolution in u. Let p be the intersection point of that arc with C_u. Then we rotate D_u and the drawing of T_u around the center of D_u until u is placed at p, see node v_7 in Figure 7. Since the stub of u for e also has an angle of $\alpha(u)/2$ with C_u, the arc a indeed realizes the edge e with the angles in both u and v_i required for perfect angular resolution. Furthermore, a does not enter the disk bounded by C_u and hence it does not intersect any part of the drawing of T_u other than u.

We can summarize our results for drawing the light children of a node as follows:

Lemma 4.3. *Let v be a node of T at level j of $H(T)$ with two incident heavy edges. For every light child $u \in L(v)$ assume there is a disk D_u of radius $r_u = 2 \cdot 4^{h(T)-j-1}|T_u|$ that contains a fixed drawing of T_u with perfect angular resolution and such that u is exposed in the outer annulus of D_u. Then we can construct a drawing of v and its light subtrees inside a disk D, potentially with an extended small zone, such that the following properties hold:*

1. *the edge between v and any light child $u \in L(v)$ is a circular arc that does not intersect any disk other than D_u;*
2. *the heavy edges do not intersect any disk D_u;*
3. *any two disks D_u and $D_{u'}$ for $u \neq u'$ are disjoint;*
4. *the angular resolution of v is $2\pi/d(v)$;*
5. *the disk D has radius $r_v = 4^{h(T)-j}l(v)$.*

By combining Lemmas 4.1 and 4.3 we obtain the following theorem:

Theorem 4.4. *Given an ordered tree T with n nodes we can find a crossing-free Lombardi drawing of T that preserves the embedding of T and fits inside a disk D of radius $2 \cdot 4^{h(T)}n$, where $h(T)$ is the height of the heavy-path decomposition of T. Since $h(T) \leq \log_2 n$ the radius of D is no more than $2n^3$.*

Figure 2b shows a drawing of an ordered tree according to our method. We note that instead of asking for perfect angular resolution, the same algorithm can be used to construct a circular-arc drawing of an ordered tree with any assignment of angles between consecutive edges around each node that add up to 2π. The drawing remains crossing-free and fits inside a disk of radius $O(n^3)$.

5   Conclusion and Closing Remarks

We have shown that straight-line drawings of trees can be performed with perfect angular resolution and polynomial area, by carefully ordering the children of each vertex and by using a style similar to balloon drawings in which the children of any vertex are placed on two concentric circles rather than on a single circle. However, using our Fibonacci caterpillar example we showed that this combination of straight lines, perfect angular resolution, and polynomial area could no longer be achieved if the children of each vertex may not be reordered. For trees with a fixed embedding, Lombardi drawings in which edges are drawn as circular arcs allow us to retain the other desirable qualities of polynomial area and perfect angular resolution. In [9] we report on a basic implementation and some practical improvements of the straight-line drawing algorithm.

Our work opens up new problems in the study of Lombardi drawings of trees, but much remains to be done in this direction. In particular, our polynomial area bounds seem unlikely to be tight, and our method is impractically complex. It would be of interest to find simpler Lombardi drawing algorithms that achieve perfect angular resolution for more limited classes of trees, such as binary trees, with better area bounds.

Acknowledgments. This research was supported in part by the National Science Foundation under grant 0830403, by the Office of Naval Research under MURI grant N00014-08-1-1015, and by the German Research Foundation under grant NO 899/1-1.

References

[1] Brandes, U., Wagner, D.: Using graph layout to visualize train interconnection data. J. Graph Algorithms Appl. 4(3), 135–155 (2000),
http://jgaa.info/accepted/00/BrandesWagner00.4.3.pdf

[2] Buchheim, C., Jünger, M., Leipert, S.: Improving Walker's algorithm to run in linear time. In: Goodrich, M.T., Kobourov, S.G. (eds.) GD 2002. LNCS, vol. 2528, pp. 344–353. Springer, Heidelberg (2002), doi:10.1007/3-540-36151-0_32

[3] Cappos, J., Estrella-Balderrama, A., Fowler, J.J., Kobourov, S.G.: Simultaneous graph embedding with bends and circular arcs. Computational Geometry 42(2), 173–182 (2009), doi:10.1016/j.comgeo.2008.05.003

[4] Carlson, J., Eppstein, D.: Trees with convex faces and optimal angles. In: Kaufmann, M., Wagner, D. (eds.) GD 2006. LNCS, vol. 4372, pp. 77–88. Springer, Heidelberg (2007), doi:10.1007/978-3-540-70904-6_9

[5] Chan, T., Goodrich, M.T., Kosaraju, S.R., Tamassia, R.: Optimizing area and aspect ratio in straight-line orthogonal tree drawings. Computational Geometry 23(2), 153–162 (2002), doi:10.1016/S0925-7721(01)00066-9

[6] Cheng, C.C., Duncan, C.A., Goodrich, M.T., Kobourov, S.G.: Drawing planar graphs with circular arcs. Discrete Comput. Geom. 25(3), 405–418 (2001), doi:10.1007/s004540010080

[7] Dickerson, M., Eppstein, D., Goodrich, M.T., Meng, J.: Confluent drawings: visualizing non-planar diagrams in a planar way. In: Liotta, G. (ed.) GD 2003. LNCS, vol. 2912, pp. 1–12. Springer, Heidelberg (2004)

[8] Duncan, C.A., Efrat, A., Kobourov, S.G., Wenk, C.: Drawing with fat edges. Int. J. Found. Comput. Sci. 17(5), 1143–1164 (2006), doi:10.1142/S0129054106004315

[9] Duncan, C.A., Eppstein, D., Goodrich, M.T., Kobourov, S.G., Nöllenburg, M.: Drawing Trees with Perfect Angular Resolution and Polynomial Area, (September 2010) ArXiv e-prints, arXiv:1009.0581

[10] Duncan, C.A., Eppstein, D., Goodrich, M.T., Kobourov, S.G., Nöllenburg, M.: Lombardi drawings of graphs. In: Proc. 18th Int. Symp. on Graph Drawing (GD 2010), Springer, Heidelberg (2010), http://arxiv.org/abs/1009.0579

[11] Eades, P.: Drawing free trees. Bull. Inst. Combinatorics and Its Applications 5, 10–36 (1992)

[12] Finkel, B., Tamassia, R.: Curvilinear graph drawing using the force-directed method. In: Pach, J. (ed.) GD 2004. LNCS, vol. 3383, pp. 448–453. Springer, Heidelberg (2005)

[13] Garg, A., Goodrich, M.T., Tamassia, R.: Planar upward tree drawings with optimal area. Int. J. Comput. Geom. Appl. 6(3), 333–356 (1996), http://www.cs.brown.edu/cgc/papers/ggt-aoutd-96.ps.gz

[14] Garg, A., Rusu, A.: Area-efficient order-preserving planar straight-line drawings of ordered trees. Int. J. Comput. Geom. Appl. 13(6), 487–505 (2003), doi:10.1142/S021819590300130X

[15] Garg, A., Rusu, A.: Straight-line drawings of binary trees with linear area and arbitrary aspect ratio. J. Graph Algorithms Appl. 8(2), 135–160 (2004), http://jgaa.info/accepted/2004/GargRusu2004.8.2.pdf

[16] Grivet, S., Auber, D., Domenger, J.P., Melançon, G.: Bubble tree drawing algorithm. In: Proc. Int. Conf. Computer Vision and Graphics, pp. 633–641. Springer, Heidelberg (2004), http://www.labri.fr/publications/is/2004/GADM04

[17] Harel, D., Tarjan, R.E.: Fast algorithms for finding nearest common ancestors. SIAM J. Comput. 13(2), 338–355 (1984), doi:10.1137/0213024

[18] Hobbs, R., Lombardi, M.: Mark Lombardi: Global Networks. Independent Curators International, New York (2003)

[19] Lin, C.-C., Yen, H.-C.: On balloon drawings of rooted trees. J. Graph Algorithms Appl. 11(2), 431–452 (2007), http://jgaa.info/accepted/2007/LinYen2007.11.2.pdf

[20] Melançon, G., Herman, I.: Circular Drawings of Rooted Trees. Tech. Rep. INS-R9817, CWI Amsterdam (1998)

[21] Reingold, E.M., Tilford, J.S.: Tidier drawings of trees. IEEE Trans. Software Engineering 7(2), 223–228 (1981)

[22] Shin, C.-S., Kim, S.K., Chwa, K.-Y.: Area-efficient algorithms for straight-line tree drawings. Computational Geometry 15(4), 175–202 (2000), doi:10.1016/S0925-7721(99)00053-X

[23] Walker, J.: A node-positioning algorithm for general trees. Software Practice and Experience 20(7), 685–705 (1990), doi:10.1002/spe.4380200705

[24] Wetherell, C., Shannon, A.: Tidy drawings of trees. IEEE Trans. Software Engineering 5(5), 514–520 (1979)

Lombardi Drawings of Graphs

Christian A. Duncan[1], David Eppstein[2], Michael T. Goodrich[2],
Stephen G. Kobourov[3], and Martin Nöllenburg[2]

[1] Department of Computer Science, Louisiana Tech. Univ., Ruston, Louisiana, USA
[2] Department of Computer Science, University of California, Irvine, California, USA
[3] Department of Computer Science, University of Arizona, Tucson, Arizona, USA

Abstract. We introduce the notion of Lombardi graph drawings, named after the American abstract artist Mark Lombardi. In these drawings, edges are represented as circular arcs rather than as line segments or polylines, and the vertices have *perfect angular resolution*: the edges are equally spaced around each vertex. We describe algorithms for finding Lombardi drawings of regular graphs, graphs of bounded degeneracy, and certain families of planar graphs.

1 Introduction

The American artist Mark Lombardi [24] was famous for his drawings of social networks representing conspiracy theories. Lombardi used curved arcs to represent edges, leading to a strong aesthetic quality and high readability. Inspired by this work, we introduce the notion of a *Lombardi drawing* of a graph, in which edges are drawn as circular arcs with *perfect angular resolution*: consecutive edges are evenly spaced around each vertex. While not all vertices have perfect angular resolution in Lombardi's work, the even spacing of edges around vertices is clearly one of his aesthetic criteria; see Fig. 1.

Traditional graph drawing methods rarely guarantee perfect angular resolution, but poor edge distribution can nevertheless lead to unreadable drawings. Additionally, while some tools provide options to draw edges as curves, most rely on straight-line edges, and it is known that maintaining good angular resolution can result in exponential drawing area for straight-line drawings of planar graphs [17,25]. Our requirement of perfect angular resolution forces us to use curved edges, since even very simple graphs such as cycle graphs cannot be drawn with perfect angular resolution and straight edges.

New Results. We define a *Lombardi drawing* of a graph G to be a drawing of G in the plane in which vertices are represented as points (or as disks or labels centered on those points), edges are represented as line segments or circular arcs between their endpoints, and every vertex has perfect angular resolution, as measured by the angle formed by the tangents to the edges at the vertex. We do not necessarily insist that the drawings are free of crossings; the drawings of Lombardi had crossings, sometimes even in cases where they could have been avoided. We also do not consider crossings when we measure the angular resolution of a drawing. However, we do require that the only vertices that intersect the arc for an edge (u, v) are its two endpoints u and v.

Several of Mark Lombardi's drawings used a circle as their overall shape. We define a *circular Lombardi drawing* to be a Lombardi drawing in which the vertices lie on a

U. Brandes and S. Cornelsen (Eds.): GD 2010, LNCS 6502, pp. 195–207, 2011.

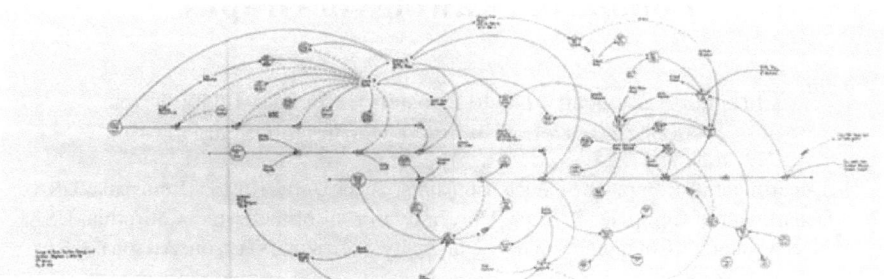

Fig. 1. Mark Lombardi, *George W. Bush, Harken Energy, and Jackson Stevens c.1979-90*, 1999. Graphite on paper, 20 × 44 inches [24].

circle. Similarly, we define a *k-circular **Lombardi drawing*** to be a Lombardi drawing in which the vertices lie on *k* concentric circles. We provide the following:

- We characterize the regular graphs that have circular Lombardi drawings, and we find efficient algorithms for constructing these drawings.
- We describe methods of finding Lombardi drawings for any 2-degenerate graph (a graph that may be reduced to the empty graph by repeated removal of vertices of degree at most 2) and many but not all 3-degenerate graphs.
- We investigate the graphs that have planar Lombardi drawings. We show that certain subclasses of the planar graphs always have such drawings, but that there exist planar graphs with no planar Lombardi drawing.
- We implement an algorithm for constructing *k*-circular Lombardi drawings and use it to draw many symmetric graphs.

Related Work. Although most previous work on angular resolution concerns straight-line drawings (e.g., see [10,17,25]) or polyline drawings (e.g., see [18,21]), the angular resolution of drawings with circular-arc edges was previously studied by Cheng *et al.* [8], who showed that maintaining bounded angular resolution in planar drawings may require exponential area even with circular-arc edges. Our circular Lombardi drawings use a circular layout of vertices that is already popular (e.g., see [3,16,30]). However, previous methods for circular layouts draw edges as straight line segments or curves perpendicular to the circle, neither of which leads to good angular resolution.

Efrat *et al.* [13] show that given a fixed placement of the vertices of a planar graph, determining whether the edges can be drawn with circular arcs so that there are no crossings is NP-Complete. For fixed position drawings with cubic Bézier curves, Brandes *et al.* [5,7] use force-directed algorithms to maximize the angular resolution and Brandes, Shubina, and Tamassia [6] rotate optimal angular resolution templates. Aicholzer *et al.* [1] show that, for a given embedded planar triangulation with fixed vertex positions, one can find a circular-arc drawing of the triangulation that maximizes the minimum angular resolution by solving a linear program. Finkel and Tamassia [14] also try to optimize angular resolution using force-directed methods for laying out graphs with curved edges. Di Battista and Vismara [10] give a nonlinear optimization characterization that can find straight-line drawings of embedded planar graphs with a prescribed assignment of angles if such drawings exist.

Any tree may be drawn with straight edges and perfect angular resolution. However, in a separate paper [12], we show that (when the order of the edges is fixed around each vertex) straight-line tree drawings with perfect angular resolution may require exponential area, whereas Lombardi drawings can achieve polynomial area.

2   Circular Lombardi Drawings of Regular Graphs

We begin by investigating *circular Lombardi drawings*, Lombardi drawings in which all vertices are placed on a circle. As we show, drawings of this type exist for many regular graphs. Our proofs use the following basic geometric observation:

Property 1. Let A be a circular arc or line segment connecting two points p and q that both lie on circle O. Then A makes the same angle to O at p that it makes at q. Moreover, for any p and q on O and any angle $0 \leq \theta \leq \pi$, there exists an arc, line segment, or pair of collinear rays A connecting p and q, making angle θ with O, and lying either inside or outside of O.

The case of two collinear rays is problematic (we only allow edges to be represented by arcs or line segments) but easily avoided by perturbing the vertices on O.

Lemma 1. *A d-regular graph G has a circular Lombardi drawing if and only if G can be decomposed into a disjoint union of 1-regular and 2-regular graphs and one of the following conditions is true: $d \not\equiv 2$ (mod 4), one of the 2-regular subgraphs is bipartite, or one of the 2-regular subgraphs is a Hamiltonian cycle.*

Proof. Suppose G has a circular Lombardi drawing on a circle O centered at o; in this drawing, define the *twist* θ_v of a vertex v to be the sharpest of the angles between line segment vo and the edges incident to v (with positive sign if one of the edges forming the sharpest angle is clockwise of the line segment, and negative sign if there is only one edge forming the sharpest angle and it is counterclockwise of v). Then if v and w are adjacent in G, $\theta_v = -\theta_w$ except when there are two equal sharpest angles at both v and w, in which case $\theta_v = \theta_w$. In each connected component either all vertices have the same twist, and have edge angles that are symmetric with respect to reflections through axis vo, or the component is bipartite, all vertices on one side of the bipartition have one twist, and all vertices on the other side of the bipartition have the opposite twist.

We can decompose each connected component of G into 1-regular and 2-regular graphs by partitioning the edges of the component according to the angle they make with circle O. For a bipartite component in which the vertices on the two sides of the bipartition have different twists, this forms a decomposition into 1-regular graphs (some of which may be combined in pairs to form bipartite 2-regular graphs). When d is 2 mod 4 and a component of G is not bipartite, the only possibilities for a symmetric twist are to make some edges parallel or perpendicular to O. Edges that are parallel to O must be drawn as arcs of O through all vertices, so they form a Hamiltonian cycle. Edges perpendicular to O must form even-length cycles that alternate between the inside and outside of O. Thus, in all cases a graph with a circular Lombardi drawing can be decomposed into 1-regular and 2-regular graphs matching the conditions of the lemma.

In the other direction, suppose that G can be decomposed into 1-regular and 2-regular graphs with the additional conditions of the lemma. By combining pairs of 1-regular graphs into a single 2-regular graph, we may assume that all but at most one of these subgraphs are 2-regular. Then we may choose an evenly spaced set of angles, draw each 2-regular graph as a set of arcs that meet O at one of these fixed angles, and draw the 1-regular graph (if it exists) as a set of arcs that are perpendicular to and interior to O. If d is divisible by four, we can choose these angles in such a way that no angle is parallel to the circle O and no angle is perpendicular to O. If d is odd, the angles can be chosen so that the 1-regular subgraph of G is perpendicular to and interior to O, and all other angles are neither perpendicular nor parallel to O. If d is congruent to 2 mod 4 and one of the 2-regular graphs is a Hamiltonian cycle, we may draw it using edges that lie on C, placing the vertices in the order of this cycle. And if d is congruent to 2 mod 4 and one of the 2-regular graphs is bipartite, we may draw it using edges that are perpendicular to O, taking care in the vertex placement to avoid using an edge that connects two diametrally opposite points on O via an exterior arc. In both of these cases where d is 2 mod 4 we then draw the other subgraphs of the decomposition using arcs that are neither parallel to nor perpendicular to O. □

Theorem 1. *Every regular graph G of degree divisible by four has a circular Lombardi drawing. A regular graph of odd degree has a circular Lombardi drawing if and only if it has a perfect matching. A regular graph of degree congruent to two modulo four has a circular Lombardi drawing if and only if it is Hamiltonian or has a 2-regular bipartite subgraph. In the cases of odd degree and degree divisible by four, when a circular Lombardi drawing exists it can be constructed in polynomial time.*

Proof. This follows from Lemma 1 together with Petersen's theorem that a regular graph of even degree can always be decomposed into 2-regular subgraphs [27,28]. □

Testing for the existence of a 2-regular bipartite subgraph in a regular graph is NP-complete (in 3-regular graphs, it is equivalent to 3-edge-coloring) but we have not determined its complexity for the case of interest to us, d-regular graphs in which d is congruent to two modulo four.

Figures 2(a–c) show drawings produced by this method for 3-regular, 4-regular, and 6-regular graphs. Figure 2(d) shows a 3-regular graph that does not have a perfect matching, and that therefore has no circular Lombardi drawing.

For bipartite regular graphs of bounded degree the method of Theorem 1 again leads to a linear-time algorithm.

Corollary 1. *Every bipartite d-regular graph has a circular Lombardi drawing that can be constructed in time $O(dn \log d)$.*

Proof. It is known that every bipartite regular graph can be decomposed into perfect matchings in the given time bound [2,9,29].[1] The result follows by applying Theorem 1 to this decomposition. □

---

[1] The fact that every regular bipartite graph has a decomposition into matchings is commonly attributed to König [22], but is equivalent to a result proved in terms of point-line configurations in the 1894 Ph.D. thesis of Ernst Steinitz.

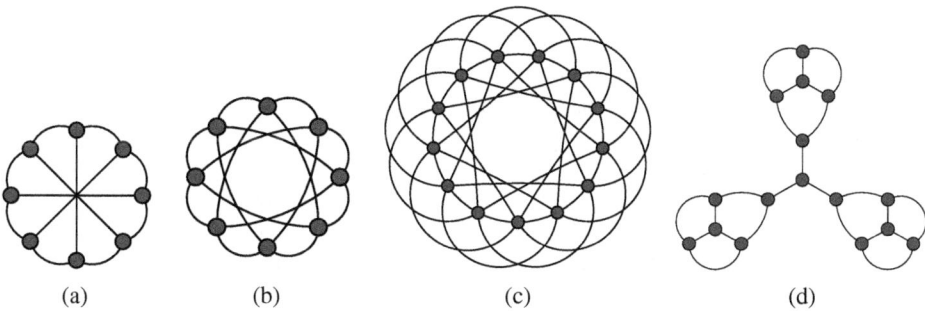

Fig. 2. (a) A circular Lombardi drawing of the 3-regular Wagner graph; (b) A circular Lombardi drawing of the 4-regular graph $K_{4,4}$; (c) The 6-regular Paley graph connecting integers modulo 13 if their difference is a quadratic residue; (d) A 3-regular graph that has no perfect matching and therefore has no circular Lombardi drawing

Corollary 2. *Every d-regular graph for which d is a power of two, with the exception of 2-regular non-bipartite disconnected graphs, has a circular Lombardi drawing that can be constructed in time $O(dn \log d)$.*

Proof. Repeatedly decompose the graph into pairs of subgraphs with half the degree by taking alternating edges of an Euler tour [15] and then apply Theorem 1 to the decomposition. □

Corollary 3. *Every 3-regular bridgeless graph has a circular Lombardi drawing that can be constructed in time $O(n \log^3 n \log \log n)$.*

Proof. The result that every 3-regular bridgeless graph has a perfect matching (equivalently, a decomposition into a 2-regular and a 1-regular subgraph) is known as Petersen's theorem [28]. Such a matching can be found in the stated time bound via an algorithm based on dynamic 2-edge-connectivity testing data structures [4,20,31]. □

3   Two-Degenerate and Three-Degenerate Graphs

The ***degeneracy*** of a graph G is the minimum number d such that G can be reduced to the empty graph by repeatedly removing a vertex of degree at most d; equivalently, it is the minimum degree in the subgraph of G that maximizes the minimum degree [23]. If a graph G has degeneracy at most d, it is known as d-degenerate. In this section we consider algorithms for drawing 2-degenerate and 3-degenerate graphs, with a specified cyclic ordering of the edges around each vertex. The main idea of these algorithms is to delete a low-degree vertex, draw the remaining graph with the appropriate angles at each of its vertices, and then find a position for the deleted vertex that allows it to be connected to the drawing of the remaining graph.

For 2-degenerate graphs, when we add back the vertices in reverse order of deletion, there is always a circle on which they can be added so we can choose one point on the circle that is not crossed by a previously drawn feature. For 3-degenerate graphs there are two points at which the point can be added to give the correct edge angles (the common intersection points of three circles) so there might be circumstances under which this addition is forced to create an undesirable edge-vertex or vertex-vertex intersection.

The results in this section rely on the following geometric property, which is proven in the full version of the paper [11].

Property 2. Suppose we are given two points p and q with associated vectors v_p and v_q and an angle θ_{pq}. Consider all pairs of circular arcs that leave p and q with tangent vectors v_p and v_q respectively and meet at an angle θ_{pq}. The locus of meeting points for these pairs of arcs is a circle.

3.1 2-Degenerate Graphs

Theorem 2. *Every 2-degenerate graph with a specified cyclic ordering of the edges around each vertex has a Lombardi drawing.*

Proof. Order the vertices by repeatedly removing a low-degree vertex. Reinsert the vertices in reverse order creating subgraphs $G_0, G_1 \ldots G_n$ with the invariant that after each insertion the drawing is a *partial* Lombardi drawing Γ_i of G_i where some vertices may not yet have all of their neighbors placed. To insert a new vertex $v = v_{i+1}$ with degree two in G_{i+1} (the case for degree one is simpler) let p and q be its two neighbors in G_{i+1}. Since there is a specified ordering around p, which has already been placed in Γ_i, there is a unique tangent vector v_p associated with the arc from p to v. Similarly, there is a unique tangent vector v_q. In addition, since the degree of v in G is known and the ordering of the neighbors at v is also given, there is a unique angle θ_{pq} associated with the two arcs from p and q to v. From Property 2, we may choose to place v at any position on the defined circle. Choosing a point v that does not coincide with any other arcs or vertices already placed guarantees we have a valid drawing Γ_{i+1}.   □

Corollary 4. *Every outerplanar or series-parallel graph has a Lombardi drawing.*

Proof. This follows from the fact that these graphs are 2-degenerate.   □

3.2 3-Degenerate Graphs

An algorithm following the same approach can be used to draw many, but not all, 3-degenerate graphs. In this case we have three points p, q, and r that we want to connect by arcs to an unplaced new vertex v. Each pair of known points yields a circle of possible choices for v. These three circles, O_{pq}, O_{pr}, O_{qr}, have to pairwise cross, and where they cross the third one must also cross because fixing the angles between two pairs of incoming arcs at the new point fixes all angles. Every graph with maximum degree four is either 4-regular or 3-degenerate, so the same algorithm applies in this case.

However, for certain graphs and certain orderings of the edges around the vertices of the graph, this algorithm can fail by placing a vertex on another edge or vertex. An example in which this occurs is the seven-vertex split graph G_7 formed by adding four

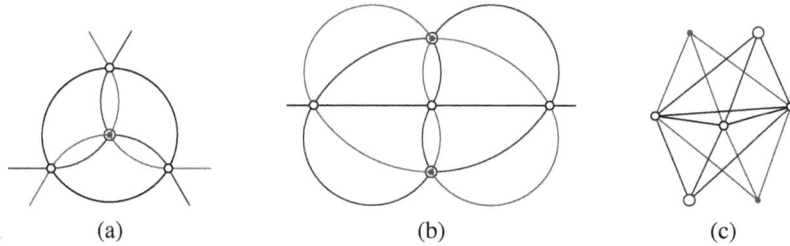

Fig. 3. A 7-vertex 3-degenerate graph that has no Lombardi drawing with the given vertex ordering. (a) A Möbius transformation makes one triangle equilateral, forcing the other 4 vertices to be placed at the centroid and the point at infinity; (b) A different transformation with finite vertex locations; (c) A straight-line drawing of the graph.

independent vertices p, q, r, and s to a triangle xyz, with an edge from each of p, q, r, and s to each of x, y, and z, as shown in Figure 3. In any Lombardi drawing of G_7 with the edge order as shown, we can assume by making an appropriate Möbius transformation of the drawing that xyz is equilateral. It follows that the only possible locations for p, q, r, and s are the centroid of the equilateral triangle and the point at infinity, so at least two vertices would have to be placed at the same point, forming an invalid drawing.

4 Non-crossing Lombardi Drawings

4.1 Planar Graphs without Planar Lombardi Drawings

Not every planar graph has a planar Lombardi drawing. To see this, consider the *k-nested triangle graphs*, maximal planar graphs with $3k$ vertices formed by k nested triangles with $k - 1$ six-cycles connecting consecutive triangles. A k-nested triangle graph may also be formed geometrically by gluing $k - 1$ octahedra end-to-end.

As can be seen in Figure 4, the 2-nested and 3-nested triangle graphs have planar Lombardi drawings. The 4-nested triangle graph, however, does not. If it did have such a drawing, its middle two triangles would form circles (the only smooth curve formed by three circular arcs). By an appropriate Möbius transformation, the outer circle O can be assumed to have its three vertices equally spaced around it. The three cirles C_1, C_2, and C_3 that (by Property 2) describe the potential positions of the vertices on the inner circle have the same radius as O and meet at the center of O, and the inner circle would have to be tangent to all three of C_1, C_2, and C_3. However, the only circle tangent to all three is exterior to O, concentric with O and having twice the radius of O. Therefore, using an edge ordering around each vertex that comes from a planar embedding but enforcing perfect angular resolution leads to a nonplanar drawing, shown in Figure 4(c).

4.2 Halin Graphs

A Halin graph [19] is a planar graph obtained from a plane tree T (with at least four vertices and with no vertices of degree 2), by connecting all the leaves of T into a

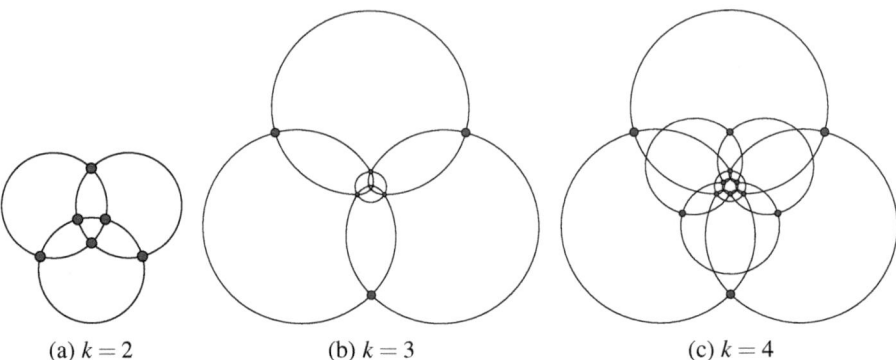

(a) $k = 2$ (b) $k = 3$ (c) $k = 4$

Fig. 4. k-nested triangle graphs. The 2-nested and 3-nested triangle graphs have planar Lombardi drawings, but the 4-nested triangle graph does not.

cycle in the order given by its embedding. As we now describe, Halin graphs (and the graphs formed in the same way from trees with degree-2 vertices) have planar Lombardi drawings that can be constructed using hyperbolic geometry.

We draw T within a Poincaré disk model of the hyperbolic plane, with its leaves on the boundary circle of the model, and then draw the cycle connecting the leaves outside this circle. If T is drawn using hyperbolic line segments, with perfect angular resolution, then its edges will form circular arcs in the Poincaré model; the conformal (angle-preserving) nature of the Poincaré model implies that the angular resolution of the hyperbolic line segments equals the angular resolution of these Euclidean arcs.

For a given straight-line drawing of a rooted tree in the hyperbolic plane, and a non-root vertex v, partition the hyperbolic plane into wedges bounded by the bisectors of the angles around the parent of v and define the ***dominance region*** of v to be the wedge containing v. Equivalently, in a Voronoi diagram generated by the rays from the parent of v to its children, the dominance region of v is the Voronoi cell containing v. We define a ***good hyperbolic drawing*** of a rooted tree T to be a drawing in which the edges are straight line segments or rays in the hyperbolic plane, the leaves are placed on the circle at infinity, and the dominance regions for two vertices v and w are either nested within each other (if one of the two vertices is an ancestor of the other) or disjoint otherwise. Two dominance regions in a good hyperbolic drawing are shown in Figure 5(a).

Lemma 2. *Every rooted tree has a good hyperbolic drawing.*

Proof. We use induction on the number of non-leaf nodes in the given tree T. As a base case, when there is one non-leaf node, it may be placed at the center of the Poincaré disk model of the hyperbolic plane with its leaves at the limit points of equally-spaced rays (radii of the disk model). Otherwise, let v be a non-leaf that is as far from the root of T as possible, and let T' be formed from T by removing all children of v. Then by induction, T' has a good hyperbolic drawing. In this drawing, v is on the circle at infinity; let R be the ray connecting the parent of v to v. For any position x along this ray, let θ_x be the maximum angle made to R by a line that stays within the dominance

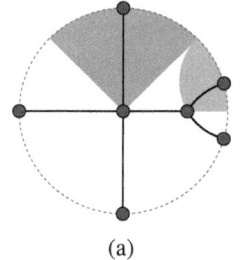 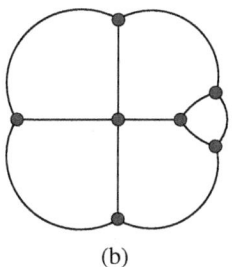

(a)   (b)

Fig. 5. (a) A good hyperbolic drawing of a seven-node tree, with the dominance regions of two leaves of the tree shown as shaded regions; (b) The Lombardi drawing formed by adding arcs outside the Poincaré model, at 30° angles to the boundary, connecting consecutive leaves

region of v. Then θ_x varies continuously along R, starting from a value of π/d at the parent of v (where d is the degree of the parent) and ending with a value of π at v itself. If the degree of v in T is d', there must be an intermediate position x on R for which $\theta_x = \pi(1 - 1/d')$. If we move v to x and place its leaf children at the limit points of equally spaced rays around x, the result is a good hyperbolic drawing of T.   □

Theorem 3. *Every Halin graph has a planar Lombardi drawing that may be constructed in linear time.*

Proof. Root the tree T at an arbitrarily chosen non-leaf node, and construct a good hyperbolic drawing of T according to Lemma 2. Draw the cycle connecting the leaves of T using circular arcs that meet the circle bounding the Poincaré model at angles of 30° as in Figure 5(b). Then each non-leaf node of T has perfect angular resolution from the tree drawing, and each leaf node has perfect angular resolution because the ray connecting it to its parent in T is perpendicular to the boundary circle and therefore at 120° angles from the two arcs connecting it to adjacent leaves.   □

4.3   Other Classes of Planar Graphs

The networks formed by two-dimensional soap bubbles naturally form 3-regular planar Lombardi drawings: they have circular arcs as their edges (the boundaries between bubbles), and 120° angles at each vertex where three arcs meet [26]. However, we do not have a precise characterization of the graphs that can be formed in this way.

The vertices of every Platonic solid, Archimedean solid, and prism lie on a common sphere. In all but two cases (the snub cube and snub dodecahedron) one may draw the edges of the polyhedron as circular arcs on the sphere with perfect angular resolution. By stereographic projection, each of these graphs has a Lombardi drawing in the plane. For instance, Figure 4(a) depicts the graph of the octahedron drawn in this way.

All outerplanar and series-parallel graphs have Lombardi drawings (Corollary 4), but we do not know whether they all have planar Lombardi drawings.

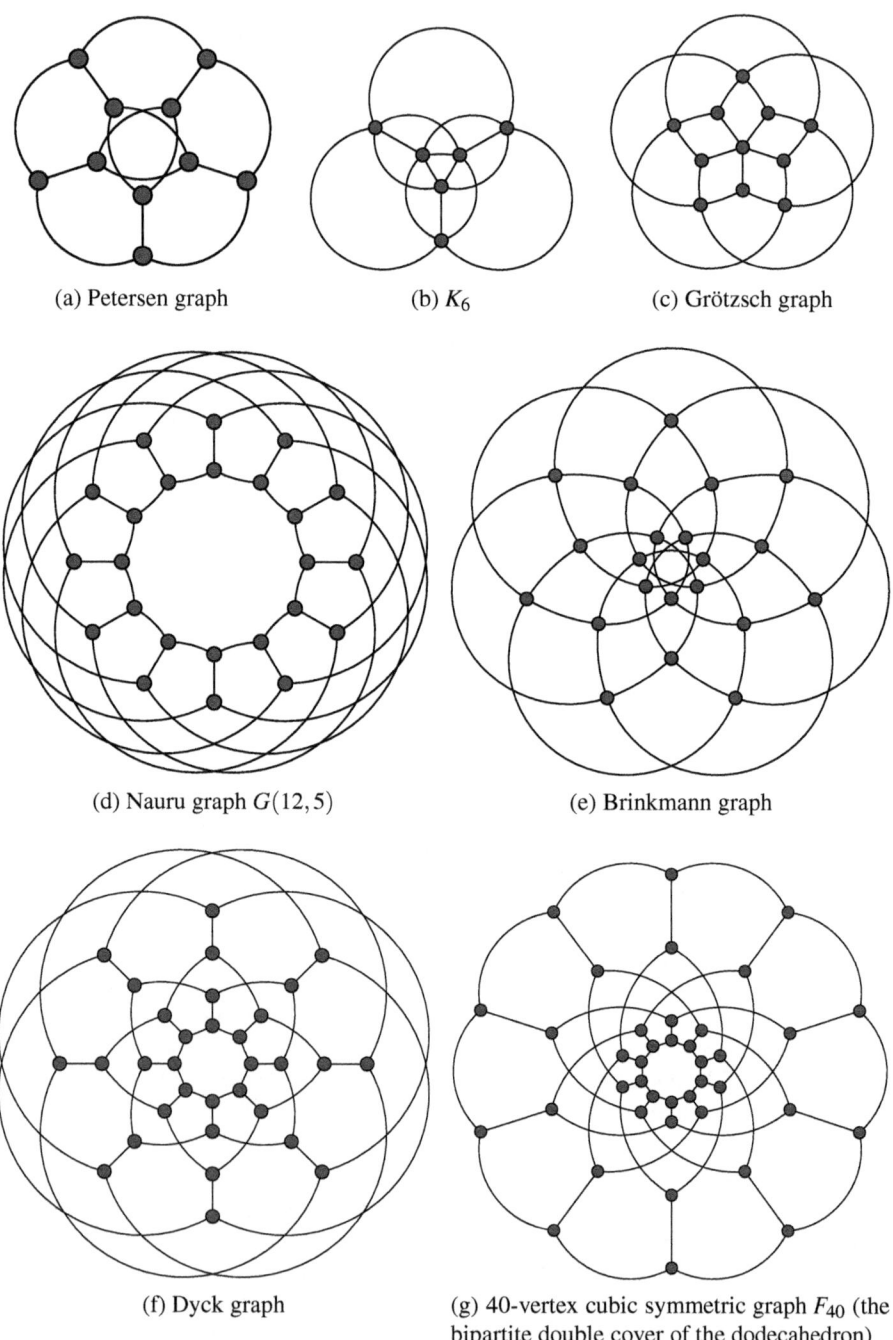

(a) Petersen graph   (b) K_6 (c) Grötzsch graph

(d) Nauru graph $G(12,5)$ (e) Brinkmann graph

(f) Dyck graph   (g) 40-vertex cubic symmetric graph F_{40} (the bipartite double cover of the dodecahedron)

Fig. 6. Sample drawings by the Lombardi Spirograph

5   The Lombardi Spirograph

We have implemented a program for constructing k-circular Lombardi drawings of graphs with dihedral symmetry; we call it the *Lombardi Spirograph*, as its drawings resemble those created by the Spirograph™ drawing toy produced by Hasbro, Inc. Our program places vertices on k concentric circles; the input specifies not only the number of vertices per circle and the set of edges to be drawn, but also the order in which those edges are incident at each vertex. Each vertex can have at most three neighbors on smaller circles; a circle on which the vertices have two or three inward neighbors has a unique radius for which the vertices have perfect angular resolution, whereas the radius for circles on which the vertices have one inner neighbor is chosen heuristically.

Figures 2 (a–c), 4 (a & b), and 6 were all drawn using this program.

6   Conclusions

We have begun an investigation into Lombardi drawings and found algorithms based on graph matching, incremental construction, hyperbolic geometry, and symmetry display for constructing drawings of this type. Based on our constructions, we can show that many regular graphs, sparse graphs, special classes of planar graphs, and symmetric graphs have Lombardi drawings, and we have found drawings of this type for many well-known graphs. In addition, we have implemented a method, called the *Lombardi Spirograph*, for producing Lombardi drawings of graphs with dihedral symmetry.

There are many related problems that remain open, including the following:

1. What are the complexities of finding circular Lombardi drawings for regular graphs with degrees that are 2 mod 4?
2. Is there an effective classification of 3-degenerate graphs according to whether they can or cannot be drawn in a way that avoids overlapping features?
3. Are there efficient methods for producing planar Lombardi drawings for outerplanar graphs, series-parallel graphs, and 3-regular planar graphs?

It would also be of interest to combine Lombardi drawing with other standard graph drawing quality criteria such as edge-length minimization. In general, we believe that Lombardi drawings will be a fruitful area for much additional research.

Acknowledgments

This research was supported in part by the National Science Foundation under grant 0830403, by the Office of Naval Research under MURI grant N00014-08-1-1015, and by the German Research Foundation under grant NO 899/1-1.

References

1. Aichholzer, O., Aigner, W., Aurenhammer, F., Dobiášová, K.Č., Jüttler, B.: Arc triangulations. In: Proc. 26th Eur. Worksh. Comp. Geometry (EuroCG 2010), Dortmund, Germany, pp. 17–20 (2010)

2. Alon, N.: A simple algorithm for edge-coloring bipartite multigraphs. Information Processing Letters 85(6), 301–302 (2003), doi:10.1016/S0020-0190(02)00446-5

3. Baur, M., Brandes, U.: Crossing reduction in circular layouts. In: Hromkovič, J., Nagl, M., Westfechtel, B. (eds.) WG 2004. LNCS, vol. 3353, pp. 332–343. Springer, Heidelberg (2004), http://www.springerlink.com/content/fepu2a3hd195ffjg/

4. Biedl, T.C., Bose, P., Demaine, E.D., Lubiw, A.: Efficient algorithms for Petersen's matching theorem. J. Algorithms 38(1), 110 (2001), doi:10.1006/jagm.2000.1132

5. Brandes, U., Schlieper, B.: Angle and distance constraints on tree drawings. In: Kaufmann, M., Wagner, D. (eds.) GD 2006. LNCS, vol. 4372, pp. 54–65. Springer, Heidelberg (2007), doi:10.1007/978-3-540-70904-6_7

6. Brandes, U., Shubina, G., Tamassia, R.: Improving angular resolution in visualizations of geographic networks. In: Data Visualization 2000. Proc. 2nd Eurographics/IEEE TCVG Symp. Visualization (VisSym 2000), pp. 23–32. Springer, Heidelberg (2000)

7. Brandes, U., Wagner, D.: Using graph layout to visualize train interconnection data. J. Graph Algorithms Appl. 4(3), 135–155 (2000)

8. Cheng, C.C., Duncan, C.A., Goodrich, M.T., Kobourov, S.G.: Drawing planar graphs with circular arcs. Discrete Comput. Geom. 25(3), 405–418 (2001), doi:10.1007/s004540010080

9. Cole, R., Ost, K., Schirra, S.: Edge-coloring bipartite multigraphs in $O(E log D)$ time. Combinatorica 21(1), 5–12 (2001), doi:10.1007/s004930170002

10. di Battista, G., Vismara, L.: Angles of planar triangular graphs. SIAM J. Discrete Math. 9(3), 349 (1996), doi:10.1137/S0895480194264010

11. Duncan, C.A., Eppstein, D., Goodrich, M.T., Kobourov, S.G., Nöllenburg, M.: Lombardi Drawings of Graphs. (September 2010) ArXiv e-prints, arXiv:1009.0579

12. Duncan, C.A., Eppstein, D., Goodrich, M.T., Kobourov, S.G., Nöllenburg, M.: Drawing trees with perfect angular resolution and polynomial area. In: Proc. 18th Int. Symp. on Graph Drawing (GD 2010), Springer, Heidelberg (2010)

13. Efrat, A., Erten, C., Kobourov, S.G.: Fixed-location circular arc drawing of planar graphs. J. Graph Algorithms Appl. 11(1), 145–164 (2007), http://jgaa.info/accepted/2007/EfratErtenKobourov2007.11.1.pdf

14. Finkel, B., Tamassia, R.: Curvilinear graph drawing using the force-directed method. In: Pach, J. (ed.) GD 2004. LNCS, vol. 3383, pp. 448–453. Springer, Heidelberg (2005), doi:10.1007/978-3-540-31843-9_46

15. Gabow, H.N.: Using Euler partitions to edge color bipartite multigraphs. Int. J. Parallel Programming 5(4), 345–355 (1976), doi:10.1007/BF00998632

16. Gansner, E.R., Koren, Y.: Improved circular layouts. In: Kaufmann, M., Wagner, D. (eds.) GD 2006. LNCS, vol. 4372, pp. 386–398. Springer, Heidelberg (2007), doi:10.1007/978-3-540-70904-6_37

17. Garg, A., Tamassia, R.: Planar drawings and angular resolution: algorithms and bounds. In: van Leeuwen, J. (ed.) ESA 1994. LNCS, vol. 855, pp. 12–23. Springer, Heidelberg (1994), doi:10.1007/BFb0049393

18. Gutwenger, C., Mutzel, P.: Planar polyline drawings with good angular resolution. In: Whitesides, S.H. (ed.) GD 1998. LNCS, vol. 1547, pp. 167–182. Springer, Heidelberg (1999), doi:10.1007/3-540-37623-2_13

19. Halin, R.: Über simpliziale Zerfällungen beliebiger (endlicher oder unendlicher) Graphen. Math. Ann. 156(3), 216–225 (1964), doi:10.1007/BF01363288

20. Holm, J., de Lichtenberg, K., Thorup, M.: Poly-logarithmic deterministic fully-dynamic algorithms for connectivity, minimum spanning tree, 2-edge, and biconnectivity. J. ACM 48(4), 723–760 (2001), doi:10.1145/502090.502095

21. Kant, G.: Drawing planar graphs using the canonical ordering. Algorithmica 16, 4–32 (1996), doi:10.1007/BF02086606

22. Kőnig, D.: Gráfok és mátrixok. Matematikai és Fizikai Lapok 38, 116–119 (1931)
23. Lick, D.R., White, A.T.: K-degenerate graphs. Canad. J. Math. 22, 1082–1096 (1970), http://www.smc.math.ca/cjm/v22/p1082
24. Lombardi, M., Hobbs, R.: Mark Lombardi: Global Networks. Independent Curators (2003)
25. Malitz, S., Papakostas, A.: On the angular resolution of planar graphs. SIAM J. Discrete Math. 7(2), 172–183 (1994), doi:10.1137/S0895480193242931
26. Morgan, F.: Soap bubbles in $\mathbb{R}^2$ and in surfaces. Pacific J. Math. 165(2), 347–361 (1994), http://projecteuclid.org/euclid.pjm/1102621620
27. Mulder, H.M.: Julius Petersen's theory of regular graphs. Discrete Mathematics 100(1-3), 157–175 (1992), doi:10.1016/0012-365X(92)90639-W
28. Petersen, J.: Die Theorie der regulären Graphs. Acta Math. 15(1), 193–220 (1891), doi:10.1007/BF02392606
29. Schrijver, A.: Bipartite edge coloring in $O(\Delta m)$ time. SIAM J. Comput. 28(3), 841–846 (1999), doi:10.1137/S0097539796299266
30. Six, J.M., Tollis, I.G.: A framework for circular drawings of networks. In: Kratochvíl, J. (ed.) GD 1999. LNCS, vol. 1731, pp. 107–116. Springer, Heidelberg (1999)
31. Thorup, M.: Near-optimal fully-dynamic graph connectivity. In: Proc. 32nd ACM Symp. Theory of Computing (STOC 2000), pp. 343–350 (2000), doi:10.1145/335305.335345

Optimal 3D Angular Resolution for Low-Degree Graphs

David Eppstein[1], Maarten Löffler[1], Elena Mumford[2], and Martin Nöllenburg[1]

[1] Department of Computer Science, University of California, Irvine, USA
[2] Eindhoven, The Netherlands

Abstract. We show that every graph of maximum degree three can be drawn in three dimensions with at most two bends per edge, and with $120°$ angles between any two edge segments meeting at a vertex or a bend. We show that every graph of maximum degree four can be drawn in three dimensions with at most three bends per edge, and with $109.5°$ angles, i.e., the angular resolution of the diamond lattice, between any two edge segments meeting at a vertex or bend.

1 Introduction

Much past research in graph drawing has shown the importance of avoiding sharp angles at vertices, bends, and crossings of a drawing, as they make the edges difficult to follow [15]. There has been much interest in finding drawings where the angles at these features are restricted, either by requiring all angles to be at most $90°$ (as in orthogonal drawings [11] and RAC drawings [1, 7, 8]) or more generally by attempting to optimize the *angular resolution* of a drawing, the minimum angle that can be found within the drawing [4, 13, 14, 16].

Three-dimensional graph drawing opens new frontiers for angular resolution in two ways. First, in three-dimensional graph drawing, there is no need for crossings, as any graph can be drawn without crossings; however, finding a compact layout that uses few bends and avoids crossings can sometimes be challenging. Second, and more importantly, in 3d there is a much greater variety in the set of ways that a collection of edges can meet at a vertex to achieve good angular resolution, and the angular resolution that may be obtained in 3d is often better than that for a two-dimensional drawing. For instance, in 3d, six edges may meet at a vertex forming angles of at most $90°$, whereas in 2d the same six edges would have an angular resolution of $60°$ at best.

The problem of optimizing the angular resolution of a collection of edges incident to a single vertex in 3d is equivalent to the well-known *Tammes' problem* of placing points on a sphere to maximize their minimum separation; this problem is named after botanist P. M. L. Tammes who studied it in the context of pores on grains of pollen [18], and much is known about it [5]. For graphs of degree five or six, the optimal angular resolution of a three-dimensional drawing is $90°$, as above, achieved by placing vertices on a grid and drawing all edges as grid-aligned polylines. The simplicity of this case has freed researchers to look for three-dimensional orthogonal drawings that, as well as optimizing the angular resolution, also optimize secondary criteria such as the number of bends per edge, the volume of the drawing, or combinations of both [3, 10, 19]. Thus, in this case, it is known that the graph may be drawn with at most 3 bends per edge in an $O(n) \times O(n) \times O(n)$ grid and with $O(1)$ bends per edge in an $O(\sqrt{n}) \times O(\sqrt{n}) \times O(\sqrt{n})$

U. Brandes and S. Cornelsen (Eds.): GD 2010, LNCS 6502, pp. 208–219, 2011.

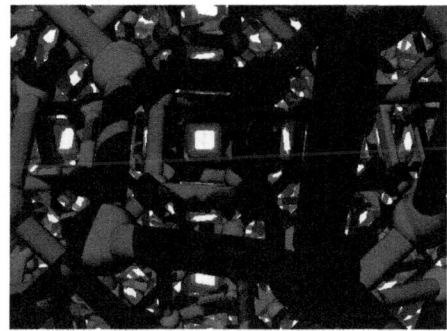

Fig. 1. Left: The three-dimensional diamond lattice, from [12]. Right: A space-filling 3-regular graph with 120° angular resolution.

grid [10]. For graphs of maximum degree five a tighter bound of two bends per edge is also known [19]; a well known open problem asks whether the same two-bend-per-edge bound may be achieved for degree six graphs [6].

In two dimensions, 90° angular resolution is optimal for graphs that may be nonplanar, because every crossing has an angle at least this sharp. However, in 3d, crossings are no longer a concern and graphs of degree three and four may have angular resolution even better than 90°. In particular, in the *diamond lattice*, a subset of the integer grid, the edges are parallel to the long diagonals of the grid cubes and meet at angles of $\arccos(-1/3) \approx 109.5°$, the optimal angular resolution for degree-four graphs (Figure 1, left). For graphs with maximum degree three, the best possible angular resolution at any vertex is clearly 120°; three edges with these angles are coplanar, but the planes of the edges at adjacent vertices may differ: for instance, Figure 1(right) shows an infinite space-filling graph in which all vertices are on integer grid points, all edges form face diagonals of the integer grid, and all vertices have 120° angular resolution.

The primary questions we study in this paper are how to achieve optimal 120° angular resolution for 3d drawings of arbitrary graphs with maximum degree three, and optimal 109.5° angular resolution for 3d drawings of arbitrary graphs with maximum degree four. We define angular resolution to be the minimum angle at any bend or vertex, matching the orthogonal drawing case, and we do not allow edges to cross. These questions are not difficult to solve without further restrictions (just place the vertices arbitrarily and use polylines with many bends to connect the endpoints of each edge) so we further investigate drawings that minimize the number of bends, align the vertices and edges of the drawing with the integer grid similarly to the alignment of the spacefilling patterns in Figure 1, and use a small total volume. We show:

- Any graph of maximum degree four can be drawn in 3d with optimal 109.5° angular resolution with at most three bends per edge, with all vertices placed on an $O(n) \times O(n) \times O(n)$ grid and with all edges parallel to the long diagonals of the grid cubes.
- Any graph of maximum degree three can be drawn in 3d with optimal 120° angular resolution with at most two bends per edge. However, our technique for achieving

this small number of bends does not use a grid placement and does not achieve good volume bounds.

- Any graph of maximum degree three has a drawing with 120° angular resolution, integer vertex coordinates, edges parallel to the face diagonals of the integer grid, at most three bends per edge, and polynomial volume.

We believe that, as in the orthogonal case, it should be possible to achieve tighter bounds on the volume of the drawing at the expense of greater numbers of bends per edge.

2   Three-Bend Drawings of Degree-Four Graphs on a Grid

Our technique for three-dimensional drawings of degree-four graphs with angular resolution 109.5° and three bends per edge is based on lifting two-dimensional drawings of the same graphs, with angular resolution 90° and two bends per edge. The three-dimensional vertex placements are all on the plane $z = 0$, essentially unchanged from their two-dimensional placements, but the edges are raised and lowered above and below the plane to avoid crossings and improve the angular resolution.

Our two-dimensional orthogonal drawing technique uses ideas from previous work on drawing degree-four graphs with bounded geometric thickness [9]. We begin by augmenting the graph with dummy edges and a constant number of dummy vertices if necessary to make it a simple 4-regular graph, find an Euler tour in the augmented graph, and color the edges alternately red and green in their order along this path. In this way, the red edges and the green edges each form 2-regular subgraphs [17] consisting of disjoint unions of cycles. We denote the number of red (green) cycles by m_{red} (m_{green}).

Next, we draw the red subgraph so that every cycle passes horizontally through its vertices with two bends per edge, and we draw the green subgraph so that every cycle passes vertically through its vertices with two bends per edge. We can do that by using the cycle ordering within each of these two subgraphs as one of the two Cartesian coordinates for each point. More precisely, we do the following.

We define the *green order* of the vertices of the graph to be an order of the vertices such that the vertices of each green path or cycle are consecutive; we define the *red order* the same way. Let $r_{green}(v) \geq 0$ be the rank of a vertex v in some green order, and $r_{red}(v)$ be its rank in some red order. We further order the red and green cycles and define $c_{red}(v) \geq 0$ and $c_{green}(v) \geq 0$ to be the ranks in the two cycle orders of the red and green cycles to which v belongs. We embed the vertices on a $(2n + 2m_{green} - 4) \times (2n + 2m_{red} - 4)$ grid such that the x-coordinate of each vertex is $2r_{green}(v) + 2c_{green}(v)$, and its y-coordinate is $2r_{red}(v) + 2c_{red}(v)$.

Let $v_1, ... v_k$ be the vertices of a green cycle C in the green order. We embed C as follows. We mark each end of each edge with a plus or a minus such that at every vertex exactly one end is marked with a plus and exactly one with a minus. We then would like to embed C in such a way that plus would correspond to the edge entering the vertex from above and a minus corresponds to the edge entering the vertex from below. Note that every edge whose two ends are marked the same can be embedded in this way with two bends. Whenever the marks alternate along the edge one can only embed it with two bends if the lower end (the end incident to a vertex with smaller y-coordinate) is marked with plus.

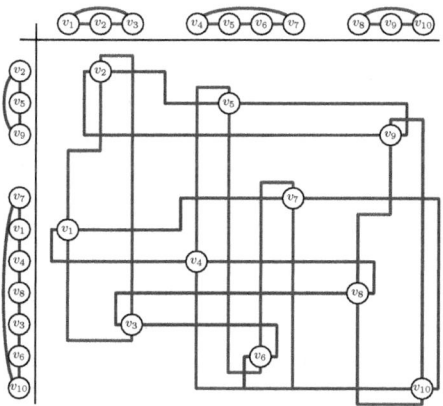

Fig. 2. A 4-regular graph with 10 vertices embedded according to the decomposition into disjoint red and green cycles

We next describe how to label C so that it has a 2-bends-per-edge embedding respecting the labeling. If k is even, we mark both ends of the edge (v_1, v_2) with pluses. If k is odd, we mark the higher end of (v_1, v_2) with a minus and its lower end with a plus. In both cases there is a unique way to label the rest of the edges such that both ends of each edge have the same signs and the labels alternate at every vertex.

To complete our 2d embedding we draw all edges consistently with the labeling as follows. Each edge (v_i, v_{i+1}) is placed such that the y-distance of its horizontal segment to one of the vertices is 1. If the last edge (v_1, v_k) is labeled negatively, its horizontal segment is drawn on the grid line one unit below the lowest vertex or bend of C. Similarly, if (v_1, v_k) is labeled positively, the horizontal segment is drawn one unit above the highest part of C. See Figure 2 for an illustration.

Lemma 1. *The embedding described above has the following properties:*

- *no two edges of the same color intersect;*
- *a vertex lies on an edge if and only if it is incident to the edge;*
- *no midpoint of an edge coincides with a bend of the edge;*
- *the embedding fits on a $(2n + 2m_{green}) \times (2n + 2m_{red})$ grid.*

Proof. Green edges connecting consecutive vertices in the green order of the same cycle C are trivially disjoint. The horizontal segment of the edge connecting the first and the last vertex of C is placed below or above all other edges of C. Two different green components are disjoint because the edges of every component are contained inside the vertical strip defined by its first and last vertices and components are ordered along the x-axis. The argument for red edges is symmetric.

Since all the vertices have distinct x-coordinates, and every green vertical segment has a vertex at one of its ends we can conclude that every vertex is incident to at most two vertical green segments. Every green horizontal segment has odd y-coordinate and every vertex has even y-coordinate hence a green horizontal segment cannot contain a vertex. The argument for red edges is symmetric.

For arbitrary red and green vertex orders it is possible that the midpoint of an edge coincides with one of its bends. We show that there are red and green vertex orders for which this is not the case. For any edge whose ends are labeled differently we can always place the horizontal segment such that the midpoint of the edge does not coincide with a bend. For edges whose ends have the same label it is easy to see that the midpoint coincides with a bend if and only if the vertical distance and the horizontal distance of its vertices are equal. Apart from the last edge in each green cycle the horizontal distance between any two adjacent vertices v_i and v_{i+1} is 2. We claim that the vertical distance between v_i and v_{i+1} is larger than 2 since otherwise v_i and v_{i+1} are adjacent in a red cycle which contradicts the assumption that the 4-regular graph is simple. Note that this is the reason why different components are spaced by at least 4 units. Finally consider the last edge (v_1, v_k) of a cycle C with vertices $v_1, \ldots, v_k$. The horizontal distance of v_1 and v_k is $2k - 2$. If their vertical distance equals $2k - 2$ as well, we cyclically shift the green order of the vertices in C by moving v_k to the vertical grid line of v_1 and shifting each of $v_1, \ldots, v_{k-1}$ two units to the right. Now (v_k, v_{k-1}) is the last edge of C. We perform this shifting until the vertices of the last edge no longer have vertical distance $2k - 2$. Since every vertex has an exclusive y-coordinate there is at least one edge with this property in C. The local shifting of C does not influence other parts of the drawing. The argument for red cycles is analogous.

The vertices lie on $(2n + 2m_{green} - 4) \times (2n + 2m_{red} - 4)$ grid, and each grid line with coordinate $2k$ contains exactly one vertex. The lowest vertex is incident to a green edge with a horizontal segment at the height -1; the highest one is incident to a green edge with a horizontal segment at the height $2n + 2m_{red} - 3$. One of the green edges connecting the first and last vertices of some cycle can lie one grid line below the height -1 or one grid lines above $2n + 2m_{red} - 3$.   □

It remains to lift the 2d drawing described above into three dimensions. We first rotate the drawing by $45°$; this expands the grid size to $(4n + 4m_{green}) \times (4n + 4m_{red})$. The vertices themselves stay in the plane $z = 0$, but we replace each edge by a path in 3d that goes below the plane for the red edges and above the plane for the green edges, eliminating all crossings between red and green edges. The path for a green edge goes upwards along the long diagonals of the diamond lattice cubes until its midpoint, where it has a bend and turns downwards again. The lifted images of the two bends in the underlying 2d edge remain bends in the 3d path and hence we get three bends per edge in total. The red edges are drawn analogously below the plane $z = 0$. Since in the original 2d drawing every edge has even length, the midpoint of every edge is a grid point and hence the lifted midpoint is also a grid point of the diamond lattice. By Lemma 1 a midpoint of an edge never coincides with a 2d bend and hence all bend angles as well as the vertex angles are $109.5°$ diamond lattice angles. Finally, we remove all the edges we added to make the graph 4-regular. Considering the longest possible red and green edges the total grid size is at most $(4n + 4m_{green}) \times (4n + 4m_{red}) \times (12n + 6m_{green} + 6m_{red})$. We note that $m_{green}, m_{red} \leq n/3$ since every component is a cycle. This yields the following theorem.

Theorem 1. *Any graph G with maximum vertex degree four can be drawn in a 3d grid of size $16n/3 \times 16n/3 \times 16n$ with angular resolution $109.5°$, three bends per edge and no edge crossings.*

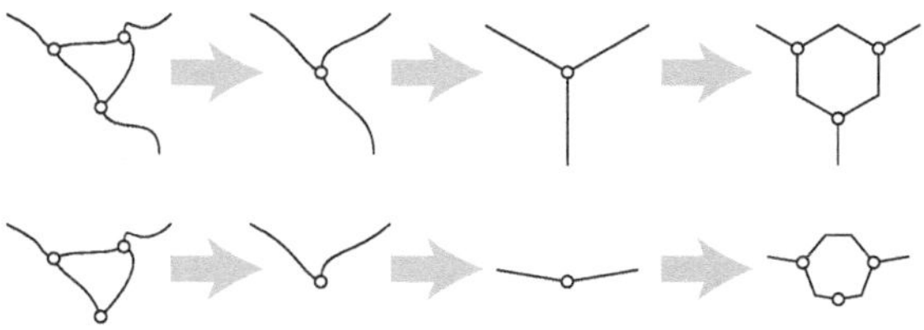

Fig. 3. Δ–Y transformation of a graph G containing a triangle, and undoing the transformation to find a drawing of G (Lemma 2). Top: the contracted vertex has degree three, and is replaced by a hexagon. Bottom: the contracted vertex has degree two, and is replaced by a heptagon.

3 Two-Bend Drawings of Degree-Three Graphs

The main idea of our algorithm for drawing degree-three graphs with optimal angular resolution and at most two bends per edge is to decompose the graph into a collection of vertex-disjoint cycles. Each cycle of length four or more can be drawn in such a way that the edges incident to the cycle all attach to it via segments that are parallel to the z axis (Lemma 4). By placing the cycles far enough apart in the z direction, these segments can be connected to each other with at most two bends per edge. However, several issues complicate this method:

- Cycles of length three cannot be drawn in the same way, and must be handled differently (Lemma 2).
- Our method for eliminating cycles of length three does not apply to the graph K_4, for which we need a special-case drawing (Lemma 3).
- Although Petersen's theorem [2, 17] can be used to decompose any bridgeless cubic graph into cycles and a matching, it is not suitable for our application because some of the matching edges may connect two vertices in a single cycle, a case that our method cannot handle. In addition, we wish to handle graphs that may contain bridges. Therefore, we need to devise a different decomposition algorithm. However, with our decomposition, the complement of the cycles is a forest rather than just a matching, and again we need additional analysis to handle this case.

Lemma 2. *Let G be a graph with maximum degree three containing a triangle uvw. If uvw is not part of any other triangle, let G' be the result of contracting uvw into a single vertex (that is, performing a Δ–Y transformation on G). Otherwise, if there is a triangle vwx, let G' be the result of contracting $uvwx$ into a single vertex. If G' can be drawn in 3d with two bends per edge and with angles of at least $120°$ between the edges at each vertex or bend, then so can G.*

Proof. First we consider the case that G' is obtained by collapsing uvw. The edges incident to the merged vertex uvw must lie in a plane in any drawing of G'. If uvw has degree zero, one, or three in G', or if it has degree two and is drawn with angular resolution exactly $120°$, then we may draw G by replacing uvw by a small regular hexagon in the same plane, with at most one bend for each of the three triangle edges

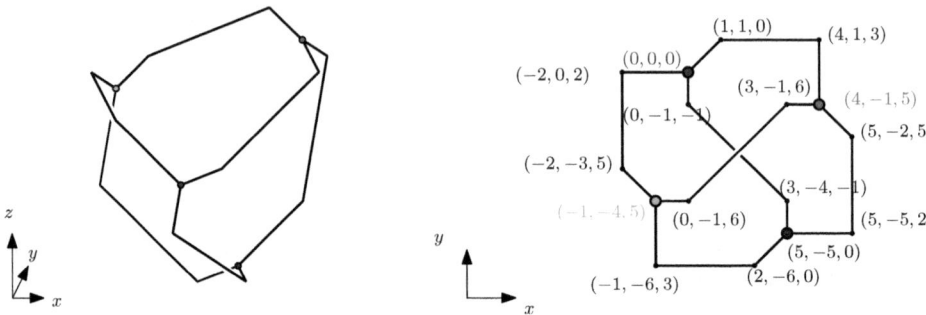

Fig. 4. A two-bend drawing of K_4 with $120°$ angular resolution (left) and its two-dimensional projection (right)

(Figure 3, top). If the merged vertex uvw has degree two in G' and is drawn with angular resolution greater than $120°$, we may replace it by a small heptagon (Figure 3, bottom).

The case that G' is obtained by collapsing four vertices $uvwx$ is similar: the collapsed vertex may be replaced by a pair of regular hexagons or irregular heptagons, meeting edge-to-edge. The four vertices $uvwx$ are placed at the points where these two polygons meet the other edges of the drawing and the two endpoints of the edge where they meet each other; the edge vw has no bends and the other edges all have one or two bends.   □

Lemma 3. *The graph K_4 may be drawn in 3d with all vertices on integer grid points, angular resolution $120°$, and at most two bends per edge.*

Proof. See Figure 4.   □

Lemma 4. *Let G be a graph with maximum degree three, consisting of a cycle C of $n \geq 4$ vertices together with some number of degree-one vertices that are adjacent to some of the vertices in C. Suppose also that each degree-one vertex in G is labeled with the number $+1$ or -1. Then, there is a drawing of G with the following properties:*

- *All vertices and bends have angular resolution at least $120°$.*
- *All edges of C have at most two bends.*
- *All edges attaching the degree-one vertices to C have no bends.*
- *Every degree-one vertex has the same x and y coordinates as its (unique) neighbor, and its z coordinate differs from its neighbor's z coordinate by its label. Thus, all edges connecting degree-one vertices to C are parallel to the z axis, all positively labeled vertices are above (in the positive z direction from) their neighbors, and all negatively labeled vertices are below their neighbors.*
- *No three vertices of C project to collinear points in the (x, y)-plane.*

Proof. As shown in Figure 5, we draw C in such a way that it projects onto a polygon P in the xy-plane, with $135°$ angles and with sides parallel to the coordinate axes and at $45°$ angles to the axes. There are polygons of this type with a number of sides that can be any even number greater than seven; we choose the number of sides of P so that at least one and at most two vertices of C can be assigned to each axis-parallel side of

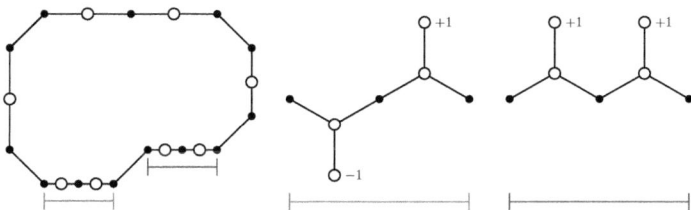

Fig. 5. The embedding of a cycle with degree-one neighboring vertices described by Lemma 4. Left: the xy-projection of the cycle; cycle vertices are indicated as large hollow circles and bends are indicated as small black disks. Right (at a larger scale): the xz-projection of the portions of the embedding corresponding to the two horizontal bottom sides of the xy-projected polygon.

the polygon. (E. g., when C has from four to eight vertices, P can have eight sides, but when C has more vertices P must be more complex.)

We assign the vertices of C consecutively to the axis-parallel sides of P, in such a way that at least one vertex of C and at most two vertices are assigned to each axis-parallel side. If one vertex is assigned to a side, it is placed at the midpoint of that side, and if two vertices are assigned to a side of length ℓ, then they are placed at distances of $\ell/4$ from one endpoint of the side, as measured in the xy plane, with a bend at the midpoint of the side.

In three dimensions, the diagonal sides of P are placed in the plane $z = 0$. For any axis-parallel side of P of length ℓ containing k vertices of C, we place the vertices with no degree-one neighbor or with a positively labeled neighbor at elevation $z = \ell/(2k\sqrt{3})$, and the vertices with a negatively labeled neighbor at elevation $z = -\ell/(2k\sqrt{3})$, so that the portion of C that projects onto a single side of P forms a polygonal curve with angles of exactly 120°. The degree-one neighbors of the vertices in C are then placed above or below them according to their signs.

With this embedding, each vertex of C gets angular resolution exactly 120°. Any two consecutive vertices of C that are assigned to the same side of P are separated either by zero bends (if their neighbors have opposite signs) or a single bend (if their neighbors have the same signs). Two consecutive vertices of C that belong to two different sides of P are separated by two bends at two of the corners of P; these bends have angles of $\arccos(-\sqrt{3/8}) \approx 127.8°$. By adjusting the lengths of the sides of P appropriately, we may ensure that no three vertices of C project to collinear points in the xy-plane. $\square$

The main idea of our drawing algorithm is to use Lemma 4, and some simpler cases for individual vertices, to repeatedly extend partial drawings of the given graph G until the entire graph is drawn. We define a *vertically extensible partial drawing* of a set S of vertices of G to be a drawing of the subgraph $G[S]$ induced in G by S, with the following properties:

- The drawing of $G[S]$ has angular resolution 120° or greater and has at most two bends per edge.
- Each vertex in S has at most one neighbor in $G \setminus S$.

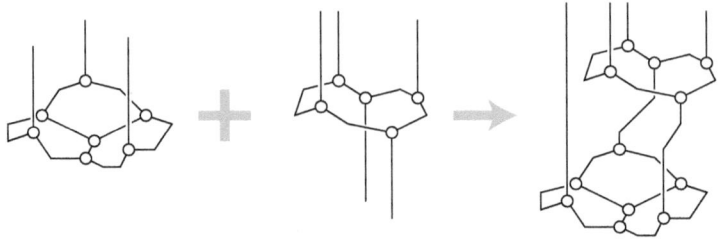

Fig. 6. Extending a vertically extensible drawing by adding a cycle

 – If a vertex v in S has a neighbor w in $G \setminus S$, then w could be placed anywhere along a ray in the positive z-direction from v, producing a drawing of $G[S \cup \{w\}]$ that remains non-crossing, continues to have angular resolution $120°$ or greater, and has no bends on edge vw. We call the ray from v the *extension ray* for edge vw.
 – No three extension rays are coplanar.

For instance, if C is a chordless cycle of length four or greater in G, then by Lemma 4 there exists a vertically extensible partial drawing of C. More, the same lemma may be used to add another cycle to an existing vertically extensible partial drawing (Figure 6):

Lemma 5. *For any vertically extensible drawing of a set S of vertices in a graph G of maximum degree three, and any chordless cycle C of length four or more in $G \setminus S$, there exists a vertically extensible drawing of $S \cup C$.*

Proof. For each vertex v in C that has a neighbor w in G, replace w with a degree-one vertex that has label -1 if $w \in S$ and $+1$ if $w \notin S$. Apply Lemma 4 to find a drawing of C that can be connected in the negative z-direction to the neighbors of C in S, and in the positive z-direction for the remaining neighbors of C. Translate this drawing of C in the xy-plane so that, among the extension rays of S and the vertices of C, there are no three points and rays whose projections into the xy-plane are collinear and so that, when projected onto the xy-plane, the extension rays of S (points in the xy-plane) are disjoint from the projection of the drawing of C.

For each extension ray of S that connects a vertex v of S to a vertex w in C, draw a two-bend path with $120°$ bends in the plane containing the extension ray and w, such that the final segment of the path has the same x and y coordinates of w. By making the transverse section of this path be far enough away from S in the positive z direction, it will not intersect any other features of the existing drawing, and it cannot cross any of the other extension rays due to the requirement that no three of these rays be coplanar. If C is translated in the positive z direction farther than all of the bends in these paths, it can be connected to S to form a vertically extensible drawing of $S \cup C$, as required. □

Lemma 6. *For any vertically extensible drawing of a set S of vertices in a graph G of maximum degree three, and any vertex v in $G \setminus S$ with at most two neighbors in S and at most one neighbor in $G \setminus S$, there exists a vertically extensible drawing of $S \cup \{v\}$.*

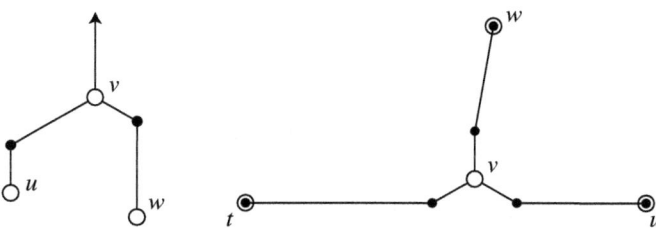

Fig. 7. Left: Adding a vertex v with two neighbors u and w in S and one neighbor in $G \setminus S$ to a vertically extensible drawing (shown in the plane of the extension rays of u and w). Right: Adding a vertex v with three neighbors t, u, and w in S (shown in the xy-plane). The three segments incident to v are parallel to the xy plane and the three remaining transverse segments form $120°$ angles to the extension rays of t, u, and w. The bends where these transverse segments meet their extension rays are shown on top of the three points t, u, and w.

Proof. If v has no neighbors in S, then v may be placed anywhere on any z-parallel line that does not pass through a feature of the existing drawing and is not coplanar with any two existing extension rays. If v has a single neighbor w in S, then v may be placed anywhere on the extension ray of wv.

In the remaining case, v connects to two extension rays of S. Within the plane of these two rays, we may connect v to these two rays by transverse segments at $120°$ angles to the rays. By placing v far enough in the positive z direction, these transverse segments can be made to avoid any existing features of the drawing. The extension ray from v can lie on any line parallel to and between the lines of the two incoming extension rays; only finitely many of these lines lead to coplanarities with other extension rays, so it is always possible to place v avoiding any such coplanarity. As shown in Figure 7(left), this construction produces one bend on each edge into v. □

Lemma 7. *If we are given a vertically extensible drawing of a set S of vertices in a graph G of maximum degree three, and a vertex v in $G \setminus S$ that has three neighbors t, u, and w in S, then there exists a vertically extensible drawing of $S \cup \{v\}$.*

Proof. Suppose that tu is the longest edge of the triangle formed by the projections of t, u, and w into the xy plane. Then, as a first approximation to the position of v in the xy-plane, let the (two-dimensional) point v' be placed on edge tu of this triangle, at the point where $v'w$ is perpendicular to tu. We adjust this position along edge tu, keeping the angle between $v'w$ and tu close to $90°$ in order to ensure that line segment $v'w$ does not pass through the two-dimensional projection of any extension ray. Then, we replace v' by three short line segments at $120°$ angles to each other meeting the three line segments $v't$, $v'u$, and $v'w$ at angles of $150°$, $150°$, and close to $180°$. Let v be the point where these three short line segments meet.

This configuration can be lifted into three-dimensional space by placing v and the three edges that attach to it in a plane perpendicular to the z axis, and by replacing the remaining portions of line segments $v't$, $v'u$, and $v'w$ by transverse segments that make $120°$ angles with the extension rays of t, u, and w. There are two bends per edge: one at the point where the extension ray of t, u, or w meets a transverse segment, and one where a transverse segment meets one of the horizontal segments incident to v.

The angles at the bends on the extension rays of t, u, and w are all exactly $120°$, and the angles at the other bends on the paths connecting t and u to w are $\arccos(3/4) \approx 138.6°$. As long as segment $v'w$ stays within $54°$ of perpendicular to tu in the xy-plane, the angle at the final remaining bend will be at least $120°$.   □

The construction of Lemma 7 is illustrated in Figure 7(right).

Theorem 2. *Any graph G of degree three has a drawing with $120°$ angular resolution and at most two bends per edge.*

Proof. While G contains a triangle, apply Lemma 2 to simplify it, resulting in either K_4 or a triangle-free graph G'. If this simplification process leads to K_4, draw it according to Lemma 3. Otherwise, starting from $S = \emptyset$, we repeatedly grow a vertically extensible drawing of a subset S of G' until all of G' has been drawn. If $G' \setminus S$ contains a vertex with at most one neighbor in $G' \setminus S$, then either Lemma 6 or Lemma 7 applies and we can add this vertex to the vertically extensible drawing. Otherwise, all vertices in $G' \setminus S$ have two or more neighbors in $G' \setminus S$, so $G' \setminus S$ contains a cycle. Let C be the shortest cycle in $G' \setminus S$; it has length at least four (because we eliminated all triangles) and no chords (because a chord would lead to a shorter cycle) so we may apply Lemma 5 to incorporate it into the vertically extensible drawing. Once we have included all vertices in the vertically extensible drawing, we have drawn all of G', and we may reverse the transformations performed according to Lemma 2 to produce a drawing of G.   □

In the full version (arXiv:1009.0045) we show that any graph of degree three has a drawing with $120°$ angular resolution, integer vertex coordinates, edges parallel to the face diagonals of the integer grid, at most three bends per edge, and polynomial volume.

4   Conclusions

We have shown how to draw degree-three graphs in three dimensions with optimal angular resolution and two bends per edge, and how to draw degree-four graphs in three dimensions with optimal angular resolution, three bends per edge, integer vertex coordinates, and cubic volume. Multiple questions remain open for investigation, however:

- It does not seem to be possible to draw K_4 or K_5 in three dimensions with optimal angular resolution and one bend per edge. Can this be proven rigorously?
- Does every degree-four graph have a drawing in three dimensions with optimal angular resolution and two bends per edge? In particular, is this possible for K_5?
- How many bends per edge are necessary to draw degree-three graphs with optimal angular resolution in an $O(n) \times O(n) \times O(n)$ grid, with all edges parallel to the face diagonals of the grid?
- It should be possible to draw degree-three and degree-four graphs with optimal angular resolution in an $O(\sqrt{n}) \times O(\sqrt{n}) \times O(\sqrt{n})$ grid. How many bends per edge are necessary for such a drawing?

Acknowledgments. This research was supported in part by the National Science Foundation under grant 0830403, by the Office of Naval Research under MURI grant N00014-08-1-1015, and by the German Research Foundation (DFG) under grant NO 899/1-1.

References

1. Angelini, P., Cittadini, L., Di Battista, G., Didimo, W., Frati, F., Kaufmann, M., Symvonis, A.: On the perspectives opened by right angle crossing drawings. In: Eppstein, D., Gansner, E.R. (eds.) GD 2009. LNCS, vol. 5849, pp. 21–32. Springer, Heidelberg (2010), doi:10.1007/978-3-642-11805-0_5
2. Biedl, T.C., Bose, P., Demaine, E.D., Lubiw, A.: Efficient algorithms for Petersen's matching theorem. Journal of Algorithms 38(1), 110–134 (2001), doi:10.1006/jagm.2000.1132
3. Biedl, T.C., Thiele, T., Wood, D.R.: Three-dimensional orthogonal graph drawing with optimal volume. Algorithmica 44(3), 233–255 (2006), doi:10.1007/s00453-005-1148-z
4. Carlson, J., Eppstein, D.: Trees with convex faces and optimal angles. In: Kaufmann, M., Wagner, D. (eds.) GD 2006. LNCS, vol. 4372, pp. 77–88. Springer, Heidelberg (2007)
5. Clare, B.W., Kepert, D.L.: The closest packing of equal circles on a sphere. Proc. Roy. Soc. London A 405(1829), 329–344 (1986), doi:10.1098/rspa.1986.0056
6. Demaine, E.D.: Problem 46: 3D minimum-bend orthogonal graph drawings. The Open Problems Project. Posed by David R. Wood at the CCCG, open-problem session (2002), http://maven.smith.edu/~orourke/TOPP/P46.html
7. Didimo, W., Eades, P., Liotta, G.: Drawing graphs with right angle crossings. In: Dehne, F., Gavrilova, M., Sack, J.-R., Tóth, C.D. (eds.) WADS 2009. LNCS, vol. 5664, pp. 206–217. Springer, Heidelberg (2009), doi:10.1007/978-3-642-03367-4_19
8. Dujmović, V., Gudmundsson, J., Morin, P., Wolle, T.: Notes on large angle crossing graphs. arXiv:0908.3545 (2009)
9. Duncan, C.A., Eppstein, D., Kobourov, S.G.: The geometric thickness of low degree graphs. In: Proc. 20th ACM Symp. Computational Geometry (SoCG 2004), pp. 340–346 (2004), doi:10.1145/997817.997868, arXiv:cs.CG/0312056
10. Eades, P., Symvonis, A., Whitesides, S.: Three-dimensional orthogonal graph drawing algorithms. Discrete Applied Mathematics 103(1-3), 55–87 (2000), doi:10.1016/S0166-218X(00)00172-4
11. Eiglsperger, M., Fekete, S.P., Klau, G.W.: Orthogonal graph drawing. In: Kaufmann, M., Wagner, D. (eds.) Drawing Graphs. LNCS, vol. 2025, pp. 121–171. Springer, Heidelberg (2001), doi:10.1007/3-540-44969-8_6
12. Eppstein, D.: Isometric diamond subgraphs. In: Tollis, I.G., Patrignani, M. (eds.) GD 2008. LNCS, vol. 5417, pp. 384–389. Springer, Heidelberg (2009), doi:10.1007/978-3-642-00219-9_37
13. Garg, A., Tamassia, R.: Planar drawings and angular resolution: algorithms and bounds. In: van Leeuwen, J. (ed.) ESA 1994. LNCS, vol. 855, pp. 12–23. Springer, Heidelberg (1994), doi:10.1007/BFb0049393
14. Gutwenger, C., Mutzel, P.: Planar polyline drawings with good angular resolution. In: Whitesides, S.H. (ed.) GD 1998. LNCS, vol. 1547, pp. 167–182. Springer, Heidelberg (1999), doi:10.1007/3-540-37623-2_13
15. Huang, W., Hong, S.-H., Eades, P.: Effects of crossing angles. In: Proc. IEEE Pacific Visualization Symp., pp. 41–46 (2008), doi:10.1109/PACIFICVIS.2008.4475457
16. Malitz, S.: On the angular resolution of planar graphs. In: Proc. 24th ACM Symp. Theory of Computing (STOC 1992), pp. 527–538 (1992), doi:10.1145/129712.129764
17. Petersen, J.: Die Theorie der regulären Graphs. Acta Math. 15(1), 193–220 (1891), doi:10.1007/BF02392606
18. Tammes, P.M.L.: On the origin of the number and arrangement of the places of exit on the surface of pollen grains. Ree. Trav. Bot. Néerl. 27, 1–82 (1930)
19. Wood, D.R.: Optimal three-dimensional orthogonal graph drawing in the general position model. Theor. Comput. Sci. 299(1-3), 151–178 (2003), doi:10.1016/S0304-3975(02)00044-0

Improved Lower Bounds on the Area Requirements of Series-Parallel Graphs*

Fabrizio Frati

Dipartimento di Informatica e Automazione, Università Roma Tre
frati@dia.uniroma3.it

Abstract. We show that there exist series-parallel graphs requiring $\Omega(n2^{\sqrt{\log n}})$ area in any straight-line or poly-line grid drawing, improving the previously best known $\Omega(n \log n)$ lower bound.

1 Introduction

Determining asymptotic bounds for the area requirements of straight-line and poly-line drawings of planar graphs is a classical topic in the Graph Drawing literature. Groundbreaking works of the beginning of the nineties [5,10] have shown that every n-vertex planar graph admits a planar straight-line drawing in an $O(n^2)$ grid. Such a bound is worst-case optimal, even for poly-line drawings [7,5]. Hence, it is natural to search for interesting sub-classes of planar graphs admitting sub-quadratic area drawings.

In this paper we deal with series-parallel graphs, a class of planar graphs that has been widely investigated in Graph Theory and Graph Drawing (see, e.g., [11,8,1,6]). Series-parallel graphs can be equivalently defined as the graphs excluding K_4 as a minor or, inductively, by series and parallel compositions of smaller series-parallel graphs.

Biedl [2,3] proved that a series-parallel graph with n vertices admits a poly-line grid drawing in $O(n^{3/2})$ area. She achieved such a bound by first constructing *visibility representations* of series-parallel graphs in $O(n^{3/2})$ area and by then turning such representations into poly-line drawings with asymptotically the same area. No sub-quadratic area upper bound is known for straight-line grid drawings of series-parallel graphs.

The author proved [9] that there exist series-parallel graphs requiring $\Omega(n \log n)$ area in any straight-line or poly-line grid drawing. To achieve such a bound, the following theorem[1] was proved in [9], improving upon previous results of Biedl et al. [4].

Theorem 1. *Every planar straight-line or poly-line grid drawing of $K_{2,n}$ in a $W \times H$ grid satisfies $\max\{W, H\} \geq c \cdot n$, for some constant $c \leq 1/2$.*

In this paper, we prove the following.

Theorem 2. *There exist series-parallel graphs with n vertices requiring $\Omega(2^{\sqrt{\log n}})$ width and $\Omega(2^{\sqrt{\log n}})$ height in any straight-line or poly-line grid drawing.*

---

* This work is partially supported by the Italian Ministry of Research, Grant number RBIP06BZW8, FIRB project "Advanced tracking system in intermodal freight transportation".

[1] Theorem 1 is stated in [9] in an equivalent form using the $\Omega(n)$ notation.

U. Brandes and S. Cornelsen (Eds.): GD 2010, LNCS 6502, pp. 220–225, 2011.
© Springer-Verlag Berlin Heidelberg 2011

Such a result is achieved by carefully constructing a graph out of several copies of $K_{2,n}$ and by then exploiting Theorem 1 and some further geometric considerations. Theorem 1, together with Theorem 2, directly implies the following.

Theorem 3. *There exist series-parallel graphs with n vertices requiring $\Omega(n2^{\sqrt{\log n}})$ area in any straight-line or poly-line grid drawing.*

We remark that the function $2^{\sqrt{\log n}}$ is asymptotically greater than any polylogarithmic function of n and smaller than any polynomial function of n.

2 Preliminaries

A *planar grid drawing* of a graph is a mapping of each vertex to a distinct point of the plane with integer coordinates and of each edge to a Jordan curve between the endpoints of the edge, so that no two edges intersect except, possibly, at common endpoints. In the following we always refer to planar grid drawings. A *straight-line* drawing is such that all edges are rectilinear segments. A *poly-line* drawing is such that the edges are sequences of rectilinear segments. In a poly-line drawing a *bend* is a point in which an edge changes its slope, i.e., a point common to two consecutive segments in the sequence of segments representing the edge. In a grid drawing bends have integer coordinates. A *polygonal path* is a poly-line grid drawing of a path. The *bounding box* of a drawing Γ is the smallest rectangle with sides parallel to the axes that covers Γ completely. The *height (width)* of Γ is the height (resp. width) of its bounding box. The *area* of Γ is the height of Γ times its width.

A drawing of the complete bipartite graph $K_{2,n}$ can be thought as a drawing of n paths that start and end at the same two vertices and that do not share any other vertex. In the following we will refer to such paths as to the *paths of $K_{2,n}$*.

In the next section, we will use the following lemmata [9].

Lemma 1. *Consider any poly-line grid drawing of $K_{2,n}$, any path π of $K_{2,n}$, and any vector $\boldsymbol{v}$. There exists a grid point $p \in \pi$ such that $\boldsymbol{v} \cdot p \geq \boldsymbol{v} \cdot p'$, for any point $p' \in \pi$.*

Lemma 2. *Let a and b be the endvertices of the paths of $K_{2,n}$. Consider any planar drawing of $K_{2,n}$. Let l be any line that does not intersect nor contain the open segment $\overline{ab}$. No three paths $\pi_1, \pi_2,$ and π_3 of $K_{2,n}$ exist such that: (i) $\pi_1, \pi_2,$ and π_3 do not intersect each other; (ii) $\pi_1, \pi_2,$ and π_3 are contained in the closed half-plane delimited by l and containing a and b; (iii) each of $\pi_1, \pi_2,$ and π_3 touches l at least once.*

3 Proof of Theorem 2

As straight-line drawings are also poly-line drawings, it suffices to prove Theorem 2 for poly-line drawings. Let $f(n)$ be a function to be computed later and let $d = c/4$, where c is the constant of Theorem 1. Observe that $d \leq 1/8$.

Graph G_1 is $K_{2,f(n)-2}$. Graph G_{i+1} is defined as follows. Consider $f(n)$ copies $G_{i,1,1}, G_{i,1,2}, G_{i,2,1}, G_{i,2,2}, \ldots, G_{i,j,1}, G_{i,j,2}, \ldots, G_{i,f(n)/2,1}, G_{i,f(n)/2,2}$ of G_i; construct $f(n)/2$ series-parallel graphs $G_{i,1}, G_{i,2}, \ldots, G_{i,j}, \ldots, G_{i,f(n)/2}$, where $G_{i,j}$ is

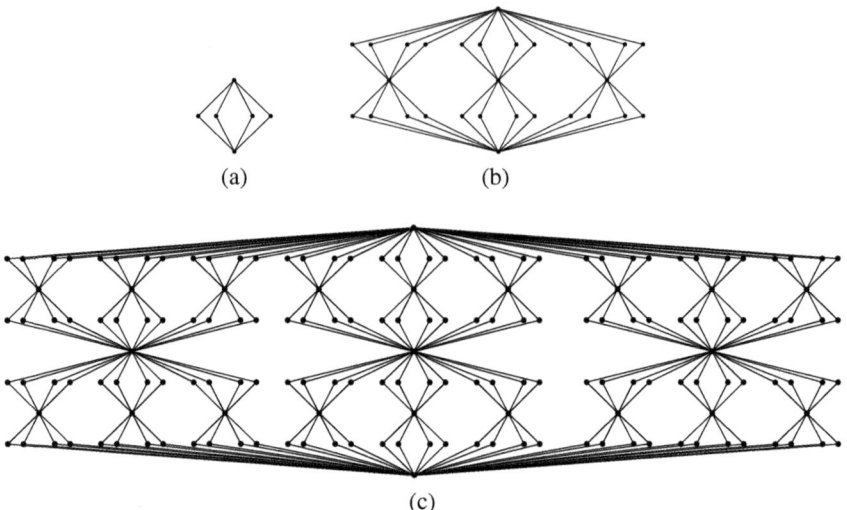

Fig. 1. Graphs G_i, with $f(n) = 6$. (a) G_1. (b) G_2. (c) G_3

the series composition of $G_{i,j,1}$ and $G_{i,j,2}$; then, G_{i+1} is the parallel composition of graphs $G_{i,1}, G_{i,2}, \ldots, G_{i,f(n)/2}$. See Fig. 1.

First, we prove Theorem 2 for sufficiently large graphs, that is, for graphs having a number of vertices that is at least some constant n_0 to be determined later. From now till it is otherwise specified, assume that $n \geq n_0$.

Suppose that $f(n) \geq 8$, $\forall n \geq n_0$. Let n be the number of vertices of graph G_k. We have the following main lemma.

Lemma 3. *Let Γ_i be any poly-line grid drawing of G_i and let a_i and b_i be the poles of G_i, for each $1 \leq i \leq k$. Then, one of the following holds:*

- *Condition 1: The height and the width of Γ_i are both greater than or equal to $d \cdot f(n)$.*
- *Condition 2: The width of Γ_i is greater than or equal to $d \cdot f(n)$ and Γ_i contains a polygonal path l_i connecting a_i to b_i that has height greater than or equal to 2^i and such that, for every point $p \in l_i$, $\min\{y(a_i), y(b_i)\} \leq y(p) \leq \max\{y(a_i), y(b_i)\}$; or the height of Γ_i is greater than or equal to $d \cdot f(n)$ and Γ_i contains a polygonal path l_i connecting a_i to b_i that has width greater than or equal to 2^i and such that, for every point $p \in l_i$, $\min\{x(a_i), x(b_i)\} \leq x(p) \leq \max\{x(a_i), x(b_i)\}$.*

Proof: We prove the statement by induction on i. In the base case, consider any poly-line grid drawing Γ_1 of G_1. By Theorem 1, one of the height and the width of Γ_1, say the width of Γ_1, is at least $c \cdot f(n)$, hence it is at least $d \cdot f(n)$.

Assume, without loss of generality, that $y(a_1) \leq y(b_1)$. Suppose that at least $2d \cdot f(n)$ paths of $G_1 = K_{2,f(n)-2}$ intersect the open half-plane $H^-(a_1)$ defined as $y < y(a_1)$ or the open half-plane $H^+(b_1)$ defined as $y > y(b_1)$. By Lemma 1 with $v = (0, -1)$, for each path π of G_1 intersecting $H^-(a_1)$, a grid point $p \in \pi$ exists whose y-coordinate is minimum among the points of π. Clearly, p belongs to $H^-(a_1)$. Hence, p belongs to a horizontal grid line h not intersecting nor containing segment $\overline{a_1 b_1}$. By Lemma 2, at most two paths of G_1 have their points with smallest y-coordinate in h. Analogously, by Lemma 1 with $v = (0, 1)$, for each path π of G_1 that intersects $H^+(b_1)$, a grid point $p \in \pi$ exists whose y-coordinate is maximum among the points of π. Clearly, p belongs to $H^+(b_1)$. Hence, p belongs to a horizontal grid line h not intersecting nor containing segment $\overline{a_1 b_1}$. By Lemma 2, at most two paths of G_1 have their points with greatest y-coordinate in h. Hence, as $2d \cdot f(n)$ paths of G_1 intersect $H^-(a_1)$ or $H^+(b_1)$, it follows that Γ_1 has height at least $d \cdot f(n)$.

Now suppose that no $2d \cdot f(n)$ paths of G_1 intersect $H^-(a_1)$ or $H^+(b_1)$. Then, since $d \leq 1/8$, at least $f(n) - 2 - 2d \cdot f(n) + 1 \geq 3f(n)/4 - 1$ paths of G_1 are such that, for every point p of any such a path, $y(a_1) \leq y(p) \leq y(b_1)$. By planarity of Γ_1 at most one path of G_1 contains a point p such that $y(p) = y(a_1)$ and $x(p) < x(a_1)$. Analogously, at most one path of G_1 contains a point p such that $y(p) = y(a_1)$ and $x(p) > x(a_1)$, at most one path of G_1 contains a point p such that $y(p) = y(b_1)$ and $x(p) < x(b_1)$, and at most one path of G_1 contains a point p such that $y(p) = y(b_1)$ and $x(p) > x(b_1)$. Since $f(n) \geq 8$, it follows that $3f(n)/4 - 1 \geq 5$, hence there is at least one path of G_1 whose only vertex $v \neq a_1, b_1$ is such that $y(v) > y(a_1)$ and $y(v) < y(b_1)$. Then, the polygonal path (a_1, v, b_1) has height at least two and is such that, for every point $p \in (a_1, v, b_1)$, $y(a_1) \leq y(p) \leq y(b_1)$, thus proving the base case.

In the inductive case, consider any poly-line grid drawing Γ_{i+1} of G_{i+1}, containing drawings $\Gamma_{i,1,1}, \Gamma_{i,1,2}, \Gamma_{i,2,1}, \Gamma_{i,2,2}, \dots, \Gamma_{i,j,1}, \Gamma_{i,j,2}, \dots, \Gamma_{i,f(n)/2,1}, \Gamma_{i,f(n)/2,2}$ of $G_{i,1,1}, G_{i,1,2}, G_{i,2,1}, G_{i,2,2}, \dots, G_{i,j,1}, G_{i,j,2}, \dots, G_{i,f(n)/2,1}, G_{i,f(n)/2,2}$, respectively. By induction, for $1 \leq j \leq f(n)/2$ and $1 \leq k \leq 2$, $\Gamma_{i,j,k}$ satisfies Condition 1 or 2.

If two indices $1 \leq j \leq f(n)/2$ and $1 \leq k \leq 2$ exist such that $\Gamma_{i,j,k}$ satisfies Condition 1, then the width and the height of $\Gamma_{i,j,k}$ are both greater than or equal to $d \cdot f(n)$, hence so are the width and the height of Γ_{i+1}.

Hence, we can assume that, for every $1 \leq j \leq f(n)/2$ and $1 \leq k \leq 2$, $\Gamma_{i,j,k}$ satisfies Condition 2. If indices $1 \leq j', j'' \leq f(n)/2$ and $1 \leq k', k'' \leq 2$ exist, where $j' = j''$ and $k' = k''$ do not hold simultaneously, such that the width of $\Gamma_{i,j',k'}$ is greater than or equal to $d \cdot f(n)$ and the height of $\Gamma_{i,j'',k''}$ is greater than or equal to $d \cdot f(n)$, then the width and the height of Γ_{i+1} are both greater than or equal to $d \cdot f(n)$.

Hence, we can assume that, for every $1 \leq j \leq f(n)/2$ and $1 \leq k \leq 2$, the width of $\Gamma_{i,j,k}$ is greater than or equal to $d \cdot f(n)$ and $\Gamma_{i,j,k}$ contains a polygonal path $l_{i,j,k}$ connecting a_i to b_i that has height greater than or equal to 2^i and such that, for every point $p \in l_{i,j,k}$, $\min\{y(a_i), y(b_i)\} \leq y(p) \leq \max\{y(a_i), y(b_i)\}$; the case in which, for every $1 \leq j \leq f(n)/2$ and $1 \leq k \leq 2$, the height of $\Gamma_{i,j,k}$ is greater than or equal to $d \cdot f(n)$ and $\Gamma_{i,j,k}$ contains a polygonal path $l_{i,j,k}$ connecting a_i to b_i that has width greater than or equal to 2^i and such that, for every point $p \in l_{i,j,k}$, $\min\{x(a_i), x(b_i)\} \leq x(p) \leq \max\{x(a_i), x(b_i)\}$ can be treated analogously.

Denote by $l_{i,j}$ the path connecting a_{i+1} and b_{i+1} composed of $l_{i,j,1}$ and $l_{i,j,2}$. Assume, without loss of generality, that $y(a_{i+1}) \leq y(b_{i+1})$. Suppose that at least $2d \cdot f(n)$ paths $l_{i,j}$ intersect the open half-plane $H^-(a_{i+1})$ defined as $y < y(a_{i+1})$ or the open half-plane $H^+(b_{i+1})$ defined as $y > y(b_{i+1})$. By Lemma 1 with $v = (0, -1)$, for each path $l_{i,j}$ that intersects $H^-(a_{i+1})$, a grid point $p \in l_{i,j}$ exists whose y-coordinate is minimum among the points of $l_{i,j}$. Clearly, p belongs to $H^-(a_{i+1})$. Hence, p belongs to a horizontal grid line h not intersecting nor containing segment $\overline{a_{i+1}b_{i+1}}$. By Lemma 2, at most two paths $l_{i,j}$ have their points with smallest y-coordinate in h. Analogously, by Lemma 1 with $\boldsymbol{v} = (0, 1)$, for each path $l_{i,j}$ that intersects $H^+(b_{i+1})$, a grid point $p \in l_{i,j}$ exists whose y-coordinate is maximum among the points of $l_{i,j}$. Clearly, p belongs to $H^+(b_{i+1})$. Hence, p belongs to a horizontal grid line h not intersecting nor contain segment $\overline{a_{i+1}b_{i+1}}$. By Lemma 2, at most two paths $l_{i,j}$ have their points with greatest y-coordinate in h. Hence, as $2d \cdot f(n)$ paths $l_{i,j}$ intersect $H^-(a_{i+1})$ or $H^+(b_{i+1})$, it follows that Γ_{i+1} has height at least $d \cdot f(n)$.

Now suppose that no $2d \cdot f(n)$ paths $l_{i,j}$ intersect $H^-(a_{i+1})$ or $H^+(b_{i+1})$. Then, since $d \leq 1/8$, at least $f(n) - 2d \cdot f(n) + 1 \geq 3f(n)/4 + 1$ paths $l_{i,j}$ are such that, for every point p of any such a path, $y(a_{i+1}) \leq y(p) \leq y(b_{i+1})$. By planarity of Γ_{i+1} at most one path $l_{i,j}$ contains a point p such that $y(p) = y(a_{i+1})$ and $x(p) < x(a_{i+1})$. Analogously, at most one path $l_{i,j}$ contains a point p such that $y(p) = y(a_{i+1})$ and $x(p) > x(a_{i+1})$, at most one path $l_{i,j}$ contains a point p such that $y(p) = y(b_{i+1})$ and $x(p) < x(b_{i+1})$, and at most one path $l_{i,j}$ contains a point p such that $y(p) = y(b_{i+1})$ and $x(p) > x(b_{i+1})$. Since $f(n) \geq 8$, it follows that $3f(n)/4 + 1 \geq 5$, hence there is at least one path $l_{i,j}$ composed of path $l_{i,j,1}$, that connects the poles a_{i+1} and v of $G_{i,j,1}$, and of path $l_{i,j,2}$, that connects the poles b_{i+1} and v of $G_{i,j,2}$, such that $y(v) > y(a_{i+1})$ and $y(v) < y(b_{i+1})$. By inductive hypothesis, $l_{i,j,1}$ has height greater than or equal to 2^i and, for every point $p \in l_{i,j,1}$, $y(a_{i+1}) \leq y(p) \leq y(v)$; further, $l_{i,j,2}$ has height greater than or equal to 2^i and, for every point $p \in l_{i,j,2}$, $y(v) \leq y(p) \leq y(b_{i+1})$; hence, $l_{i,j}$ has height greater than or equal to 2^{i+1} and, for every point $p \in l_{i,j}$, $y(a_{i+1}) \leq y(p) \leq y(b_{i+1})$, thus completing the induction. □

Corollary 1. *Any poly-line grid drawing of G_k has height and width that are both greater than or equal to $\min\{d \cdot f(n), 2^k\}$.*

Let $f(n) = n^{x(k)}$. By construction $|G_1| = n^{x(k)}$; since G_i is composed of $f(n) = n^{x(k)}$ copies of G_{i-1}, $|G_i| \leq n^{x(k)} \cdot |G_{i-1}|$; inductively, we obtain $|G_k| \leq n^{k \cdot x(k)}$. Assuming $|G_k| = n$, we get $x(k) \geq 1/k$ and $f(n) \geq n^{1/k}$.

We now choose k in such a way that $n^{1/k}$ and 2^k are equal. This is done as follows. $2^k = n^{1/k} \Rightarrow \log_2(2^k) = \log_2(n^{1/k}) \Rightarrow k \log_2(2) = 1/k \log_2(n) \Rightarrow k^2 = \log_2(n) \Rightarrow k = \sqrt{\log_2(n)}$. By Corollary 1, both the height and the width of Γ_k, with $k = \sqrt{\log_2(n)}$, are greater than or equal to $\min\{d \cdot n^{1/\sqrt{\log_2(n)}}, 2^{\sqrt{\log_2(n)}}\} = d \cdot 2^{\sqrt{\log_2(n)}} = \Omega(2^{\sqrt{\log_2(n)}})$, and Theorem 2 follows if $n \geq n_0$.

As we need $f(n) = 2^{\sqrt{\log_2(n)}} \geq 8, \forall n \geq n_0$, then $n_0 = 512$. However, $d \cdot 2^{\sqrt{\log_2(n)}} < 1$, for all $n < 512$, as $d \leq 1/8$. Since every drawing of a graph that is not a collection of paths has height and width at least one, the $d \cdot 2^{\sqrt{\log_2(n)}}$ lower bound holds also for graphs with less than 512 nodes, thus completing the proof of Theorem 2.

4   Conclusions and Open Problems

In this paper we have shown that there exist series-parallel graphs with n vertices requiring $\Omega(n2^{\sqrt{\log n}})$ area in any straight-line or poly-line grid drawing.

The best known area upper bound for poly-line grid drawings of series-parallel graphs is $O(n^{3/2})$ [2,3], while no sub-quadratic area upper bound is known in the case of straight-line grid drawings. Hence, in both cases, closing the gap between upper and lower bound is an intriguing challenge.

Concerning straight-line drawings, David Wood [12] conjectures the following: Let $p_1, \ldots, p_k$ be positive integers. Let $G(p_1)$ be the graph obtained from K_3 by adding p_1 new vertices adjacent to v and w for each edge (v, w) of K_3. For $k \geq 2$, let $G(p_1, p_2, \ldots, p_k)$ be the graph obtained from $G(p_1, p_2, \ldots, p_{k-1})$ by adding p_k new vertices adjacent to v and w for each edge (v, w) of $G(p_1, p_2, \ldots, p_{k-1})$. Observe that $G(p_1, p_2, \ldots, p_k)$ is a series-parallel graph.

Conjecture 1. (D. R. Wood) Every straight-line grid drawing of $G(p_1, p_2, \ldots, p_k)$ requires $\Omega(n^2)$ area for some choice of k and $p_1, p_2, \ldots, p_k$.

Acknowledgments

Thanks to Patrizio Angelini and Giuseppe Di Battista for very useful discussions. Also thanks to David Wood for sharing his strong conjecture.

References

1. Bertolazzi, P., Cohen, R.F., di Battista, G., Tamassia, R., Tollis, I.G.: How to draw a series-parallel digraph. International Journal of Computational Geometry & Applications 4(4), 385–402 (1994)
2. Biedl, T.C.: Small poly-line drawings of series-parallel graphs. Tech. Report CS-2007-23, School of Computer Science, University of Waterloo, Canada (2005)
3. Biedl, T.C.: On small drawings of series-parallel graphs and other subclasses of planar graphs. In: Eppstein, D., Gansner, E.R. (eds.) GD 2009. LNCS, vol. 5849, pp. 280–291. Springer, Heidelberg (2010)
4. Biedl, T.C., Chan, T.M., López-Ortiz, A.: Drawing $K_{2,n}$: A lower bound. Information Processing Letters 85(6), 303–305 (2003)
5. de Fraysseix, H., Pach, J., Pollack, R.: How to draw a planar graph on a grid. Combinatorica 10(1), 41–51 (1990)
6. Di Giacomo, E., Didimo, W., Liotta, G., Wismath, S.K.: Book embeddability of series-parallel digraphs. Algorithmica 45(4), 531–547 (2006)
7. Dolev, D., Leighton, T., Trickey, H.: Planar embeddings of planar graphs. Advances in Computing Research 2, 147–161 (1984)
8. Eppstein, D.: Parallel recognition of series-parallel graphs. Information and Computation 98(1), 41–55 (1992)
9. Frati, F.: A lower bound on the area requirements of series-parallel graphs. In: Broersma, H., Erlebach, T., Friedetzky, T., Paulusma, D. (eds.) WG 2008. LNCS, vol. 5344, pp. 159–170. Springer, Heidelberg (2008)
10. Schnyder, W.: Embedding planar graphs on the grid. In: ACM-SIAM Symposium on Discrete Algorithms (SODA 1990), pp. 138–148 (1990)
11. Valdes, J., Tarjan, R.E., Lawler, E.L.: The recognition of series parallel digraphs. SIAM Journal on Computing 11(2), 298–313 (1982)
12. Wood, D.R.: Private Communication (2008)

A Computational Approach to Conway's Thrackle Conjecture*

Radoslav Fulek[1] and János Pach[2]

[1] Ecole Polytechnique Fédérale de Lausanne
`radoslav.fulek@epfl.ch`
[2] Ecole Polytechnique Fédérale de Lausanne and City College, New York
`pach@cims.nyu.edu`

Abstract. A drawing of a graph in the plane is called a *thrackle* if every pair of edges meets precisely once, either at a common vertex or at a proper crossing. Let $t(n)$ denote the maximum number of edges that a thrackle of n vertices can have. According to a 40 years old conjecture of Conway, $t(n) = n$ for every $n \geq 3$. For any $\varepsilon > 0$, we give an algorithm terminating in $e^{O((1/\varepsilon^2)\ln(1/\varepsilon))}$ steps to decide whether $t(n) \leq (1+\varepsilon)n$ for all $n \geq 3$. Using this approach, we improve the best known upper bound, $t(n) \leq \frac{3}{2}(n-1)$, due to Cairns and Nikolayevsky, to $\frac{167}{117}n < 1.428n$.

1 Introduction

A *drawing* of a graph (or a *topological graph*) is a representation of the graph in the plane such that the vertices are represented by distinct points and the edges by (possibly crossing) simple continuous curves connecting the corresponding point pairs and not passing through any other point representing a vertex. If it leads to no confusion, we make no notational distinction between a drawing and the underlying abstract graph (i.e. its set representation) G. In the same vein, $V(G)$ and $E(G)$ will stand for the vertex set and edge set of G as well as for the sets of points and curves representing them.

A drawing of G is called a *thrackle* if every pair of edges meets precisely once, either at a common vertex or at a proper crossing. (A crossing p of two curves is *proper* if at p one curve passes from one side of the other curve to its other side.) More than *forty* years ago Conway [19,2,15] conjectured that every thrackle has at most as many edges as vertices, and offered a bottle of beer for a solution. Since then the prize went up to one thousand dollars. In spite of considerable efforts, Conway's thrackle conjecture is still open. It is believed to represent the tip of an "iceberg", obstructing our understanding of crossing patterns of edges in topological graphs. If true, the inequality of Conway's conjecture would be tight as any cycle of length at least *five* can be drawn as a thrackle, see [18]. Two thrackle drawings of C_5 and C_6 are shown in Figure 1.

---

* Research partially supported by NSF grant CCF-08-30272, grants from OTKA, SNF, and PSC-CUNY.

U. Brandes and S. Cornelsen (Eds.): GD 2010, LNCS 6502, pp. 226–237, 2011.

Obviously, the property that G can be drawn as a thrackle is *hereditary*: if G has this property, then any subgraph of G does. It is very easy to verify (cf. [18]) that C_4, a cycle of length *four*, cannot be drawn as a thrackle. Therefore, every "thrackleable" graph is C_4-free, and it follows from extremal graph theory that every thrackle of n vertices has at most $O(n^{3/2})$ edges [6]. The

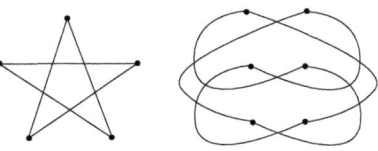

Fig. 1. C_5 and C_6 drawn as thrackles

first linear upper bound of $2n-3$ on the maximum number of edges of a thrackle of n vertices was given by Lovász et al. [12]. This was improved to $\frac{3}{2}(n-1)$ by Cairns and Nikolayevsky [3].

The aim of this note is to provide a finite approximation scheme for estimating the maximum number of edges that a thrackle of n vertices can have. We apply our technique to improve the best known upper bound for this maximum.

To state our results, we need a definition. Given three integers $c', c'' > 2$, $l \geq 0$, the *dumbbell* $\mathrm{DB}(c', c'', l)$ is a simple graph consisting of two edge disjoint cycles of length c' and c'', connected by a path of length l. For $l = 0$, the two cycles share a vertex. It is natural to extend this definition to negative values of l, as follows. For

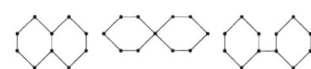

Fig. 2. Dumbbells $\mathrm{DB}(6,6,-1)$, $\mathrm{DB}(6,6,0)$, and $\mathrm{DB}(6,6,1)$

any $l > -\min(c', c'')$, let $\mathrm{DB}(c', c'', l)$ denote the graph consisting of two cycles of lengths c' and c'' that share a path of length $-l$. That is, for any $l > -\min(c', c'')$, we have

$$|V(\mathrm{DB}(c', c'', l))| = c' + c'' + l - 1.$$

The three types of dumbbells (for $l < 0$, $l = 0$, and $l > 0$) are illustrated in Figure 2.

Our first theorem shows that for any $\varepsilon > 0$, it is possible to prove Conway's conjecture up to a multiplicative factor of $1 + \varepsilon$, by verifying that no dumbbell smaller than a certain size depending on ε is thrackleable.

Theorem 1. *Let $c \geq 6$ and $l \geq -1$ be two integers, such that c is even, with the property that no dumbbell $\mathrm{DB}(c', c'', l')$ with $-c'/2 \leq l' \leq l$ and with even c', c'', for which $6 \leq c', c'' \leq c$, can be drawn in the plane as a thrackle. Let $r = \lfloor l/2 \rfloor$. Then the maximum number of edges $t(n)$ that a thrackle on n vertices can have satisfies $t(n) \leq \tau(c, l)n$, where*

$$\tau(c, l) = \begin{cases} \dfrac{47c^2 + 116c + 80}{35c^2 + 68c + 32} & \text{if } l = -1, \\[2ex] 1 + \dfrac{2c^2r + 4cr^2 + 22cr + 7c^2 + 22c + 8r^2 + 24r + 16}{2c^2r^2 + 14c^2r + 4cr^2 + 16cr + 24c^2 + 12c} & \text{if } l \geq 0, \end{cases}$$

as n tends to infinity.

As both c and l get larger, the constant $\tau(c, l)$ given by the second part of Theorem 1 approaches 1. On the other hand, assuming that Conway's conjecture is true for all bipartite graphs with up to 10 vertices, which will be verified in Section 4, the first part of the theorem applied with $c = 6, l = -1$ yields that $t(n) \leq \frac{617}{425}n < 1.452n$. This bound is already better than the bound $\frac{3}{2}n$ established in [3].

By a more careful application of Theorem 1, i.e. taking $c = 6$ and $l = 0$, we obtain an even stronger result.

Theorem 2. *The maximum number of edges $t(n)$ that a thrackle on n vertices can have satisfies the inequality $t(n) \leq \frac{167}{117}n < 1.428n$.*

Our method is algorithmic. We design an $e^{O((1/\varepsilon^2)\ln(1/\varepsilon))}$ time algorithm by means of which we can prove, for any $\varepsilon > 0$, that $t(n) \leq (1+\varepsilon)n$ for all n, or which provides a counterexample to Conway's conjecture. The proof of Theorem 2 is computer assisted: it requires testing the planarity of certain relatively small graphs.

For thrackles drawn by straight-line edges, Conway's conjecture has been settled in a slightly different form by Hopf and Pannwitz [10] and by Sutherland [17] before Conway was even born, and later, in the above form, by Erdős and Perles. Assuming that Conway's conjecture is true, Woodall [18] gave a complete characterization of all graphs that can be drawn as a thrackle. He also observed that it is sufficient to verify the conjecture for dumbbells. This observation is one of the basic ideas behind our arguments.

Several interesting special cases and variants of the conjecture are discussed in [3,4,5,9,12,13,14].

In Section 2, we describe a crucial construction by Conway and summarize some earlier results needed for our arguments. The proofs of Theorems 1 and 2 are given in Sections 3 and 4. The analysis of the algorithm for establishing the $(1 + \varepsilon)n$ upper bound for the maximum number of edges that a thrackle of n vertices can have is also given in Section 4 (Theorem 3). In the last section, we discuss some related Turán-type extremal problems for planar graphs.

2   Conway's Doubling and Preliminaries

In this section, we review some earlier results that play a key role in our arguments.

A *generalized thrackle* is a drawing of a graph in the plane with the property that any pair of edges share an odd number of points at which they properly cross or which are their common endpoints. Obviously, every thrackle is a generalized thrackle but not vice versa: although C_4 is not thrackleable, it can be drawn as a generalized thrackle, which is not so hard to see (Figure 3(a)).

We need the following simple observation based on the Jordan curve theorem.

Lemma 1. [12] *A (generalized) thrackle cannot contain two vertex disjoint odd cycles.*

Lovász, Pach, and Szegedy [12] gave a somewhat counterintuitive characterization of generalized thrackles containing no odd cycle: a bipartite graph is a

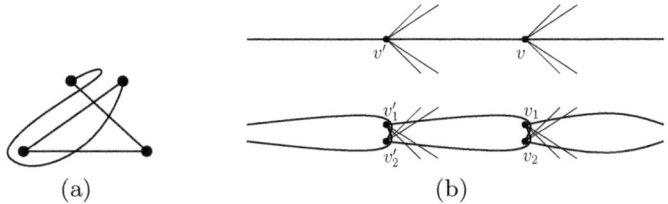

Fig. 3. (a) C_4 drawn as generalized thrackle, (b) Conway's doubling of a cycle

generalized thrackle if and only if it is *planar*. Moreover, it follows immediately from Lemma 3 and the proof of Theorem 3 in Cairns and Nikolayevsky [3] that this statement can be strengthened as follows.

Lemma 2. [3] *Let G be a bipartite graph with vertex set $V(G) = A \cup B$ and edge set $E(G) \subseteq A \times B$. If G is a generalized thrackle then it can be redrawn in the plane without crossing so that the cyclic order of the edges around any vertex $v \in V(G)$ is preserved if $v \in A$ and reversed if $v \in B$.*

We recall a construction by Conway for transforming a thrackle into another one. It can be used to eliminate odd cycles.

Let G be a thrackle or a generalized thrackle that contains an *odd* cycle C. In the literature, the following procedure is referred to as *Conway's doubling*: First, delete from G all edges incident to a vertex belonging to C, including all edges of C. Replace every vertex v of C by two vertices v_1 and v_2 which lie near to each other and to the former position of v. For any edge vv' of C, connect v_1 to v'_2 and v_2 to v'_1 by two edges running very close to the original edge vv', as depicted in Figure 3(b). For any vertex v belonging to C, the set of edges incident to v but not belonging to C can be divided into two classes, $E_1(v)$ and $E_2(v)$: the set of all edges whose initial arcs around v lie on one side and the other side of C, respectively. In the resulting topological graph G', connect all edges in $E_1(v)$ to v_1 and all edges in $E_2(v)$ to v_2 so that every edge connected to v_1 crosses all edges connected to v_2 exactly once in their small neighborhood. See Figure 3(b). All other edges of G remain unchanged. Denote the vertices of the original odd cycle C by $v^1, v^2, \ldots, v^k$, in this order. In the resulting drawing G', we obtain an *even* cycle $C' = v_1^1 v_2^2 v_1^3 v_2^4 \ldots v_2^1 v_1^2 v_2^3 v_1^4 \ldots$ instead of C. It is easy to verify that G' is drawn as a thrackle, which is stated as part (ii) of the following lemma (see also Lemma 2 in [3]).

Lemma 3. (Conway, [18,3]) *Let G be a (generalized) thrackle with at least one odd cycle C. Then the topological graph G' obtained from G by Conway's doubling of C is*

    (i) bipartite, and
    (ii) a (generalized) thrackle.

Finally, we recall an observation of Woodall [18] mentioned in the introduction, which motivated our investigations.

As thrackleability is a hereditary property, a minimal counterexample to the thrackle conjecture must be a connected graph G with exactly $|V(G)| + 1$ edges and with no vertex of degree *one*. Such a graph G is necessarily a dumbbell $DB(c', c'', l)$. If $l \neq 0$, then G consists of two cycles that share a path or are connected by a path uv. In both cases, we can "double" the path uv, as indicated in Figure 4, to obtain another thrackle G'. It is easy to see that G' is a dumbbell consisting of two cycles that share precisely one vertex (the vertex v in the figure). Moreover, if any of these two cycles is not even, then we can double it and repeat the above procedure, if necessary, to obtain a dumbbell $DB(b', b'', 0)$ drawn as a thrackle, where b' and b'' are even numbers. Thus, in order to prove the thrackle conjecture, it is enough to show that no dumbbell $DB(c', c'', 0)$ consisting of two even cycles that share a vertex is thrackleable.

3   Proof of Theorem 1

Let $c \geq 6$ and $l \geq -1$ be two integers, and suppose that no dumbbell $DB(c', c'', l')$ with $-c'/2 \leq l' \leq l$ and with even c', c'', for which $6 \leq c', c'' \leq c$, can be drawn in the plane as a thrackle. For simpler notation, let $r = \lfloor l/2 \rfloor$.

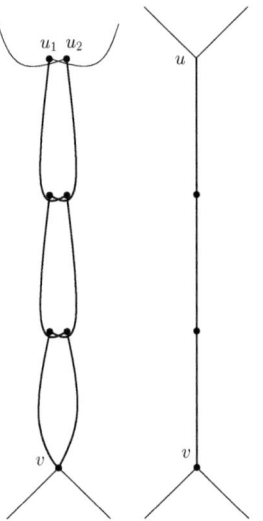

Let $G = (V, E)$ be a thrackleable graph with n vertices and m edges. We assume without loss of generality that G is connected and that it has no vertex of degree *one*. Otherwise, we can successively delete all vertices of degree *one*, and argue for each connected component of the resulting graph separately.

As usual, we call a graph *two-connected* if it is connected and it has no *cut vertex*, i.e., it cannot be separated into two or more parts by the removal of a vertex [6].

We distinguish three cases:

Fig. 4.   Doubling   the path uv

(A) G is bipartite;

(B) G is not bipartite, and the graph G' obtained by performing Conway's doubling of a shortest odd cycle $C \subset G$ is 2-connected;

(C) G is not bipartite, and the graph G' obtained by performing Conway's doubling of a shortest odd cycle $C \subset G$ is not 2-connected.

In each case, we will prove that $m \leq \tau(c, l)n$.

(A) By Lemma 2, in this case G is planar. We fix an embedding of G in the plane. According to the assumption of our theorem, G contains no subgraph that is a dumbbell $DB(c', c'', l')$, for any even $6 \leq c' \leq c'' \leq c$, and $-c'/2 \leq l' \leq l$. We also know that G has no C_4. We are going to use these conditions to bound the number of edges $m = |E(G)|$.

Notice that we also exclude dumbbells $DB(c', c'', l')$ with $-c' \le l' < -c'/2$. Indeed, in this case $DB(c', c'', l')$ is isomorphic to $DB(c', d, k)$, where $d = (c' + c'' + 2l')$, $k = (-c' - l')$, and $d < c'' \le c$, $\max(-c'/2, -d/2) \le k < 0$.

Suppose first that G is *two-connected*. Let f denote the number of faces, and let f_c stand for the number of faces with at most c sides. By double counting the edges, we obtain

$$2m \ge 6f_c + (c+2)(f - f_c). \tag{1}$$

If $l = -1$, then applying the condition of forbidden dumbbells, we obtain that no two faces of size at most c share an edge, so that $6f_c \le m$. If $l \ge 0$, Menger's theorem implies that any two faces of size at most c are connected by two vertex disjoint paths. Since any such path must be longer than l, to each face we can assign its vertices as well as the $r = \lfloor l/2 \rfloor$ closest vertices along two vertex disjoint paths leaving the face, and these sets are disjoint for distinct faces. Thus, we have $f_c(2r + 6) \le n$. In either case, we have

$$f_c \le \begin{cases} \frac{m}{6} & \text{if } l = -1, \\ \frac{n}{2r+6} & \text{if } l \ge 0. \end{cases} \tag{2}$$

Combining the last two inequalities, we obtain

$$f \le \frac{(c-4)f_c + 2m}{c+2} \le \begin{cases} \frac{(c-4)\frac{m}{6}+2m}{c+2} & \text{if } l = -1, \\ \frac{(c-4)\frac{n}{2r+6}+2m}{c+2} & \text{if } l \ge 0. \end{cases}$$

Using Euler's polyhedral formula $m + 2 = n + f$ we obtain

$$m \le \begin{cases} \frac{6c+12}{5c+4}n - \frac{12c+24}{5c+4} & \text{if } l = -1, \\ \frac{2cr+4r+7c+8}{2cr+6c}n - \frac{2c+4}{c} & \text{if } l \ge 0. \end{cases} \tag{3}$$

It can be shown by routine calculations that the last estimates, even if we ignore their negative terms independent of n, are stronger than the ones claimed in the theorem. (In fact, they are also stronger than the corresponding bounds (5) and (4) in Case (B); see below.) This concludes the proof of the case (A) when G is 2-connected.

If G is not 2-connected, then consider a block decomposition of G, and proceed by induction on the number of blocks. The base case, i.e when G is 2-connected, is treated above. Otherwise G can be obtained as a union of two bipartite graphs $G_1 = (V_1, E_1)$ and $G_2 = (V_2, E_2)$ sharing exactly one vertex. By induction hypothesis we can use (3) to bound the number of edges in G_i, for $i = 1, 2$, by substituting $|E_i|$ and $|V_i|$ for m and n, respectively. We obtain the claimed bound on the maximum number of edges in G by adding up the bounds on $|E_1|$ and $|E_2|$ as follows. $|E(G)| = |E(G_1)| + |E(G_2)| \le k_1|V(G_1)| + k_1|V(G_2)| - 2k_2 = k_1|V(G)| + k_1 - 2k_2$ where $k_1 = k_1(c, l)$ and $k_2 = k_2(c, l)$ represent the constants in (3). The induction step follows analogously, because $k_1 < k_2$ for all considered values of c and l.

(B) In this case, we establish two upper bounds on the maximum number of edges in G: one that decreases with the length of the shortest odd cycle $C \subseteq G$ and one that increases. Finally, we balance between these two bounds.

By doubling a shortest odd cycle $C \subseteq G$, as before, we obtain a bipartite thrackle G' (see Lemma 3). Let C' denote the doubled cycle in G'. By Lemma 2, G' is a planar graph. Since G' is bipartite, it is trivially two-colorable. Moreover, it can be embedded in the plane without crossing so that the cyclic order of the edges around each vertex in one color class is preserved, and for each vertex in the other color class reversed. A closer inspection of the way how we double C shows that as we traverse C' in G', the edges incident to C' start on alternating sides of C'. This implies that, after redrawing G' as a plane graph, all edges incident to C' lie on one side, that is, C' is a *face*.

Slightly abusing the notation, from now on let G' denote a crossing-free drawing with the above property, which has a $2|C|$-sided face C'. Denoting the number of vertices and edges of G' by n' and m', the number of faces and the number of faces of size at most c by f' and f'_c, respectively, we have $n' = n + |C| = |V(G')|$, $m' = m + |C| = |E(G')|$, and, as in Case (A), inequality (2),

$$ f'_c \leq \begin{cases} \frac{1}{6}m' & \text{if } l = -1, \\[2mm] \frac{n'}{2r+6} & \text{if } l \geq 0. \end{cases} $$

Double counting the edges of G', we obtain

$$ 2m' \geq 6f'_c + (c+2)(f' - 1 - f'_c) + 2|C|. $$

In case $l \geq 0$, combining the last two inequalities, we have

$$ f' \leq \frac{(c-4)f'_c + 2(m' - |C|) + c + 2}{c+2} \leq \frac{(c-4)\frac{n'}{2r+6} + 2(m' - |C|) + c + 2}{c+2}. $$

By Euler's polyhedral formula, $f' = m' - n' + 2$. Thus, after ignoring the negative term, which depends only on c and l, the last inequality yields

$$ |E(G)| \leq \frac{2cr + 4r + 7c + 8}{2cr + 6c}n + |C|\frac{c-4}{2cr + 6c}. \tag{4} $$

The case $l = -1$ can be treated analogously, and the corresponding bound on $E(G)$ becomes

$$ |E(G)| \leq \frac{6c + 12}{5c + 4}n + |C|\frac{c-4}{5c+4}. \tag{5} $$

We now establish another upper bound on the number of edges in G: one that decreases with the length of the shortest odd cycle C in G. As in [12], we remove from G the vertices of C together with all edges incident to them. Let G'' denote the resulting thrackle. By Lemma 1, G'' is bipartite. By Lemma 2, it is a planar graph. From now on, let G'' denote a fixed (crossing-free) embedding of this graph. According to our assumptions, G'' has no subgraph isomorphic to

$DB(c', c'', l')$, for any even numbers c' and c'' with $6 \leq c' \leq c'' \leq c$, and for any integer l' with $-c'/2 \leq l' \leq l$.

We can bound $|E(G'')|$, as follows. By the minimality of C, each vertex $v \in V(G)$ that does not belong to C is adjacent in G to at most one vertex on C. Indeed, otherwise, v would create either a C_4 or an odd cycle shorter than C. Hence, if $l \geq 0$, inequality (3) implies that

$$|E(G)| \leq |E(G'')| + |C| + (n - |C|) \leq \frac{2cr + 4r + 7c + 8}{2cr + 6c}(n - |C|) + n. \qquad (6)$$

In the case $l = -1$, we obtain

$$|E(G)| \leq |E(G'')| + |C| + (n - |C|) \leq \frac{6c + 12}{5c + 4}(n - |C|) + n. \qquad (7)$$

It remains to compare the above upper bounds on $|E(G)|$ in order to find the value of $|C|$ for which the minimum of our two bounds is maximal. If $l > -1$, then the value of $|C|$ for which the right-hand sides of (4) and (6) coincide is

$$|C| = \frac{cr + 3c}{cr + 2r + 4c + 2}n.$$

The claimed bound follows by inserting this value into (4) or (6).

In the case $l = -1$, the critical value of $|C|$, obtained by comparing the bounds (5) and (7), is

$$|C| = \frac{5c + 4}{7c + 8}n.$$

Inserting this value into (5) or (7), the claimed bound follows.

(C) As before, let C be a shortest odd cycle in G, and let G' be the graph obtained from G after doubling C. The doubled cycle is denoted by $C' \subset G'$. Let $G_0 \supseteq C$ denote a *maximal* subgraph of G, which is turned into a *two*-connected subgraph of G' after performing Conway's doubling on C. Let G_1 stand for the graph obtained from G by the removal of all *edges* in G_0.

It is not very hard to see that G_1 is bipartite, and each of its connected components shares exactly one vertex with G_0. Indeed, if a connected component $G_2 \subseteq G_1$ were not bipartite, then, by Lemma 1, G_2 would share at least one vertex with C, which belongs to an odd cycle of G_2. By the maximal choice of G_0, after doubling C, the component G_2 must turn into a subgraph $G_2' \subset G'$, which shares precisely *one* vertex with the doubled cycle C'. Thus, G_2 must also share precisely *one* vertex with C, which implies that $G_2' \subseteq G'$ has an odd cycle. This contradicts Lemma 3(i), according to which G' is a bipartite graph.

Therefore, G_1 is the union of all blocks of G, which are not entirely contained in G_0. Since each connected component G_2 of G_1 is bipartite, the number of edges of G_2 can be bounded from above by (3), just like in Case (A).

In order to bound the number of edges of G, we proceed by adding the connected components of G_1 to G_0, one by one. As at the end of the discussion of Case (3), using the fact that the last terms in (3), which do not depend on n, are smaller than -2, we can complete the proof by induction on the number of connected components of G_1.

4   A Better Upper Bound

As was pointed out in the introduction, if we manage to prove that for any l', $-3 \le l' \le -1$, the dumbbell $DB(6, 6, l')$ is not thrackleable, then Theorem 1 yields that the maximum number of edges that a thrackle on n vertices can have is at most $\frac{617}{425}n < 1.452n$. This estimate is already better than the currently best known upper bound $\frac{3}{2}n$ due to Cairns and Nikolayevsky [3].

In order to secure this improvement, we have to exclude the subgraphs $DB(6, 6, -1)$, $DB(6, 6, -2)$, and $DB(6, 6, -3)$. The fact that $DB(6, 6, -3)$ cannot be drawn as a thrackle was proved in [12] (Theorem 5.1). Here we present an algorithm that can be used for checking whether a "reasonably" small graph G can be drawn as a thrackle. We applied our algorithm to verify that $DB(6, 6, -1)$ and $DB(6, 6, -2)$ are indeed not thrackleable. In addition, we show that $DB(6, 6, 0)$ cannot be drawn as a thrackle, which leads to the improved bound in Theorem 2.

Let $G = (V, E)$ be a thrackle. Direct the edges of G arbitrarily. For any $e \in E$, let $E_e \subseteq E$ denote the set of all edges of G that do not share a vertex with e, and let $m(e) = |E_e|$. Let $\pi_e = (\pi_e(1), \pi_e(2), \ldots, \pi_e(m(e)))$ stand for the $m(e)$-tuple (permutation) of all edges belonging to E_e, listed in the order of their crossings along e.

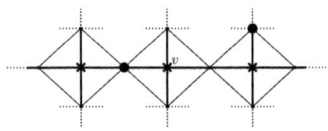

Fig. 5. 4-cycle around a vertex v of G', which was a crossing point in G

Construct a planar graph G' from G, by introducing a new vertex at each crossing between a pair of edges of G, and by replacing each edge by its pieces. In order to avoid G' having an embedding in which two paths corresponding to a crossing pair of edges of G do not *properly* cross, we introduce a new vertex in the interior of every edge of G', whose both endpoints are former crossings. For each former crossing point v, we add a cycle of length *four* to G', connecting its neighbors in their cyclic order around v, as illustrated in Figure 5. In the figure, the thicker lines and points represent edges and vertices or crossings of G, while the thinner lines and points depict the *four*-cycles added at the second stage.

Obviously, G' is completely determined by the directed abstract graph of G and by the set of permutations $\Pi(G) := \{\pi_e \in E_e^{m(e)} | e \in E\}$. Thus, a graph $G = (V, E)$ can be drawn as a thrackle if and only if there exists a set Π of $|E|$ permutations of $E_e, e \in E$, such that the abstract graph G' corresponding to the pair (G, Π) is *planar*. In other words, to decide whether a given abstract graph $G = (V, E)$ can be drawn as a thrackle, it is enough to consider all possible sets of permutations Π of $E_e, e \in E$, and to check if the corresponding graph $G' = G'(G, \Pi)$ is planar for at least one of them. The first deterministic linear time algorithm for testing planarity was found by Hopcroft and Tarjan [11]. However, in our implementation we used an improved algorithm for planarity testing by Fraysseix et al. [7], in particular, its implementation in the library P.I.G.A.L.E. [8]. The source code can be found in [20].

It was shown in [12] (Lemma 5.2) that in every drawing of a directed cycle C_6 as a thrackle, either every oriented path $e_1e_2e_3e_4$ is drawn in such a way that $\pi_{e_1} = (e_4, e_3)$ and $\pi_{e_4} = (e_1, e_2)$, or every oriented path $e_1e_2e_3e_4$ is drawn in such a way that $\pi_{e_1} = (e_3, e_4)$ and $\pi_{e_4} = (e_2, e_1)$. Using this observation (which is not crucial, but saves computational time), we ran a backtracking algorithm to rule out the existence of a set of permutations Π, for which $G'(DB(6, 6, 0), \Pi)$, $G'(DB(6, 6, -1), \Pi)$, or $G'(DB(6, 6, -2), \Pi)$ is planar. Our algorithm attempts to construct larger and larger parts of a potentially good set Π, and at each step it verifies if the corresponding graph still has a chance to be extended to a planar graph.

Summarizing, we have the following

Lemma 4. *None of the dumbbells* $DB(6, 6, l')$, $-3 \le l' \le 0$ *can be drawn as a thrackle.*

According to Lemma 4, Theorem 1 can be applied with $c = 6, l = 0$, and Theorem 2 follows.

For any $\varepsilon > 0$, our Theorem 1 and the above observations provide a deterministic algorithm with bounded running time to prove that all thrackles with n vertices have at most $(1 + \varepsilon)n$ edges or to exhibit a counterexample to Conway's conjecture.

In what follows we give an estimate of the dependence of the running time of our algorithm on ε (the proof is omitted due to lack space). The analysis uses the standard random access machine model. In particular, we assume that all basic arithmetic operations can be carried out in constant time.

Theorem 3. *For any* $\varepsilon > 0$, *there is a deterministic algorithm with running time* $e^{O((1/\varepsilon^2)\ln(1/\varepsilon))}$ *to prove that all thrackles with* n *vertices have at most* $(1 + \varepsilon)n$ *edges or to exhibit a counterexample to Conway's conjecture.*

5   Concluding Remarks

We say that two cycles C_1 and C_2 of a graph are at distance $l \ge 0$, if the length of a shortest path joining a vertex of C_1 to a vertex of C_2 is l. The following Turán-type questions were motivated by the proof of Theorem 1.

(1) Given two integers c_1, c_2, with $3 \le c_1 \le c_2$, what is the maximum number of edges that a planar graph on n vertices can have, if its girth (i.e. the length of the shortest cycle) is at least c_1, and no two cycles of length at most c_2 share an edge?

(2) Given three integers c_1, c_2, and l, with $3 \le c_1 \le c_2$ and $l \ge 0$, what is the maximum number of edges that a planar graph on n vertices can have, if its girth is at least c_1, and any two of its cycles of length at most c_2 are at distance larger than l ?

The inequalities (3) provide nontrivial upper bounds for restricted versions of the above problem for bipartite graphs.

We have the following general result.

Theorem 4. *Let c_1, c_2, and l denote three non-negative natural numbers with $3 \leq c_1 \leq c_2$. Let G be a planar graph with n vertices and girth at least c_1.*

(i) If no two cycles of length at most c_2 share an edge, then
$$|E(G)| \leq \frac{c_1 c_2 + c_1}{c_1 c_2 - c_2 - 1} n.$$
(ii) If no two cycles of length at most c_2 are at distance at most l, then
$$|E(G)| \leq \frac{c_1 c_2 + 2\lfloor l/2 \rfloor c_2 + 2\lfloor l/2 \rfloor + c_2 + 1}{2\lfloor l/2 \rfloor c_2 - 2\lfloor l/2 \rfloor + c_1 c_2 - c_1} n.$$

Proof. (Outline.) Without loss of generality, we can assume in both cases that G is connected, it has no vertex of degree *one*, and it is not a cycle. To establish part (i), consider an embedding of G in the plane. Let $m = |E(G)|$, and let f and f_{c_2} stand for the number of faces of G and for the number of faces of length at most c_2. We follow the idea of the proof of Case (A), Theorem 1, with $f_{c_2} \leq \frac{1}{c_1} m$ instead of $f_c \leq \frac{1}{6} m$, and with the inequality

$$2m \geq c_1 f_{c_2} + (c_2 + 1)(f - f_{c_2}) \tag{8}$$

replacing (1). Analogously, in the proof of part (ii), we use $f_{c_2} \leq \frac{1}{2\lfloor l/2 \rfloor + c_1} n$ instead of the inequality $f_c \leq \frac{1}{2r+6} n$.

It is possible that the constant factor in the part (i) of Theorem 4 is tight for all values of c_1 and c_2. It is certainly tight for all values of the form $c_1 = ml$ and $c_2 = m(l+1) - 1$, where m and l are natural numbers, as is shown by the following result, the proof of which is omitted.

Theorem 5. *For any positive integers n_0, $m \geq 1$, and $l \geq 3$, one can construct a plane graph $G = (V, E)$ with at least n_0 vertices with girth ml such that all of its inner faces are of size ml or $m(l+1)$, its outer face is of size $2ml$, and each edge of G not on its outer face belongs to exactly one cycle of size ml, which is a face of G. The second smallest length of a cycle in G is $m(l+1)$.*

If we slightly relax the conditions in Theorem 4 by forbidding only dumbbells determined by *face cycles*, we obtain some tight bounds. For instance, it is not hard to prove the following.

Theorem 6. *Let c_1 and c_2 be two nonnegative integers with $3 \leq c_1 \leq c_2$. Let G be a plane graph on n vertices that has no face shorter than c_1 and no two faces of length at most c_2 that share an edge. Then we have $|E(G)| \leq \frac{c_1 c_2 + c_1}{c_1 c_2 - c_2 - 1} n$, and the inequality does not remain true with any smaller constant.*

References

1. Bollobás, B.: Modern Graph Theory. Springer, New York (1998)
2. Brass, P., Moser, W., Pach, J.: Research Problems in Discrete Geometry. Springer, New York (2005)
3. Cairns, G., Nikolayevsky, Y.: Bounds for generalized thrackles. Discrete Comput. Geom. 23, 191–206 (2000)

4. Cairns, G., McIntyre, M., Nikolayevsky, Y.: The thrackle conjecture for K_5 and $K_{3,3}$. In: Towards a theory of Geometric Graphs, Contemp. Math. Amer. Math. Soc., vol. 342, pp. 35–54. RI, Providence (2004)
5. Cairns, G., Nikolayevsky, Y.: Generalized thrackle drawings of non-bipartite graphs. Discrete Comput. Geom. 41, 119–134 (2009)
6. Diestel, R.: Graph Theory. Springer, New York (2008)
7. de Fraysseix, H., de Mendez, P.O., Rosenstiehl, P.: Trémaux Trees and Planarity. Internat. J. Found. of Comput. Sc. 17, 1017–1030
8. de Fraysseix, H., de Mendez, P.O.: Public Implementation of a Graph Algorithm Library and Editor, http://pigale.sourceforge.net/
9. Green, J.E., Ringeisen, R.D.: Combinatorial drawings and thrackle surfaces. In: Graph Theory, Combinatorics, and Algorithms, Kalamazoo, MI, vol. 2, pp. 999–1009. Wiley-Intersci. Publ., New York (1995)
10. Hopf, H., Pannwitz, E.: Aufgabe Nr. 167. Jahresbericht Deutsch. Math.-Verein. 43, 114 (1934)
11. Hopcroft, J., Tarjan, R.E.: Efficient planarity testing. Journal of the Association for Computing Machinery 21(4), 549–568
12. Lovász, L., Pach, J., Szegedy, M.: On Conway's thrackle conjecture. Discrete Comput. Geom. 18, 369–376 (1998)
13. Perlstein, A., Pinchasi, R.: Generalized thrackles and geometric graphs in R^3 with no pair of strongly avoiding edges. Graphs Combin. 24, 373–389 (2008)
14. Piazza, B.L., Ringeisen, R.D., Stueckle, S.K.: Subthrackleable graphs and four cycles. Graph theory and Applications (Hakone, 1990), Discrete Math 127, 265–276 (1994)
15. Ringeisen, R.D.: Two old extremal graph drawing conjectures: progress and perspectives. Congressus Numerantium 115, 91–103 (1996)
16. Rosenstiehl, P.: Solution algebrique du probleme de Gauss sur la permutation des points d'intersection d'une ou plusieurs courbes fermees du plan. C. R. Acad. Sci. Paris Ser. A-B 283, A551–A553 (1976)
17. Sutherland, J.W.: Lösung der Aufgabe 167. Jahresbericht Deutsch. Math.-Verein. 45, 33–35 (1935)
18. Woodall, D.R.: Thrackles and deadlock. In: Welsh, D.J.A. (ed.) Combinatorial Mathematics and Its Applications, vol. 348, pp. 335–348. Academic Press, London (1969)
19. Unsolved problems. Chairman: P. Erdős, in: Combinatorics Proc. Conf. Combinatorial Math., Math. Inst., Oxford, Inst. Math. Appl., Southend-on-Sea, 351–363 (1972)
20. http://dcg.epfl.ch/webdav/site/dcg/users/183292/public/Thrackle.zip

Optimal k-Level Planarization and Crossing Minimization

Graeme Gange[1], Peter J. Stuckey[1], and Kim Marriott[2]

[1] Department of Computer Science and Software Engineering
The University of Melbourne, Vic. 3010, Australia
{ggange,pjs}@csse.unimelb.edu.au
[2] Clayton School of IT
Monash University, Vic. 3800, Australia
Kim.Marriott@infotech.monash.edu.au

Abstract. An important step in laying out hierarchical network diagrams is to order the nodes on each level. The usual approach is to minimize the number of edge crossings. This problem is NP-hard even for two layers when the first layer is fixed. Hence, in practice crossing minimization is performed using heuristics. Another suggested approach is to maximize the planar subgraph, i.e. find the least number of edges to delete to make the graph planar. Again this is performed using heuristics since minimal edge deletion for planarity is NP-hard. We show that using modern SAT and MIP solving approaches we can find *optimal* orderings for minimal crossing or minimal edge deletion for planarization on reasonably sized graphs. These exact approaches provide a benchmark for measuring quality of heuristic crossing minimization and planarization algorithms. Furthermore, we can straightforwardly extend our approach to minimize crossings followed by maximizing planar subgraph or vice versa; these hybrid approaches produce noticeably better layout then either crossing minimization or planarization alone.

1 Introduction

The standard approach for drawing hierarchical network diagrams is a three phase approach in which (a) nodes in the graph are assigned levels producing a k-level graph; (b) nodes are assigned an order so as to minimize edge crossings in the k-level graph; and (c) the edge routes and node positions are computed. There has been considerable research into step (b) which is called *k-level crossing minimization*. Unfortunately this step is NP-hard even for two layers ($k = 2$) where the ordering on one layer is given. Thus, research has focussed on developing heuristics to solve it. In practice the approach is to iterate through the levels, re-ordering the nodes on each level using heuristic techniques such as the barycentric method [1].

An alternative to performing crossing minimization in phase (b) is *k-level planarization* problem. This was introduced by Mutzel [2] and is the problem of finding the minimal set of edges that can be removed which allow the remaining edges in the k-level graph to be drawn without any crossings. Mutzel has argued convincingly that for hierarchical network diagrams with many crossings,

U. Brandes and S. Cornelsen (Eds.): GD 2010, LNCS 6502, pp. 238–249, 2011.

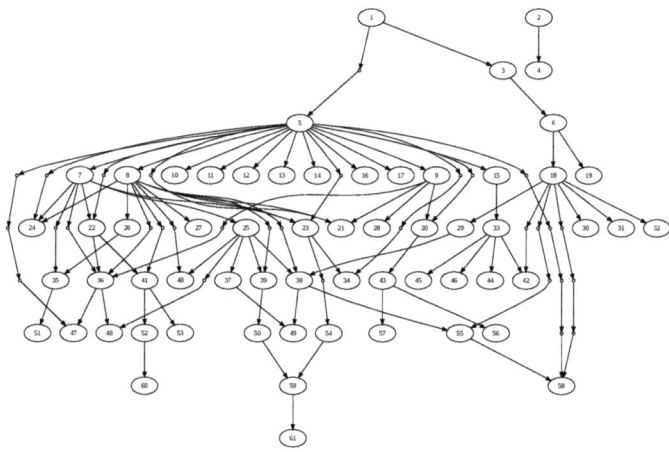

Fig. 1. Graphviz heuristic layout for the *profile* example graph

this leads to better drawings than those obtained by simply minimizing the total number of edge crossings. While in some sense simpler than k-level crossing minimization (since the problem is tractable for $k = 2$ with one side fixed) it is still NP-hard for $k > 2$. A disadvantage of k-level planarization is that it does not take into account the number of crossings that the non-planar edges generate and so a poor choice of which edges to remove can give rise to unnecessary edge crossings.

Here we introduce a combination of the two approaches we call *k-level planarization and crossing minimization*. This minimizes the weighted sum of the number of crossings *and* the number of edges that need to be removed to give a planar drawing. We believe that this gives rise to nicer drawings than either k-level planarization or k-level crossing minimization while providing a natural generalization of both.

As some evidence for this consider the drawings shown in Figures 1 and Figure 2 of the example graph `profile` from the GraphViz gallery [3]. Figure 1 shows the layout from GraphViz using its heuristic for edge crossing minimization. It has 54 edge crossing and requires removal of 17 edges to become planar.

The layout resulting from minimizing edge crossings is shown in Figure 2(a). It has 38 crossings, significantly less than the heuristic layout. The layout resulting from maximizing the planar subgraph is shown in Figure 2(b) with deleted edges dotted. It requires only 9 edges to be deleted but has 81 crossings. The layout clearly shows that maximizing the planar subgraph in isolation is not enough, leading to many unnecessary crossings.

The combined model allows us to minimize both crossings and edge deletions for planarity simultaneously. Figure 2(c) shows the result of minimizing crossings and then maximizing the planar subset. It yields 38 crossings and 11 edge deletions. Figure 2(d) shows the results of of maximizing the planar subset and the minimize crossings. It yields 9 edge deletions and 57 edge

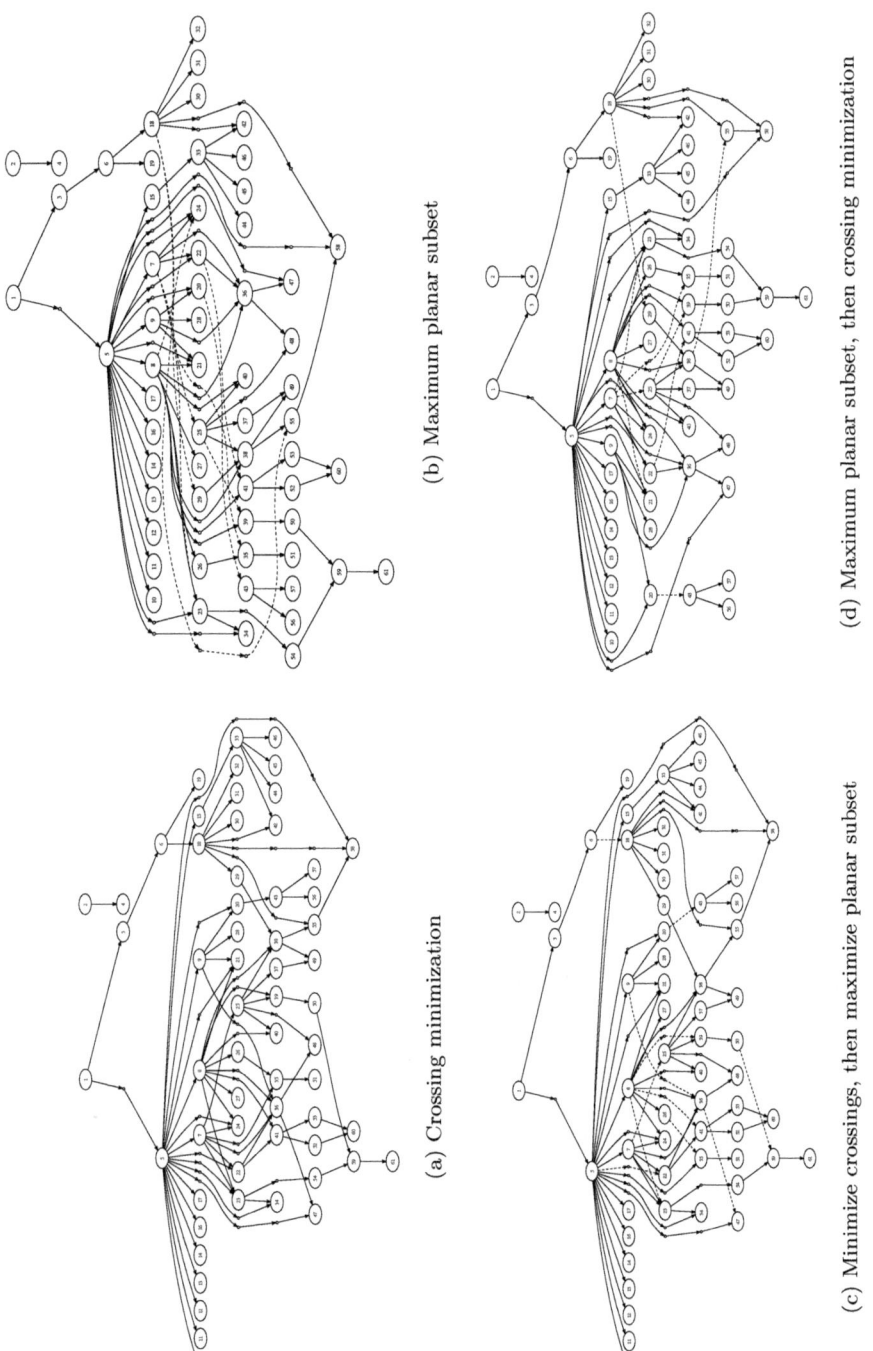

(a) Crossing minimization

(b) Maximum planar subset

(c) Minimize crossings, then maximize planar subset

(d) Maximum planar subset, then crossing minimization

Fig. 2. Different layouts of the *profile* example graph

crossings, a substantial improvement over the maximal planar subgraph layout of Figure 2(b).

We believe these combined layouts clearly illustrate that some combination of minimal edge crossing and minimal edge deletions for planarity leads to better layout than either individually. Part of the advantage is simply that displaying edges deleted for planarity differently makes the layout much clearer.

Apart from introducing these combined layouts, the paper has two main technical contributions. The first is to give a binary program for the combined k-level planarization and crossing minimization problem. By appropriate choice of the weighting factor this model reduces to either k-level planarization or k-level crossing minimization. Our basic model is reasonably straightforward but we use some tricks to reduce symmetries, handle leaf nodes in trees and improve bounds for edge cycles.

Our second technical contribution is to evaluate performance of the binary program using both a generic MIP solver and a generic SAT solver. While MIP techniques are not uncommon in graph drawing the use of SAT techniques is quite unusual. Our reason for considering MIP is that MIP is well suited to combinatorial optimization problems in which the linear relaxation of the problem is close to the original problem. However this does not seem true for k-level planarization and/or k-level crossing minimization. Hence it is worth investigating the use of other generic optimization techniques. Over the last decade there has been considerable improvement in SAT solving techniques and they are now capable of solving problems with thousands of variables in a few seconds. Part of this improvement arises by learning combinations of assignments to the Boolean variables that lead to unsatisfiability (called "no goods") as the search for the optimum solution proceeds. The no goods are used to prune the search space leading to orders of magnitude performance improvement.

We find that modern SAT solving with learning, and modern MIP solvers (which now have special routines to handle SAT style models) are able to handle the k-level planarization and crossing minimization problems and their combination for quite large k, meaning that we can solve step (b) to optimality. They are fast enough to find the optimal ordering of nodes on all layers for graphs with hundreds of nodes in a few seconds, so long as the graph is reasonably narrow (less than 10 nodes on each level) and for larger graphs they find reasonable solutions within one minute.

The significance of our research is twofold. First it provides a benchmark for measuring the quality of heuristic methods for solving k-level crossing minimization and/or k-level planarization. Second, the method is practical for small to medium graphs and leads to significantly fewer edge crossings involving fewer edges than is obtained with the standard heuristic approaches. As computers increase in speed and SAT solving and MIP solving techniques continue to improve we predict that optimal solution techniques based on MIP and SAT will replace the use of heuristics for step (b) in layout of hierarchical networks.

Furthermore, our research provides support for the use of generic optimization techniques for exploring different aesthetic criteria. The use of generic techniques allows easy experimentation with, for instance, our hybrid objective function. As another example rather than k-level planarization we might wish to minimize the total number of edges involved in crossings. This is simple to do with generic optimization. Another advantage of generic optimization techniques is that they also readily handle additional constraints on the layout, such as placing some nodes on the outside or clustering nodes together.

The task of reducing crossings in k-layered graphs has received considerable attention, particularly due to the layout algorithm of Sugiyama [4]. However, most of these approaches, such as [5] tend to use heuristics, rather than finding the global optimum.

The most closely related work is on the use of MIP and branch-and-bound techniques for solving k-level crossing minimization. Jünger and Mutzel [6] compared heuristic methods for two layer crossing minimization with a MIP encoding solved using a specialized branch-and-cut algorithm to solve to optimality. They found that the MIP encoding for the case when one layer is fixed is practical for reasonably sized graphs. In another paper, Jünger et al [7] gave a 0-1 model for k-level crossing minimization and solved it using a generic MIP solver. They found that at that time MIP techniques were impractical except for quite small graphs. We differ from this in considering planarization as well and in investigating SAT solvers. Randerath et al [8] gave a partial-MAXSAT model of crossing minimization, however did not provide any experiments. We show that SAT solving with learning, and more recent MIP solvers (which now have special routines to handle SAT style models) are now practical for reasonably sized graphs.

Also related is Mutzel [2] which describes the results of using a MIP encoding with branch-and-cut for the 2-level planarization problem. Here we give a binary program model for k-level planarization and show that SAT with learning and modern MIP solvers can solve the k-level planarization problem for quite large k. We use a similar model to that of Jünger and Mutzel but examine both MIP and SAT techniques to solving it.

The paper is organized as follows. In the next section we give our model for combined planarity and crossing minimization. In Section 3 we show how to improve the model by taking into account graph properties. In Section 4 we give results of experiments comparing the different measures, and finally in Section 5 we conclude.

2   Model

A general framework for generating layouts of hierarchical data was presented by [4]. This proceeds in three stages. First, the vertices of the graph are partitioned into horizontal layers. Then, the ordering of vertices within these horizontal layers is permuted to reduce the number of edge crossings. Finally, these layers are positioned to straighten long edges and minimize edge length. Our focus is on the second stage of this process – permuting the vertices on each layer.

Consider a graph with nodes divided into k layers, with edges restricted to adjacent layers, ie. edges from layer i to $i + 1$. Denote the nodes in the $k - th$

layer by *nodes*[k], and the edges from layer k to layer $k + 1$ by *edges*[k]. For a given edge e, denote the start and end nodes by $e.s$ and $e.d$ respectively.

The combined model for maximal planar subgraph and crossing minimization is defined by the binary program:

$$\min \sum_{k \in levels} C \sum_{e,f \in edges[k]} c_{(e,f)} + P \sum_{e \in edges[k]} r_e \tag{1}$$

s.t.

$$\bigwedge_{k \in levels} \bigwedge_{i,j,k \in nodes[k]} l_{(i,j)} \wedge l_{(j,k)} \rightarrow l_{(i,k)} \tag{2}$$

$$\bigwedge_{k \in levels} \bigwedge_{e,f \in edges[k]} c_{(e,f)} \leftrightarrow l_{(e.s,f.s)} \oplus l_{(e.d,f.d)} \tag{3}$$

$$\bigwedge_{k \in levels} \bigwedge_{e,f \in edges[k]} r_e \vee r_f \vee \neg c_{(e,f)} \tag{4}$$

The variable $l_{(i,j)}$ indicates node i is before j in the level ordering. The variable $c_{(e,f)}$ indicates that edge e crosses edge f. The variable r_e indicates that edge e is deleted to make the graph planar. The constants C and P define the relative weights of crossing minimization and edge deletion for planarity. The 3-cycle constraints of Equation 2 ensures that the order variables are assigned to a consistent ordering. Equation 3 defines the edge crossings variables in terms of the ordering: the edges cross if the relative order of the start and end nodes are reversed. It is encoded in clauses as

$$c_{(e,f)} \vee l_{(e.s,f.s)} \vee \neg l_{(e.d,f.d)}, \quad c_{(e,f)} \vee \neg l_{(e.s,f.s)} \vee l_{(e.d,f.d)},$$
$$\neg c_{(e,f)} \vee l_{(e.s,f.s)} \vee l_{(e.d,f.d)}, \quad \neg c_{(e,f)} \vee \neg l_{(e.s,f.s)} \vee \neg l_{(e.d,f.d)}.$$

The planarity requirement is encoded in Eq. 4 which states that for each pair either one is removed, or they don't cross. The combined model uses $O(k.(e^2 + n^2))$ Boolean variables and is $O(k.(n^3 + e^2))$ in size.

We can convert this clausal model to a MIP binary program by converting each clause $b_1 \vee \cdots b_l \vee \neg b_{l+1} \vee \cdots \vee \neg b_m$ to the linear constraint $b_1 + \cdots + b_l - b_{l-1} - \cdots - b_m \geq m - l + 1$.

Long edges are handled by adding intermediate nodes in the levels that the long edges cross and breaking the edge into components. For crossing minimization each of these new edges is treated like an original edge. For the minimal deletion of edges each component edge in a long edge e is encoded using the same deletion variable r_e.

By adjusting the relative weights for crossing C, and planarization P, we can create and evaluate new measures of clarity of the graph. With $C = 1 + \sum_{k \in levels} |edges[k]|$ and $P = 1$ we first minimize crossings, then minimize edge deletions for planarity. With $C = 1$ and $P = \sum_{k \in levels} |edges[k]|^2$ we first minimize edge deletions and then crossings.

3   Additional Constraints

While the basic model described in Section 2 are sufficient to ensure correctness, finding the optimum still requires a great deal of search. We can modify the model to significantly improve performance.

First note that we add symmetry breaking by fixing the order of the first two nodes appearing on the same level. If the graph to be layed out has more

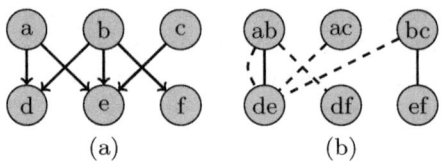

Fig. 3. (a) A graph, with an initial ordering (b) The corresponding vertex-exchange graph

symmetries than this left-to-right symmetry we could use this to fix more variables (although we don't do this in the experiments). Next, we can improve edge crossing minimization by using as an upper bound the number of crossings in a heuristic layout. We could also use heuristic solutions to bound planarity but doing so requires computing how many edges need deletion, which is non-trivial.

3.1 Cycle Parity

Healy and Kuusik introduced the vertex-exchange graph [9] for analyzing layered graphs. Each edge in the vertex-exchange graph corresponds to a potential crossing in the initial graph; each node corresponds to a pair of nodes within a level.

Consider the graph shown in Figure 3(a), its vertex-exchange graph is shown in Figure 3(b). Note there are two edges (ab, de) corresponding to the two pairs $((a, d), (b, e))$ and $((a, e), (b, d))$. Edges corresponding to crossings in Figure 3(a) are shown as solid, the rest are dashed.

For any given cycle in the vertex exchange graph, permuting nodes within a layer will maintain parity in the number of crossings in the cycle. For cycles with an odd number of crossings, this means that at least one of the pairs of edges in the cycle will be crossing. This can be represented by the clause $\bigvee_{(e,f) \in cycle} c_{(e,f)}$. When finding the maximal planar subgraph, we then know that at least one edge involved in the cycle must be removed from the subgraph. Similarly since the cycle is even in length we know that not all edges can cross, represented by $\bigvee_{(e,f) \in cycle} \neg c_{(e,f)}$. Both these constraints can be added to the model.

A special case of cycle parity is the $K_{2,2}$ subgraph. This subgraph always produces exactly one crossing, irrespective of the relative orderings of the nodes in the subgraph. When minimizing crossings, the corresponding $c_{(e,f)}$ variables need not be included in the objective function, which considerably simplifies the problem structure. Note that, for example, a $K_{3,3}$ subgraph contains 9 $K_{2,2}$ subgraphs, and each of the 9 $c_{e,f}$ variables arising can be ommitted from the problem. For the experiments we add constraints for cycles of length 6 or less, since the larger cycles did not improve performance.

3.2 Leaves

It is not difficult to prove that if a node on layer k has m child leaf nodes (unconnected to any other node) on layer $k + 1$, then all of these leaf nodes can be ordered together.

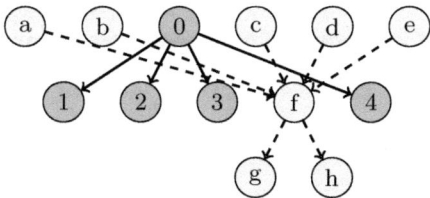

Fig. 4. A partial layout with respect to some leaf nodes 1,2,3,4

Consider the partial layout illustrated in Figure 4, where each node 1,2,3 and 4 is a leaf node with no outgoing arcs. If we place a node f in between nodes 1,2,3 and 4 (as illustrated) there is always at least as good a crossing solution by placing f either before or after all of them. Here since there are 2 parents before 0 and 3 after, f should be placed after 4, leading to 8 crossings rather than the 9 illustrated.

Similarly maximizing planarity always requires that all edges to siblings left of f be removed or all edges from parents before 0, and all edges to siblings right of f or all edges from parents after 0. An optimal solution always results by either deleting all edges to leaf nodes (which makes the leaf positions irrelevant), or ordering f after all leaves and deleting all edges from parents before 0, or ordering f before all leaves and deleting all edges from parents after 0.

Since there is no benefit in splitting leaf siblings we can treat them as a single node, but note we must appropriately weight the edge resulting, since it represents multiple crossings and multiple edge deletions.

Let N be a set of m leaf nodes from a single parent node i. We replace N by a new node j', and replace all edges $\{(i,j) \mid j \in N\}$ by the single edge (i,j'). We replace each m terms $c_{((i,j),f)}, j \in N$ in the objective function by one term $m \times c_{((i,j'),f)}$ and replace each of the m terms $r_{(i,j)}, j \in N$ in the objective by the term $m \times r_{(i,j')}$.

4   Experimental Results

We tested the binary model on a variety of graphs, using the pseudo-Boolean constraint solver MiniSAT+[10], and the Mixed Integer Programming solver CPLEX 12.0. All experiments were performed on a 3.0GHz Xeon X5472 with 32 Gb of RAM running Debian GNU/Linux 4.0. We ran for a maximum of 60s, and all times are given in seconds. We compared 4 different objective functions:

- crossing minimization: $C = 1$, $P = 0$;
- maximal planar subgraph $C = 0$, $P = 1$;
- crossing minimization then maximal planar subgraph $C = 1 + \sum_{k \in levels} |edges[k]|$, $P = 1$; and
- maximal planar subgraph then minimize crossings $C = 1$, $P = \sum_{k \in levels} |edges[k]|^2$.

246   G. Gange, P.J. Stuckey, and K. Marriott

Table 1. Time to find and prove the minimal crossing layout and maximal planar subgraph for Graphviz examples using MIP and SAT

Problem	graphviz	Crossing minimization				Maximal planar subgraph			
		MIP		SAT		MIP		SAT	
	best	best	solved	best	solved	best	solved	best	solved
crazy	3	2	0.04	2	**0.01**	1	0.03	1	**0.01**
datastruct	2	2	0.00	2	0.00	1	0.02	1	0.00
fsm	0	0	0.00	0	0.00	0	0.00	0	0.00
lion_share	7	4	**0.04**	4	0.11	2	0.05	2	0.02
profile	54	38	**6.81**	54	—	12	—	9	5.39
switch	20	20	0.75	20	**0.64**	17	—	17	34.49
traffic_lights	0	0	0.00	0	0.00	0	0.00	0	0.00
unix	3	2	0.05	2	**0.01**	1	0.04	1	0.02
world	50	**46**	—	50	—	15	—	13	—

To compare speed and effectiveness of the model we ran it on two sets of graphs. The first set of graphs are all the hierarchical network diagrams appearing in the GraphViz gallery [3]. The second set of graphs are random graphs of k-levels with n nodes per level and a fixed edge density of 20% (that is each node is connected on average to 20% of the nodes on the next layer); these do not include any long edges. Problem class gk_n is a suite of 10 randomly generated instances with k levels and n nodes per level.

Table 1 shows the results of minimizing edge crossings and maximizing planar subgraphs with MIP and SAT solvers, as well as the crossings resulting in the Graphviz heuristic layout for graphs from the GraphViz gallery. We use the best variation of our model for each solver: for the MIP solver this is with all improvements described in the previous section, while for the SAT solver we omit the leaf optimization since it slows down the solver. For each solver we show the solution, with least edge crossings or minimal number of edges deleted for planarity, found in 60s and the time to prove optimality or '—' if it was not proved optimal. The results show that for realistic graphs we can find better solutions than the heuristic method, even when there are very few crossings. The best found crossing and edges deleted for planarization and problems solved/time combination are highlighted in bold. We can find optimal crossing solutions for 9 out of ten examples, and maximal planar subgraph solutions for 7 out of 10 examples. The MIP approach is clearly superior for minimizing edge crossings, while SAT is superior for maximizing planarity.

Table 2 shows the results of crossing minimization and maximal planar subgraph for the second data set of random graphs using the MIP and SAT solver. The table shows: the total number of crossings when the graphs are laid out using GraphViz then for each solver: the total number of crossings or edge deletions in the best solutions found in 60s for the suite (a '—' indicates that for at least one instance the method found no solution better than the Graphviz

Table 2. Time to find and prove the minimal crossing layout and maximal planar subgraph, using MIP and SAT for random examples

| Problem | graphviz | Crossing minimization | | | | Maximal planar subgraph | | | |
| | | MIP | | SAT | | MIP | | SAT | |
	best	best	solved	best	solved	best	solved	best	solved
g3_7	41	35	10 / 0.00	35	10 / 0.02	18	10 / 0.01	18	10 / 0.02
g3_8	103	86	10 / 0.03	86	10 / 0.10	31	10 / 0.15	31	10 / 0.04
g3_9	204	195	10 / 0.07	195	10 / 6.67	63	10 / 4.60	63	10 / 0.12
g3_10	399	373	10 / 0.97	380	8 / 14.44	91	5 / 15.53	91	10 / 3.37
g4_7	70	55	10 / 0.01	55	10 / 0.04	32	10 / 0.11	32	10 / 0.04
g4_8	187	169	10 / 0.09	169	10 / 0.68	61	10 / 3.83	61	10 / 0.16
g4_9	351	342	10 / 0.89	345	8 / 13.69	94	6 / 26.85	94	10 / 2.82
g4_10	703	681	9 / 2.37	—	—	161	—	152	—
g5_7	101	95	10 / 0.03	95	10 / 0.11	47	10 / 0.43	47	10 / 0.08
g5_8	284	245	10 / 0.32	245	10 / 4.01	93	10 / 25.46	93	10 / 2.78
g5_9	474	450	10 / 0.99	—	5 / 30.47	139	1 / 35.37	138	3 / 47.46
g6_7	141	131	10 / 0.03	131	10 / 0.18	57	10 / 1.51	57	10 / 0.18
g6_8	357	324	10 / 0.53	324	10 / 13.34	112	4 / 11.64	111	10 / 28.81
g6_9	684	637	10 / 3.28	—	—	190	—	197	—
g7_7	159	148	10 / 0.08	148	10 / 0.58	67	10 / 3.91	67	10 / 0.78
g7_8	390	366	10 / 0.72	372	6 / 12.35	134	1 / 47.75	140	—
g7_9	813	786	9 / 12.54	—	—	238	—	233	—
g8_7	249	235	10 / 0.16	235	10 / 4.06	92	8 / 13.38	92	10 / 4.91
g8_8	466	431	10 / 1.51	—	1 / 10.73	154	—	165	—
g9_7	269	238	10 / 0.30	238	10 / 2.60	108	7 / 17.72	108	8 / 15.18
g9_8	572	541	10 / 2.95	—	3 / 28.71	197	—	200	—
g10_7	334	304	10 / 0.27	304	10 / 9.67	119	8 / 24.05	121	4 / 19.39
g10_8	733	661	10 / 10.68	—	—	216	—	225	—

bound in 60s) and the number of instances where optimal solutions were found and proved and the average time to prove optimality.

The results are in accord with those for the first dataset and show that the MIP solver can almost always find optimal minimal crossing solutions within this time bound (only two instances failed). The Graphviz solutions can be substantially improved, the best solutions found have 10-20% fewer crossings.

For maximal planar subgraph, in contrast to edge crossings, the SAT solver is better than the MIP solver, although as the number of levels increases the advantage decreases.

Tables 3 and 4 show the results for the mixed objective functions: minimizing crossings then maximizing planar subgraph and the reverse. For minimizing crossings first MIP dominates as before, and again is able to solve almost all problems optimally within 60s. For the reverse objective SAT is better for the small instances, but suffers as the instances get larger. This problem is significantly harder than the minimizing crossings first.

Results not presented demonstrate that the improvements presented in the previous section make a substantial difference. The elimination of $K_{2,2}$ cycles is

Table 3. Time to find and prove optimal mixed objective solutions for Graphviz examples using MIP and SAT

Problem	Crossing then planarization				Planarization then crossing			
	MIP		SAT		MIP		SAT	
	best	solved	best	solved	best	solved	best	solved
crazy	**(2, 1)**	**0.07**	**(2, 1)**	10.51	**(1, 2)**	**0.08**	**(1, 2)**	0.31
datastruct	**(2, 1)**	**0.02**	**(2, 1)**	0.26	**(1, 2)**	**0.03**	**(1, 2)**	0.23
fsm	**(0, 0)**	**0.00**	**(0, 0)**	0.00	**(0, 0)**	**0.00**	**(0, 0)**	0.00
lion_share	**(4, 3)**	**0.15**	**(4, 3)**	3.55	**(2, 5)**	**0.52**	**(2, 5)**	0.60
profile	**(38, 11)**	**29.96**	(281, 34)	—	**(12, 66)**	—	(13, 145)	—
switch	**(20, 17)**	**1.36**	**(20, 17)**	3.61	**(17, 20)**	—	**(17, 20)**	—
traffic_lights	**(0, 0)**	**0.00**	**(0, 0)**	0.00	**(0, 0)**	**0.00**	**(0, 0)**	0.01
unix	**(2, 1)**	**0.07**	**(2, 1)**	10.52	**(1, 2)**	**0.09**	**(1, 2)**	0.32
world	**(47, 14)**	—	(108, 19)	—	(18, 79)	—	**(15, 106)**	—

Table 4. Time to find and prove optimal mixed objective solutions for random examples using MIP and SAT

Problem	Crossing then planarization				Planarization then crossing			
	MIP		SAT		MIP		SAT	
	best	solved	best	solved	best	solved	best	solved
g3_7	**(35, 22)**	10 / 0.01	**(35, 22)**	10 / 0.04	**(18, 39)**	10 / 0.02	**(18, 39)**	10 / 0.04
g3_8	**(86, 41)**	10 / 0.13	**(86, 41)**	10 / 0.37	**(31, 102)**	10 / 0.31	**(31, 102)**	10 / 0.21
g3_9	**(195, 78)**	10 / 0.48	**(195, 78)**	10 / 3.09	**(63, 231)**	9 / 10.31	**(63, 231)**	10 / 1.04
g3_10	**(373, 115)**	10 / 2.67	(564, 166)	2 / 40.99	**(91, 444)**	2 / 13.84	**(91, 419)**	9 / 10.64
g4_7	**(55, 35)**	10 / 0.05	**(55, 35)**	10 / 0.23	**(32, 58)**	10 / 0.21	**(32, 58)**	10 / 0.23
g4_8	**(169, 76)**	10 / 0.57	**(169, 76)**	10 / 1.58	**(61, 223)**	10 / 6.66	**(61, 223)**	10 / 1.91
g4_9	**(342, 116)**	10 / 1.66	(418, 162)	5 / 33.21	**(94, 386)**	3 / 31.03	(95, 382)	8 / 14.76
g4_10	**(681, 195)**	9 / 8.17	(1249, 338)	—	**(158, 933)**	—	(160, 928)	—
g5_7	**(95, 55)**	10 / 0.08	**(95, 55)**	10 / 0.35	**(47, 104)**	10 / 0.76	**(47, 104)**	10 / 0.53
g5_8	**(245, 108)**	10 / 0.66	**(245, 108)**	10 / 8.38	(95, 269)	4 / 29.48	(94, 290)	8 / 25.24
g5_9	**(450, 174)**	10 / 3.83	(694, 210)	1 / 42.43	**(142, 612)**	—	(146, 656)	—
g6_7	**(131, 64)**	10 / 0.25	**(131, 64)**	10 / 1.98	**(57, 153)**	10 / 2.74	**(57, 153)**	10 / 2.32
g6_8	**(324, 136)**	10 / 1.16	(357, 150)	6 / 22.50	**(112, 419)**	2 / 22.62	(117, 413)	2 / 31.29
g6_9	**(637, 228)**	10 / 8.15	(1353, 513)	—	**(192, 881)**	—	(212, 967)	—
g7_7	**(148, 83)**	10 / 0.34	**(148, 83)**	10 / 23.94	**(67, 168)**	10 / 10.66	**(67, 168)**	10 / 10.59
g7_8	**(366, 159)**	10 / 3.00	(454, 236)	2 / 20.16	**(136, 472)**	—	(148, 500)	—
g7_9	**(778, 255)**	8 / 18.96	(1372, 481)	—	**(236, 1031)**	—	(258, 1303)	—
g8_7	**(235, 116)**	10 / 0.50	**(235, 116)**	10 / 14.06	(92, 272)	5 / 15.54	(93, 277)	8 / 22.09
g8_8	**(431, 195)**	10 / 5.06	(641, 345)	1 / 33.37	**(154, 552)**	—	(182, 639)	—
g9_7	**(238, 123)**	10 / 0.77	(241, 126)	9 / 25.00	**(108, 260)**	6 / 16.29	(112, 283)	2 / 57.22
g9_8	**(541, 229)**	10 / 6.17	(981, 464)	—	**(198, 757)**	—	(216, 871)	—
g10_7	**(304, 144)**	10 / 1.59	(329, 201)	7 / 33.19	**(119, 362)**	4 / 32.01	(126, 415)	1 / 58.29
g10_8	**(661, 256)**	9 / 15.15	(1216, 546)	—	**(199, 832)**	—	(224, 987)	—

highly beneficial to both solvers. Constraints for larger cycles can have significant benefit for the MIP solver but rarely benefit the SAT solver. The leaf optimization is good for the MIP solver, but simply slows down the SAT solver. We believe this is because it complicates the MiniSAT+ translation of the objective function to clauses. Overall the optimizations improve speed by around 2-5×.

They allow 6 more instances to find optimal solutions for minimizing crossing, 5 for maximal planar subgraph, 19 for crossing minimization then maximal planar subgraph, and 9 for maximal planar subgraph then crossing minimization.

5 Conclusion

This paper demonstrates that maximizing clarity of heirarchical network diagrams by edge crossing minimization or maximal planar subgraph or their combination can be solved optimally for reasonable sized graphs using modern SAT and MIP software. Using this generic solving technology allows us to experiment with other notions of clarity combining or modifying these notions. It also gives us the ability to accurately measure the effectiveness of heuristic methods for solving these problems.

References

1. Di Battista, G., Eades, P., Tamassia, R., Tollis, I.: Graph Drawing: Algorithms for the Visualization of Graphs. Prentice Hall, Englewood Cliffs (1999)
2. Mutzel, P.: An alternative method to crossing minimization on hierarchical graphs. In: Graph Drawing, pp. 318–333 (1997)
3. Gansner, E., North, S.: An open graph visualization system and its applications to software engineering. Software: Practice and Experience 30(11), 1203–1233 (2000)
4. Sugiyama, K., Tagawa, S., Toda, M.: Methods for visual understanding of hierarchical system structures. IEEE Trans. Syst. Man Cybern. 11(2), 109–125 (1981)
5. Matuszewski, C., Schönfeld, R., Molitor, P.: Using sifting for k-layer straightline crossing minimization. In: Kratochvíl, J. (ed.) GD 1999. LNCS, vol. 1731, pp. 217–224. Springer, Heidelberg (1999)
6. Jünger, M., Mutzel, P.: 2-layer straightline crossing minimization: Performance of exact and heuristic algorithms. Journal of Graph Algorithms and Applications 1(1), 1–25 (1997)
7. Jünger, M., Lee, E.K., Mutzel, P., Odenthal, T.: A polyhedral approach to the multi-layer crossing minimization problem. In: DiBattista, G. (ed.) GD 1997. LNCS, vol. 1353, pp. 13–24. Springer, Heidelberg (1997)
8. Randerath, B., Speckenmeyer, E., Boros, E., Hammer, P., Kogan, A., Makino, K., Simeone, B., Cepek, O.: A satisfiability formulation of problems on level graphs. Electronic Notes in Discrete Mathematics 9, 269–277 (2001)
9. Healy, P., Kuusik, A.: The vertex-exchange graph: A new concept for multi-level crossing minimisation. In: Kratochvíl, J. (ed.) GD 1999. LNCS, vol. 1731, pp. 205–216. Springer, Heidelberg (1999)
10. Eén, N., Sörensson, N.: Translating pseudo-boolean constraints into SAT. Journal on Satisfiability, Boolean Modeling and Computation 2, 1–26 (2006)

On Touching Triangle Graphs

Emden R. Gansner[1], Yifan Hu[1], and Stephen G. Kobourov[2]

[1] AT&T Labs - Research, Florham Park, NJ
{erg,yifanhu}@research.att.com
[2] University of Arizona, Tucson, AZ
kobourov@cs.arizona.edu

Abstract. In this paper, we consider the problem of representing graphs by triangles whose sides touch. We present linear time algorithms for creating touching triangles representations for outerplanar graphs, square grid graphs, and hexagonal grid graphs. The class of graphs with touching triangles representations is not closed under minors, making characterization difficult. We do show that pairs of vertices can only have a small common neighborhood, and we present a complete characterization of the subclass of biconnected graphs that can be represented as triangulations of some polygon.

1 Introduction

Planar graphs are a widely studied class that includes naturally occurring subclasses such as trees and outerplanar graphs. Typically planar graphs are drawn using the node-link model, where vertices are represented by points and edges are represented by line segments. Alternative representations, such as contact circles [5] and contact triangles [8] have also been explored. In these representations, a vertex is a circle or triangle, and an edge is represented by pairwise contact at a common point.

In this paper, we explore the case where vertices are polygons, with an edge whenever the sides of two polygons touch. Specifically, given a planar graph $G = (V, E)$, we would like to find a set of polygons R such that there is bijection between V and R, and two polygons touch non-trivially if and only if the corresponding vertices are adjacent in G.

Note that, unlike the case of contact circle and contact triangle representations, two polygons that share a common point are not considered adjacent. In the sequel, we use "contact" to refer to edges touching non-trivially.

A theorem of Thomassen [20] implies that all planar graphs can be represented using convex hexagons and this also follows from results by Kant [15] and de Fraysseix *et al.* [7]. Gansner *et al.* [10] have shown that six sides are not only sufficient but also necessary, and gave a linear time construction. This leads us to consider which planar graphs can be represented by polygons with fewer than six sides.

This paper presents some initial results for the case of touching triangles. We assume we are dealing with connected planar graphs $G = (V, E)$. We let TTG denote the class of graphs that have a touching triangles representation. Concerning how to attack the problem, we can start with some simple observations. First, unlike such classes as planar graphs, TTG graphs are not closed under homeomorphisms or minors. On the

U. Brandes and S. Cornelsen (Eds.): GD 2010, LNCS 6502, pp. 250–261, 2011.
© Springer-Verlag Berlin Heidelberg 2011

other hand, as Corollary 1 shows, we can sometimes find a subclass of TTG graphs which can be extended by homeomorphism. In addition, there is the special subclass of *filled* TTG graphs, i.e., those for which the polygon formed by the union of triangles is simply-connected. At times, the filled version can be more tractable than the general version, and may lead to a solution of the general problem [6].

In Section 2, we show that all outerplanar graphs can be represented as filled TTG. Similarly, we show in Section 3 that all subgraphs of a square or hexagonal grid are in TTG. Section 4 characterizes the special case of graphs arising from filled triangulations of polygons. Finally, in Section 5, we show that, for graphs in TTG, pairs of vertices can have very limited common neighborhoods. This allows us to identify concrete examples of graphs not in TTG.

1.1   Related Work

In the limiting case, one can date results on representing planar graphs as touching polygons to Koebe's 1936 theorem [16] which states that any planar graph can be represented as a contact graph of disks in the plane. Kant's linear time algorithm for drawing degree-3 planar graphs on a hexagonal grid [15] can be used to obtain hexagonal drawings for planar graphs. Gansner *et al.* [10] show that at least six sides are necessary and that the lower bound is matched by an upper bound of six sides with a linear time algorithm for representing any planar graph by touching convex hexagons.

The problem restricting the polygons to isothetic rectangles has been extensively studied, starting with Ungar [21]. Rahman *et al.* [18] describe a linear time algorithm for constructing rectangular contact graphs, if one exists. Buchsbaum *et al.* [6] provide a characterization of the class of graphs that admit rectangular contact graph representation. The version of the problem where it is further required that the rectangles partition a rectangle is known as the rectangular dual problem. Bhasker and Sahni[4] and He [12] describe linear time algorithms for constructing a rectangular dual of a planar graph, if one exists.

In VLSI floor-planning it is often required to partition a rectangle into rectilinear regions so that non-trivial region adjacencies correspond to a given planar graph. It is natural to try to minimize the complexities of the resulting regions and the best known results are due to He [13] and Liao *et al.* [17] who show that regions need not have more than 8 sides. Both of these algorithms run in $O(n)$ time and produce layouts on an integer grid of size $O(n) \times O(n)$, where n is the number of vertices.

Rectilinear cartograms can be defined as rectilinear contact graphs for vertex weighted planar graphs, where the area of a rectilinear region must be proportional to the weight of its corresponding node. Even with this extra condition, de Berg *et al.* [2] show that rectilinear cartograms with constant region complexity can be constructed in $O(n \log n)$ time. Specifically, a rectilinear cartogram with region complexity 40 can always be found.

2   Outerplanar Graphs

In this section, we show that any outerplanar graph can be represented by a set of touching triangles, that is, outerplanar graphs belong to the class TTG. Here we assume that

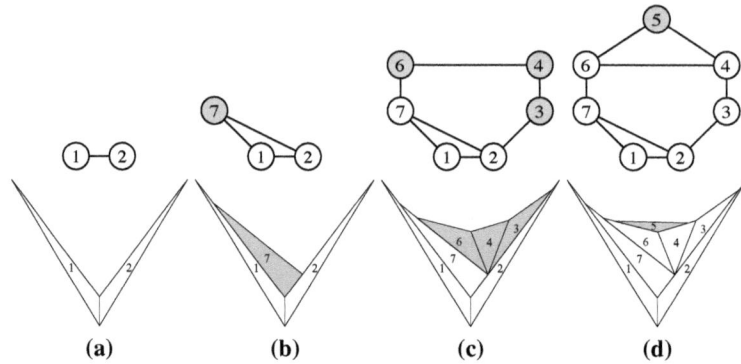

Fig. 1. Incremental construction of the TTG representation for outerplanar graphs. The shaded vertices on the top row and shaded regions on the bottom row are the ones processed at the current step.

we are given an outerplanar graph $G = (V, E)$ and the goal is to represent G as a set of touching triangles. We describe a linear time algorithm based on inserting the vertices of G is an easy-to-compute "peeling" order.

2.1   Algorithm Overview

1. Compute an outerplanar embedding of G.
2. Compute a reverse "peeling" order of chains of vertices of G.
3. Insert region(s) corresponding to the current set of vertices in the peeling order, while maintaining a concave upper envelope.

We now look at each step in more detail. First we compute an outerplanar embedding of the graph, that is, an embedding in which all the vertices are on the outer face. For a given planar graph $G = (V, E)$, this can be easily done in linear time as follows. Let w be a new vertex and let $G' = (V', E')$, where $V' = V \cup \{w\}$ and $E' = E \cup \{(v, w)$ for all $v \in V\}$. Note that G' is planar: if it contained a subgraph homeomorphic to K_5 or $K_{3,3}$, then G would contain a subgraph homeomorphic to K_4 or $K_{3,2}$, which would imply that G was not outerplanar to begin with as these are forbidden graphs for outerplanar graphs (Theorem 11.10, [11]). We can then compute a planar embedding for G' with w on the outer face. Removing w and all its edges yields the desired outerplanar embedding, with all vertices on the outer face.

The second step of the algorithm is to compute a reverse "peeling" order of the vertices of G. Such an order is defined by peeling off one face at a time and keeping track of the set of removed vertices. If G is a single edge, the result is trivial, so we may assume that $|V| > 2$. In addition, we may assume that G is biconnected. If not, we can traverse the outer face $v_1, v_2, \ldots, v_n$. If we encounter a node $v_i = v_j, 1 < j < i$, we add a new node w_i and edges (v_{i-1}, w_i) and (w_i, v_{i+1}). The graph remains outerplanar, and we continue the traversal from v_{i+1}. This yields a biconnected graph G'. If we can construct a TTG for G', we need only remove the triangles corresponding to the added vertices to get a TTG for G.

The dual of an outerplanar graph restricted to the interior faces is a tree. We pick a face with at least one edge (v_1, v_2) on the outer face and make that the root of the tree. We then remove the faces in depth-first order. At each step, a face consists of vertices $v, u_1, u_2, \ldots, u_j, w$, where $j \geq 1$ and only the edge (w, v) is part of another face. We then remove the path $u_1, u_2, \ldots, u_j$ and continue the process until we come to the root face. We then remove the path connecting v_1 and v_2.

The third step of the algorithm is to create the touching triangles representation of G, by processing the graph using the peeling order from the second step. We begin by placing the vertices v_1 and v_2 as shown in Fig. 1(a). We then recreate each face in the reverse order in which it was removed by adding triangles corresponding to the path removed from the face. We assume that, at each step, we have the following two invariants:

1. each pair of adjacent triangles corresponding to the path from v_1 to v_2 form a concave angle
2. each triangle has part of its upper side forming part of the boundary.

This is clearly true for the first step.

Suppose the path being added consists of a single vertex w connecting to adjacent vertices v_i and v_k. Let p be the point where triangles v_i and v_k meet concavely, and let q and r be any two points of the exposed upper sides of the two triangles v_i and v_k. We can then add w as the triangle p, q, r, giving us the next face and maintaining the invariants. This is illustrated in Figures 1(b) and (d).

If the path to be added consists of multiple vertices $u_1, u_2, \ldots, u_j$, with u_1 and u_j connecting to adjacent vertices v_i and v_k, respectively, we again let the points p, q, r be as defined in the previous paragraph. We then pick points $s_1, s_2, \ldots, s_{j-1}$ so that path $q, s_1, s_2, \ldots, s_{j-1}, r$ is a concave path of line segments. We can then add this face using the triangles

$$(q, p, s_1), (s_1, p, s_2), \ldots, (s_{j-2}, p, s_{j-1}), (s_{j-1}, p, r)$$

while maintaining the invariants. Figure 1(c) shows a sample of this.

Figure 1 provides an example of the algorithm. We start with the outerplanar embedding shown in the top line of Figure 1(d), and progressively remove chains until we are left with a single edge. This is used to create the configuration shown at the bottom of Fig. 1(a). The chains are added as fans of triangles until we finish with the TTG shown at the bottom right.

The first step of this algorithm can be done in linear time as it is a slight modification of a standard planar embedding algorithm such as that by Hopcroft and Tarjan [14]. The second step can also be done in linear time as computing the "peeling ordering" requires constant time per face, given the embedding of the graph from the previous step. In the third step, we record the three edges of each triangle corresponding to each processed vertex. Inserting a new chain of vertices involves finding, say, the midpoint of the exposed edges, and forming the "fan" of new triangles, all tasks which require constant time per vertex and add up to linear overall time. Thus, we have the following theorem:

Theorem 1. *A touching triangles representation can be computed in linear time for any outerplanar graph.*

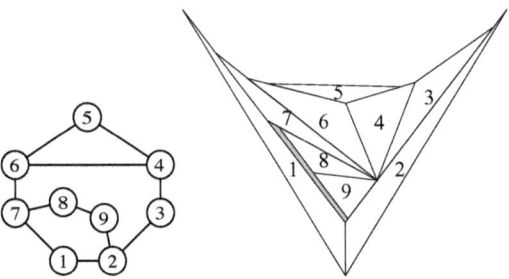

Fig. 2. Replacing a chord in an outerplanar graph with a path

Given that the above construction relies on fitting chains of triangles into smaller and smaller areas with each face, the area bounds are likely to be poor.

Corollary 1. *Any graph homeomorphic to an outerplanar graph has a touching triangles representation.*

Proof. (Sketch) Without loss of generality, we may assume that G is biconnected with an embedding such that all vertices are on the outer face except for paths of nodes connecting two nodes on the outer cycle. We then replace these interior chains by chords, yielding an outerplanar graph, and use the algorithm described above. By the construction, any chord is represented by two triangles, one of whose sides is totally within a side of the other. The shorter side can then be rotated, breaking the chord but leaving all other adjacencies intact. It is then simple to insert a fan of triangles corresponding to replacing the chord by a path of nodes. Figure 2 shows how the edge between nodes 2 and 7 in Figure 1(d) can be replaced by a chain of two nodes.

3   Grid Graphs

In this section, we show that any subgraph of a square or hexagonal grid graph is in TTG. We describe a linear time algorithm based on inserting the vertices of the graph in an outward fashion starting from an interior square/hexagon. We illustrate the algorithm with examples in Figure 3.

3.1   Algorithm Overview

We first consider TTG representations for grid graphs. We assume we have a canonical embedding of the graph.

1. Compute a "spiral" order of the vertices of G.
2. Insert region(s), corresponding to a vertex or a path of vertices in the spiral order, while maintaining a concave upper envelope in each quadrant (in the case of square grid), or by carving out triangles out of trapezoids that correspond to the current spiral segment (in the case of hexagonal grid).

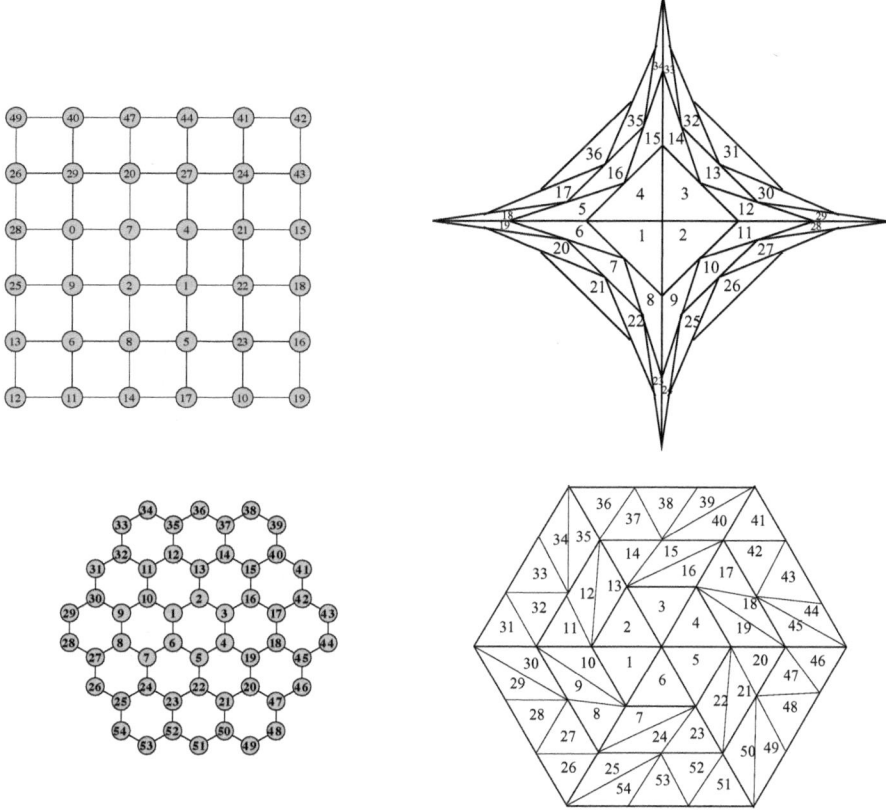

Fig. 3. Grid graphs (left) as touching triangles (right). The Hamiltonian path that visits all the vertices in the spiral order is given by the labels of the vertices.

The "spiral" order is constructed as a Hamiltonian path which starts at a center node and visits all the vertices as shown in Fig. 3.

In the case of square grids, the plane is partitioned into four quadrants and in each quadrant the spiral order introduces vertices in paths of increasing lengths $(1, 3, 5, \ldots)$. In general these paths can be introduced recursively, provided that the upper envelope of the quadrant remains concave. The insertion of regions is similar to the process described for outerplanar graphs above.

In the case of hexagonal grids the plane is partitioned into six sectors and in each sector the spiral order introduces vertices in paths of increasing lengths $(1, 3, 5, \ldots)$. In general, these paths can be introduced directly by adding an adjacent trapezoidal region and carving it into triangles.

The above algorithms show how to construct a TTG representation for any square or hexagonal grid graph. To get a TTG representation for any subgraph, given a canonical embedding, we need to add vertices and edges to create a grid graph, construct the TTG representation for that, and then remove the triangles corresponding to vertices unused in the subgraph, and adjust the remaining triangles to remove any contacts corresponding to unused edges. Note that, in degenerate cases, extending the subgraph and then removing unnecessary triangles and contacts may be quadratic in the size of the subgraph. Thus, we have the following theorem:

Theorem 2. *A touching triangles representation can be computed for any subgraph of a square or hexagonal grid graph.*

4   Triangulations

In a triangle representation, if we require that a vertex of one triangle cannot touch the interior of the side of another, we get the special case of TTGs we call *triangulation graphs*. These representations clearly correspond to creating a triangular mesh [3,1], allowing Steiner points within the interior of a polygon. For example, the representation in the bottom right of Fig. 3 is a triangulation graph and the representation in the top right of Fig. 3 is not.

It is easy to see that triangulation graphs form a strict subset of TTGs. For example, K_4 is a TTG but not a triangulation graph. It is also immediate that a triangulation graph has maximum degree 3, because by the definition of triangulation graphs, the vertex of one triangle cannot touch the side of another.

Lemma 1. *If G is a triangulation graph with no nodes of degree 1, G has at least 3 nodes of degree 2.*

Proof. The only triangles that can contribute to the polygon's boundary or outer face must have degree 2 in the graph, each contributing exactly 1 edge to the boundary. Since the polygon has at least 3 edges, the result follows.

Here we focus on the *filled triangulation graphs*, those whose TTG representation is filled. It is possible to fully characterize the biconnected subset of these graphs.

Theorem 3. *A biconnected graph G is a filled triangulation graph if and only if it has:*

1. *only nodes of degree 2 or 3*
2. *an embedding in the plane such that:*
   (a) *every internal node has degree 3;*
   (b) *there are at least 3 nodes of degree 2 on the boundary;*
   (c) *if there are any degree 3 nodes on the boundary, all of the degree 2 nodes cannot be consecutive; and*
   (d) *if the degree 2 nodes on both ends of a chain of degree 3 boundary nodes are removed, the graph remains connected.*

Proof. We first prove necessity. Let G be a filled triangulation graph. Since it is biconnected, it cannot have any vertices of degree 1. Its triangulation representation yields an embedding with all internal nodes of degree 3. Lemma 1 shows we have at least 3 nodes of degree 2 on the boundary.

Suppose there are degree 3 nodes on the boundary and the degree 2 nodes are consecutive. The chain of degree 2 nodes cannot connect at a single vertex, because this would be a cut vertex. Thus, if we remove all triangles corresponding to degree 2 nodes, we would have a triangulation representation of a graph with exactly 2 vertices of degree two, which is not allowed by Lemma 1.

To finish the proof of necessity, we note that for two degree 2 triangles to disconnect the triangulation, they would have to share an interior vertex. On the other hand, if all intervening triangles on the boundary have degree 3, they can contribute nothing to the polygon boundary, so the two degree 2 must share another vertex. But then, they share a side, so there can't be any intervening degree 3 triangles.

Next, we prove sufficiency. We assume G is biconnected, all of its vertices have degree 2 or 3, and it has the specified embedding. We construct a graph G' which is a special kind of dual of G. G' contains the dual of the interior faces and edges of G. In addition, G' has a vertex for each maximal sequence of degree 3 nodes on the boundary, and a vertex for each boundary edge connecting two degree 2 nodes. These are placed in the external face of G, near the corresponding nodes or edges. These vertices are connected in a cycle of G' following the ordering induced by the boundary nodes and edges of G. Finally, for each boundary edge e of G, we add an edge from the node of G' corresponding to the interior face of G containing e to one of the vertices on the external cycle of G'. If e is adjacent to a vertex of degree 3, we connect the edge to the node of G' corresponding to the degree 3 vertex. Otherwise, we connect to the node of G' corresponding to e.

It is immediate from the construction that G' is a planar embedding of nodes and edges; all interior faces are triangles; and there is a 1-1 correspondence between faces of G' and vertices of G and between edges in G and G'. We need to show that G' is a simple graph.

As G is biconnected, G' can have no loops. Property 2(d) of the embedding implies that each interior face is connected to at most one of the nodes associated with the exterior face. The only way that multiedges could then occur would be if G' has a boundary consisting of two nodes and two edges. We know G has as least $n_2 \geq 3$ nodes of degree 2 on the boundary. If there are only degree 2 nodes on the boundary, G' has a boundary of n_2 nodes. Assume G has some degree 3 nodes on the boundary. If these nodes split into 3 or more paths, the construction creates at least 3 nodes on the boundary of G'. If not, they must split into 2 paths, since the degree 2 nodes must be separated. One group of degree 2 nodes must contain at least 2 nodes. The construction then creates one node for each group of degree 3 nodes, and at least one node for the path of more than 2 degree nodes, again given G' at least 3 boundary nodes.

As G' is simple, by using one of the algorithms (e.g, [9]) for making the edges of planar graph into line segments while retaining the embedding, we derive a triangulation representation of G, completing the proof.

Perhaps not surprisingly, the conditions of the theorem have a similar feel to those for rectangular drawings [18]. It is also not hard to see that the result can probably be derived from the duality between planar, cubic, 3-connected graphs and triangulations of the plane [19], but our proof seems more straightforward. Lastly, we note that Theorem 3 gives another proof that the hexagonal grid graphs of Section 3 are in TTG.

Figure 4 illustrates the algorithm. Figure 4(a) shows a graph satisfying the conditions of the theorem. In Figure 4(b), we have added a node for each internal face, and node

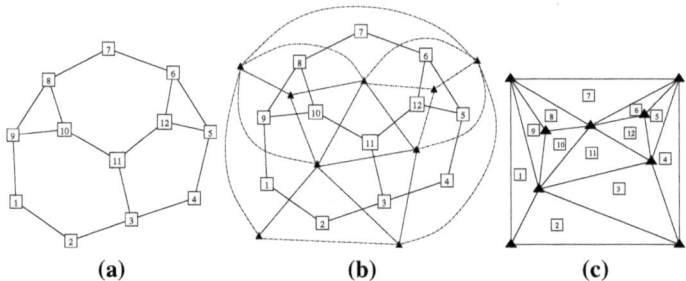

Fig. 4. Constructing a triangulation graph. **(a)** Original graph; **(b)** Creating the "dual" graph; **(c)** Straightening the edges.

on the outside for each sequence of degree 3 nodes or for each edge both of whose nodes have degree 2. This gives us a planar graph with each face having three sides and associated with a node of the original graph. Straightening the sides of the faces makes each face a triangle.

5   Necessary Conditions

Thus far, we have shown that various categories of graphs are in TTG. Now, we wish to pursue some necessary conditions which will eliminate many graphs from TTG. Specifically, we show that pairs of vertices in any graph that can be represented by touching triangles must have a small common neighborhood: if a pair of vertices is connected by an edge, they cannot have more than 3 common neighbors, and if they are not directly connected, they cannot have more than 4 common neighbors. We start with some definitions.

Given triangles T_0 and T_1, pick two sides s_0 and s_1, one from each triangle, and orient the side counter-clockwise around the interior of the triangle. Extend the sides into directed lines L_0 and L_1. If the lines intersect at a unique point, the intersection is *feasible* if a non-trivial portion of s_0 lies to the right of L_1 and a non-trivial portion of s_1 lies to the right of L_0. Considering the four rays induced by the intersection, only one of the four angles corresponds to a ray pointing into the intersection followed by right turn to a ray point out. We call this a *feasible angle*. Two sides are *collinear* if the directed lines L_0 and L_1 are identical.

Lemma 2. *If a triangle T touches both T_0 and T_1, using two distinct sides, one of its angles must be a feasible angle of T0 and T1.*

Proof. If α is the angle of T determined by the two touching sides of T_0 and T_1, it immediate that α is a feasible angle. See Figure 5.

This lemma already greatly reduces the possible TTG graphs. If two triangles have no collinear sides, there can be at most nine triangles touching both of them, since any such triangle uses at least one of the feasible angles. If two sides are collinear, one triangle can touch those two sides. Any other triangles must correspond to feasible angles, and

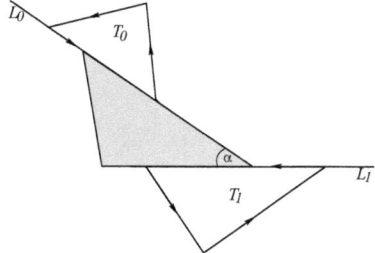

Fig. 5. A triangle T touching two other triangles T_0 and T_1. The angle α is a feasible angle of T_0 and T_1.

since the remaining sides of both triangles are all to the left of the two collinear sides, there can be at most 4 feasible angles. We next work at tightening these bounds.

For a node u in G, we let N_u be the nodes in G joined to u by an edge. If u and v are two nodes in a graph G, define N_{uv} as the mutual neighbors of u and v, that is, $N_{uv} = N_u \cap N_v$. Finally, define E_{uv} be the subset of edges of G induced by N_{uv}.

Theorem 4. *Let G be a TTG, and let u and v be two nodes in G joined by an edge. Then $|N_{uv}| \leq 3$ and $|E_{uv}| \leq 1$.*

Proof. Let T_u and T_v be the two triangles corresponding to nodes u and v. Since the two nodes share an edge, T_u and T_v must touch. There are basically two possibilities: one side is totally contained in the other or not.

In the first case, we have the situation represented in Figure 6. We immediately note that there can be no feasible angle associated with **12** and **ab**. In addition, **ab** is to the left of both **23** and **31**. On the other hand, there are feasible angles formed by **12** with **bc** and **ca**. So, we only have to consider pairings of **23** and **31** with **bc** and **ca**.

If point c is placed in region II, both **bc** and **ca** are to the left of **23** and **31**, so there are no more feasible angles, giving a total of two.

If c is in region III, we get a new feasible angle formed by **31** and **bc**. In this case, though, we are left with **bc** and **ca** to the left of **23**, and **31** to the left of **ca**. Thus, we have at most three feasible points. We also note that any triangle associated with the feasible angle formed by **12** and **ca** cannot share an edge with any triangle of the other two feasible angles, so there can be at most one edge among the neighbors of u and v. The argument is similar if c is in region I.

If points 1 and b are identical, the same arguments hold except, in addition, we no longer have a feasible angle formed by **12** and **bc** because **12** is to the left of **bc**. Thus, we have at most two mutual neighbors and no edge between them. If points 2 and a are the same, the same arguments hold. Putting these two cases together, we find that if 1 and b are identical and 2 and a are identical, there can be at most one feasible angle.

The remaining case occurs when neither shared side is contained in the other. This is the situation represented by Figure 7.

As previously, there can be no feasible angle associated with **12** and **ab**, but now we have feasible angles formed by **12** and **ca**, and by **31** and **ab**. In addition, **12** is to the left of **bc** and **ab** is to the left of **23**. Again, we are reduced to considering the

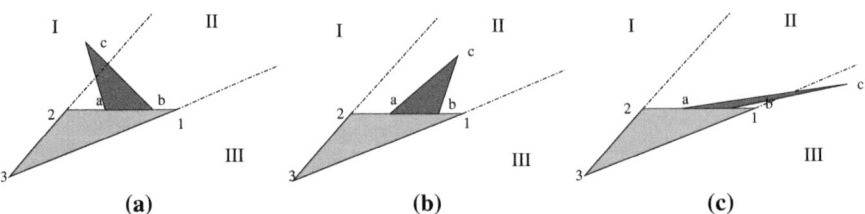

Fig. 6. Touching triangles with one side contained in the other **(a)** Node c in region I **(b)** Node c in region II **(c)** Node c in region III

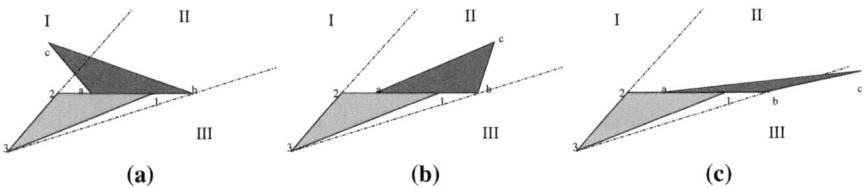

Fig. 7. Touching triangles with touching sides overlapping **(a)** Node c in region I **(b)** Node c in region II **(c)** Node c in region III

four pairings of **23** and **31** with bc and ca. If ca is to the right of **31**, then **31** is to the left of ca, and vice versa, so that pairing is not possible. Finally, we note that if c is in regions I or II, then **23** and **31** are to the left of bc, while if c is in regions II or III, bc and ca are to the left of **23**. So, if c is in region II, there are at most two feasible angles. Otherwise, there can be three but, as above, at most two of the associated triangles can touch.

We next consider what happens to the set of common neighbors if we relax the condition that there is an edge between two nodes.

Theorem 5. *Let G be a TTG, and let u and v be any two nodes in G. Then $|N_{uv}| \leq 4$ and $|E_{uv}| \leq 2$.*

Proof. The proof is similar to that of Theorem 4, and is omitted for lack of space.

6 Conclusion and Future Work

We considered the class of graphs TTG that can be represented as side-touching triangles, and showed that this includes outerplanar graphs, as well as subgraphs of square and hexagonal grids. We derived some necessary conditions for such TTG graphs, and described a complete characterization of biconnected triangulation graphs.

A complete characterization of graphs in TTG, as well as contact graphs of 4-gons and 5-gons, remains open. This is true even for the typically simpler case of filled graphs. Theorem 3 does solve the filled problem for a restricted version of TTG graphs. Can this be extended to allow for holes? Finally, the work on hexagonal contact graphs [10] gives a small ($|V| \times |V|$) area bound. Are small areas possible in the triangular or, at least, the outerplanar case?

Acknowledgments

We thank Stephen North and Stephen Wismath for useful discussions about the problem, and the referees for valuable suggestions and references.

References

1. de Berg, M., van Kreveld, M., Overmars, M.H., Schwarzkopf, O.: Computational Geometry: Algorithms and Applications, 2nd edn. Springer, Heidelberg (2000)
2. de Berg, M., Mumford, E., Speckmann, B.: On rectilinear duals for vertex-weighted plane graphs. Discrete Mathematics 309(7), 1794–1812 (2009)
3. Bern, M.: Triangulations. In: Goodman, J.E., O'Rourke, J. (eds.) Handbook of Discrete and Computational Geometry. CRC Press, Boca Raton (1997)
4. Bhasker, J., Sahni, S.: A linear algorithm to find a rectangular dual of a planar triangulated graph. Algorithmica 3, 247–278 (1988)
5. Brightwell, G.R., Scheinerman, E.R.: Representations of planar graphs. SIAM Journal on Discrete Mathematics 6(2), 214–229 (1993)
6. Buchsbaum, A.L., Gansner, E.R., Procopiuc, C.M., Venkatasubramanian, S.: Rectangular layouts and contact graphs. ACM Transactions on Algorithms 4(1) (2008)
7. de Fraysseix, H., de Mendez, P.O., Rosenstiehl, P.: On triangle contact graphs. Combinatorics, Probability and Computing 3, 233–246 (1994)
8. de Fraysseix, H., de Mendez, P.O., Rosenstiehl, P.: Representation of planar hypergraphs by contacts of triangles. In: Hong, S.-H., Nishizeki, T., Quan, W. (eds.) GD 2007. LNCS, vol. 4875, pp. 125–136. Springer, Heidelberg (2008)
9. de Fraysseix, H., Pach, J., Pollack, R.: Small sets supporting Fary embeddings of planar graphs. In: 20th Symposium on Theory of Computing (STOC), pp. 426–433 (1988)
10. Gansner, E., Hu, Y., Kaufmann, M., Kobourov, S.: Optimal polygonal representation of planar graphs. In: 9th LATIN Sympoisum, pp. 417–432 (2010)
11. Harary, F.: Graph Theory. Addison-Wesley, Reading (1972)
12. He, X.: On finding the rectangular duals of planar triangular graphs. SIAM Journal of Computing 22(6), 1218–1226 (1993)
13. He, X.: On floor-plan of plane graphs. SIAM Journal of Computing 28(6), 2150–2167 (1999)
14. Hopcroft, J., Tarjan, R.E.: Efficient planarity testing. Journal of the ACM 21(4), 549–568 (1974)
15. Kant, G.: Hexagonal grid drawings. In: 18th Workshop on Graph-Theoretic Concepts in Computer Science, pp. 263–276 (1992)
16. Koebe, P.: Kontaktprobleme der konformen Abbildung. Berichte über die Verhandlungen der Sächsischen Akademie der Wissenschaften zu Leipzig. Math.-Phys. Klasse 88, 141–164 (1936)
17. Liao, C.C., Lu, H.I., Yen, H.C.: Compact floor-planning via orderly spanning trees. Journal of Algorithms 48, 441–451 (2003)
18. Rahman, M., Nishizeki, T., Ghosh, S.: Rectangular drawings of planar graphs. Journal of Algorithms 50(1), 62–78 (2004)
19. Steinitz, E., Rademacher, H.: Vorlesungen über die Theorie der Polyeder. Springer, Berlin (1934)
20. Thomassen, C.: Plane representations of graphs. In: Bondy, J.A., Murty, U.S.R. (eds.) Progress in Graph Theory, pp. 43–69. Academic Press, Canada (1984)
21. Ungar, P.: On diagrams representing maps. Journal of the London Mathematical Society 28, 336–342 (1953)

Triangle Contact Representations and Duality[*]

Daniel Gonçalves, Benjamin Lévêque, and Alexandre Pinlou

LIRMM, CNRS & Univ. Montpellier 2
161 rue Ada 34392 Montpellier Cedex 5

Abstract. A contact representation by triangles of a graph is a set of triangles in the plane such that two triangles intersect on at most one point, each triangle represents a vertex of the graph and two triangles intersects if and only if their corresponding vertices are adjacent. de Fraysseix, Ossona de Mendez and Rosenstiehl proved that every planar graph admits a contact representation by triangles. We strengthen this in terms of a simultaneous contact representation by triangles of a planar map and of its dual.

A primal-dual contact representation by triangles of a planar map is a contact representation by triangles of the primal and a contact representation by triangles of the dual such that for every edge uv, bordering faces f and g, the intersection between the triangles corresponding to u and v is the same point as the intersection between the triangles corresponding to f and g. We prove that every 3-connected planar map admits a primal-dual contact representation by triangles. Moreover, the interiors of the triangles form a tiling of the triangle corresponding to the outer face and each contact point is a node of exactly three triangles. Then we show that these representations are in one-to-one correspondence with generalized Schnyder woods defined by Felsner for 3-connected planar maps.

1 Introduction

A *contact system* is a set of curves (closed or not) in the plane such that two curves cannot cross but may intersect tangentially. A *contact point* of a contact system is a point that is in the intersection of at least two curves. A *contact representation* of a graph $G = (V, E)$ is a contact system $\mathcal{C} = \{c(v) : v \in V\}$, such that two curves intersect if and only if their corresponding vertices are adjacent.

The Circle Packing Theorem of Koebe [14] states that every planar graph admits a contact representation by circles.

Theorem 1 (Koebe [14]). *Every planar graph admits a contact representation by circles.*

Theorem 1 implies that every planar graph has a contact representation by convex polygons, and de Fraysseix et al. [8] strengthened this by showing that every planar graph admits a contact representation by triangles. A contact representation by triangles is *strict* if each contact point is a node of exactly one triangle. de Fraysseix et al. [8] proved the following:

---
[*] This work was partially supported by the grant ANR-09-JCJC-0041.

U. Brandes and S. Cornelsen (Eds.): GD 2010, LNCS 6502, pp. 262–273, 2011.
© Springer-Verlag Berlin Heidelberg 2011

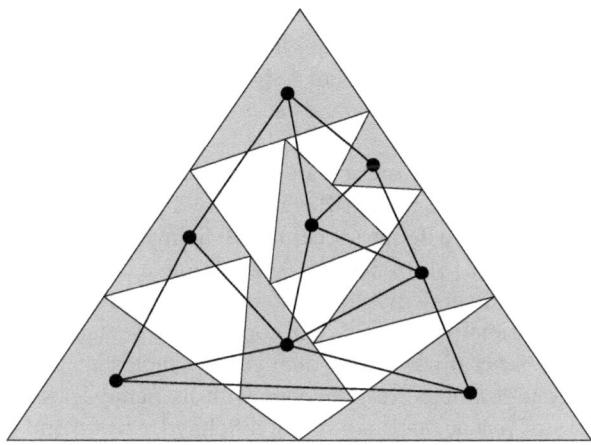

Fig. 1. A strict tiling primal-dual contact representation by triangles

Theorem 2 (de Fraysseix et al. [8]). *Every planar graph admits a strict contact representation by triangles.*

Moreover, de Fraysseix et al. [8] proved that strict contact representations by triangles of a planar triangulation are in one-to-one correspondence with its Schnyder woods defined by Schnyder [17].

Andre'ev [1] strengthen Theorem 1 in terms of a simultaneous contact representation of a planar map and of its dual. The *dual* of a planar map $G = (V, E)$ is noted $G^* = (V^*, E^*)$. A *primal-dual contact representation* $(\mathcal{V}, \mathcal{F})$ of a planar map G is two contact systems $\mathcal{V} = \{c(v) : v \in V\}$ and $\mathcal{F} = \{c(f) : f \in V^*\}$, such that $\mathcal{V}$ is a contact representation of G, and $\mathcal{F}$ is a contact representation of G^*, and for every edge uv, bordering faces f and g, the intersection between $c(u)$ and $c(v)$ is the same point as the intersection between $c(f)$ and $c(g)$. A *contact point* of a primal-dual contact representation is a contact point of V or a contact point of F. Andre'ev [1] proved the following:

Theorem 3 (Andre'ev [1]). *Every 3-connected planar map admits a primal-dual contact representation by circles.*

Our main result is an analogous strengthening of Theorem 2. We say that a primal-dual contact representation by triangles is *tiling* if the triangles corresponding to vertices and those corresponding to bounded faces form a tiling of the triangle corresponding to the outer face (see Figure 1). We say that a primal-dual contact representation by triangles is *strict* if each contact point is a node of exactly three triangles corresponding to vertices or faces (see Figure 1). We prove the following :

Theorem 4. *Every 3-connected planar map admits a strict tiling primal-dual contact representation by triangles.*

In [12], Gansner et al. study representation of graphs by triangles where two vertices are adjacent if and only if their corresponding triangles are intersecting on a side (touching representation by triangles). Theorem 4 shows that for 3-connected planar graphs, the incidence graph between vertices and faces admits a touching representation by triangles.

The tools needed to prove Theorem 4 are introduced in section 2. In section 2.1, we present a result of de Fraysseix et al. [10] concerning the stretchability of a contact system of arcs. In section 2.2, we define (generalized) Schnyder woods and present related results obtained by Felsner [4]. In Section 3, we define a contact system of arc, based on a Schnyder wood, and show that this system of arc is stretchable. When stretched, this system gives the strict tiling primal-dual contact representation by triangles. In Section 4, we show that strict tiling primal-dual contact representations by triangles of a planar map are in one-to-one correspondence with its Schnyder woods. In Section 5, we define the class of planar maps admitting a Schnyder wood and thus a strict tiling primal-dual contact representation by triangles. In Section 6, we discuss possible improvements of Theorem 4.

2 Tools

2.1 Stretchability

An *arc* is a non-closed curve. An *internal point* of an arc is a point of the arc distinct from its extremities. A contact system of arcs is *strict* if each contact points is internal to at most one arc. A contact system of arcs is *stretchable* if there exists a homeomorphism which transforms it into a contact system whose arcs are straight line segments. An *extremal point* of a contact system of arcs is a point on the outer-boundary of the system and which is internal to no arc.

We define in Section 3 a contact system of arcs such that when stretched it gives a strict tiling primal-dual contact representation by triangles. To prove that our contact system of arcs is stretchable, we need the following theorem of de Fraysseix et al. [10].

Theorem 5 (de Fraysseix et al. [10]). *A strict contact system of arcs is stretchable if and only if each subsystem of cardinality at least two has at least three extremal points.*

2.2 Schnyder Woods

The contact system of arcs defined in Section 3 is constructed from a Schnyder wood.

Schnyder woods where introduced by Schnyder [17] and then generalized by Felsner [4]. Here we use the definition from [4] except if explicitly mentioned. We refer to *classic Schnyder woods* defined by Schnyder [17] or *generalized Schnyder woods* defined by Felsner [4] when there is a discussion comparing both.

Given a planar map G. Let x_0, x_1, x_2 be three distinct vertices occurring in clockwise order on the outer face of G. The *suspension* G^σ is obtained by attaching a half-edge that reaches into the outer face to each of these special vertices. A *Schnyder wood* rooted at x_0, x_1, x_2 is an orientation and coloring of the edges of G^σ with the colors 0, 1, 2 satisfying the following rules (see Figures 2 and 3):

- Every edge e is oriented in one direction or in two opposite directions. We will respectively say that e is uni- or bi-directed. The directions of edges are colored such that if e is bi-directed the two directions have distinct colors.
- The half-edge at x_i is directed outwards and colored i.
- Every vertex v has out-degree one in each color. The edges $e_0(v)$, $e_1(v)$, $e_2(v)$ leaving v in colors 0, 1, 2, respectively, occur in clockwise order. Each edge entering v in color i enters v in the clockwise sector from $e_{i+1}(v)$ to $e_{i-1}(v)$ (where $i+1$ and $i-1$ are understood modulo 3).
- There is no interior face the boundary of which is a directed monochromatic cycle.

The difference with the original definition of Schnyder [17] it that edges can be oriented in two opposite directions.

A Schnyder wood of G^σ defines a labelling of the angles of G^σ where every angle in the clockwise sector from $e_{i+1}(v)$ to $e_{i-1}(v)$ is labeled i.

A *Schnyder angle labellings* of G^σ is a labeling of the angles of G^σ with the labels 0, 1, 2 satisfying the following rules (see Figures 2 and 3):

- The two angles at the half-edge of the special vertex x_i have labels $i+1$ and $i-1$ in clockwise order.
- Rule of vertices: The labels of the angles at each vertex form, in clockwise order, a nonempty interval of 0's, a nonempty interval of 1's and a nonempty interval of 2's.
- Rule of faces: The labels of the angles at each interior face form, in clockwise order, a nonempty interval of 0's, a nonempty interval of 1's and a nonempty interval of 2's. At the outer face the same is true in counterclockwise order.

Felsner [5] proved the following correspondence:

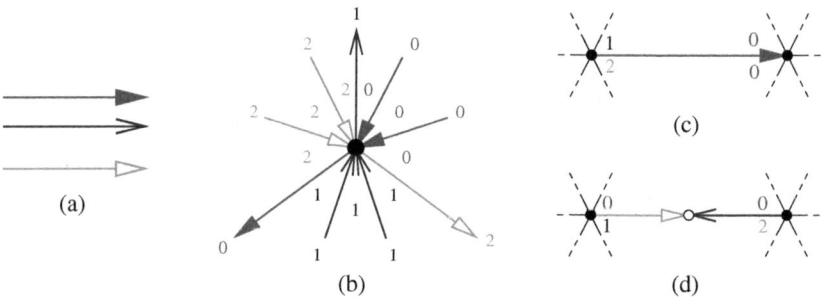

Fig. 2. (a) Edge colored respectively with color 0, 1, and 2. We use distinct arrow types to distinguish those colors. (b) Rules for Schnyder woods and angle labellings. (c) Example of angle labelling around an uni-directed egde colored with color 0. (d) Example of angle labelling around a bi-directed edge colored with colors 2 and 1.

Theorem 6 (Felsner [5]). *Schnyder woods of G^σ are in one-to-one correspondence with Schnyder angle labellings.*

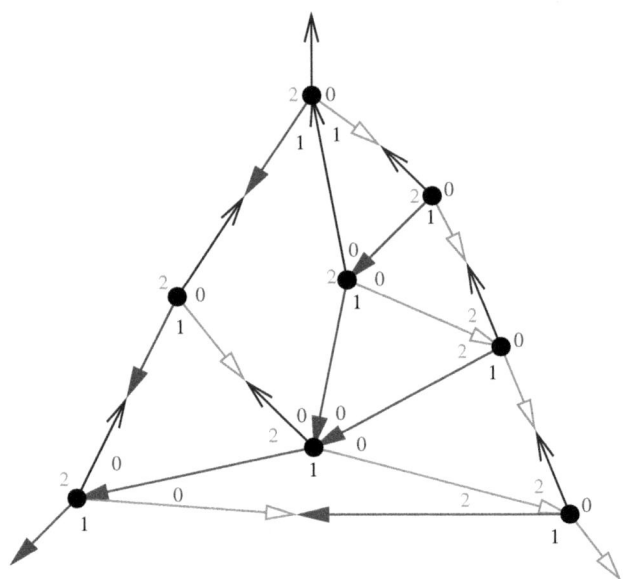

Fig. 3. A Schnyder wood with its corresponding angle labeling

3   Mixing Tools

Given a planar map G and a Schnyder wood of G rooted at x_0, x_1, x_2 we construct a contact system of arcs $\mathcal{A}$ *corresponding* to the Schnyder wood by the following method (see Figure 4):

Each vertex v is represented by three arcs $a_0(v), a_1(v), a_2(v)$, where the arc $a_i(v)$ is colored i and represent the interval of angles labeled i of v. It may be the case that $a_i(u) = a_i(v)$ for some values of i, u and v. For every edge e of G, we choose a point on its interior that we note $p(e)$. There is also such a point on the half-edge leaving x_i, for $i \in \{0, 1, 2\}$. The points $p(e)$ are the contact points of the contact system of arcs.

Actually the arcs of $\mathcal{A}$ are completely defined by the following subarcs : For each angle labeled i at a vertex v in-between the edges e and e', there is a subarc of $a_i(v)$ going from $p(e)$ to $p(e')$ along e and e'. Each contact point $p(e)$ is the end of 4 such subarcs. The Schnyder labelling implies that the three colors are represented at $p(e)$ and so the two subarcs with the same color are merged and form a longer arc.

One can easily see that this defines a contact system (there is no crossing arcs) of arcs (there is no closed curve) whose contact points are the points $p(e)$. It is also clear that the arcs satisfy the following rules:

- For every edge $e = vw$ uni-directed from v to w in color i: The arcs $a_{i+1}(v)$ and $a_{i-1}(v)$ end at $p(e)$ and the arc $a_i(w)$ goes through $p(e)$.
- For every edge $e = vw$ bi-directed, leaving v in color i and leaving w in color j: Let k be such that $\{i, j, k\} = \{0, 1, 2\}$. The arcs $a_j(v)$ and $a_i(w)$ ends at $p(e)$, and the arcs $a_k(v)$ and $a_k(w)$ are equal and go through $p(e)$.

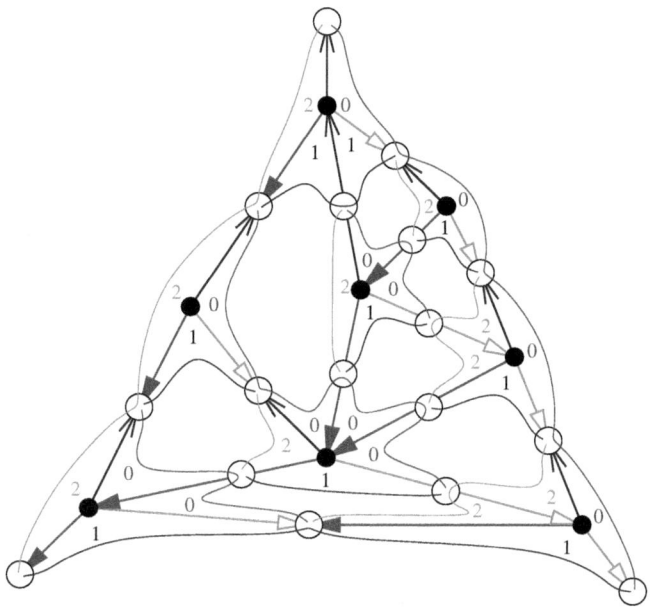

Fig. 4. A Schnyder wood with its corresponding angle labeling and contact system of arcs

The following lemma will be used to transform the contact system of arcs into a strict tiling primal-dual contact representation by triangles.

Lemma 1. *The contact system of arcs corresponding to a Schnyder wood is stretchable.*

Proof. Let G be a planar map, given with a Schnyder wood rooted at x_0, x_1, x_2. Let $\mathcal{A}$ be the contact system of arcs corresponding to the Schnyder wood as defined before. By definition of $\mathcal{A}$, every point $p(e)$, corresponding to an edge e uni- or bi-directed, is interior to one arc and is the end of two other arcs, so the contact system of arcs $\mathcal{A}$ is strict. By Theorem 5, we have to prove that each subsystem of $\mathcal{A}$, of cardinality at least two, has at least three extremal points. Let $\mathcal{B}$ be a subsystem of arcs of cardinality at least two. We have to prove that $\mathcal{B}$ has at least three extremal points.

The rest of this technical proof is omitted due to lack of space.

4   One-to-One Correspondence

De Fraysseix et al. [8] already proved that strict contact representations by triangles of a planar triangulation are in one-to-one correspondence with its Schnyder woods defined by Schnyder [17]. In this section, we are going to prove a similar result for primal-dual contact representations.

De Fraysseix et al. [9] proved that classic Schnyder woods of a planar triangulation are in one-to-one correspondence with orientation of the edges of the graph where each interior vertex has out-degree 3. This shows that it is possible to retrieve the coloring of the edges of a classic Schnyder wood from the orientation of all the edges of this Schnyder wood.

For generalized Schnyder woods (with some edges bi-directed) such a property is not true: it is not always possible to retrieve the coloring of the edges of a generalized Schnyder wood from the orientation of the edges (see for example the graph of Figure 8 in [6]). But Felsner proved that a Schnyder wood of a planar map uniquely defines a Schnyder wood of the dual and when both the orientation of the edges of the primal and the dual are given, then the coloring of the Schnyder wood can be retrieved. We will use this to obtain the one-to-one correspondence with strict tiling primal-dual contact representations by triangles. To this purpose, we need to introduce some formalism from [6].

The *suspension dual* $G^{\sigma*}$ is obtained from the dual G^* by the following: The dual-vertex corresponding to the unbounded face is replaced by a triangle with vertices y_0, y_1, y_2. More precisely, let X_i be the set of edges on the boundary of the outer face of G between vertices x_j and x_k, with $\{i, j, k\} = \{0, 1, 2\}$. Let Y_i be the set of dual edges to the edges in X_i, i.e. $Y_0 \cup Y_1 \cup Y_2$ is the set of edges containing the vertex f_∞ of G^* which corresponds to the unbounded face of G. Exchange f_∞ by y_i at all the edges of Y_i, add three edges $y_0 y_1$, $y_1 y_2$, $y_2 y_0$, and finally add a half-edge at each y_i inside the face $y_0 y_1 y_2$. The resulting graph is the suspension dual $G^{\sigma*}$. Felsner [5,6] proved that Schnyder woods of G^σ are in one-to-one correspondence with Schnyder woods of $G^{\sigma*}$.

The *completion* of a plane suspension G^σ and its dual $G^{\sigma*}$ is obtain by the following: Superimpose G^σ and $G^{\sigma*}$ so that exactly the primal dual pairs of edges cross (the half-edge at x_i cross the dual edge $y_j y_k$, for $\{i, j, k\} = \{0, 1, 2\}$). The common subdivision of each crossing pair of edges is a new edge-vertex. Add a new vertex v_∞ which is the second endpoint of the six half-edges reaching into the unbounded face. The resulting graph is the completion $\widetilde{G^\sigma}$.

A *s-orientation* of $\widetilde{G^\sigma}$ is an orientation of the edges of $\widetilde{G^\sigma}$ satisfying the following out-degrees :

- $d^+(v) = 3$ for all primal- and dual-vertices v
- $d^+(e) = 1$ for all edge-vertices e.
- $d^+(v_\infty) = 0$ for the special vertex v_∞.

Felsner [6] proved the following:

Theorem 7 (Felsner [6]). *Schnyder woods of G^σ are in one-to-one correspondence with s-orientations of $\widetilde{G^\sigma}$.*

We are now able to prove the following correspondence:

Theorem 8. *The non-isomorphic strict tiling primal-dual contact representations by triangles of a planar map are in one-to-one correspondence with its Schnyder woods.*

Proof. Given a strict tiling primal-dual contact representation by triangles $(\mathcal{V}, \mathcal{F})$ of a graph G, one can associate a corresponding suspension G^σ, its suspension dual $G^{\sigma*}$, the completion $\widetilde{G^\sigma}$ and a s-orientation of the completion. The three vertices x_0, x_1, x_2 that define the suspension G^σ are, in clockwise order, the three triangles of $\mathcal{V}$ that share a node with the triangle corresponding to the outer face. We modify our contact system by exchanging the triangle $c(f_\infty)$, representing the outer face f_∞, by three triangles $c(y_0), c(y_1), c(y_2)$ each one representing y_0, y_1, y_2 of the suspension dual. Each $c(y_i)$ share a side with $c(f_\infty)$ and two $c(y_i)$ have parallel and intersecting sides. The interiors of the triangles of this new system still form a tiling of a triangle $c(v_\infty)$ representing the vertex v_∞ of the completion. The edge-vertices of the completion corresponds to the nodes of the triangles of the new system.

The s-orientation of $\widetilde{G^\sigma}$ is obtained by the following. For a primal- or dual-vertex v, represented by a triangle $c(v)$, all edges ve of $\widetilde{G^\sigma}$ are directed from v to e if e corresponds to a node of $c(v)$ and from e to v otherwise. For the special vertex v_∞, all its incident edges are directed towards itself. Clearly, for every primal- or dual-vertex v, we have $d^+(v) = 3$ as $c(v)$ is a triangle and for v_∞ we have $d^+(v_\infty) = 0$. As the primal-dual contact representation $(\mathcal{V}, \mathcal{F})$ is strict, i.e. each contact point is a node of exactly three triangles, we have $d^+(e) = 1$ for every edge-vertex that is a contact point of $(\mathcal{V}, \mathcal{F})$. For edge-vertices between special vertices x_i, y_j and v_∞ one can check that the out-degree constraint is also satisfied.

One can remark that two non-isomorphic triangle contact systems representing the same planar map G define two distinct orientations of $\widetilde{G^\sigma}$ and thus two different Schnyder woods of G^σ by Theorem 7.

Conversely, let G be a planar map, given with a Schnyder wood rooted at x_0, x_1, x_2 and the corresponding s-orientation of $\widetilde{G^\sigma}$. Let $\mathcal{A}$ be the contact system of arcs corresponding to the Schnyder wood as defined in Section 3. For each vertex $v \in V$, we note $c(v)$ the closed curve that is the union, for $i \in \{0, 1, 2\}$, of the part of the arc $a_i(v)$ between the contact point with $a_{i-1}(v)$ and $a_{i+1}(v)$. The set of curves $\mathcal{V} = (c(v))_{v \in V}$ is a contact representation of G by closed curves. For each interior face F, the labels of its angles form a nonempty interval of 0's, a nonempty interval of 1's and a nonempty interval of 2's by Theorem 6. By definition of the arcs, each interval of i's corresponds to only one arc, noted $a_i(f)$. We note $c(f)$ the closed curve that is the union, for $i \in \{0, 1, 2\}$, of the part of the arc $a_i(f)$ between the contact point with $a_{i-1}(f)$ and $a_{i+1}(f)$. For the outer face f_∞, the curve $c(f_\infty)$ is the union, for $i \in \{0, 1, 2\}$, of $a_{i+1}(x_i)$. The set of curves $\mathcal{F} = (c(f))_{f \in V^*}$ is a contact representation of G^* by closed curves.

By Lemma 1, the contact system of arcs $\mathcal{A}$ is stretchable. For each $v \in V \cup V^*$, the closed curves $c(v)$ is the union of three part of arcs of $\mathcal{A}$, so when stretched it becomes a triangle. Thus, we obtain a primal-dual contact representation by triangles $(\mathcal{V}, \mathcal{F})$ of G. By definition of $(\mathcal{V}, \mathcal{F})$ the interiors of the triangles form a tiling of the triangle corresponding to the outer face. Thus, the primal-dual contact representation by triangles $(\mathcal{V}, \mathcal{F})$ is tiling. By definition of $\mathcal{A}$, every contact point, corresponding to an uni- or bi-directed edge, is interior to one arc and is the extremity of two arcs. So each contact point of $(\mathcal{V}, \mathcal{F})$ is a node of exactly three triangles. Thus, the primal-dual contact representation by triangles $(\mathcal{V}, \mathcal{F})$ is strict. The strict tiling primal-dual contact representation by triangles $(\mathcal{V}, \mathcal{F})$ corresponds to the s-orientation of $\widetilde{G^\sigma}$ and thus to the Schnyder wood by Theorem 7.

5   Internally 3-Connected Planar Maps

A planar map G is *internally 3-connected* if there exists three vertices on the outer face such that the graph obtain from G by adding a vertex adjacent to the three vertices is 3-connected. Miller [16] proved the following (see also [4] for existence of Schnyder woods for 3-connected planar maps and [3] were the following result is stated in this form):

Theorem 9 (Miller [16]). *A planar map admits a Schnyder wood if and only if it is internally 3-connected.*

As a corollary of Theorems 8 and 9, we obtain the following:

Corollary 1. *A planar map admits a strict tiling primal-dual contact representation by triangles if and only if it is internally 3-connected.*

A 3-connected planar map is obviously internally 3-connected, so we obtain Theorem 4 as a consequence of Corollary 1.

6   Particular Types of Triangles

The construction given by de Fraysseix et al. [8] to obtain a strict contact representation by triangles of a planar triangulation can be slightly modified to give a strict tiling primal-dual contact representation by triangles (the three triangles corresponding to the outer face have to be modified to obtain the tiling property). In de Fraysseix et al.'s construction all the triangles have a horizontal side at their bottom and moreover it is possible to require that all the triangles are right (with the right angle on the left extremity of the horizontal side). This leads us to propose the following conjecture.

Conjecture 1. Every 3-connected planar map admits a strict tiling primal-dual contact representation by right triangles where all triangles have a horizontal and a vertical side and where the right angle is bottom-left for primal vertices and the outer face and top-right otherwise.

One may wonder if further requirements can be asked. Is it possible to obtain primal-dual contact representation by homothetic triangles ? The 4-connected planar triangulation of Figure 5 has a unique contact representation by homothetic triangles (for a fixed size of the external triangles). The central face corresponds to an empty triangle and thus this graph has no primal-dual contact representation by homothetic triangles. Moreover if one add a vertex in the central face adjacent to all the vertices of this face, then, there is no contact representation by homothetic triangles. In this case, the planar triangulation that is obtain is not 4-connected anymore. This leads Kratochvil [15] (see also [2]) to conjecture that every 4-connected planar triangulation admits a contact representation by homothetic triangles. Actually this conjecture holds by an application of the following theorem of O. Schramm [18] that is a generalization of Theorem 1 (in the sense that circle are replaced by convex bodies).

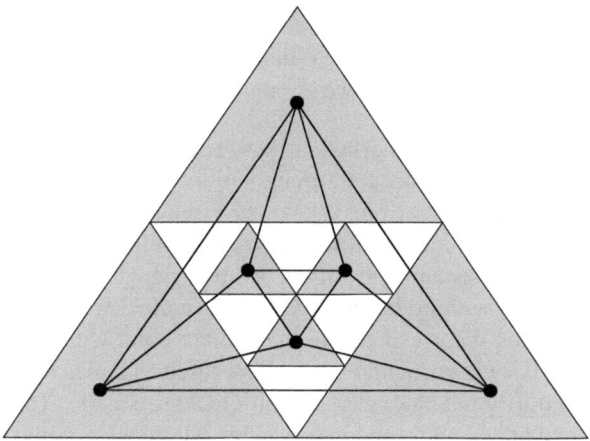

Fig. 5. A contact representation by homothetic triangles

Theorem 10 (Convex Packing Theorem). *Let T be a planar triangulation with outerface abc. Let C be a simple closed curve in the plane, and let $\mathcal{P}_a$, $\mathcal{P}_b$, $\mathcal{P}_c$ be three arcs composing C, which are determined by three distinct points of C. For each vertex $v \in V(T) \setminus \{a, b, c\}$, let there be a prototype $\mathcal{P}_v$, which is a convex set in the plane containing more than one point. Then there is a contact system in the plane $\mathcal{Q} = \{\mathcal{Q}_v : v \in V(T)\}$, where $\mathcal{Q}_a = \mathcal{P}_a$, $\mathcal{Q}_b = \mathcal{P}_b$, $\mathcal{Q}_c = \mathcal{P}_c$ and each $\mathcal{Q}_v$ (for $v \in V(T) \setminus \{a, b, c\}$ is either a point or (positively) homothetic to $\mathcal{P}_v$, and such that T is a subgraph of the graph induced by $\mathcal{Q}$.*

This theorem makes an intersecting link between Theorem 1 and Theorem 2.

Theorem 11. *Every 4-connected planar triangulation T admits a contact representation by homothetic triangles.*

Proof. Indeed, in the Convex Packing Theorem if we let the prototypes be homothetic triangles and the curves $\mathcal{P}_a$, $\mathcal{P}_b$, $\mathcal{P}_c$ be segments with appropriate slopes (in such a way that those segment can be the sides of homothetic triangles added in the outer-region), we obtain a contact system of homothetic triangles $\mathcal{Q}$, where the triangles may be reduced to a point, and that induces a graph $G \supseteq T$. Thus, to prove the theorem we just have to show that (a) none of the triangles are reduced to a point and (b) that $E(G) = E(T)$.

(a) If there was a vertex v such that its triangle $\mathcal{Q}_v$ is reduced to a point p then by taking a sufficiently small circle C around p we intersect at most three non-degenerated triangles. Since a path P from x to y in H clearly corresponds to a curve in $\cup_{z \in P} \mathcal{Q}_z$ from $\mathcal{Q}_x$ to $\mathcal{Q}_y$, the triangles intersecting C correspond to a set of vertices separating v to some $u \in \{a, b, c\}$ in G, contradicting the 4-connectedness of G and T.

(b) Since none of the triangles is degenerated, the contact points are either the intersection of two or three triangles. In those cases the contact point respectively correspond to one or three edges of H. Then, since these contact points are respectively the nodes of at least one, or exactly three triangles, and according to the position of the segments $\mathcal{P}_a$, $\mathcal{P}_b$ and $\mathcal{P}_c$, we have that $|E(G)| \leq 3 + 3(n-3) = 3n - 6 = |E(T)|$. Thus $E(G) = E(T)$ and we are done.

It is still an open question to know whether these representations by homothetic triangles are unique for a given 4-connected triangulation. These representations being not strict (three triangles can meet at one point, see Figure 5) we can not always derive a unique Schnyder wood as in [8]. However, we can define a set of Schnyder woods corresponding to the representation as follow. All the triangles of the representation are homothetic to a triangle with nodes colored $0, 1, 2$ in clockwise order. The out-going arc of color i of a vertex v corresponds to the contact point with the node i of its corresponding triangle. For the particular case where three triangles meet in one point we have to choose arbitrarily the clockwise or anti-clockwise cycle. This set of Schnyder woods can be embedded on an orthogonal drawing where edge-points are coplanar (by allowing a degenerate patterns for each point that is the intersection of three triangles, see Felsner and Zickfeld [7]). Another interesting conjecture concerning contact system of triangles is the following.

Conjecture 2. Every planar graph has a contact representation by equilateral (not necessarily homothetic) triangles.

Concerning intersection systems (not contact systems) of triangles, M. Kaufmann *et al.* [13] proved that Max-tolerance graphs are exactly those graphs that have an intersection representation by homothetic triangles. Then K. Lehmann conjectured that every planar graph has such a representation. We can derive from Theorem 11 that her conjecture holds.

Theorem 12. *A graph G is planar if and only if it has an intersection representation by homothetic triangles where no three triangles intersect.*

It is interesting to notice that this theorem implies a result of Gansner et al. [11], that planar graphs have a representation by touching hexagons. Indeed consider an intersection model of G by homothetic triangles where no three triangles intersect, and where two intersecting triangles intersect in more than one point (inflate the triangles if necessary). Then remove, for each pair of intersecting triangles $T(u)$ and $T(v)$ (where $T(u)$ has a node strictly inside $T(v)$), the triangle $T(u) \cap T(v)$ from $T(u)$. However, Gansner et al.'s construction also provides, for triangulations, a model of touching polygons (with at most six sides) that form a tilling.

References

1. Andreev, E.: On convex polyhedra in Lobachevsky spaces. In: Mat. Sbornik, Ser., vol. 81, pp. 445–478 (1970)
2. Badent, M., Binucci, C., Di Giacomo, E., Didimo, W., Felsner, S., Giordano, F., KratochvÃl, J. Palladino, P., Patrignani, M., Trotta., F.: Homothetic triangle contact representations of planar graphs. In: Proceedings of the 19th Canadian Conference on Computational Geometry CCCG 2007, pp. 233–236 (2007)

3. Bonichon, N., Felsner, S., Mosbah, M.: Convex Drawings of 3-Connected Plane Graphs. Algorithmica 47, 399–420 (2007)
4. Felsner, S.: Convex Drawings of Planar Graphs and the Order Dimension of 3-Polytopes. Order 18, 19–37 (2001)
5. Felsner, S.: Geodesic Embeddings and Planar Graphs. Order 20, 135–150 (2003)
6. Felsner, S.: Lattice structures from planar graphs. Electron. J. Combin. 11 (2004)
7. Felsner, S., Zickfeld, F.: Schnyder Woods and Orthogonal Surfaces. Discrete Comput. Geom. 40, 103–126 (2008)
8. de Fraysseix, H., Ossona de Mendez, P., Rosenstiehl, P.: On Triangle Contact Graphs. Combinatorics, Probability and Computing 3, 233–246 (1994)
9. de Fraysseix, H., Ossona de Mendez, P.: On topological aspects of orientations. Discrete Mathematics 229, 57–72 (2001)
10. de Fraysseix, H., de Mendez, P.O.: Barycentric systems and stretchability. Discrete Applied Mathematics 155, 1079–1095 (2007)
11. Gansner, E.R., Hu, Y., Kaufmann, M., Kobourov, S.G.: Optimal Polygonal Representation of Planar Graphs. In: López-Ortiz, A. (ed.) LATIN 2010. LNCS, vol. 6034, pp. 417–432. Springer, Heidelberg (2010)
12. Gansner, E.R., Hu, Y., Kobourov, S.G.: On Touching Triangle Graphs. In: Proc. Graph Drawing (2010), http://arxiv1.library.cornell.edu/abs/1001.2862v1
13. Kaufmann, M., Kratochvíl, J., Lehmann, K.A., Subramanian, A.R.: Max-tolerance graphs as intersection graphs: Cliques, cycles and recognition. In: Proc. SODA 2006, pp. 832–841 (2006)
14. Koebe, P.: Kontaktprobleme der konformen Abbildung. Berichte Äuber die Verhandlungen d. SÄachs. Akad. d. Wiss., Math.-Phys. Klasse 88, 141–164 (1936)
15. Kratochvíl, J.: Bertinoro Workshop on Graph Drawing (2007)
16. Miller, E.: Planar graphs as minimal resolutions of trivariate monomial ideals. Documenta Mathematica 7, 43–90 (2002)
17. Schnyder, W.: Planar graphs and poset dimension. Order 5, 323–343 (1989)
18. Schramm, O.: Combinatorically Prescribed Packings and Applications to Conformal and Quasiconformal Maps, Modified version of PhD thesis from (1990), http://arxiv.org/abs/0709.0710v1

On Maximum Differential Graph Coloring

Yifan Hu[1], Stephen Kobourov[2], and Sankar Veeramoni[2]

[1] AT&T Labs Research, Florham Park, NJ, USA
yifanhu@research.att.com
[2] Computer Science Dept., University of Arizona, Tucson, AZ, USA
{kobourov,sankar}@cs.arizona.edu

Abstract. We study the *maximum differential graph coloring problem*, in which the goal is to find a vertex labeling for a given undirected graph that maximizes the label difference along the edges. This problem has its origin in map coloring, where not all countries are necessarily contiguous. We define the differential chromatic number and establish the equivalence of the maximum differential coloring problem to that of k-Hamiltonian path. As computing the maximum differential coloring is NP-Complete, we describe an exact backtracking algorithm and a spectral-based heuristic. We also discuss lower bounds and upper bounds for the differential chromatic number for several classes of graphs.

1   Introduction

The Four Color Theorem states that only four colors are needed to color any map so that no neighboring countries share the same color. This theorem assumes that each country forms a contiguous region in the map. However, if countries in the map are not all contiguous then the result no longer holds [6]. Instead, this necessitates the use of a unique color for each country to avoid ambiguity. As a result, the number of colors needed is equal to the number of countries.

Given a map, we define the country graph $G = \{V, E\}$ to be the undirected graph where countries are nodes and two countries are connected by an edge if they share a nontrivial boundary. We then consider the problem of assigning colors to nodes of G so that the color distance between nodes that share an edge is maximized. Figure 1 gives an illustration of this map coloring problem.

As not all colors make suitable choices for country colors, a good color palette is often a gradation of certain "map-like colors". Furthermore, if the final result is to be printed in black and white, then the color space is strictly 1D. We therefore make the assumption that the colors form a line in the color space, and model our map coloring problem as one of node labeling of the graph, where the available labels are from set C of all permutations $\{1, 2, \cdots |V|\}\}$. This brings us to the *maximum differential graph coloring problem*, in which we aim to find a vertex labeling for a given undirected graph that maximizes the label difference along the edges in the graph.

U. Brandes and S. Cornelsen (Eds.): GD 2010, LNCS 6502, pp. 274–286, 2011.

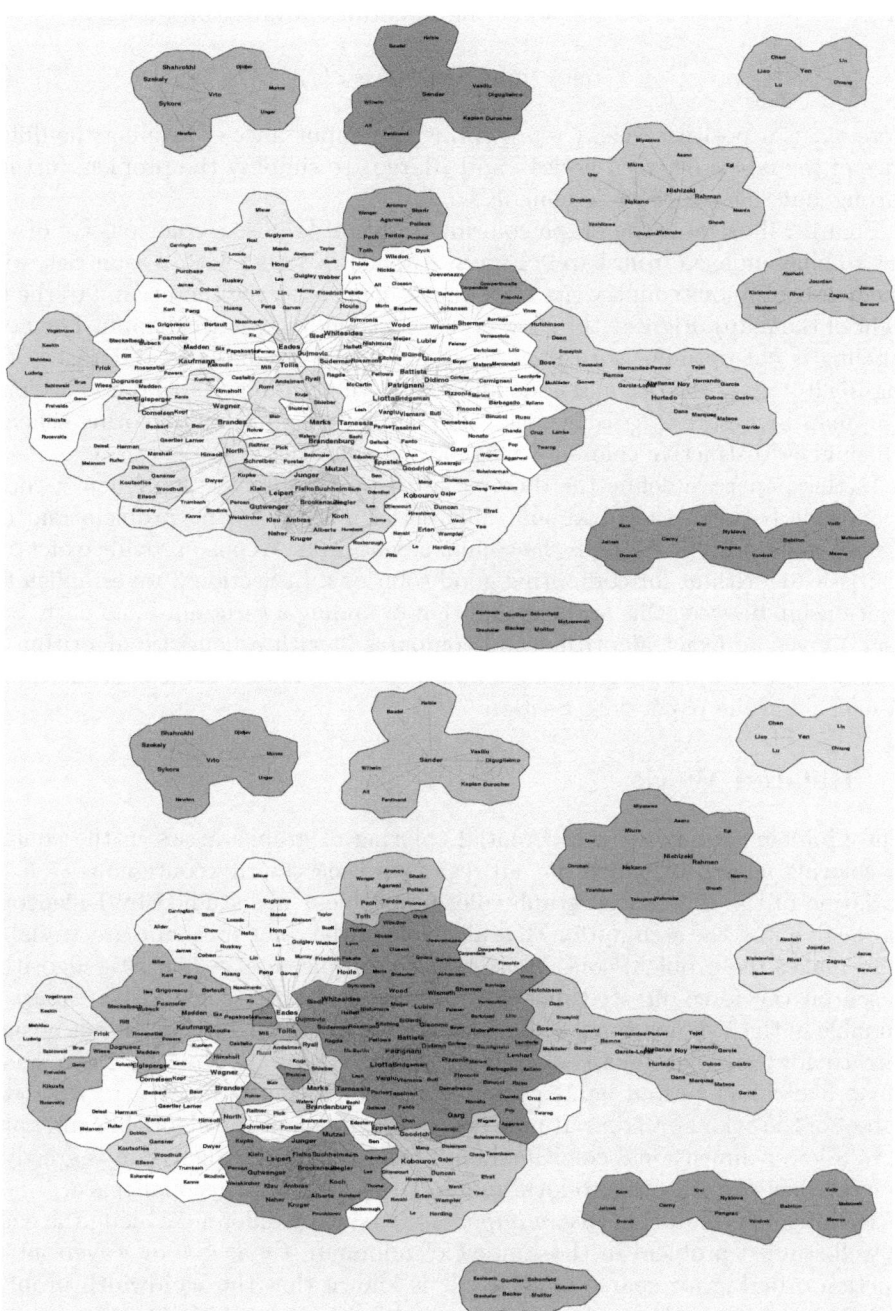

Fig. 1. Graph Drawing Symposia (1994-2004) co-authorship map with a single-hue color palette from ColorBrewer [2]: a random color assignment above and optimal below

More formally, we are looking for a labeling function, a bijection $c : V \to \{1, 2, \ldots, |V|\}$, that solves the following MaxMin optimization problem:

$$\max_{c \in C} \min_{\{i,j\} \in E} w_{ij} |c(i) - c(j)| \qquad (1)$$

Here w_{ij} is a positive weight representing the importance of keeping the difference of the labels between nodes i and j large. To simplify the problem further, throughout this paper we assume $w_{ij} = 1$.

Figure 2 illustrates the graph coloring problem. We use a color palette of yellow to blue, indexed from 1 to 9. Figure 2(a) shows a map with 9 countries, with one non-contiguous country (its two components are at the center, and to the far right of the map). Figure 2(b) shows the corresponding country graph. The node labeling is not optimal, with many adjacent nodes having label difference of 1. Figure 2(c) gives the optimal node labeling, with a minimal label difference of 3. The map in Figure 2(a) is in fact colored using this optimal coloring scheme, which gives distinctive colors for neighboring countries.

In this paper we define the differential chromatic number and show a correspondence between the maximum differential graph coloring problem and the Hamiltonian path problem in the complement graph. We also provide exact and heuristic algorithms for computing good solutions. In Section 3 we establish the relationship between this problem and that of finding a k-Hamiltonian path. Section 4 gives an exact algorithm and compares it with a heuristic algorithm on a number of well known graphs. Section 5 gives results for some special graphs. We conclude the paper with Section 6.

2   Related Work

The problem of maximum differential coloring of graphs arises in the context of coloring a map in which not all regions are necessarily contiguous [7, 6]. A variation of the differential graph coloring problem was studied by Dillencourt *et al.* [4], under the assumption that all colors in the color spectrum are available. This makes the problem continuous rather than discrete. A heuristic algorithm based on the force-directed model is used to select $|V|$ colors as far apart as possible in the 3-dimensional color space. However this algorithm cannot be used directly for the general map coloring problem, as maps are typically colored with "map-like colors", often light, pastel colors which come from a very restricted subset of the color spectrum. It may be possible to adapt the algorithm and apply it to a lower-dimensional color manifold, but because the algorithm is greedy it is more likely to converge to local minimum in lower dimensional space.

Finding a permutation that *minimizes* the labeling differences along the edges is well-studied problem in the context of minimum *bandwidth* or wavefront reduction ordering for sparse matrices. It is known that the bandwidth problem is NP-complete, with a reduction from 3SAT dating to 1975 [11]. Moreover, it is NP-complete to find any constant approximation, even when restricting the problem to trees [1, 15]. A number of effective heuristics for that problem have been proposed [8, 9, 13].

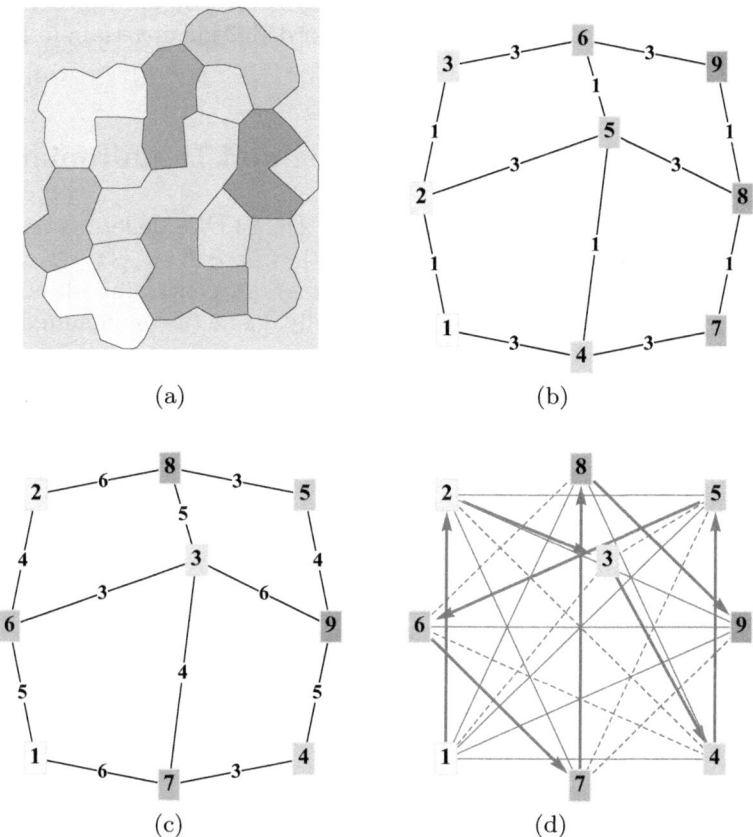

Fig. 2. An illustration of the graph coloring problem and its relation to a k-Hamiltonian path. A yellow to blue coloring scheme (indexed from 1 to 9) is used; nodes are labeled by color indices and edges are labeled with the absolute difference of of the adjacent nodes. (a) An input map. (b) The country graph G corresponding to the input map; note that many adjacent nodes have very similar colors. (c) An optimal 3-differential coloring of the country graph. (d) The complement graph $\bar{G}$ corresponding to G and a 2-Hamiltonian path. The Hamiltonian path $\{1, 2, \ldots, 9\}$ is shown with orange arrows, and edges with labeling difference of 2 are shown as blue dashed lines. Note that the 2-Hamiltonian path of $\bar{G}$ is such that any two nodes of labeling difference ≤ 2 form an edge of $\bar{G}$.

The complement of the bandwidth problem, that of *maximizing* the labeling difference along the edges, is the less well-known *antibandwidth* problem. Not surprisingly, it is also NP-Complete as shown in 1984 [10]. Exact values for the antibandwidth are known for some Hamming graphs [5], as well as for meshes [14], hypercubes [12,16], and complete k-ary trees for odd values of k [3].

There have been definitions for a graph being k-*edge Hamiltonian* and k-*vertex Hamiltonian* in the literature [17], which are defined as the graph having a Hamiltonian cycle after removing any k edges or vertices, respectively. These concepts, though related, are different from our definition of k-Hamiltonian path, introduced in the next section.

3   Maximum Differential Coloring and Hamiltonian Path

Let the undirected graph of interest be $G = \{V, E\}$. We denote $\bar{G}$ as the complement of graph G, defined as $\bar{G} = \{V, \bar{E}\}$, where $\bar{E} = \{\{i, j\} | i \neq j, i, j \in V, \text{ and } \{i, j\} \notin E\}$. In other words, $\bar{G}$ is the graph containing all nodes in G, and all edges that are not in G. We now formally define the maximum differential coloring problem.

Definition 1. *A* coloring *of the nodes of G is a bijection $c : V \to \{1, 2, \ldots, |V|\}$. We denote the set of all colorings $C(G)$.*

Definition 2. *A k-differential coloring of G is one in which the absolute coloring difference of the endpoints for any edge is k or more. We denote the set of all such k-differential colorings $DC(G, k) = \{c \mid c \in C(G), |c(i) - c(j)| \geq k \text{ for all } \{i, j\} \in E\}$.*

Definition 3. *A graph is k-differential colorable if $DC(G, k) \neq \emptyset$.*

Definition 4. *If a graph is k-differential colorable, but not $(k + 1)$-differential colorable, it has a differential chromatic number k, denoted as $dc(G) = k$.*

Definition 5. *A Hamiltonian path of G is a bijection $p : \{1, 2, \ldots, |V|\} \to V$, such that $\{p(i), p(i + 1)\} \in E$ for all $i = 1, 2, \ldots, |V| - 1$. We denote the set of all Hamiltonian paths $H(G)$.*

The key insight in understanding the maximum differential coloring comes from observing a good differential coloring scheme. For example, in Figure 2(c), the minimum labeling difference between any two adjacency nodes is 3. This means that any two nodes with labeling difference of ≤ 2 can not form an edge in G, in other word they must form an edge in the complement of the graph, shown in Figure 2(d). Therefore the list of nodes induced by this coloring scheme of G, with labels $\{1, 2, \ldots, |V|\}$, forms a Hamiltonian path in $\bar{G}$, shown in Figure 2(d) with orange arrows. Furthermore, this Hamiltonian path is such that any two nodes along the path with labeling difference of 2 are also connected by an edge of $\bar{G}$, shown in Figure 2(d) with dashed lines.

This observation leads to a natural extension of the concept of Hamiltonian path, which we call a k-Hamiltonian path.

Definition 6. *A k-Hamiltonian path $(k \geq 1)$ of G is a Hamiltonian path, such that each i-th node on the path is connected to the j-th node, if $|i - j| \leq k$. We define the set of k-Hamiltonian paths as $H(G, k) = \{p | p \in H(G), \text{ and } \{p(i), p(j)\} \in E \text{ if } 0 \leq i, j \leq |V| \text{ and } |i - j| \leq k\}$. Clearly $H(G) = H(G, 1)$.*

Based on our previous discussion, we can relate the k-differential coloring problem to that of finding a $(k-1)$-Hamiltonian path.

Theorem 1. *A k-differential coloring $(k \geq 2)$ of a graph G exists if and only if a $(k-1)$-Hamiltonian path exists in the complement of G, i.e., $DC(G, k) \neq \emptyset$ if and only if $H(\bar{G}, k-1) \neq \emptyset$. Furthermore, the inverse function of each element of $DC(G, k)$ is in $H(\bar{G}, k-1)$, and vice visa.*

Proof. Suppose a k-differential coloring c exists for G. Define the path, $p = c^{-1}$, so that it visits the vertices of the graph in order of their color index (i.e., the first vertex is that with color 1, the second is that with color 2 and so on). Consider two nodes $u = p(i)$ and $v = p(j)$ such that $|i - j| \leq k - 1$. Since the color difference of these two nodes in the original graph, $|c(u) - c(v)| = |i - j| \leq k - 1$, by the definition of k-differential coloring (u, v) is not an edge of G, hence $\{u, v\} = \{p(i), p(j)\}$ is an edge of $\bar{G}$. It follows that $p = c^{-1}$ is a $(k-1)$-Hamiltonian path.

Conversely, suppose a $(k-1)$-Hamiltonian path p exists for $\bar{G}$. Define a coloring $c = p^{-1}$, where the color index is given by the order in which a vertex appears in the path (i.e., color 1 is assigned to the first vertex along the path, color 2 to the second, and so on). Consider any edge $\{u, v\} \in E$. We prove that $|c(u) - c(v)| \geq k$. Assume that $|c(u) - c(v)| < k$. Let $i = c(u)$ and $j = c(v)$, then $|i - j| < k$, and $u = c^{-1}(i) = p(i)$ and $v = c^{-1}(j) = p(j)$. By the definition of a $(k-1)$-Hamiltonian path, (u, v) must be an edge of $\bar{G}$, which is a contradiction with the fact that $\{u, v\} \in E$. It follows that $|c(u) - c(v)| \geq k$, and so $c = p^{-1}$ is a k-differential coloring of G. □

This theorem immediately gives an upper bound for the differential chromatic number of a graph based on its maximum degree:

Corollary 1. *A graph of maximum degree $\Delta(G)$ has a differential chromatic number of at most $|V| - \Delta(G)$.*

Proof. $\bar{G}$ must have a node with degree $|V| - 1 - \Delta(G)$, therefore $H(\bar{G}, |V| - \Delta(G)) = \emptyset$, or by Theorem 1 $DC(G, |V| - \Delta(G) + 1) = \emptyset$. □

A special case of differential coloring is finding a scheme with maximum difference of 2 or more. Clearly:

Corollary 2. *Finding a 2-differential coloring of a graph G is equivalent to finding a Hamiltonian path of $\bar{G}$.*

Given the equivalence between 2-differential coloring and Hamiltonian path, a number of well known results for Hamiltonian path can immediately be used to give results on 2-differential coloring.

Theorem 2. *(Ore's Theorem) A graph with more than 2 nodes is Hamiltonian if, for each pair of non-adjacent nodes, the sum of their degrees is $|V|$ or greater.*

Corollary 3. *A graph with more than 2 nodes is 2-differential colorable if, for each pair of adjacent nodes, the sum of their degrees is $|V| - 1$ or less.*

Theorem 3. *(Dirac's Theorem). A graph with more than 2 nodes is Hamiltonian if each node has degree $|V|/2$ or greater.*

Corollary 4. *A graph with more than 2 nodes is 2-differential colorable if each node has degree $|V|/2 - 1$ or less.*

Corollaries 3-4 confirm the intuition that a sparser graph has a better chance of being more differential colorable.[1] The flip side of this intuition is that the complement graph needs to be denser. For a graph to be k-colorable, the complement graph must be well connected. The next theorem follows from a result in [12]:

Theorem 4. *The complement of a k-differential colorable graph is $(k - 1)$-connected.*

Given the connection between finding a 2-differential coloring and a Hamiltonian path, it is not difficult to prove directly that the maximum differential coloring problem is NP-Complete. An equivalent result was established in 1984 in the context of the antibandwidth problem [10].

4   Algorithms for Maximum Differential Coloring

Theorem 1 provides a way to check whether a graph is k-differential colorable. The following `kpath` algorithm attempts to find a k-Hamiltonian path of $\bar{G} = \{V, \bar{E}\}$ from a given node i. Before calling `kpath` the k-Hamiltonian path is initialized as $p = \emptyset$. Each call to `kpath` recursively tries to add a neighbor to the last node in the path, and checks that this maintains the k-path condition. If the condition is violated, the next neighbor is explored, or the algorithm backtracks. If a k-Hamiltonian path of $\bar{G}$ is found by the algorithm, we know that the graph G is $(k + 1)$-colorable.

As the exact algorithm requires exponential time, it is impractical for large graphs, where heuristic algorithms might be a better choice. Gansner *et al.* [6] proposed a heuristic based on a relaxation of the discrete MaxMin problem (1) into a continuous maximization problem of 2-norm:

$$\max \sum_{\{i,j\} \in E} w_{ij}(c_i - c_j)^2, \text{ subject to } \sum_{i \in V} c_i^2 = 1 \qquad (2)$$

where $c \in R^{|V|}$. This continuous problem is solved when c is the eigenvector corresponding to the largest eigenvalue of the weighted Laplacian of the graph. Once (2) is solved, the ordering defined by the eigenvector is used as an approximate solution for the MaxMin problem. This is followed by a greedy refinement algorithm which repeatedly swaps pairs of vertices, provided that the

---

[1] Even planar graphs, which have average degree less than 6, may be tough to color well. Consider a star graph, where one vertex is connected to every other vertex in the graph. Regardless of the coloring, at least one pair of nodes will have a difference of at most one, thus showing that $dc(G) = 1$.

Input: $\bar{G}$, i, p (initialized to $\emptyset$ on first entry)
Output: p
if $i \in p$, or i is not connected with the last k nodes in p **then**
    return $\emptyset$;
end
$p = p \cup \{i\}$;
if $|p| = |V|$ **then**
    return p;
end
foreach neighbor node j of i **do**
    if kpath$(\bar{G}, j, p) \neq \emptyset$ **then**
        return p;
    end
end
$p = p - \{i\}$;
return $\emptyset$;

Algorithm 1. kpath

swap improves the coloring scheme. We call this algorithm GSpectral (Greedy
Spectral). Table 1 gives the chromatic number of some well-known graphs found
by kpath, as well as an estimate of the differential chromatic number obtained
by GSpectral. The exact algorithm kpath can be prohibitively expensive even
for small graphs. For example, on the 60-node football graph (skeleton graph
of a truncated icosahedron), one week of CPU time was not enough to find the
exact differential chromatic number, whereas the greedy spectral algorithm gives
a lower bound of 18 in a few milli-seconds. We note that the GSpectral algorithm often finds good solutions, though it rarely matches the optimal solution.
We also tested GSpectral on larger grid graphs for which the differential chromatic number is known; see Section 5. For grid10 and grid20 graph, it gives
an estimate of 30 and 124, and the actual differential chromatic numbers are 45
and 190, respectively. In both case the CPU time for GSpectral is less than 0.1
seconds.

5   Differential Chromatic Numbers of Special Graphs

Theorem 5. *A line graph on n nodes has differential chromatic number of*
$\lfloor n/2 \rfloor$. *A cycle graph on n nodes has differential chromatic number of* $\lfloor (n-1)/2 \rfloor$.

Proof. Consider a line graph with even number of nodes labeled in order with
$n/2+1$, $1, n/2+2$, $2, \ldots$, n, $n/2$; see Fig. 3(top). This labeling is clearly a $n/2$-
differential coloring and it is easy to show that this is the best possible coloring.
Take any labeling of this graph and consider the node labeled $n/2$. Regardless of
the labels of its neighbor(s) this node must induce an edge difference of at most
$n/2$ achieved if the neighbor(s) is labeled n. Hence the differential chromatic
number can not be more than $n/2$, that is, $dc(G) = n/2 = \lfloor n/2 \rfloor$. A similar
argument also works when n is odd; see Fig. 3 (bottom).

Now consider a cycle graph with an even number of vertices labeled in order
with 1, $n/2+1$, 2, $n/2+2$, $\ldots, n/2$, n; see Fig. 4(left). This labeling is clearly

Table 1. Differential chromatic numbers given by `kpath` and by `GSpectral`

name	drawing	$dc(G)$	GSpectral
petersen		3	3
hypercube		4	4
binarytree3		7	5
grid4		6	3
grid5		10	7
football		-	18

On Maximum Differential Graph Coloring 283

6–5–1–6–7–5–2–6–8–5–3–6–**9**–5–4–6–**10**–5–5
1–6–**7**–5–2–6–**8**–5–3–6–**9**–5–4–6–**10**–5–5–6–**11**–5–6

Fig. 3. Optimal coloring of line graphs with 10 and 11 nodes

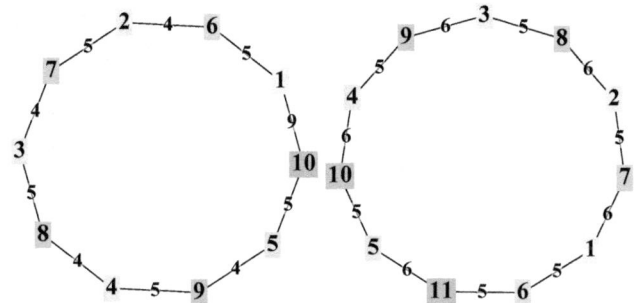

Fig. 4. Optimal coloring of ring graphs with 10 and 11 nodes

$n/2 - 1 = \lfloor (n-1)/2 \rfloor$ differential coloring and no better solution exists. Take any labeling of this graph and consider the node labeled $n/2$. Regardless of the labels of its two neighbors this node must induce an edge difference of at most $n/2 - 1$, achieved if the first neighbor is labeled n and the second neighbor is 1 or $n - 1$. A similar argument also works when n is odd; see Fig. 4 (right). □

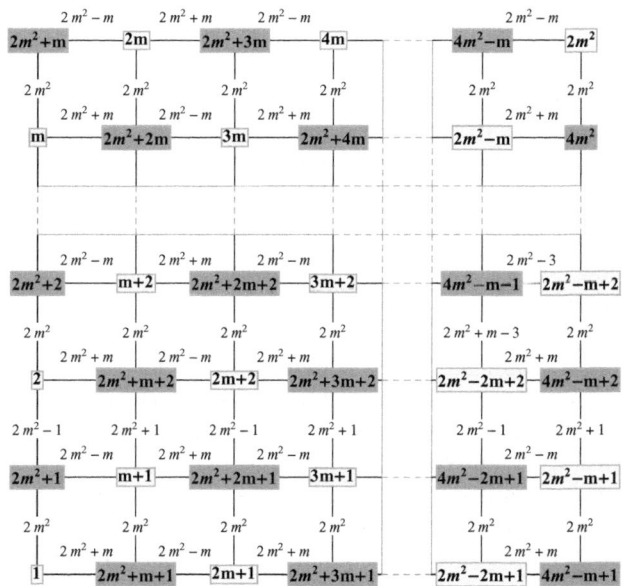

Fig. 5. Optimal coloring of $n \times n$-grids for even n

Fig. 6. Optimal grid map coloring (left) and one induced by the largest eigenvector (right)

Theorem 6. *A grid graph of $n \times n$ nodes has differential chromatic number $\geq \frac{1}{2}n(n-1)$.*

Proof. If n is even, let $m = n/2$, we can color the grid as shown in Fig. 5. Due to the symmetries in the numbering scheme, only five labeling differences are possible: $2m^2 - m$, $2m^2 - 1$, $2m^2$, $2m^2 + 1$, $2m^2 + m$. The smallest is $2m^2 - m = \frac{1}{2}n(n-1)$. For odd n there is also solution with only four label differences.   □

It turns out that $\frac{1}{2}n(n-1)$ is also an upper bound for the chromatic number of the $n \times n$ grid graph, making this result tight [12]. It is informative to contrast an optimally colored grid map with a grid map that uses the coloring induced by the largest eigenvector; see Fig. 6. The eigenvector coloring provides good contrast between neighboring countries, particularly in the center. This indicates that GSpectral might ineed be a good practical heuristic for large graphs.

6   Conclusion

In this paper we introduced the maximum differential graph coloring problem which arises in the context of map coloring. We described exact and heuristic algorithms for this problem, and considered some special classes of graphs for which good solutions can be computed. We showed that this problem is related to that of finding a k-Hamiltonian path. There is also a close relationship between this problem and the antibandwidth problem.

We note that the results of this paper extend easily to the case when there are more than $|V|$ colors available: we can augment the graph with the same number of isolated "dummy" nodes as there are extra colors. We further note that countries that are not neighbors, but are nevertheless close (e.g., neighbor's neighbor), can also be forced to have distinctive colors by adding additional edges in the country graph linking these countries.

Throughout our paper we have been assuming that the edge weights in the country graph are uniform. That is, the importance of having a different color between any pair of adjacent countries in the map is the same. We would also like to investigate ways to handle non-uniform weights, where more importance can be placed on certain countries or certain types of adjacencies (e.g., long vs. short common borders).

Acknowledgments

We thank L'ubomír Török for pointing out the related work on the antibandwidth problem.

References

1. Blache, G., Karpinski, M., Wirtgen, J.: On approximation intractability of the bandwidth problem. Technical Report TR98-014, University of Bonn (1998)
2. Brewer, C.: ColorBrewer - Color Advice for Maps, http://www.colorbrewer.org
3. Calamoneri, T., Massini, A., Török, L., Vrt'o, I.: Antibandwidth of complete k-ary trees. Electronic Notes in Discrete Mathematics 24, 259–266 (2006)
4. Dillencourt, M.B., Eppstein, D., Goodrich, M.T.: Choosing colors for geometric graphs via color space embeddings. In: Kaufmann, M., Wagner, D. (eds.) GD 2006. LNCS, vol. 4372, pp. 294–305. Springer, Heidelberg (2007)
5. Dobrev, S., Královic, R., Pardubská, D., Török, L., Vrt'o, I.: Antibandwidth and cyclic antibandwidth of hamming graphs. Electronic Notes in Discrete Mathematics 34, 295–300 (2009)
6. Gansner, E.R., Hu, Y.F., Kobourov, S.G.: GMap: Visualizing graphs and clusters as maps. In: IEEE Pacific Visualization Symposium (PacVis), pp. 201–208 (2010)
7. Gansner, E.R., Hu, Y.F., Kobourov, S.G., Volinsky, C.: Putting recommendations on the map: visualizing clusters and relations. In: 3rd ACM Conference on Recommender Systems (RecSys), pp. 345–348 (2009)
8. Hu, Y.F., Scott, J.A.: A multilevel algorithm for wavefront reduction. SIAM Journal on Scientific Computing 23, 1352–1375 (2001)
9. Kumfert, G., Pothen, A.: Two improved algorithms for envelope and wavefront reduction. BIT 35, 1–32 (1997)
10. Leung, J., Vornberger, O., Witthoff, J.: On some variants of the bandwidth minimization problem. SIAM Journal on Computing 13, 650 (1984)
11. Papadimitriou, C.: The NP-Completeness of the bandwidth minimization problem. Computing 16, 263–270 (1975)
12. Raspaud, A., Schröder, H., Sýkora, O., Török, L., Vrt'o, I.: Antibandwidth and cyclic antibandwidth of meshes and hypercubes. Discrete Mathematics 309(11), 3541–3552 (2009)
13. Sloan, S.W.: An algorithm for profile and wavefront reduction of sparse matrices. International Journal for Numerical Methods in Engineering 23, 239–251 (1986)
14. Török, L., Vrt'o, I.: Antibandwidth of three-dimensional meshes. Discrete Mathematics 310(3), 505–510 (2010)

15. Unger, W.: The complexity of the approximation of the bandwidth problem. In: Proceedings of the 39th Symposium on Foundations of Computer Science (FOCS), pp. 82–91 (1998)
16. Wang, X., Wu, X., Dumitrescu, S.: On explicit formulas for bandwidth and antibandwidth of hypercubes. Discrete Applied Mathematics 157(8), 1947–1952 (2009)
17. Wong, W.W., Wong, C.K.: Minimum K-Hamiltonian graphs. Journal of Graph Theory 8, 155–165 (2006)

Dot Product Representations of Planar Graphs

Ross J. Kang[1],[*] and Tobias Müller[2],[**]

[1] Durham University
ross.kang@gmail.com
[2] Centrum Wiskunde & Informatica
tobias@cwi.nl

Abstract. A graph G on n vertices is a k-dot product graph if there are vectors $u_1, \ldots, u_n \in \mathbb{R}^k$, one for each vertex of G, such that $u_i^T u_j \geq 1$ if and only if $ij \in E(G)$. Fiduccia, Scheinerman, Trenk and Zito (1998) asked whether every planar graph is a 3-dot product graph. We show that the answer is "no". On the other hand, every planar graph is a 4-dot product graph.

1 Introduction and Statement of Results

We study a type of geometric representation of graphs using vectors from $\mathbb{R}^k$ for some $k \in \mathbb{N}$. Let G be a graph with n vertices. We say G is a *k-dot product graph* if there exist vectors $u_1, \ldots, u_n \in \mathbb{R}^k$ such that $u_i^T u_j \geq 1$ if and only if $ij \in E(G)$. An explicit set of vectors in $\mathbb{R}^k$ that exhibits G in this way is called a *k-dot product representation* of G. The *dot product dimension* of G is the least k such that there is a k-dot product representation of G. (Notice that G can be trivially represented in $\mathbb{R}^{|E(G)|}$, so that the dot product dimension is finite for all graphs.)

The well-studied class of threshold graphs coincides with the 1-dot product graphs; consult the monograph by Mahadev and Peled [6] for a comprehensive survey of results on these and related structures.

Partially motivated by the striking application by Lovász of a similar geometric representation to an important problem on Shannon capacity [5], Reiterman, Rödl and Siňajová [7,8,9] studied the dot product dimension extensively and obtained several bounds in terms of threshold dimension, sphericity, chromatic number, maximum degree, maximum average degree, and maximum complementary degree; they also detailed various examples. Arriving from a different direction Fiduccia, Scheinerman, Trenk and Zito [3] also considered dot product dimension and, for example, analysed bipartite, complete multipartite, and interval graphs. Both Reiterman et al. and Fiduccia et al. proved that every forest is a 3-dot product graph [8,3].

---

[*] Research partially supported by the Engineering and Physical Sciences Research Council (EPSRC), grant EP/G066604/1.
[**] Research partially supported by a VENI grant from Netherlands Organisation for Scientific Research (NWO).

U. Brandes and S. Cornelsen (Eds.): GD 2010, LNCS 6502, pp. 287–292, 2011.
© Springer-Verlag Berlin Heidelberg 2011

Seeing a potential extension to this result, Fiduccia et al. asked whether every planar graph is a 3-dot product graph. Here we will answer this in the negative by describing a counterexample. In contrast, we show that any planar graph has dimension at most 4.

Theorem 1. *Every planar graph is a 4-dot product graph, and there exist planar graphs which are not 3-dot product graphs.*

It remains open to characterise which planar graphs have dot product dimension exactly 4.

The structure of the paper is as follows. In the next section, we develop some notation and review some spherical geometry. In Section 3, we present our counterexample. We show how every planar graph has a 4-dot product representation in Section 4.

2   Preliminaries

For $u, v \in S^2$, let us denote by $[u, v]$ the (shortest) spherical arc between u and v. Let $\mathrm{dist}_{S^2}(u, v)$ denote the length of $[u, v]$. Then one can see that $\mathrm{dist}_{S^2}(u, v)$ equals the angle between the two vectors $u, v \in S^2$. It can thus be expressed as

$$\mathrm{dist}_{S^2}(u, v) = \arccos(v^T u).$$

For $r \geq 0$, let the *spherical cap* of radius r around $v \in S^2$ be defined as

$$\mathrm{cap}(v, r) := \{u \in S^2 : \mathrm{dist}_{S^2}(u, v) \leq r\}.$$

Suppose that $u, v, w \in S^2$ are three points on the sphere in general position. We shall call the union of the three circular arcs $[u, v], [v, w], [u, w]$ a *spherical triangle*. Similarly one can define a *spherical polygon*.

Let us write $a := \mathrm{dist}_{S^2}(u, v), b := \mathrm{dist}_{S^2}(u, w), c := \mathrm{dist}_{S^2}(v, w)$, and let γ denote the angle between $[u, v]$ and $[u, w]$. See Figure 1.

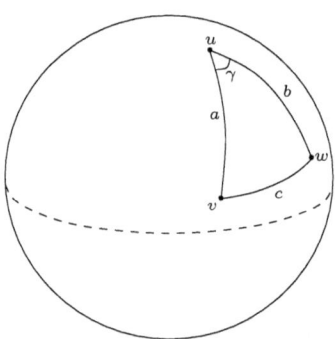

Fig. 1. A spherical triangle

Recall the spherical law of cosines:

$$\cos(c) = \cos(a)\cos(b) + \sin(a)\sin(b)\cos(\gamma). \tag{1}$$

The spherical law of cosines can be rephrased as:

$$v^T w = (u^T v) \cdot (u^T w) + \cos(\gamma)\sqrt{(1 - (u^T v)^2)(1 - (u^T w)^2)}. \tag{2}$$

This second form will be more useful for our purposes.

3 Planar Graphs That Are Not 3-Dot Product Graphs

We will construct graphs F, G, H as follows.

(i) We start with K_4, the complete graph on the four vertices t_1, t_2, t_3, t_4.

(ii) To obtain F, we replace each edge $t_i t_j$ of K_4 by a path $t_i t_{ij} t_{ji} t_j$ of length 3.

(iii) An embedding of the graph F divides the plane into four faces. To obtain G, we place an additional vertex inside each face of F and connect it to all vertices on the outer cycle of the face. Here, f_i will denote the vertex in the face whose limiting cycle does not contain t_i.

(iv) Finally, to obtain H we attach four leaves to each vertex of G.

The graphs G and H are depicted in Figure 2. For $k \in \mathbb{N}$, let the graph H_k consist of k disjoint copies of H. Clearly H_k is planar for all k. In the rest of this section, we shall prove the following.

Theorem 2. *The graph H_k is not a 3-dot product graph, for sufficiently large k.*

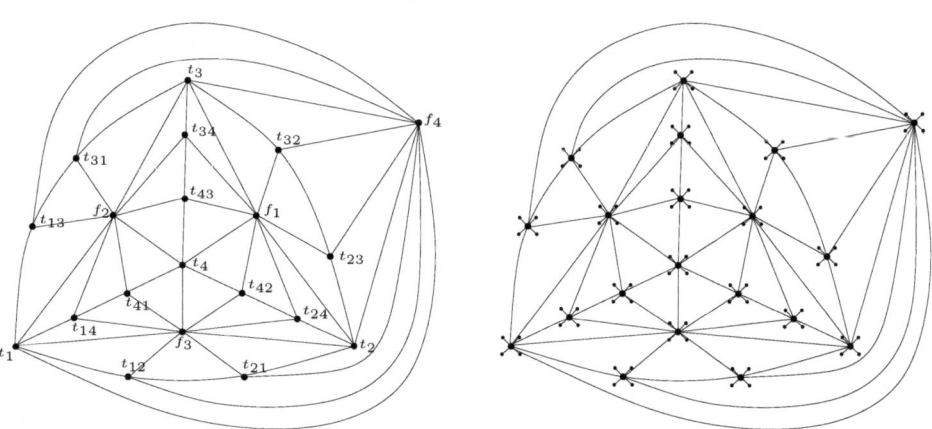

Fig. 2. The graphs G, H

Proof of Theorem 2. The proof is by contradiction. Let us assume that every H_k has a 3-dot product representation. The proof is divided into a number of intermediate steps. The details of the proofs of Claims 3 through 8 below can be found in the journal version of this paper.

Claim 3. For every $\eta > 0$ there is a 3-dot product representation of H, with $\|u(t)\| < 1 + \eta$ for all $t \in V(H)$.   □

Let us fix a small η (say $\eta := 10^{-10}$), and let $u : V(H) \to \mathbb{R}^3$ be the representation provided by Claim 3. For $s \in V(H)$ let us write $l(s) := \|u(s)\|, v(s) := u(s)/\|u(s)\|$. Let us observe that

$$st \in E(H) \quad \text{if and only if} \quad v(s)^T v(t) \geq 1/l(s)l(t).$$

Recall that $G \subseteq H$ is the subgraph induced by all non-leaf vertices.

Claim 4. For every $s \in V(G)$ we have $l(s) > 1$.   □

Claim 5. Suppose that $st, s't' \in E(G)$ are edges with s, s', t, t' distinct and suppose that the arcs $[v(s), v(t)]$ and $[v(s'), v(t')]$ cross. Then at least one of ss', st', ts', tt' is also an edge of G.   □

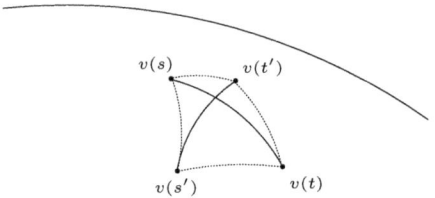

Claim 6. Suppose that s_1, s_2, s_3 form a clique in G, and $v(s)$ lies inside the (smaller of the two areas defined by the) spherical triangle defined by $v(s_1)$, $v(s_2)$, $v(s_3)$. Then either $s_1 s \in E(G)$ or $s_2 s \in E(G)$ or $s_3 s \in E(G)$.   □

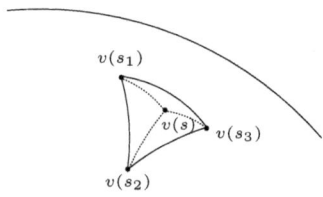

From now on, let us write $v_i := v(t_i)$, $l_i = l(t_i)$ and $v_{ij} = v(t_{ij})$, $l_{ij} = l(t_{ij})$. By Claim 5, the arcs $[v_{ij}, v_{ji}]$ and $[v_{kl}, v_{lk}]$ do not cross each other (for $\{i, j\} \neq \{k, l\}$). However, the arc $[v_{ij}, v_{ji}]$ could cross an arc of the form $[v_i, v_{ik}]$ or $[v_j, v_{jk}]$.

Let C denote the cycle $t_1 t_{12} t_{21} t_2 t_{23} t_{32} t_3 t_{31} t_{13} t_1$ in G, and let P denote the corresponding spherical polygon. Because each circular arc corresponding to an edge has length at most ρ, we have $P \subseteq \text{cap}(v_1, 5\rho)$. Also note that $S^2 \backslash P$ consists of at least two path-connected components (by the Jordan Curve Theorem). As

ρ is small, exactly one of these components has area $> 3.9\pi$. We shall refer to this component as the "outside" of P, and the union of the other components we will call the "inside".

Claim 7. We can assume without loss of generality that v_4 lies inside the polygon P.   □

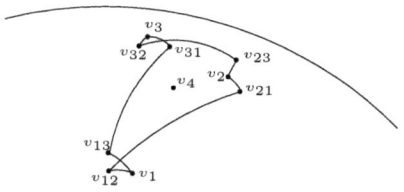

Claim 8. v_4 lies inside the spherical triangle defined by $v(f_4)$ and two consecutive points on P.   □

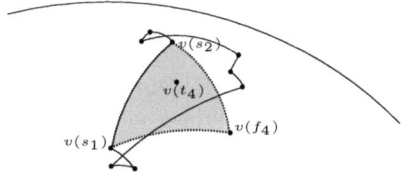

Since f_4, s_1, s_2 form a triangle in G, Claim 6 implies that t_4 must be adjacent to at least one of them. This is a contradiction, since t_4 is neither adjacent to f_4 nor to any vertex of C. This completes the proof of Theorem 2.   □

4 All Planar Graphs Are 4-Dot Product Graphs

The Colin de Verdière parameter $\mu(G)$ of a graph G is the maximum co-rank over all matrices M that satisfy

(i) $M_{ij} < 0$ if $ij \in E(G)$;
(ii) $M_{ij} = 0$ if $ij \notin E(G)$ and $i \neq j$;
(iii) M has exactly one negative eigenvalue;
(iv) if X is symmetric with $X_{ij} = 0$ for all $ij \in E(G)$ and $X_{ii} = 0$, for all i, and $MX = 0$, then we must have $X = 0$.

This parameter was introduced Y. Colin de Verdiére in [1,2], where it is shown that planar graphs are exactly the graphs G with $\mu(G) \leq 3$.

Kotlov, Lovasz and Vempala [4] introduced the following related parameter. Let $\nu(G)$ denote the smallest d such that there exist vectors $u_1, \ldots, u_n \in \mathbb{R}^d$ that satisfy

(i) $u_i^T u_j = 1$ if $ij \in E(G)$;
(ii) $u_i^T u_j < 1$ if $ij \notin E(G)$ and $i \neq j$;

(iii) if X is a symmetric $n \times n$ matrix such that $X_{ij} = 0$ for all $ij \in E(G)$ and $X_{ii} = 0$ for all i, and $\sum_j X_{ij} u_j = 0$ for all i, then $X = 0$.

Clearly, every graph G is a $\nu(G)$-dot product graph. However, because **(i)** asks for equality and because of the extra demand **(iii)**, G might also be a k-dot product graph for some $k < \nu(G)$. The relation between $\nu(G)$ and $\mu(G)$ is given by the following result.

Theorem 9 ([4]). *If $G \neq K_2$ then $\nu(G) = n - 1 - \mu(\overline{G})$.*

That K_2 is a 4-dot product graph is obvious. That every other planar graph is a 4-dot product graph is a direct consequence of Theorem 9 and the following result.

Theorem 10 ([4]). *If G is the complement of a planar graph then $\mu(G) \geq n-5$.*

References

1. Colin de Verdière, Y.: Sur un nouvel invariant des graphes et un critère de planarité. J. Combin. Theory Ser. B 50(1), 11–21 (1990)
2. de Verdière, Y.C.: On a new graph invariant and a criterion for planarity. In: Graph structure theory, Contemp. Math. Amer. Math. Soc., Seattle, WA, vol. 147, pp. 137–147. RI, Providence (1991)
3. Fiduccia, C.M., Scheinerman, E.R., Trenk, A., Zito, J.S.: Dot product representations of graphs. Discrete Math. 181(1-3), 113–138 (1998)
4. Kotlov, A., Lovász, L., Vempala, S.: The Colin de Verdière number and sphere representations of a graph. Combinatorica 17(4), 483–521 (1997)
5. Lovász, L.: On the Shannon capacity of a graph. IEEE Trans. Inform. Theory 25(1), 1–7 (1979)
6. Mahadev, N.V.R., Peled, U.N.: Threshold graphs and related topics. Annals of Discrete Mathematics, vol. 56, North-Holland Publishing Co, Amsterdam (1995)
7. Reiterman, J., Rödl, V., Šiňajová, E.: Embeddings of graphs in Euclidean spaces. Discrete Comput. Geom. 4(4), 349–364 (1989)
8. Reiterman, J., Rödl, V., Šiňajová, E.: Geometrical embeddings of graphs. Discrete Math. 74(3), 291–319 (1989)
9. Reiterman, J., Rödl, V., Šiňajová, E.: On embedding of graphs into Euclidean spaces of small dimension. J. Combin. Theory Ser. B 56(1), 1–8 (1992)

Drawing Planar Graphs of Bounded Degree with Few Slopes

Balázs Keszegh[1,3], János Pach[1,3], and Dömötör Pálvölgyi[1,2]

[1] Ecole Politechnique Fédérale de Lausanne
[2] Eötvös University, Budapest
[3] A. Rényi Institute of Mathematics, Budapest

Abstract. We settle a problem of Dujmović, Eppstein, Suderman, and Wood by showing that there exists a function f with the property that every planar graph G with maximum degree d admits a drawing with noncrossing straight-line edges, using at most $f(d)$ different slopes. If we allow the edges to be represented by polygonal paths with *one* bend, then $2d$ slopes suffice. Allowing *two* bends per edge, every planar graph with maximum degree $d \geq 3$ can be drawn using segments of at most $\lceil d/2 \rceil$ different slopes. There is only one exception: the graph formed by the edges of an octahedron is 4-regular, yet it requires 3 slopes. These bounds cannot be improved.

Keywords: Graph drawing, Slope number, Planar graphs.

1 Introduction

A planar layout of a graph G is called a *drawing* if the vertices of G are represented by distinct points in the plane and every edge is represented by a continuous arc connecting the corresponding pair of points and not passing through any other point representing a vertex [3]. If it leads to no confusion, in notation and terminology we make no distinction between a vertex and the corresponding point and between an edge and the corresponding arc. If the edges are represented by line segments, the drawing is called a *straight-line drawing*. The *slope* of an edge in a straight-line drawing is the slope of the corresponding segment.

In this paper, we will be concerned with drawings of planar graphs. Unless it is stated otherwise, all *drawings* will be *noncrossing*, that is, no two arcs that represent different edges have an interior point in common.

Every planar graph admits a straight-line drawing [9]. From the practical and aesthetical point of view, it makes sense to minimize the number of slopes we use [23]. The *planar slope number* of a planar graph G is the smallest number s with the property that G has a straight-line drawing with edges of at most s distinct slopes. If G has a vertex of degree d, then its planar slope number is at least $\lceil d/2 \rceil$, because in a straight-line drawing no two edges are allowed to overlap.

Dujmović, Eppstein, Suderman, and Wood [4] raised the question whether there exists a function f with the property that the planar slope number of

U. Brandes and S. Cornelsen (Eds.): GD 2010, LNCS 6502, pp. 293–304, 2011.

every planar graph with maximum degree d can be bounded from above by $f(d)$. Jelinek et al. [13] have shown that the answer is yes for *outerplanar* graphs, that is, for planar graphs that can be drawn so that all of their vertices lie on the outer face. In Section 2, we answer this question in full generality. We prove the following.

Theorem 1. *Every planar graph with maximum degree d admits a straight-line drawing, using segments of $O(d^2(3 + 2\sqrt{3})^{12d}) \leq K^d$ distinct slopes.*

The proof is based on a paper of Malitz and Papakostas [18], who used Koebe's theorem [14] on disk representations of planar graphs to prove the existence of drawings with relatively large angular resolution. As the proof of Malitz and Papakostas, our argument is nonconstructive; it only yields a nondeterministic algorithm with running time $O(dn)$.

For $d = 3$, much stronger results are known than the one given by our theorem. Dujmović at al. [4] showed that every planar graph with maximum degree 3 admits a straight-line drawing using at most 3 different slopes, except for at most 3 edges of the outer face, which may require 3 additional slopes. This complements Ungar's old theorem [22], according to which 3-regular, 4-edge-connected planar graphs require only 2 slopes and 4 extra edges.

The exponential upper bound in Theorem 1 is probably far from being optimal. However, we were unable to give any superlinear lower bound for the largest planar slope number of a planar graph with maximum degree d. The best constructions we are aware of are presented in Section 4.

It is perhaps somewhat surprising that if we do not restrict our attention to planar graphs, then no result similar to Theorem 1 holds. For every $d \geq 5$, Barát, Matoušek, and Wood [1] and, independently, Pach and Pálvölgyi [20] constructed graphs with maximum degree d with the property that no matter how we draw them in the plane with (possibly crossing) straight-line edges, we must use an arbitrarily large number of slopes. (See also [5].) The case $d \leq 3$ is different: Keszegh et al. [15] proved that every graph with maximum degree 3 can be drawn with 5 slopes. Moreover, Mukkamala and Szegedy [19] showed that 4 slopes suffice if the graph is connected. The case $d = 4$ remains open.

Returning to planar graphs, we show that significantly fewer slopes are sufficient if we are allowed to represent the edges by short noncrossing polygonal paths. If such a path consists of $k + 1$ segments, we say that the edge is drawn by k *bends*. If we allow one bend per edge, then every planar graph can be drawn using segments with $O(d)$ slopes. The proof of Theorem 2 is based on a result of Fraysseix et al. [10], according to which every planar graph can be represented as a contact graph of T-shapes. See the full version of this paper for details.

Theorem 2. *Every planar graph G with maximum degree d can be drawn with at most 1 bend per edge, using at most $2d$ slopes.*

Allowing *two* bends per edge yields an optimal result: almost all planar graphs with maximum degree d can be drawn with $\lceil d/2 \rceil$ slopes. In Section 3, we establish

Theorem 3. *Every planar graph G with maximum degree $d \geq 3$ can be drawn with at most 2 bends per edge, using segments of at most $\lceil d/2 \rceil$ distinct slopes. The only exception is the graph formed by the edges of an octahedron, which is 4-regular, but requires 3 slopes. These bounds are best possible.*

It follows from the proof of Theorem 3 that in the cyclic order of directions, the slopes of the edges incident to any given vertex form a contiguous interval. Moreover, the $\lceil d/2 \rceil$ directions we use can be chosen to be equally spaced in $[0, 2\pi)$. We were unable to guarantee such a nice property in Theorem 2: even for a fixed d, as the number of vertices increases, the smallest difference between the $2d - 2$ slopes we used tends to zero. We suspect that this property is only an unpleasant artifact of our proof technique.

2   Straight-Line Drawings–Proof of Theorem 1

Note that it is sufficient to prove the theorem for triangulated planar graphs, because any planar graph can be triangulated by adding vertices and edges so that the degree of each vertex increases only by a factor of at most three [21], so at the end we will lose this factor.

We need the following result from [18], which is not displayed as a theorem there, but is stated right above Theorem 2.2.

Lemma 1. (Malitz-Papakostas) *The vertices of any triangulated planar graph G with maximum degree d can be represented by nonoverlapping disks in the plane so that two disks are tangent to each other if and only if the corresponding vertices are adjacent, and the ratio of the radii of any two disks that are tangent to each other is at least α^{d-2}, where $\alpha = \frac{1}{3+2\sqrt{3}} \approx 0.15$.*

Lemma 1 can be established by taking any representation of the vertices of G by tangent disks, as guaranteed by Koebe's theorem, and applying a conformal mapping to the plane that takes the disks corresponding to the three vertices of the outer face to disks of the same radii. The lemma now follows by the observation that any internal disk is surrounded by a ring of at most d mutually touching disks, and the radius of none of them can be much smaller than that of the central disk.

The idea of the proof of Theorem 1 is as follows. Let G be a triangulated planar graph with maximum degree d, and denote its vertices by $v_1, v_2, \ldots$. Consider a disk representation of G meeting the requirements of Lemma 1. Let D_i denote the disk that represents v_i, and let O_i be the center of D_i. By properly scaling the picture if necessary, we can assume without loss of generality that the radius of the smallest disk D_i is sufficiently large. Place an integer grid on the plane, and replace each center O_i by the nearest grid point. Connecting the corresponding pairs of grid points by segments, we obtain a straight-line drawing of G. The advantage of using a grid is that in this way we have control of the slopes of the edges. The trouble is that the size of the grid, and thus the number of slopes used, is very large. Therefore, in the neighborhood of each disk D_i, we

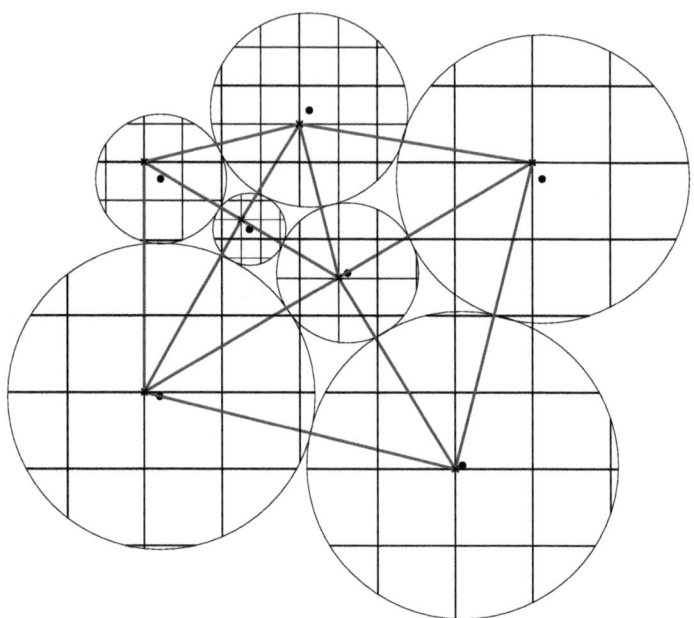

Fig. 1. Straight-line graph from disk representation

use a portion of a grid whose side length is proportional to the radius of the disk. These grids will nicely fit together, and each edge will connect two nearby points belonging to grids of comparable sizes. Hence, the number of slopes used will be bounded. See Figure 1.

Now we work out the details. Let r_i denote the radius of D_i ($i = 1, 2 \ldots$), and suppose without loss of generality that r^*, the radius of the smallest disk is

$$r^* = min_i r_i = \sqrt{2}/\alpha^{d-2} > 1,$$

where α denotes the same constant as in Lemma 1.

Let $s_i = \lfloor \log_d(r_i/r^*) \rfloor \geq 0$, and represent each vertex v_i by the integer point nearest to O_i such that both of its coordinates are divisible by d^{s_i}. (Taking a coordinate system in general position, we can make sure that this point is unique.) For simplicity, the point representing v_i will also be denoted by v_i. Obviously, we have that the distance between O_i and v_i satisfies

$$\overline{O_i v_i} < \frac{d^{s_i}}{\sqrt{2}}.$$

Since the centers O_i of the disks induce a (crossing-free) straight-line drawing of G, in order to prove that moving the vertices to v_i does not create a crossing, it is sufficient to verify the following statement.

Lemma 2. *For any three mutually adjacent vertices, v_i, v_j, v_k in G, the orientation of the triangles $O_i O_j O_k$ and $v_i v_j v_k$ are the same.*

Proof. By Lemma 1, the ratio between the radii of any two adjacent disks is at least α^{d-2}. Suppose without loss of generality that $r_i \geq r_j \geq r_k \geq \alpha^{d-2}r_i$. For the orientation to change, at least one of $\overline{O_iv_i}$, $\overline{O_jv_j}$, or $\overline{O_kv_k}$ must be at least half of the smallest altitude of the triangle $O_iO_jO_k$, which is at least $\frac{r_k}{2}$.

On the other hand, as we have seen before, each of these numbers is smaller than

$$\frac{d^{s_i}}{\sqrt{2}} \leq \frac{r_i/r^*}{\sqrt{2}} = \frac{\alpha^{d-2}r_i}{2} \leq \frac{r_k}{2}$$

which completes the proof.

Now we are ready to complete the proof of Theorem 1. Take an edge v_iv_j of G, with $r_i \geq r_j \geq \alpha^{d-2}r_i$. The length of this edge can be bounded from above by

$$\overline{v_iv_j} \leq \overline{O_iO_j} + \overline{O_iv_i} + \overline{O_jv_j} \leq r_i + r_j + \frac{d^{s_i}}{\sqrt{2}} + \frac{d^{s_j}}{\sqrt{2}} \leq 2r_i + \sqrt{2}d^{s_i} \leq 2r_i + \sqrt{2}r_i/r^*$$

$$\leq r_i/r^*(2r^* + \sqrt{2}) \leq \frac{r_j/r^*}{\alpha^{d-2}}(2r^* + \sqrt{2}) < \frac{d^{s_j+1}}{\alpha^{d-2}}\left(\frac{2\sqrt{2}}{\alpha^{d-2}} + \sqrt{2}\right).$$

According to our construction, the coordinates of v_j are integers divisible by d^{s_j}, and the coordinates of v_i are integers divisible by $d^{s_i} \geq d^{s_j}$, thus also by d^{s_j}.

Thus, shrinking the edge v_iv_j by a factor of d^{s_j}, we obtain a segment whose endpoints are integer points at a distance at most $\frac{d}{\alpha^{d-2}}\left(\frac{2\sqrt{2}}{\alpha^{d-2}} + \sqrt{2}\right)$. Denoting this number by $R(d)$, we obtain that the number of possible slopes for v_iv_j, and hence for any other edge in the embedding, cannot exceed the number of integer points in a disk of radius $R(d)$ around the origin. Thus, the planar slope number of any triangulated planar graph of maximum degree d is at most roughly $R^2(d)\pi = O(d^2/\alpha^{4d})$, which completes the proof. $\square$

Our proof is based on the result of Malitz and Papakostas that does not have an algorithmic version. However, with some reverse engineering, we can obtain a nondeterministic algorithm for drawing a triangulated planar graph of bounded degree with a bounded number of slopes. Because of the enormous constants in our expressions, this algorithm is only of theoretical interest. Here is a brief sketch.

Nondeterministic algorithm. First, we guess the three vertices of the outer face and their coordinates in the grid scaled according to their radii. Then embed the remaining vertices one by one. For each vertex, we guess the radius of the corresponding disk as well as its coordinates in the proportionally scaled grid. This algorithm runs in nondeterministic $O(dn)$ time.

3   Two Bends per Edge–Proof of Theorem 3

In this section, we draw the edges of a planar graph by polygonal paths with at most *two* bends. Our aim is to establish Theorem 3.

Note that the statement is trivially true for $d = 1$ and is false for $d = 2$. It is sufficient to prove Theorem 3 for even values of d. For $d = 4$, the assertion was first proved by Liu et al. [17] and later, independently, by Biedl and Kant [2] (also that the only exception is the octahedral graph). The latter approach is based on the notion of *st*-ordering of biconnected (2-connected) graphs from Lempel et al. [16]. We will show that this method generalizes to higher values of $d \geq 5$. As it is sufficient to prove the statement for even values of d, from now on we suppose that $d \geq 6$ even. We will argue that it is enough to consider *biconnected* graphs. Then we review some crucial claims from [2] that will enable us to complete the proof. We start with some notation.

Take $d \geq 5$ lines that can be obtained from a vertical line by clockwise rotation by $0, \pi/d, 2\pi/d, \ldots, (d-1)\pi/d$ degrees. Their slopes are called the d *regular slopes*. We will use these slopes to draw G. Since these slopes depend only on d and not on G, it is enough to prove the theorem for connected graphs. If a graph is not connected, its components can be drawn separately.

In this section we always use the term "slope" to mean a regular slope. The *directed slope* of a directed line or segment is defined as the angle (mod 2π) of a clockwise rotation that takes it to a position parallel to the upward directed y-axis. Thus, if the directed slopes of two segments differ by π, then they have the same slope. We say that the slopes of the segments incident to a point p form a *contiguous interval* if the set $S \subset \{0, \pi/d, 2\pi/d, \ldots, (2d-1)\pi/d\}$ of directed slopes of the segments directed away from p, has the property that for all but at most one $\alpha \in S$, we have that $\alpha + \pi/d \bmod 2\pi \in S$ (see Figure 3). Finally, we say that G admits a *good drawing* if G has a planar drawing such that every edge has at most 2 bends, every segment of every edge has one of the $\lceil d/2 \rceil$ regular slopes, and the slopes of the segments incident to any vertex form a contiguous interval. To prove Theorem 3, we show by induction that every planar graph with maximum degree d admits a good drawing.

Lemma 3. *Let G be a connected planar graph of maximum degree d, let $t \in V(G)$ be a vertex whose degree is strictly smaller than d, and let $v \in V(G)$ be a cut vertex. Suppose that for any connected planar graph G' of maximum degree d, which has fewer than $|V(G)|$ vertices, and for any vertex $t' \in V(G')$ whose degree is strictly smaller than d, there is a good drawing of G' with t' on its outer face. Then G also admits a good drawing with t on its outer face.*

Proof. Let $G_1, G_2, \ldots$ denote the connected components of the graph obtained from G after the removal of the cut vertex v, and let G_i^* be the subgraph of G induced by $V(G_i) \cup \{v\}$. If $t = v$ is a cut vertex, then each G_i^* has a good drawing with $t = v$ on its outer face. After performing a suitable rotation and scaling for each of these drawings, and identifying their vertices corresponding to v, the lemma follows because the slopes of the segments incident to v form a contiguous interval in each component. If $t \neq v$, then let G_j be the component containing t. Using the hypothesis, G_j^* has a good drawing with t on its outer face. Also, each G_i^* for $i \geq 2$ has a good drawing with v on its outer face. As in the previous case, the lemma follows by scaling down and rotating the components for $i \neq j$ and again identifying the vertices corresponding to v.

In view of Lemma 3, in the sequel we consider only biconnected graphs. We need the following definition.

Definition 1. *An ordering of the vertices of a graph, $v_1, v_2, \ldots, v_n$, is said to be an st-ordering if $v_1 = s$, $v_n = t$, and if for every $1 < i < n$ the vertex v_i has at least one neighbor that precedes it and a neighbor that follows it.*

In [16], it was shown that any biconnected graph has an st-ordering, for any choice of the vertices s and t. In [2], this result was slightly strengthened for planar graphs, as follows.

Lemma 4. (Biedl-Kant) *Let D_G be a drawing of a biconnected planar graph, G, with vertices s and t on the outer face. Then G has an st-ordering for which $s = v_1$, $t = v_n$ and v_2 is also a vertex of the outer face and $v_1 v_2$ is an edge of the outer face.*

We define G_i to be the subgraph of G induced by the vertices $v_1, v_2, \ldots, v_i$. Note that G_i is connected. If i is fixed, we call the edges between $V(G_i)$ and $V(G) \setminus V(G_i)$ the *pending edges*. For a drawing of G, D_G, we denote by D_{G_i} the drawing restricted to G_i and to an initial part of each pending edge connected to G_i.

Proposition 1. *In the drawing D_G guaranteed by Lemma 4, $v_{i+1}, \ldots v_n$ and the pending edges are in the outer face of D_{G_i}.*

Proof. Suppose for contradiction that for some i and $j > i$, v_j is not in the outer face of D_{G_i}. We know that v_n is in the outer face of D_{G_i} as it is on the outer face of D_G, thus v_n and v_j are in different faces of D_{G_i}. On the other hand, by the definition of st-ordering, there is a path in G between v_j and v_n using only vertices from $V(G) \setminus V(G_i)$. The drawing of this path in D_G must lie completely in one face of D_{G_i}. Thus, v_j and v_n must also lie in the same face, a contradiction. Since the pending edges connect $V(G_i)$ and $V(G) \setminus V(G_i)$, they must also lie in the outer face.

By Lemma 4, the edge $v_1 v_2$ lies on the boundary of the outer face of D_{G_i}, for any $i \geq 2$. Thus, we can order the pending edges connecting $V(G_i)$ and $V(G) \setminus V(G_i)$ by walking in D_G from v_1 to v_2 around D_{G_i} on the side that does not consist of only the $v_1 v_2$ edge, see Figure 2(a). We call this the *pending-order* of the pending edges between $V(G_i)$ and $V(G) \setminus V(G_i)$ (this order may depend on D_G). Proposition 1 implies

Proposition 2. *The edges connecting v_{i+1} to vertices preceding it form an interval of consecutive elements in the pending-order of the edges between $V(G_i)$ and $V(G) \setminus V(G_i)$.*

For an illustration see Figure 2(a).

Two drawings of the same graph are said to be *equivalent* if the circular order of the edges incident to each vertex is the same in both drawings. Note that in this order we also include the pending edges (which are differentiated with respect to their yet not drawn end).

Now we are ready to prove Theorem 3.

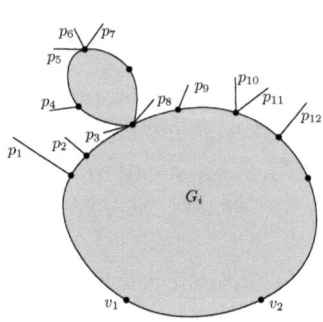

(a) The pending-order of the pending edges in D_{G_i}

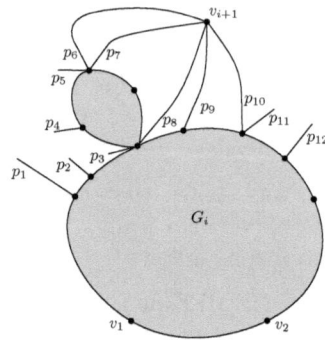

(b) The preceding neighbors of v_{i+1} are consecutive in the pending-order

Fig. 2. Properties of the st-ordering

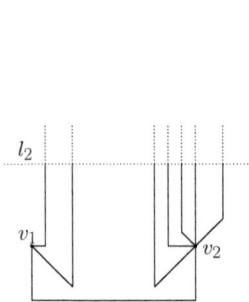

(a) Drawing v_1, v_2 and the edges incident to them

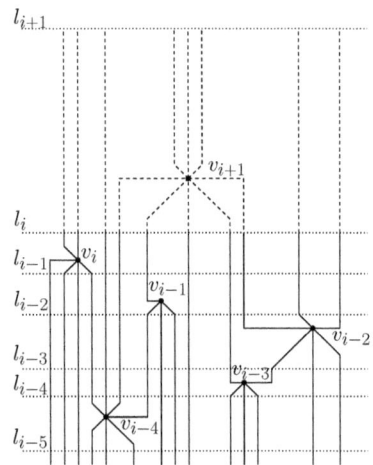

(b) Adding v_i; partial edges added in this step are drawn with dashed lines

Fig. 3. Drawing with at most two bends

Lemma 5. *For any biconnected planar graph G with maximum degree $d \geq 6$ and for any vertex $t \in V(G)$ with degree strictly less then d, G admits a good drawing with t on its outer face.*

For $d \geq 6$, it follows from Euler's polyhedral formula that G has a vertex t of degree at most $5 < d$. Thus, Theorem 3 is a direct consequence of the lemma.

Proof (Proof of Theorem 3.). Take a planar drawing D_G of G such that t is on the outer face and pick another vertex, s, from the outer face. Apply Lemma 4 to obtain an st-ordering with $v_1 = s, v_2$, and $v_n = t$ on the outer face of D_G

such that v_1v_2 is an edge of the outer face. We will build up a good drawing of G by starting with v_1 and then adding $v_2, v_3, \ldots, v_n$ one by one to the outer face of the current drawing. As soon as we add a new vertex v_i, we also draw the initial pieces of the pending edges, and we make sure that the resulting drawing is equivalent to the drawing D_{G_i}.

Another property of the good drawing that we maintain is that every edge consists of precisely three pieces. (Actually, an edge may consist of fewer than 3 segments, because two consecutive pieces are allowed to have the same slope and form a longer segment) The middle piece will always be vertical, except for the middle piece of v_1v_2.

Suppose without loss of generality that v_1 follows directly after v_2 in the clockwise order of the vertices around the outer face of D_G. Place v_1 and v_2 arbitrarily in the plane so that the x–coordinate of v_1 is smaller than the x–coordinate of v_2. Connect v_1 and v_2 by an edge consisting of three segments: the segments incident to v_1 and v_2 are vertical and lie below them, while the middle segment has an arbitrary non-vertical regular slope. Draw a horizontal auxiliary line l_2 above v_1 and v_2. Next, draw the initial pieces of the other (pending) edges incident to v_1 and v_2, as follows. For $i = 1, 2$, draw a short segment from v_i for each of the edges incident to it (except for the edge v_1v_2, which has already been drawn) so that the directed slopes of the edges (including v_1v_2) form a contiguous interval and their circular order is the same as in D_G. Each of these short segments will be followed by a vertical segment that reaches above l_2. These vertical segments will belong to the middle pieces of the corresponding pending edges. Clearly, for a proper choice of the lengths of the short segments, no crossings will be created during this procedure. So far this drawing, including the partially drawn pending edges between $V(G_2)$ and $V(G) \setminus V(G_2)$, will be equivalent to the drawing D_{G_2}. As the algorithm progresses, the vertical segments will be further extended above l_2, to form the middle segments of the corresponding edges. For an illustration, see Figure 3(a).

The remaining vertices $v_i, i > 2$, will be added to the drawing one by one, while maintaining the property that the drawing is equivalent to D_{G_i} and that the pending-order of the actual pending edges coincides with the order in which their vertical pieces reach the auxiliary line l_i. At the beginning of step $i + 1$, these conditions are obviously satisfied. Now we show how to place v_{i+1}.

Consider the set X of intersection points of the vertical (middle) pieces of all pending edges between $V(G_i)$ and $V(G) \setminus V(G_i)$ with the auxiliary line l_i. By Proposition 2, the intersection points corresponding to the pending edges incident to v_{i+1} must be consecutive elements of X. Let m be (one of) the median element(s) of X. Place v_{i+1} at a point above m, so that the x-coordinates of v_{i+1} and m coincide, and connect it to m. (In this way, the corresponding edge has only one bend, because its second and third piece are both vertical.) We also connect v_{i+1} to the upper endpoints of the appropriately extended vertical segments passing through the remaining elements of X, so that the directed slopes of the segments leaving v_{i+1} form a contiguous interval of regular slopes. For an illustration see Figure 3(b). Observe that this step can always be performed, because, by the definition of st-orderings, the number of edges leaving v_{i+1} is strictly smaller than d. This is not necessarily true in the last step, but then we have $v_n = t$, and we assumed that the degree of t was smaller than d. To complete this step, draw a horizontal auxiliary line l_{i+1} above v_{i+1} and extend the vertical portions of those pending edges between $V(G_i)$ and $V(G) \setminus V(G_i)$ that were not incident to v_{i+1} until they hit the line l_{i+1}. (These edges remain pending in the next step.) Finally, in a small vicinity of v_{i+1}, draw as many short segments from v_{i+1} using the remaining directed slopes as many pending edges connect v_{i+1} to $V(G) \setminus V(G_{i+1})$. Make sure that the directed slopes used at v_{i+1} form a contiguous interval and the circular order is the same as in D_G. Continue each of these short segments by adding a vertical piece that hits the line l_{i+1}. The resulting drawing, including the partially drawn pending edges, is equivalent to $D_{G_{i+1}}$.

In the final step, we place v_n and we obtain a drawing that meets the requiremenets.

4   Lower Bounds

In this section, we construct a sequence of planar graphs, providing a nontrivial lower bound for the planar slope number of bounded degree planar graphs. They also require more than the trivial number ($\lceil d/2 \rceil$) slopes, even if we allow one bend per edge. Remember that if we allow *two* bends per edge, then, by Theorem 3, for all graphs with maximum degree $d \geq 3$, except for the octahedral graph, $\lceil d/2 \rceil$ slopes are sufficient, which bound is optimal.

Theorem 4. *For any $d \geq 3$, there exists a planar graph G_d with maximum degree d, whose planar slope number is at least $3d - 6$. In addition, any drawing of G_d with at most one bend per edge requires at least $\frac{3}{4}(d-1)$ slopes.*

Proof. The construction of the graph G_d is as follows. Start with a graph of 6 vertices, consisting of two triangles, abc and $a'b'c'$, connected by the edges aa', bb', and cc' (see Figure 4(a)). Add to this graph a cycle C of length $3(d-3)$, and connect $d-3$ consecutive vertices of C to a, the next $d-3$ of them to b, and the remaining $d-3$ to c. Analogously, add a cycle C' of length $3(d-3)$, and connect one third of its vertices to a', one third to b', one third to c'. In the resulting graph, G_d, the maximum degree of the vertices is d.

In any crossing-free drawing of G_d, either C lies inside the triangle abc or C' lies inside the triangle $a'b'c'$. Assume by symmetry that C lies inside abc, as in Figure 4(a).

If the edges are represented by straight-line segments, the slopes of the edges incident to a, b, and c are all different, except that aa', bb', and cc' may have the same slope as some other edge. Thus, the number of different slopes used by any straight-line drawing of G_d is at least $3d - 6$.

Suppose now that the edges of G_d are represented by polygonal paths with at most one bend per edge. Assume, for simplicity, that every edge of the triangle

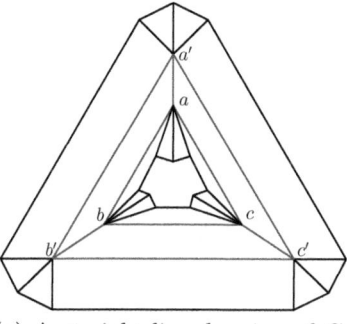

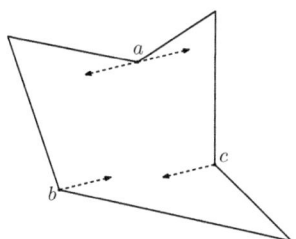

(a) A straight line drawing of G_6

(b) At most four segments starting from a, b, c can use the same slope in a drawing of G_d with one bend per edge

Fig. 4. Lower bounds

abc is represented by a path with exactly one bend (otherwise, an analogous argument gives an even better result). Consider the $3(d - 3)$ polygonal paths connecting a, b, and c to the vertices of the cycle C. Each of these paths has a segment incident to a, b, or c. Let S denote the set of these segments, together with the 6 segments of the paths representing the edges of the triangle abc.

We claim that the number of segments in S with any given slope is at most 4. The sum of the degrees of any polygon on k vertices is $(k - 2)\pi$. Every direction is covered by exactly $k - 2$ angles of a k-gon (counting each side $1/2$ times at its endpoints). Thus, if we take every other angle of a hexagon, then, even including its sides, every direction is covered at most 4 times (See Figure 4(b)).

The claim now implies that for any drawing of G with at most one bend per edge, we need at least $(3(d - 3) + 6)/4 = \frac{3}{4}(d - 1)$ different slopes.

References

1. Barát, J., Matoušek, J., Wood, D.: Bounded-degree graphs have arbitrarily large geometric thickness. Electronic J. Combinatorics 13(1) (2006) R3
2. Biedl, T., Kant, G.: A better heuristic for orthogonal graph drawings, Comput. Comput. Geom. 9, 159–180 (1998)
3. Di Battista, G., Eades, P., Tamassia, R., Tollis, I.G.: Graph Drawing. Prentice Hall, Upper Saddle River (1999)
4. Dujmović, V., Eppstein, D., Suderman, M., Wood, D.R.: Drawings of planar graphs with few slopes and segments. Comput. Geom. 38(3), 194–212 (2007)
5. Dujmović, V., Suderman, M., Wood, D.R.: Graph drawings with few slopes. Comput. Geom. 38, 181–193 (2007)
6. Duncan, C.A., Eppstein, D., Kobourov, S.G.: The geometric thickness of low degree graphs. In: Proc. 20th ACM Symp. on Computational Geometry (SoCG 2004), pp. 340–346. ACM Press, New York (2004)

7. Engelstein, M.: Drawing graphs with few slopes, Research paper submitted to the Intel Competition for high school students, New York (October 2005)
8. Eppstein, D.: Separating thickness from geometric thickness. In: Pach, J. (ed.) Towards a Theory of Geometric Graphs, Contemporary Mathematics, Amer. Math. Soc., Providence, pp. 75–86 (2004)
9. Fáry, I.: On straight line representation of planar graphs. Acta Univ. Szeged. Sect. Sci. Math. 11, 229–233 (1948)
10. de Fraysseix, H., de Mendez, P.O., Rosenstiehl, P.: On triangle contact graphs. Combinatorics, Probability and Computing 3, 233–246 (1994)
11. de Fraysseix, H., Pach, J., Pollack, R.: How to draw a planar graph on a grid. Combinatorica 10(1), 41–51 (1990)
12. Garg, A., Tamassia, R.: Planar drawings and angular resolution: Algorithms and bounds. In: van Leeuwen, J. (ed.) ESA 1994. LNCS, vol. 855, pp. 12–23. Springer, Heidelberg (1994)
13. Jelínek, V., Jelínková, E., Kratochvíl, J., Lidický, B., Tesař, M., Vyskočil, T.: The planar slope number of planar partial 3-trees of bounded degree. In: Eppstein, D., Gansner, E.R. (eds.) GD 2009. LNCS, vol. 5849, pp. 304–315. Springer, Heidelberg (2010)
14. Koebe, P.: Kontaktprobleme der konformen Abbildung. Berichte Verhand. Sächs. Akad. Wiss. Leipzig, Math.-Phys. Klasse 88, 141–164 (1936)
15. Keszegh, B., Pach, J., Plvlgyi, D., Tth, G.: Drawing cubic graphs with at most five slopes, Drawing cubic graphs with at most five slopes. Comput. Geom. 40(2), 138–147 (2008)
16. Lempel, A., Even, S., Cederbaum, I.: An algorithm for planarity testing of graphs. In: Rosenstiehl, P. (ed.) Theory of Graphs, pp. 215–232. Gordon and Breach, New York (1967)
17. Liu, Y., Morgana, A., Simeone, B.: General theoretical results on rectilinear embeddability of graphs. Acta Math. Appl. Sinica 7, 187–192 (1991)
18. Malitz, S., Papakostas, A.: On the angular resolution of planar graphs. SIAM J. Discrete Math. 7, 172–183 (1994)
19. Mukkamala, P., Szegedy, M.: Geometric representation of cubic graphs with four directions. Comput. Geom. 42, 842–851 (2009)
20. Pach, J., Pálvölgyi, D.: Bounded-degree graphs can have arbitrarily large slope numbers. Electronic J. Combinatorics 13(1) (2006) N1
21. Pach, J., Tóth, G.: Crossing number of toroidal graphs. In: Healy, P., Nikolov, N.S. (eds.) GD 2005. LNCS, vol. 3843, pp. 334–342. Springer, Heidelberg (2006)
22. Ungar, P.: On diagrams representing maps. J. London Math. Soc. 28, 336–342 (1953)
23. Wade, G.A., Chu, J.H.: Drawability of complete graphs using a minimal slope set. The Computer J. 37, 139–142 (1994)

Complexity of Finding Non-Planar Rectilinear Drawings of Graphs

Ján Maňuch[1,2,*], Murray Patterson[1,**],
Sheung-Hung Poon[3,***], and Chris Thachuk[1,†]

[1] Dept. of Computer Science, University of British Columbia, Vancouver BC, Canada
[2] Dept. of Mathematics, Simon Fraser University, Burnaby BC, Canada
[3] Dept. of Computer Science, National Tsing Hua University, Taiwan, R.O.C.

Abstract. We study the complexity of the problem of finding non-planar rectilinear drawings of graphs. This problem is known to be NP-complete. We consider natural restrictions of this problem where constraints are placed on the possible orientations of edges. In particular, we show that if each edge has prescribed direction "left", "right", "down" or "up", the problem of finding a rectilinear drawing is polynomial, while finding such a drawing with the minimum area is NP-complete. When assigned directions are "horizontal" or "vertical" or a cyclic order of the edges at each vertex is specified, the problem is NP-complete. We show that these two NP-complete cases are fixed parameter tractable in the number of vertices of degree 3 or 4.

1 Introduction

In this paper, we study the rectilinear (or bendless orthogonal) drawing of graphs, where each edge is drawn either as a horizontal or vertical line segment. Such drawings are important for various applications such as VLSI circuit design, entity-relationship diagrams for databases, flow chart drawings in software engineering, and subway-map design. This work is also motivated by the increasing research interest in *RAC (Right Angle Crossing)* drawing [1,7]. Note that a rectilinear drawing of a graph is a RAC-drawing with the additional property that the angles between adjacent incident edges around a vertex are multiples of $\pi/2$.

Formally, a *rectilinear drawing/embedding* of a graph $G = (V, E)$ is the pair of mappings $x, y : V \to \mathbb{Z}$, where $x(u)$ and $y(u)$ represent the x and y coordinates of vertex u on the rectangular grid, such that each edge $\{u, v\} \in E$ is mapped to endpoints of a horizontal or vertical line segment that does not contain any other mapped vertices but its endpoints u and v. Observe that if the maximum degree of G is larger than 4, then it does not have a rectilinear drawing. Such a

* Research supported by NSERC Discovery Grant.
** Research supported by NSERC PGS-D.
*** Research supported by NSC Grant 97-2221-E-007-054-MY3 in Taiwan.
† Research supported by NSERC PGS-D and MSFHR Trainee Award.

U. Brandes and S. Cornelsen (Eds.): GD 2010, LNCS 6502, pp. 305–316, 2011.
© Springer-Verlag Berlin Heidelberg 2011

drawing/embedding is called *planar* if none of embedded edges cross; otherwise, it is called *non-planar*. We remark here that all embedded edges are straight, and thus do not contain any bends.

There are several variants of the rectilinear drawing problem which put restrictions on how each edge is drawn. The most studied is the following variant. Associated with the input graph G is a function λ which assigns each oriented edge $\vec{e} = (u, v) \in \vec{E}$ of G one of the following four labels: L, R, D, and U, where $\lambda(u, v) = $ L (R) means edge $\vec{e}$ should be drawn horizontally to the left (right) of the source vertex u, and $\lambda(u, v) = $ D (U) means edge $\vec{e}$ should be drawn vertically below (above) the source vertex u. A graph with edges labeled in this way will be called an *LRDU-restricted graph* and is specified as $G = (V, E, \lambda)$. An *HV-restricted graph* $G = (V, E, \lambda)$ is a graph with each of its oriented edges $\vec{e} = (u, v)$ labeled with H or V. An edge $\vec{e} = (u, v)$ labeled with $\lambda(u, v) = $ H should be drawn on the plane horizontally; whereas an edge labeled $\lambda(u, v) = $ V should be drawn vertically. Furthermore, we also consider a different kind of restriction on the edges of the input graph, in which the cyclic ordering of the incident edges around each vertex is fixed in every rectilinear drawing of this graph. Such graphs will be called *cyclic-restricted*. Cyclic-restricted graphs with a planar embedding are exactly the so-called graphs with a fixed embedding. On the other hand, a fixed cyclic ordering of edges around a vertex is an important constraint in the definition of a fixed embedding condition for a non-planar graph [1]. This motivates us to investigate rectilinear drawings of such a kind of graphs. A graph with no restriction on the edges will be called *unrestricted*.

Planar rectilinear drawings have been extensively investigated in the literature of graph drawing. Garg and Tamassia [5] showed that deciding whether a graph is rectilinear planar is NP-complete. However, there are efficient algorithms, in fact linear-time algorithms, to find (construct) planar rectilinear drawings of *plane* graphs of maximum vertex degree three [13], subdivisions of planar triconnected cubic graphs [11], and series-parallel graphs of maximum vertex degree three [12]. These algorithms apply to unrestricted graphs. Some other research also considered the restricted variants. Vijayan and Wigderson [14] present a linear-time decision algorithm and a $O(n^2)$-time algorithm for finding a planar rectilinear drawing of an LRDU-restricted graph. Following this work, Hoffman and Kriegel [6] present an improved linear-time algorithm that finds such a drawing. However, Patrignani [10] showed that it is NP-complete to find a planar rectilinear drawing with minimum area for an LRDU-restricted graph. Recently, Eppstein [3] investigated the rectilinear drawing problem in three dimensions, and showed that it is NP-complete to determine whether an unrestricted graph has a rectilinear drawing with the constraint that at most two vertices lie on the same axis-parallel line.

On the other hand, non-planar rectilinear drawing has not been so well studied. Formann et al. [4] showed that it is NP-hard to decide whether a graph of maximum vertex degree 4 has a straight-line drawing with angular resolution $\frac{\pi}{2}$. This is equivalent to say that it is NP-hard to decide whether there is a drawing/embedding of an unrestricted graph. Further, Eades, Hong and Poon [2] showed that the problem is NP-hard even for an unrestricted graph consisting of 4-cycle blocks connected by single edges. In this paper, we investigate variants of the existence and the area-minimization problems of non-planar rectilinear drawings of given graphs. In particular, we show that the problem of deciding whether an LRDU-restricted graph has a rectilinear drawing (and finding such a drawing) is polynomial (Section 2.3), while the problem is NP-complete for HV-restricted graphs (Section 2.1) and cyclic-restricted graphs (Section 2.2). We then show that the NP-complete cases are fixed parameter tractable (FPT) in the number of vertices of degree 3 or 4 (Section 2.4). In addition, we show that finding a rectilinear drawing of an LRDU-restricted graph with minimum area is NP-complete as well (Section 3.1). The following table summarizes these results (the results marked with * follow immediately from the NP-completeness of the existence version):

Input Graph	Existence		Area-minimization
unrestricted	NP-c ([4])	FPT (Th. 4)	NP-c*
HV-restricted	NP-c (Th. 1)	FPT (Th. 4)	NP-c*
cyclic-restricted	NP-c (Th. 2)	FPT (Th. 4)	NP-c*
LRDU-restricted	P (Th. 3)		NP-c (Th. 5)

2   Existence of Rectilinear Drawings

2.1   Rectilinear Drawings of HV-Restricted Graphs

Theorem 1. *Given an HV-restricted graph G, the problem of deciding if G has a rectilinear drawing is NP-complete.*

Proof. It is clear that this problem is in NP. We show it is NP-hard by a reduction from the Betweenness (BTW) problem, proved to be NP-complete by Opatrny [8]. The input for the BTW problem is a set $S = \{1, \ldots, n\}$ and set of triples $t_i = (t_{i,1}, t_{i,2}, t_{i,3}) \in S^3$ for $i \in \{1, \ldots, k\}$, and the problem is to determine if there is an injective mapping $f : S \to \mathbb{Z}$ on the elements S, such that for every $i \in \{1, \ldots, k\}$, either $f(t_{i,1}) < f(t_{i,2}) < f(t_{i,3})$ or $f(t_{i,3}) < f(t_{i,2}) < f(t_{i,1})$.

Given an instance I of the BTW problem we construct an HV-restricted graph $G = (V, E, \lambda)$ on the following $3k$ vertices: for $i \in \{1, \ldots, k\}$ and every triple $t_i = (t_{i,1}, t_{i,2}, t_{i,3}) \in S^3$, we add the corresponding vertices $v_{i,1}$, $v_{i,2}$ and $v_{i,3}$. The idea of the construction is that the x-coordinates of the vertices of any rectilinear drawing x, y of G will correspond to values of their corresponding elements of S assigned by a solution f. The goal is then to add edges to the graph in such a way that (i) each triple constraint is enforced and that (ii) for every $s \in S$, the set $V_s = \{v_{i,j} \mid t_{i,j} = s\}$ of vertices in G that correspond to s are all assigned to the same x-coordinate.

To ensure (i), for $i \in \{1, \ldots, k\}$, for the set of vertices $v_{i,1}, v_{i,2}, v_{i,3}$ corresponding to each triple t_i we set $\lambda(v_{i,1}, v_{i,2}) = \lambda(v_{i,2}, v_{i,3}) = \mathrm{H}$. This ensures that for every rectilinear drawing x, y of G, that $x(v_{i,2})$ is between $x(v_{i,1})$ and

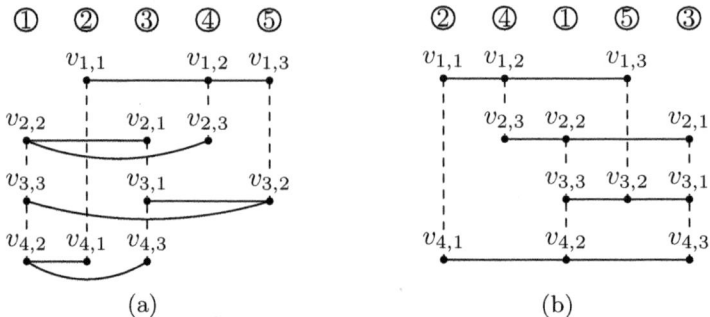

Fig. 1. (a) The input graph constructed for the instance of the BTW problem with triples $(2, 4, 5)$, $(3, 1, 4)$, $(3, 5, 1)$ and $(2, 1, 3)$. The dashed lines represent edges labeled with V. (b) Rectilinear drawing of this graph corresponding to the solution $f(2) < f(4) < f(1) < f(5) < f(3)$ of this instance.

$x(v_{i,3})$. Now to ensure (ii), for every $s \in S$, let $v_{i_1,j_1}, v_{i_2,j_2}, \ldots, v_{i_p,j_p}$ be the elements of V_s ordered such that $i_1 < i_2 < \cdots < i_p$. For $\ell \in \{1, \ldots, p-1\}$, we set $\lambda(v_{i_\ell,j_\ell}, v_{i_{\ell+1},j_{\ell+1}}) = $ V. This path containing vertical edges through the set V_s ensures that they have the same x-coordinate for every rectilinear drawing x, y of G. An example of a graph constructed for an instance of the BTW problem is shown in Figure 1(a). Figure 1(b) shows a rectilinear drawing of this graph. We now show that instance I has a solution if and only if G has a rectilinear drawing.

If I has a solution f, then we draw G as follows. The x-coordinates of vertices in V are set according to the injective mapping f and y-coordinates are set according to the order on the triples: $x(v_{i,j}) = f(t_{i,j})$ and $y(v_{i,j}) = i$. Obviously, the pair x, y satisfies all restrictions on edges. To show that it is a rectilinear drawing, we need the following lemma. For any $v \in V$, we denote $(x, y)(v) = (x(v), y(v))$.

Lemma 1. *For edge $\{v_{i,j}, v_{i',j'}\} \in E$, there is no vertex $v_{i'',j''} \in V \setminus \{v_{i,j}, v_{i',j'}\}$ such that $(x, y)(v_{i'',j''})$ lies on the line segment $L = [(x, y)(v_{i,j}), (x, y)(v_{i',j'})]$.*

Proof. Assume for contradiction that the image of vertex $v_{i'',j''} \in V \setminus \{v_{i,j}, v_{i',j'}\}$ lies on line segment L. If $\lambda(v_{i,j}, v_{i',j'}) = $ H, then $i = y(v_{i,j}) = y(v_{i',j'}) = i'$ and $j = j' \pm 1$. Since $(x, y)(v_{i'',j''})$ lies on L, we have $i'' = y(v_{i'',j''}) = i = i'$, i.e., $t_{i,j}$, $t_{i',j'}$ and $t_{i'',j''}$ are from the same triple t_i, i.e., $\{j, j', j''\} = \{1, 2, 3\}$. Furthermore, $x(v_{i'',j''}) = f(t_{i'',j''})$ is between $x(v_{i,j}) = f(t_{i,j})$ and $x(v_{i',j'}) = f(t_{i',j'})$. Hence, $j'' = 2$ which contradicts the fact that $j = j' \pm 1$.

If $\lambda(v_{i,j}, v_{i',j'}) = $ V, then we have $t_{i,j} = t_{i',j'} = s$ and $v_{i,j}, v_{i',j'} \in V_s$. Since $(x, y)(v_{i'',j''})$ lies on L, $f(t_{i'',j''}) = f(s)$. Since f is injective, $t_{i'',j''} = s$, i.e., $v_{i'',j''} \in V_s$. It follows that vertices $v_{i,j}, v_{i',j'}, v_{i'',j''}$ lie on the path of V-edges, hence, $(x, y)(v_{i'',j''})$ cannot lie on L in any rectilinear drawing of G. □

Conversely, assume G has a rectilinear drawing x, y. Such a drawing must satisfy conditions (i) and (ii) discussed above. By condition (ii), for every $s \in S$, every vertex in V_s has the same x-coordinate. We will construct mapping f by assigning this x-coordinate of vertices in V_s to s. To ensure that f is injective, it might be necessary to modify the drawing x, y slightly: if for two different values $s, s' \in S$, points in V_s and points in $V_{s'}$ are mapped to the same x-coordinate, we will slightly offset the x-coordinate of points in one of the two sets. After this modification, the pair x, y will remain a rectilinear drawing which satisfies all edge constraints. Now, condition (i) will guarantee that f is a solution to I.

Given an instance I to the BTW problem, we have constructed an HV-restricted graph G in time polynomial in the size of I that has a rectilinear drawing if and only if the instance I has a solution. Thus the problem of deciding if G has a rectilinear drawing is NP-hard.   □

2.2 Rectilinear Drawings of Cyclic-Restricted Graphs

Theorem 2. *Given a cyclic-restricted graph G, the problem of deciding if G has a rectilinear drawing is NP-complete.*

Proof. Clearly, the problem is in NP. In order to show that the problem is NP-hard, we give a reduction from the 3SAT problem. The input instance of the 3SAT problem is a set $\{x_1, x_2, \ldots, x_n\}$ of n variables, and a collection $\{c_1, c_2, \ldots, c_m\}$ of m clauses, where each clause consists of exactly three literals. The 3SAT problem is to determine whether there exists a truth assignment to the variables so that each clause has at least one true literal. In the following, we will describe our linear-time reduction, which is based on the construction of Formann *et al.* [4].

First we construct an L-shaped skeleton (denoted by K) by connecting a series of 4-cycles together and attaching ports that connect to variable towers and clause gadgets as depicted in Figure 2. The upward spikes sitting on the base of this L-shaped skeleton are the ports that connect to the variable towers, and the 2-edge paths hanging on the right hand side of the vertical column of the skeleton are ports that connect to the clause gadgets.

The variable tower for variable x_i is constructed and connected to the skeleton as shown in Figure 3. Since the cyclic order of the incident edges to every vertex is fixed, there are only two possible configurations of this tower as depicted in Figure 3(a) and (b). These will represent the *true* and *false* states, respectively, of variable x_i. The spikes $x_{i,j}$ and $\overline{x_{i,j}}$ on this variable tower will represent the literals x_i and $\overline{x_i}$, respectively, of clause c_j (if c_j contains x_i), where the truth value in c_j of this literal will correspond to the state of variable x_i, as determined by the configuration of its variable tower. Note that in each of these two configurations, the literals pointing to the right always have the *true* values.

Each clause gadget consists of three 3-edge paths connecting the port of this clause to spikes (on the corresponding variable towers) of the literals it contains. We illustrate the construction of clause gadgets with an example of constructing the gadget for clause $c_j = x_i \vee x_l \vee \overline{x_k}$, as depicted in Figure 4. Since one of these 3 paths incident to port c_j must contain an edge pointing to the right (depending on which direction this port is bent), this path forces its corresponding variable

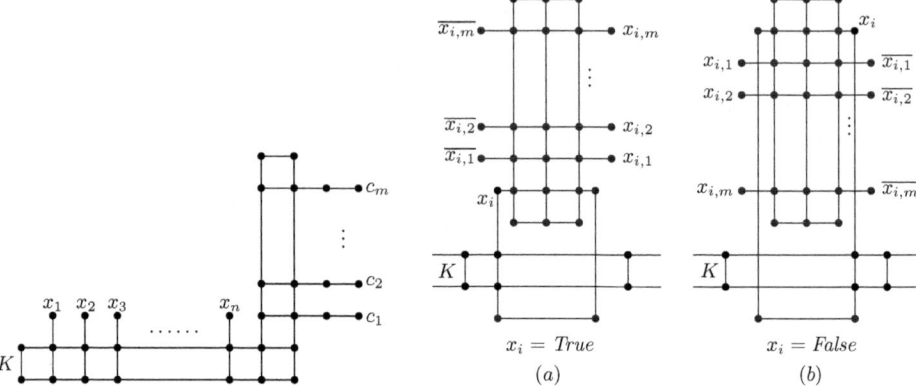

Fig. 2. L-shaped skeleton with ports that connect to variable towers and clause gadgets

Fig. 3. Variable tower for x_i and the representation of its truth values. Note that the literals on the right have value *true*.

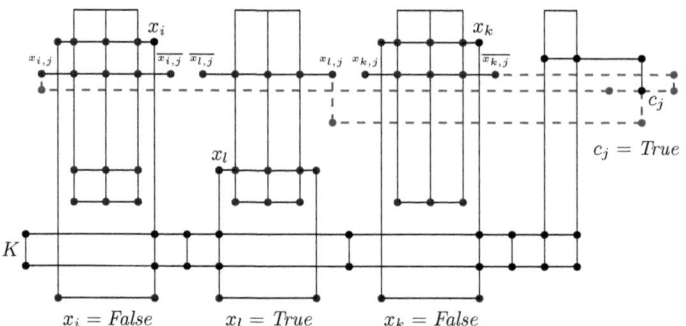

Fig. 4. An example of a clause gadget for clause $c_j = x_i \vee x_l \vee \overline{x_k}$. The port of c_j is bent down which forces the last literal $\overline{x_k}$ to have value *true*.

tower to be in the configuration that sets its literal in c_j to true, in this example, literal $\overline{x_k}$. Finally, it is easy to see that the 3SAT formula has a satisfying assignment if and only if the constructed graph has a rectilinear drawing. Thus this problem is NP-hard.   □

2.3   Rectilinear Drawings of LRDU-Restricted Graphs

Theorem 3. *Given an LRDU-restricted graph $G = (V, E, \lambda)$, the problem of deciding if G has a rectilinear drawing and finding such a drawing can be done in time $O(|V||E|)$.*

Proof. We will give a polynomial-time algorithm for the problem. Given LRDU-restricted graph $G = (V, E, \lambda)$, we first check to see if G satisfies the following necessary conditions: for every $u, v, w \in V$ and $X \in \{L, R, D, U\}$, (i) $\lambda(u, v) \in \{L, R, D, U\}$ iff $\{u, v\} \in E$; (ii) $\lambda(u, v) = L$ iff $\lambda(v, u) = R$ and $\lambda(u, v) = D$

iff $\lambda(v, u) = $ U; and (iii) if $\lambda(u, v) = X$ then $\lambda(w, v) \neq X$, i.e., that two edges cannot start at the same vertex and have the same direction. Checking (i) and (ii) each takes time $O(|E|)$, while checking (iii) takes time $O(|V||E|)$.

If this check succeeds, we will define equivalence relations E_x and E_y on the vertices of G, and construct partial orders P_x and P_y on equivalence classes of E_x and E_y, respectively, if possible. To construct E_x (E_y), for vertices $u, v \in V$, we set $u \equiv_{E_x} v$ ($u \equiv_{E_y} v$) when $\lambda(u, v) \in \{$D, U$\}$ ($\lambda(u, v) \in \{$L, R$\}$). Each class $V|_{E_x}$ ($V|_{E_y}$) of equivalence relation E_x (E_y) specifies then a set of vertices of G which must have the same x-coordinate (y-coordinate) in any rectilinear drawing of G. To construct P_x (P_y), for classes $A, B \in V|_{E_x}$ ($V|_{E_y}$), we set $A < B$ when there exist $u \in A$ and $v \in B$ such that $\lambda(u, v) = $ R ($\lambda(u, v) = $ U). The partial order P_x (P_y) on the classes of E_x (E_y) then specifies the partial ordering that the x-coordinates (y-coordinates) of these classes must have on the x-axis (y-axis) of the rectangular grid of this rectilinear drawing. We say that such a rectilinear drawing models (E_x, E_y, P_x, P_y). Orders P_x and P_y can be built (if they exist) independently of each other and in time $O(|V||E|)$. We will show that P_x and P_y exist if and only if G has a rectilinear drawing, i.e., to solve the problem it is enough to decide if partial orders P_x and P_y exist.

Since (E_x, E_y, P_x, P_y) express necessary conditions on any rectilinear drawing of G, if there exists a rectilinear drawing of G, then it models (E_x, E_y, P_x, P_y), i.e., P_x and P_y must exist. On the other hand, given (E_x, E_y, P_x, P_y) we can easily construct mappings x, y which satisfy all edge restrictions as follows. We first extend P_x (P_y) to any two total orders (this can be done in time $O(|V| + |E|)$). Then we assign to equivalence classes in $V|_{E_x}$ ($V|_{E_y}$) unique x (y) coordinates which respect these total orders, and for any $u \in A \in V|_{E_x}$ ($u \in B \in V|_{E_y}$) set $x(u)$ ($y(u)$) equal to the coordinate assigned to the class. Finally, we draw each edge $\{u, v\} \in E$ as line segment $L = [(x(u), y(u)), (x(v), y(v))]$. To show that mappings x, y form a rectilinear drawing, we need the following lemma.

Lemma 2. *For any pair of vertices $\{u, v\} \in E$, there is no vertex $w \in V \setminus \{u, v\}$ such that $(x(w), y(w))$ lies on the line segment $L = [(x(u), y(u)), (x(v), y(v))]$.*

Proof. Since $\{u, v\} \in E$, then $\lambda(u, v) \in \{$L, R, D, U$\}$, so it would be assigned to one of the equivalence classes in $V|_{E_x}$ or $V|_{E_y}$, i.e., $x(u) = x(v)$ or $y(u) = y(v)$. We can assume without loss of generality that $y(u) = y(v)$. Assume, for contradiction, that vertex $(x(w), y(w))$ lies on the line segment L. Thus $y(w) = y(u) = y(v)$, and the only way this can happen is when $w \equiv_{E_y} u \equiv_{E_y} v$, otherwise w would have been placed above or below u and v in the drawing. Now, if $(x(w), y(w))$ lies on the line segment L then either $x(u) \leq x(w) \leq x(v)$ or $x(v) \leq x(w) \leq x(u)$; without loss of generality, we assume the former. Since $x(u) \leq x(w) \leq x(v)$ and $w \equiv_{E_y} u \equiv_{E_y} v$, either there is a $u' \equiv_{E_y} u$ such that $x(u') \leq x(u)$ and $\lambda(u', w) = $ R or there is a $v' \equiv_{E_y} v$ such that $x(v) \leq x(v')$ and $\lambda(w, v') = $ R. In either case, this contradicts the fact that the check for property (iii) succeeded. □

By the above lemma, it follows that if P_x and P_y exist, we can find a rectilinear drawing of G in time $O(|V||E|)$. □

2.4 Fixed-Parameter Algorithms

In previous work, Eades, Hong and Poon showed that the problem of finding non-planar rectilinear drawings is fixed parameter tractable, where the parameter k is the number of vertices of degree 3 or 4, more precisely, they obtained that it can be solved in $O(24^k \cdot k^{2k} \cdot n)$-time [2]. In this work, we build on their idea to improve the runtime complexity of the algorithm and in addition consider rectilinear drawings for cyclic-restricted, and HV-restricted graphs. We sketch our proof here, and leave its detailed proof for the full version of this paper.

Theorem 4. *Given an unrestricted, cyclic-restricted, or HV-restricted degree-4 graph G of order n, a rectilinear drawing of G can be found in linear-time, or more precisely, in $O(24^k \cdot k^{2k+1} + n)$, $O(12^k \cdot k^{2k+1} + n)$, and $O(4^k \cdot k^{2k+1} + n)$-time, respectively, where k, the number of vertices of degree 3 or 4, is a constant.*

Proof (Sketch). Let K be the set of these k degree-3 or degree-4 vertices. We refer to any vertex $v \in K$ as a *high degree vertex*, or simply an *hd-vertex*. We call a path of degree-2 vertices connecting two *hd*-vertices an *hd-path*.

We first consider the case where G is unrestricted. Consider an *hd*-path p going from vertex u to vertex v. Suppose e_u and e_v are the two end edges of p incident to u and v, respectively. Depending on the current embedding of K, there are at most four choices of orientation for each of the two edges e_u and e_v. Thus there are at most 16 combinations. For each of these combinations, we know exactly how many edges, say m, path p needs to possess so that the connection between u and v can be built and routed around the other *hd*-vertices. We further know that m is at most five: if p possesses at least five edges, then no matter what the orientations of e_u and e_v are, the connection between u and v can be built. Hence, we perform a prepossessing step by traversing input graph G, to compute the lengths of all *hd*-paths. Since the maximum number of *hd*-paths is $2k$, this can be done in time proportional to the size of G, i.e., $O(n)$.

Since finding a routing of all *hd*-paths is sufficient for finding a rectilinear embedding of the entire graph G, we simply enumerate the number of possibilities to be checked to give us the running time of this algorithm. For any embedding of the vertices in K on the plane, it is possible that no two share a common coordinate, thus resulting in k unique horizontal and vertical coordinates. Therefore, any embedding in the plane can be considered a distribution of the vertices of K on a $k \times k$ grid. Hence, there are no more than $(k^2)^k = k^{2k}$ possible embeddings to consider. Since each vertex $v \in K$ is incident to at most four edges, each having one of four possible orientations, there are at most $4! = 24$ possible orientations of edges incident to v, and at most 24^k possible orientations for all vertices in K. Therefore, considering this number of possibilities, the time to check them and the initial length calculation of all $O(k)$ *hd*-paths, a rectilinear drawing for G (if one exists) can be found in $O(24^k \cdot k^{2k+1} + n)$-time.

Finding a rectilinear drawing when G is cyclic-restricted (resp. HV-restricted) needs to consider at most 12 (resp. 4) possible orientations of edges incident to each hd-vertex resulting in $O(12^k \cdot k^{2k+1} + n)$ (resp. $O(4^k \cdot k^{2k+1} + n)$) time overall. Note that the HV-restricted case also considers the sequence of H and V transitions along each hd-path in the same time bound.   □

3   Area-Minimization Drawings

3.1   Rectilinear Drawings of LRDU-Restricted Graphs

Theorem 5. *Given an LRDU-restricted graph G, the problem of deciding if G has a rectilinear drawing with minimum area is NP-complete.*

Proof. We show that this problem is NP-complete by reduction from 3SAT(3), a restricted version of 3SAT, proved NP-complete by Papadimitriou [9], where each variable appears exactly twice positive and once negated in the clauses. Given an instance ϕ of 3SAT(3) with n variables and m clauses, we construct a graph G which has a rectilinear drawing on a $10m \times 6n$ grid if and only if ϕ is satisfiable. Graph G will consist of $m \times n$ blocks, where each block is of three different types depicted in Figure 5. The i-th row of blocks corresponds to variable x_i and the j-th column of blocks corresponds to clause c_j. We will refer to the block in the i-th row and the j-th column, as the block at position (i, j). If neither x_i nor $\bar{x}_i$ appear in c_j, then the block at position (i, j) is the no-occurrence block depicted in Figure 5(a). If x_i appears in c_j, the positive-occurrence block in Figure 5(b) is used. If $\bar{x}_i$ appears in c_j, the negative-occurrence block in Figure 5(c) is used.

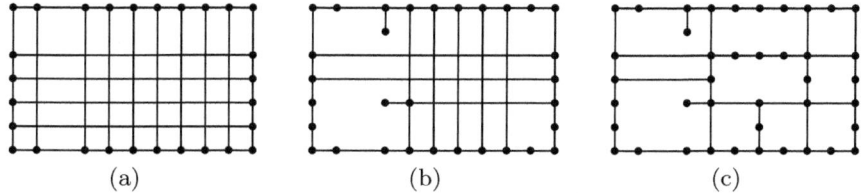

(a)   (b)   (c)

Fig. 5. Blocks: (a) a no-occurrence block — used if the variable does not occur in the clause; (b) a positive-occurrence block — used if the variable has a positive occurrence in the clause; (c) a negative-occurrence block — used if the variable has a negative occurrence in the clause

In addition, G contains a clause line and a variable line for each clause and each variable, respectively. The clause line for clause c_j starts at position $(2, 0)$ of the last block in the j-th column and ends at position $(2, 6)$ of the first block in this column. It contains $k + 2$ vertical segments, where k is the number of literals it contains. Hence, it is a vertical line spanning the whole grid with $k + 1$ internal points on it. The variable line for variable x_i starts at position $(0, 5)$ of the first block in the i-th row and ends at position $(10, 5)$ of the last block in this row. It contains the following segments RDRURDRUR (two "bumps").

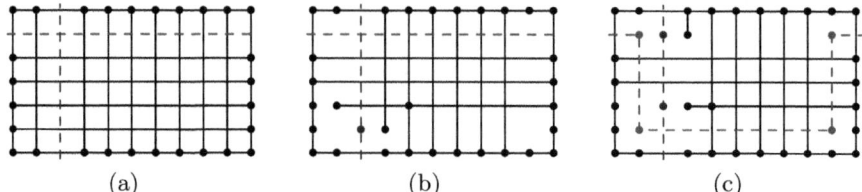

Fig. 6. (a) The clause and variable lines passing through the no-occurrence block at position (i,j), i.e., variable x_i has no occurrence in c_j. Note that if the block is in the first column (last column, first row, last row) then the variable (variable, clause, clause) line is attached to the left (right, top, bottom) of the frame. (b-c) The clause and variable lines passing through the positive-occurrence block at position (i,j), i.e., variable x_i has a positive occurrence in c_j.

Now, let us analyze how the clause line for c_j and the variable line for x_i can pass through the block at position (i,j). This will depend on the type of the block. For the no-occurrence block depicted in Figure 5(a) there is only one way that the clause and variable lines can pass through the block; see Figure 6(a). Note that no internal vertex can be placed on the clause line and no bump can occur on the variable line inside this block.

For the positive-occurrence block depicted in Figure 5(b) there are two ways that the clause and variable lines can pass through the block. They are depicted in Figure 6(b)–(c). Note that the variable line can contain at most one bump inside this block. In the case where it contains no bump (passes through directly), at most one internal vertex can be placed on the clause line inside this block. If it contains a bump, then at most two internal vertices can be placed on the clause line inside this block. Figure 6(b)–(c) depicts the possibilities with the maximal number of internal vertices on the clause line.

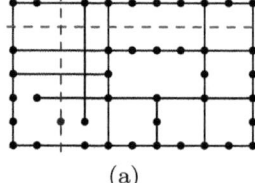 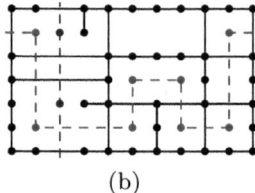

Fig. 7. The clause and variable lines passing through the negative-occurrence block at position (i,j), i.e., variable x_i has a negative occurrence in c_j

For the negative-occurrence block depicted in Figure 5(c) there are two ways that the clause and variable lines can pass through the block. They are depicted in Figure 7. Note that the variable line contains either zero or two bumps inside this block. In the case where it contains no bump (passes through directly), similarly as for the positive-occurrence block, at most one internal vertex can be placed on the clause line inside this block. If it contains two bumps, then at most two internal vertices can be placed on the clause line inside this block. Figure 7 depicts the possibilities with the maximal number of internal vertices on the clause line.

Now, we are ready to show that the constructed graph has a rectilinear drawing if and only if the 3SAT(3) instance is satisfiable. First, let us consider the variable lines. Each variable line contains two bumps. Since the variable line for x_i cannot make any bumps in the no-occurrence blocks, it must make bumps in the remaining three blocks. Let c_{j_1}, c_{j_2} (c_{j_3}) be the clauses containing a positive (negative) occurrence of x_i. The variable line for x_i can contain zero or one bump in the blocks at positions (i, j_1) and (i, j_2), and zero or two bumps in the block at position (i, j_3). It follows that we have two mutually-exclusive possibilities: either the variable line makes bumps in its positive-occurrence blocks (one in each of them) or it contains two bumps in its negative-occurrence block. In the first case, we set the value of variable x_i to *true*, in the second case, to *false*.

Next, we show that each clause is satisfied in the constructed assignment. Recall that each clause line contains $k+1$ internal vertices, where k is the number of literals in the clause. Consider a clause c_j. Since the clause line cannot contain any internal vertices inside the no-occurrence blocks it passes through, the $k+1$ internal points have to appear in the remaining k blocks. It follows that in at least one of them the clause line will contain two internal vertices. However, by the above analysis and definition of the assignment, the corresponding literal must be set to *true*, hence, clause c_j is satisfied.

Given a truth assignment for the 3SAT(3) instance, we can construct a rectilinear drawing for the graph as follows. First, for the variable line for x_i, we place the bumps in the blocks at position (i, j), where c_j contains an occurrence of the variable x_i and this occurrence (positive or negative) has value *true*. Second, for the clause line for c_j, we place two internal vertices on the clause line inside the block at position (i, j), where c_j contains an (positive or negative) occurrence of x_i which makes c_j satisfied. We might have several choices for i, but we pick one of them. Then we place one internal vertex on the clause line in the remaining blocks at positions (i', j), where c_j contains an occurrence of $x_{i'}$ and $i' \neq i$. Now, each variable line contains exactly two bumps and each clause line for a k-clause contains exactly $k + 1$ internal vertices, hence, we have a rectilinear drawing of the graph.   □

4   Conclusions

Previous work has shown the problem of finding non-planar rectilinear drawings for graphs to be NP-complete [2,4]. In this work, we have resolved the complexity of a number of natural restrictions where constraints are placed on the possible orientations of edges. In particular, we show that determining the existence of a non-planar rectilinear drawing for a graph when directions are prescribed to each edge is polynomial, while determining a minimum-area drawing for the same case is NP-complete. When edges are prescribed to be either horizontal or vertical, or when a cyclic order of the incedent edges around each vertex is prescribed, we show that the existence problem, and thus the area-minimization problem, is NP-complete. Finally, we have shown the NP-complete existence problems to be fixed parameter tractable in the number of vertices of degree 3 or 4.

It remains open whether or not the corresponding minimum-area drawing cases are also fixed parameter tractable. Since there are polynomial time algorithms for finding planar rectilinear drawings for several different classes of maximum degree 3 graphs [13,11,12], it would be interesting to find such classes for the non-planar case. Further interesting open questions are the complexity of finding planar rectilinear drawings of HV-restricted and cyclic-restricted graphs.

References

1. Didimo, W., Eades, P., Liotta, G.: Drawing graphs with right angle crossings. In: Dehne, F., Gavrilova, M., Sack, J.-R., Tóth, C.D. (eds.) WADS 2009. LNCS, vol. 5664, pp. 206–217. Springer, Heidelberg (2009)
2. Eades, P., Hong, S.-H., Poon, S.-H.: On rectilinear drawing of graphs. In: Eppstein, D., Gansner, E.R. (eds.) GD 2009. LNCS, vol. 5849, pp. 232–243. Springer, Heidelberg (2010)
3. Eppstein, D.: The topology of bendless three-dimensional orthogonal graph drawing. In: Tollis, I.G., Patrignani, M. (eds.) GD 2008. LNCS, vol. 5417, pp. 78–89. Springer, Heidelberg (2009)
4. Formann, M., Hagerup, T., Haralambides, J., Kaufmann, M., Leighton, F.T., Symvonis, A., Welzl, E., Woeginger, G.: Drawing graphs in the plane with high resolution. SIAM Journal on Computing 22(5), 1035–1052 (1993)
5. Garg, A., Tamassia, R.: On the computational complexity of upward and rectilinear planarity testing. SIAM Journal of Computing 31(2), 601–625 (2001)
6. Hoffman, F., Kriegel, K.: Embedding rectilinear graphs in linear time. Information Processing Letters 29(2), 75–79 (1988)
7. Huang, W., Hong, S., Eades, P.: Effects of crossing angles. In: Proc. of IEEE Pacific Visualization Symposium (PacificVis 2008), pp. 41–46 (2008)
8. Opatrny, J.: Total ordering problem. SIAM Journal of Computing 8(1), 111–114 (1979)
9. Papadimitriou, C.H.: Computational Complexity. Addison-Wesley, Reading (1994)
10. Patrignani, M.: On the complexity of orthogonal compaction. Computational Geometry 19, 47–67 (2001)
11. Rahman, M.S., Egi, N., Nishizeki, T.: No-bend orthogonal drawings of subdivisions of planar triconnected cubic graphs. In: Liotta, G. (ed.) GD 2003. LNCS, vol. 2912, pp. 387–392. Springer, Heidelberg (2004)
12. Rahman, M.S., Egi, N., Nishizeki, T.: No-bend orthogonal drawings of series-parallel graphs. In: Healy, P., Nikolov, N.S. (eds.) GD 2005. LNCS, vol. 3843, pp. 409–420. Springer, Heidelberg (2006)
13. Rahman, M.S., Naznin, M., Nishizeki, T.: Orthogonal drawings of plane graphs without bends. Journal of Graph Algorithms and Applications 7(4), 335–362 (2003)
14. Vijayan, G., Wigderson, A.: Rectilinear graphs and their embeddings. SIAM Journal of Computing 14(2), 355–372 (1985)

Point-Set Embeddings of Plane 3-Trees
(Extended Abstract)

Rahnuma Islam Nishat, Debajyoti Mondal, and Md. Saidur Rahman

Graph Drawing and Information Visualization Laboratory,
Department of Computer Science and Engineering, Bangladesh University of
Engineering and Technology (BUET), Dhaka - 1000, Bangladesh
nishat.buet@gmail.com, debajyoti_mondal_cse@yahoo.com,
saidurrahman@cse.buet.ac.bd

Abstract. A straight-line drawing of a plane graph G is a planar drawing of G, where each vertex is drawn as a point and each edge is drawn as a straight-line segment. Given a set S of n points on the Euclidean plane, a point-set embedding of a plane graph G with n vertices on S is a straight-line drawing of G, where each vertex of G is mapped to a distinct point of S. The problem of deciding if G admits a point-set embedding on S is NP-complete in general and even when G is 2-connected and 2-outerplanar. In this paper we give an $O(n^2 \log n)$ time algorithm to decide whether a plane 3-tree admits a point-set embedding on a given set of points or not, and find an embedding if it exists. We prove an $\Omega(n \log n)$ lower bound on the time complexity for finding a point-set embedding of a plane 3-tree. Moreover, we consider a variant of the problem where we are given a plane 3-tree G with n vertices and a set S of $k > n$ points, and give a polynomial time algorithm to find a point-set embedding of G on S if it exists.

Keywords: Point-set embedding, Plane 3-tree, Lower bound.

1 Introduction

A *straight-line drawing* Γ of a plane graph G is a planar drawing of G, where each vertex is drawn as a point and each edge is drawn as a straight-line segment. The problem of computing a straight-line drawing of a graph where the vertices are constrained to be located at integer grid points is a classical problem in the graph drawing literature [6,12]. One of the variants of this problem is to compute a planar embedding of a graph G on a set of points S where the points are located on the Euclidean plane [3,9,11].

Let G be a plane graph of n vertices and S be a set of n points on the Euclidean plane. A *point-set embedding of G on S* is a straight-line drawing of G, where each vertex of G is mapped to a distinct point of S. We do not restrict the points of S to be in general position. In other words, three or more points in S may be collinear. Figure 1(a) and (b) depict two sets S and S' of 10 points, respectively. Figure 1(c) depicts a plane graph G of 10 vertices. One can easily

U. Brandes and S. Cornelsen (Eds.): GD 2010, LNCS 6502, pp. 317–328, 2011.
© Springer-Verlag Berlin Heidelberg 2011

observe that G admits a point-set embedding on S', as illustrated in Fig. 1(d). But G does not admit a point-set embedding on S since the convex hull of S contains four points whereas the outer face of G has three vertices.

A rich body of literature has been published on the point-set embeddings when the input graph G is restricted to trees or outerplanar graphs [8,3]. Cabello [4] proved that the problem is NP-complete for planar graphs in general and even when the input graph is 2-connected and 2-outerplanar. Recently, Garcia *et al.* have given a characterization of a set of points S such that there exists a 3-connected cubic plane graph that admits a point-set embedding on S [7].

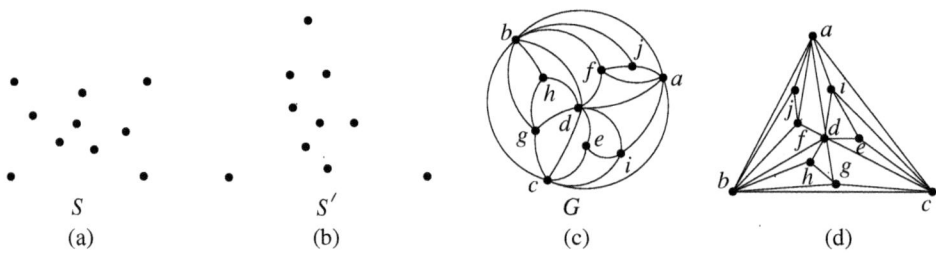

Fig. 1. (a) A set S of 10 points, (b) a set S' of 10 points, (c) a plane graph G of 10 vertices, and (d) a point-set embedding of G on S'

In this paper we consider the problem of obtaining point-set embeddings of "plane 3-trees". A *plane 3-tree* G of $n \geq 3$ vertices is a plane graph for which the following (a) and (b) hold: (a) G is a triangulated plane graph; (b) if $n > 3$, then G has a vertex whose deletion gives a plane 3-tree G' of $n - 1$ vertices. We give an $O(n^2 \log n)$ time algorithm that decides whether a plane 3-tree G admits a point-set embedding on a given set S of n points or not; and computes a point-set embedding of G if such an embedding exists. We prove an $\Omega(n \log n)$ lower bound on the time complexity for obtaining a point-set embedding of a plane 3-tree of n vertices on a set of n points. Furthermore we give a polynomial-time algorithm to decide whether a plane 3-tree G of n vertices admits a point-set embedding on a set of $k > n$ points.

The rest of this paper is organized as follows. Section 2 presents some definitions and preliminary results. Section 3 gives an $O(n^2 \log n)$ time algorithm to obtain a point-set embedding of a plane 3-tree of n vertices, if it exists. Section 4 shows an $\Omega(n \log n)$ lower bound on the running time for computing point-set embeddings of plane 3-trees. Section 5 gives an $O(nk^8)$ time algorithm to decide whether a plane 3-tree G of n vertices admits a point-set embedding on a set of $k > n$ points. Finally, Section 6 concludes the paper suggesting future works.

2   Preliminaries

In this section we give some relevant definitions that will be used throughout the paper and present some preliminary results.

For the graph theoretic definitions which have not been described here, see [10].

A graph is *planar* if it can be embedded in the plane without edge crossings except at the vertices where the edges are incident. A *plane graph* is a planar graph with a fixed planar embedding. If all the faces of a plane graph G are triangles, then G is called a *triangulated plane graph*. A plane graph G with $n \geq 3$ vertices is called a *plane 3-tree* if the following (a) and (b) hold. (a) G is a triangulated plane graph. (b) if $n > 3$, then G has a vertex whose deletion gives a plane 3-tree G' of $n - 1$ vertices. We denote by G_n a plane 3-tree of n vertices. The following results are known on plane 3-trees [2].

Lemma 1. *Let G_n be a plane 3-tree of n vertices where $n > 3$. Then the following (a) and (b) hold. (a) G_n has an inner vertex x of degree three such that the removal of x gives the plane 3-tree G_{n-1} of $n - 1$ vertices. (b) G_n has exactly one inner vertex p which is the common neighbor of all the three outer vertices of G_n.*

We call p the *representative vertex* of G_n. Let G be a plane graph. For a cycle C in G, we denote by $G(C)$ the plane subgraph of G inside C (including C). A *separating triangle* of G is a triangle in G whose interior and exterior contain at least one vertex each. We now have the following lemma.

Lemma 2. *Let G_n be a plane 3-tree of $n > 3$ vertices and C be any triangle of G_n. Then the subgraph $G_n(C)$ is a plane 3-tree.*

Let T be a rooted tree and i be any vertex of T. Then we define a *subtree $T(i)$ rooted at i* as a subgraph of T induced by the vertex i and all the descendants of i. An *ordered rooted tree* is a rooted tree where the children of any vertex are ordered counter-clockwise.

Let G_n be a plane 3-tree of n vertices. Let p be the representative vertex and a, b, c be the outer vertices of G_n. The vertex p, along with the three outer vertices a, b and c, form three triangles abp, bcp and cap. We call those three triangles the *nested triangles around p*. We now define a tree T_{n-3} with the inner vertices of G_n which we call the "representative tree" of G_n. The *representative tree* of G_n is an ordered rooted tree T_{n-3} satisfying the following two conditions (a) and (b).

(a) T_{n-3} consists of a single vertex if $n = 3$.
(b) if $n > 3$, then the root p of T_{n-3} is the representative vertex of G_n and the subtrees rooted at the three counter-clockwise ordered children p_1, p_2 and p_3 of p in T_{n-3} are the representative trees of $G_n(C_1)$, $G_n(C_2)$ and $G_n(C_3)$, respectively, where C_1, C_2 and C_3 are the three nested triangles around p in counter-clockwise order.

Figure 2(a) depicts a plane 3-tree G_n and Fig. 2(b) depicts the representative tree T_{n-3} of G_n. We now have the following lemma whose proof has been omitted.

Lemma 3. *Let G_n be any plane 3-tree of $n \geq 3$ vertices. Then G_n has a unique representative tree T_{n-3} with exactly $n - 3$ internal vertices and $2n - 5$ leaves. Moreover, T_{n-3} can be found in time $O(n)$.*

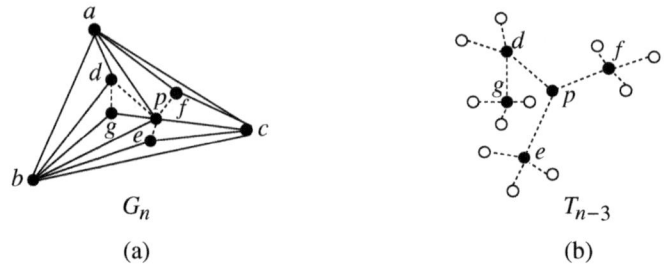

Fig. 2. (a) A plane 3-tree G_n and (b) the representative tree T_{n-3} of G_n

We now have the following lemma whose proof is immediate from the definition of the representative tree and Lemma 3.

Lemma 4. *Let T_{n-3} be the representative tree of a plane 3-tree G_n of $n \geq 3$ vertices and let $T(i)$ be an ordered subtree rooted at a vertex i of T_{n-3}. Then there exists a unique triangle C in G_n such that $T(i)$ is the representative tree of $G_n(C)$.*

By Lemma 4, for any vertex p of T_{n-3}, there is a unique triangle in G_n which we denote as C_p. For the rest of this paper, we shall often use an internal vertex p of T_{n-3} and the representative vertex of $G_n(C_p)$ interchangeably.

3 Point-Set Embeddings of Plane 3-Trees

In this section we give an $O(n^2 \log n)$ time algorithm to decide whether a plane 3-tree G of n vertices has a point-set embedding on a set S of n points or not, and obtain a point-set embedding of G on S if it exists.

Before presenting the detail of our algorithm we focus on some properties of the point-set embeddings of plane 3-trees. Let S be a set of n points on the Euclidean plane. The *convex-hull* of S is the smallest convex polygon that encloses all the points in S. Let G be a plane 3-tree of n vertices. In any point-set embedding of G the outer face of G is drawn as a triangle and hence the following fact holds.

Fact 1. *Let G be a plane 3-tree of n vertices and S be a set of n points. If G admits a point-set embedding on S, then the convex-hull of S contains exactly three points in S.*

By Fact 1, G has no point-set embedding on S if the convex-hull of S does not contain exactly three points in S. We thus assume that the convex-hull of S contains exactly three points. One may observe that the outer vertices of G can be mapped to the three points on the convex-hull of S in six ways, and hence we need to check whether G admits a point-set embedding on S or not for each of the six mappings. In the remaining of this section we give an algorithm to check whether G admits a point-set embedding on S for a given mapping of the outer vertices of G to the three points of the convex-hull of S.

Let T_{n-3} be the representative tree of G and let p be the root of T_{n-3}. Let C_{p_1}, C_{p_2}, C_{p_3} be the three nested triangles around p. By lemma 2, $G_n(C_{p_1})$, $G_n(C_{p_2})$ and $G_n(C_{p_3})$ are three plane 3-trees which have the corresponding unique representative trees $T_{n_1}(p_1)$, $T_{n_2}(p_2)$ and $T_{n_3}(p_3)$ where n_1, n_2 and n_3 are the number of internal vertices of $T_{n_1}(p_1)$, $T_{n_2}(p_2)$ and $T_{n_3}(p_3)$, respectively. Let C be a cycle in G. We denote by $\Gamma(C)$ the embedding of C on some points of S. We call a mapping of p to a point $x \in S$ a *valid mapping* of p if the proper interiors of $\Gamma(C_{p_1})$, $\Gamma(C_{p_2})$ and $\Gamma(C_{p_3})$ contain n_1, n_2 and n_3 points, respectively. We now have the following lemma whose proof is omitted in this extended abstract.

Lemma 5. *Let G be a plane 3-tree of n vertices, let a, b and c be the three outer vertices of G, and let p be the representative vertex of G. Let S be a point set of n points such that the convex hull of S contains exactly three points. Assume that G has a point-set embedding $\Gamma(G)$ on S for a given mapping of a, b and c to the three points of the convex-hull of S. Then p has a unique valid mapping for the given mapping of a, b and c.*

Based on Fact 1 and Lemma 5 we now describe an algorithm to find a point-set embedding of a plane 3-tree G on a point-set S. The algorithm first computes the convex-hull of S and then recursively finds valid mappings of the representative vertices. A naive approach to find a valid mapping of a representative vertex p is as follows. For each point $u \in S$ interior to $\Gamma(C_p)$, we assume u as the representative vertex and check whether $\Gamma(C_{p_1})$, $\Gamma(C_{p_2})$ and $\Gamma(C_{p_3})$ contains the required number of points. One can easily compute a valid mapping of p in $O(n^2)$ time. Now, if we compute the mappings of the representative vertices for the plane 3-trees $G_n(C_{p_1})$, $G_n(C_{p_2})$ and $G_n(C_{p_3})$ in a recursive fashion, we can obtain a point-set embedding of G. Since there are $n - 3$ representative vertices, computation of the final embedding takes $O(n) \times O(n^2) = O(n^3)$ time. We call this algorithm **Point-set-embedding**, and in the rest of this section we give a faster method to find a valid mapping of a representative vertex and use it to implement Algorithm **Point-set-embedding** in $O(n^2 \log n)$ time.

We first compute a convex-hull of S in $O(n \log h)$ time [5], where h is the number of points on the hull. If G admits a point-set embedding on S, then the convex-hull contains exactly three points in S. For a given mapping of the three outer vertices a, b, c of G to the three points on the convex hull of S, we sort the points interior to the triangle abc by increasing polar angle around a in clockwise order. Then we put the sorted list in an array A_a as illustrated in Fig. 3. For the points with the same slope, we keep a pointer to the point closest to a from the other points with the same slope. The point closest to a is the *parent* of the other points with the same slope. Let x and y be any two points where y has a pointer to x. Then y is a *child* of x. We also maintain a count of the children of a parent in A_a. In a similar way we construct A_b and A_c for b and c, respectively. Clearly, the construction of these three arrays takes $O(n \log n)$ time.

We next take any point $u \in S$, interior to the triangle abc, and draw straight lines through (u, a), (u, b) and (u, c) which intersect bc, ac, ab at p, q and r, respectively as illustrated in Fig 4. Thus the region inside the triangle abc gets

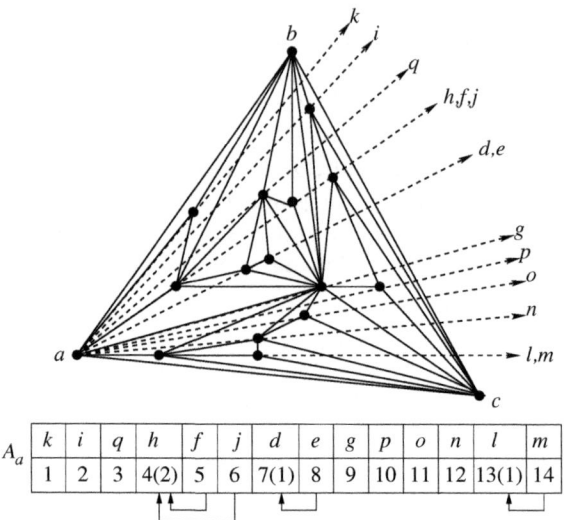

Fig. 3. Illustration for the construction of the array A_a

split into six disjoint regions which we denote as $x_1, x_2, \ldots, x_6$. The regions x_1, x_2, x_3, x_4, x_5, x_6 are bounded by the triangles aur, auq, cuq, cup, bup, bur, respectively. Moreover, by x_7, x_8 and x_9 we denote the three lines shared by region x_1 and x_6, x_2 and x_3, x_4 and x_5, respectively. In the remaining of this section we also denote by $x_i, 1 \leq i \leq 9$, the number of points of S in the region x_i.

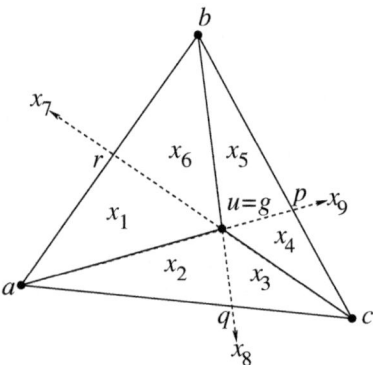

Fig. 4. Computation of the mapping of the representative vertex

Let T_{n-3} be the representative tree of G and p be the root of T_{n-3}. Let n_1, n_2 and n_3 be the number of vertices of the three subtrees rooted at the three children of p. We now formulate a set of nine linear equations and solve them to check whether u is a valid mapping for p. The three constraints are $x_2 + x_8 + x_3 = n_1$, $x_4 + x_9 + x_5 = n_2$ and $x_1 + x_7 + x_6 = n_3$ which can be obtained easily. For example, the constraints are $x_2 + x_8 + x_3 = 4$, $x_4 + x_9 + x_5 = 3$ and $x_1 + x_7 + x_6 = 6$ for the graph of Fig. 3.

The nine equations can be obtained using the three straight lines (u, a), (u, b) and (u, c) as follows. The straight line (u, a) splits the triangle abc into two disjoint regions $x_1 + x_5 + x_6 + x_7$ and $x_2 + x_3 + x_4 + x_8$. The number of points in those regions are $x_1 + x_5 + x_6 + x_7 = u_i - 1$ and $x_2 + x_3 + x_4 + x_8 = |A_a| - u_i - u_a$ where u_i is the index of u in A_a and u_a is the count of children of u in A_a. The number of points on x_9 is equal to u_a which gives another equation $x_9 = u_a$. Let u_j and u_k be the indices of a point u in A_b and A_c, respectively and let u_b and u_c be the counts of children of u in A_b and A_c, respectively. Then we can derive six other equations using (u, b) and (u, c) in a similar way as described above. The nine equations for any point u, interior to the triangle abc, are $x_1 + x_5 + x_6 + x_7 = u_i - 1$, $x_2 + x_3 + x_4 + x_8 = |A_a| - u_i - u_a$, $x_9 = u_a$, $x_5 + x_9 + x_4 + x_3 = u_j - 1$, $x_2 + x_1 + x_7 + x_6 = |A_b| - u_j - u_b$, $x_8 = u_b$, $x_3 + x_8 + x_2 + x_1 = u_k - 1$, $x_4 + x_9 + x_5 + x_6 = |A_c| - u_k - u_c$ and $x_7 = u_c$. When the vertex g is mapped to the point u in the graph of Fig. 3, the equations become $x_1 + x_5 + x_6 + x_7 = 8$, $x_2 + x_3 + x_4 + x_8 = 5$, $x_9 = 0$, $x_5 + x_9 + x_4 + x_3 = 3$, $x_2 + x_1 + x_7 + x_6 = 10$, $x_8 = 0$, $x_3 + x_8 + x_2 + x_1 = 7$, $x_4 + x_9 + x_5 + x_6 = 5$ and $x_7 = 1$.

If we get a unique solution of the set of linear equations, then u is a valid mapping of p by Lemma 5. A simple "Gaussian elimination" to solve a system of t equations for t unknowns requires $O(t^3)$ time. Here $t = 9$ and hence we can verify whether u is a valid mapping of p in $O(1)$ time. Similarly we check the points inside the triangle abc, other than u, to obtain a valid mapping of p. Since there are $O(n)$ points inside the triangle abc, this step takes $O(n) \times O(1) = O(n)$ time.

Finally, we find the valid mappings of representative vertices for the smaller plane 3-trees in a recursive fashion and obtain a point-set embedding of G. At each recursive step we construct three sorted arrays in $O(n \log n)$ time and find the mapping of the representative vertex in $O(n)$ time. Since there are $O(n)$ representative vertices, computation of the final embedding takes $O(n) \times (O(n \log n) + O(n)) = O(n^2 \log n)$ time. We now recall that the computation of the convex-hull takes $O(n \log h)$ time and there are six ways of mapping the three outer vertices of G to the three points on the convex-hull of S. Therefore the time taken for deciding whether G admits a point-set embedding on S is $(O(n \log h) + 6 \times O(n^2 \log n)) = O(n^2 \log n)$ time. We now have the following theorem.

Theorem 1. *Given a plane 3-tree G of n vertices and a point-set S of n-points, Algorithm **Point-set-embedding** computes a point-set embedding of G on S in $O(n^2 \log n)$ time if such an embedding exists.*

One can observe that the bottle-neck of Algorithm **Point-set-embedding** is to construct three sorted arrays at each recursive step; which takes $O(n \log n)$ time at each step and $O(n^2 \log n)$ time in total. But if we assume that the points are in general position, we can construct the sorted arrays for all the points collectively at the initial step of the algorithm in $O(n^2)$ time using "arrangements" [1]. At each recursive step we can update those arrays in such a way that the total number of updates after all the recursive steps becomes $O(n^2)$. Thus, when the points are assumed to be in general position, the time required to decide whether G admits a point-set embedding on S becomes $O(n^2) + O(n \log h) + 6 \times O(n^2) = O(n^2)$.

4   Lower Bound

In this section, we first analyze the time complexity of Algorithm **Point-Set-Embedding** in some restricted cases. We then show that the lower bound on time complexity for computing a point-set embedding of a plane 3-tree of n vertices is $\Omega(n \log n)$.

Let G be a plane 3-tree of n-vertices and let T be the representative tree of G. Suppose that, for each internal vertex u of T the ratio of the number of vertices in the three subtrees rooted at three children of u is $x : y : z$. Without loss of generality we assume that $x \geq y \geq z$. One can easily find that running time of the Algorithm **Point-Set-Embedding** can be written as $O(n) + T(n)$ where the term $O(n)$ is to obtain the embedding of the outer face of G and the term $T(n) \geq T(n-3)$ is the time to obtain the embedding of $n - 3$ internal vertices of G. Since at each recursive step we take $O(n \log n)$ time to obtain a valid mapping of a representative vertex, $T(n)$ can be defined recursively as follows. $T(n) \leq T(\frac{nx}{x+y+z}) + T(\frac{ny}{x+y+z}) + T(\frac{nz}{x+y+z}) + cn \log n = 3T(\frac{n}{b}) + cn \log n$. Here c is a constant hidden in $O(n \log n)$ term and $b = \frac{x+y+z}{x}$. We observe that, for $b = \sqrt{3}$, $T(n) = 3T(\frac{n}{\sqrt{3}}) + cn \log n = O(n^2)$; for $b = 2$, $T(n) = 3T(\frac{n}{2}) + cn \log n = O(n^{1.58})$; and for $b = 3$, $T(n) = 3T(\frac{n}{3}) + cn \log n = O(n \log^2 n)$. Therefore, we obtain the following theorem.

Theorem 2. *Let G be a plane 3-tree of n vertices and S be a point-set of n points. If the representative tree of G is a complete ternary tree, it can be decided whether G admits a point-set embedding on S in $O(n \log^2 n)$ time.*

We now prove a lower bound on the running time of the problem of obtaining a point-set embedding of a plane 3-tree with n vertices as in the following theorem.

Theorem 3. *The lower bound on the running time of the problem of computing point-set embeddings of plane 3-trees with n vertices is $\Omega(n \log n)$.*

Proof. We reduce sorting problems into the problem of computing point-set embeddings of plane 3-trees in the sense that point-set embedding algorithm can be used to solve sorting problems with little additional work.

Let $L = (x_1, x_2, \ldots, x_n)$ be a list of n unsorted numbers to be sorted and let the smallest number in L be x_{min}. Without loss of generality we assume that $x_i > 0$, $1 \leq i \leq n$, since if there exists an $x_i \leq 0$ we can obtain a list L' of nonzero positive numbers by adding $1 - x_{min}$ to all the x_i, $1 \leq i \leq n$. One can observe that the sorted order of the numbers in L' yields a sorted order of the numbers in L. Suppose that we have an algorithm **X** that computes a point-set embedding of a plane 3-tree of n vertices in $f(n)$ time. We show that Algorithm

X can be used to solve a sorting problem of n numbers in time $f(n) + O(n)$ where the $O(n)$ represents additional time to convert the solution of **X** to the solution of the sorting problem.

Let x_{max} be the maximum number in L which can be found in $O(n)$ time. We make a set S of two-dimensional points (x_i, x_i^2); $1 \le i \le n$ and let $S' = S \cup \{(x_{max} + 1, 0), (0, 0)\}$. Let G be a plane 3-tree of $n + 2$ vertices such that its representative tree T has the following properties (a) and (b). (a) The left child and the right child of each internal vertex of T are leaves. (b) The subgraph induced by the internal vertices of T is a path of $n - 1$ vertices. Figure 5(a), (b) and (c) illustrate S', G and T, respectively. We now use Algorithm **X** to compute a point-set embedding of G on S.

In a point-set embedding $\Gamma(G)$ of G on S', the outer vertices of G is mapped to the convex-hull of S' by Fact 1. The convex-hull of S' contains the points $(0, 0)$, $(x_{max} + 1, 0)$ and (x_{max}, x_{max}^2) which we denote by a, b and c, respectively. Let the representative vertex p of G be mapped to a point $z \in S'$ in $\Gamma(G)$ and let the proper interiors of the triangles $abz = C_{p_1}$, $bcz = C_{p_2}$, $caz = C_{p_3}$ be R_{z_1}, R_{z_2}, R_{z_3}, respectively. Since p is also the root of T with two leaves, two of the regions R_{z_1}, R_{z_2} and R_{z_3} do not contain any point of S. One can observe that such two regions can be obtained if z is the second smallest (or the second largest) number of L. Suppose that the region R_{z_2} (or R_{z_1}) contains all the points of S' other than a, b, c and z. We now consider the point-set embedding of $G(C_{p_2})$ (or $G(C_{p_1})$). Let c_1, c_2 and c_3 be the left, middle and right child of p in T. If the children are leaves, we have no vertices left to consider. Otherwise c_2 is an internal vertex. Since the left child and the right child of c_1 are leaves, c_1 must be mapped to the next smallest number or (next largest number) to ensure two regions which do not contain any points of S', maintaining the plane embedding of G.

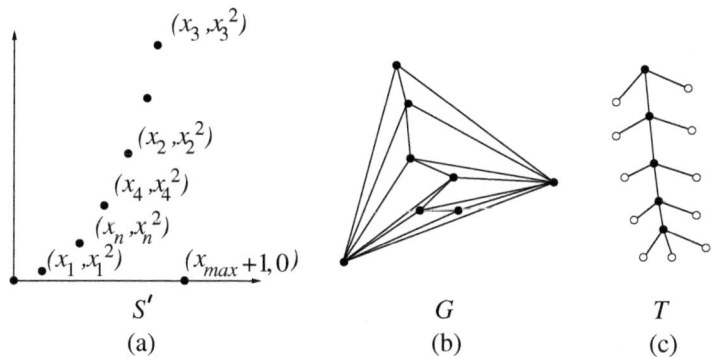

Fig. 5. Illustration for the proof of Theorem 3

Thus the sequence of x-coordinates of the mappings of the internal vertices of T from the root gives an increasing (or decreasing) order of the numbers in $L - \{x_{max}\}$. Moreover, we can check whether the obtained order is increasing or decreasing and place x_{max} in the correct position in constant time. Thus, we can use Algorithm **X** to solve the sorting problem which implies that the lower bound of Algorithm **X** is equal to the lower bound $\Omega(n \log n)$ on the running time of sorting problems.   □

5   Generalized Case

In this section we consider the problem of computing a point-set embedding of a plane 3-tree G, when the number of given points is greater than the number of vertices of G. We use dynamic programming to solve the problem.

Let G be a plane 3-tree of n vertices and S be a point-set of $k > n$ points. We obtain a grid by drawing a vertical line and a horizontal line passing through each of the k points. We assign x-coordinates for the vertical lines according to the count from left to right. Similarly we assign y-coordinates for the horizontal lines according to the count from bottom to top. The grid points are the intersections of the horizontal and the vertical lines. A *valid grid point* is a grid point which is also a point in S. The *width* W and *height* H of the grid are measured by the width and the height of the smallest rectangle with sides parallel to the axes which encloses all the points in S. The *size* of the grid is usually described as $W \times H$. Since the number of points in S is k, the size of the grid is at most $k \times k$. A graph G is *drawable on a grid* obtained from a point-set S if G admits a straight-line drawing such that the vertices of G are mapped on to the valid points on the grid obtained from S.

We assume a height h and a width w, iterate h from 1 to k and for each h, we iterate w from 1 to k. At each iteration we check whether G is drawable on a $w \times h$ grid or not. We give two algorithms Algorithm **Point-Set** and Algorithm **Feasibility-Checking** to serve this purpose. Let T be the representative tree of G. At each iteration we traverse T in preorder. For each internal vertex i of T, Algorithm **Point-Set** generates all possible (x, y)-coordinate assignments for the outer vertices a, b and c of $G(C_i)$ within area $w \times h$. For each such (x, y)-coordinate assignment of a, b and c, Algorithm **Feasibility-Checking** is called to check whether $G(C_i)$ is drawable. Here we formally define the input and output of the problem *Feasibility-Checking*.

Input: A plane 3-tree G and a mapping of the three outer vertices a, b and c of G to the three different valid points on a grid obtained from S.

Output: If G is drawable with the given mapping of a, b and c, then the output is *True*. Otherwise, the output is *False*.

We denote the x-coordinate and y-coordinate of a vertex v by v_x and v_y, respectively. We denote by $F_p(a_y^x, b_y^x, c_y^x)$ the Feasibility Checking problem of any vertex p of T where a_y^x, b_y^x, c_y^x are the (x, y)-coordinates of the three outer vertices a, b and c of $G(C_p)$. We solve this decision problem by showing in Lemma 6 that the optimal solution of the problem consists of the optimal solutions of the subproblems. Theorem 4 states the recursive solution of the *Feasibility Checking* problem.

Lemma 6. *Let G be a plane 3-tree with the representative tree T. Let p be any internal vertex of T with the three children p_1, p_2, p_3 in T and a, b, c be the outer vertices of $G(C_p)$. Then, for an assignment of the vertices a, b, c and p to valid grid points, the Feasibility Checking problems of p_1, p_2 and p_3 are the three subproblems of the Feasibility Checking problem of p.*

Theorem 4. *Let G be a plane 3-tree with the representative tree T and p be any vertex of T. Let a, b, c be the three outer vertices of $G(C_p)$ and p_1, p_2, p_3 be the three children of p when p is an internal vertex of T. Let $F_p(a_y^x, b_y^x, c_y^x)$ be the Feasibility Checking problem of p. Then $F_p(a_y^x, b_y^x, c_y^x)$ has the following recursive formula.*

$$
F_p(a_y^x, b_y^x, c_y^x) = \begin{cases}
False \ if \ (max\{a_x, b_x, c_x\} - min\{a_x, b_x, c_x\} = 0) \\
\qquad \vee \ (max\{a_y, b_y, c_y\} - min\{a_y, b_y, c_y\} = 0); \\
True \ if \ (max\{a_x, b_x, c_x\} - min\{a_x, b_x, c_x\} \geq 1) \\
\qquad \wedge \ (max\{a_y, b_y, c_y\} - min\{a_y, b_y, c_y\} \geq 1) \\
\qquad \wedge \ p \ is \ a \ leaf; \\
False \ if \ p \ is \ an \ internal \ vertex \ and \ there \ is \ no \ valid \\
\qquad grid \ point \ inside \ the \ triangle \ abc; \\
\bigvee_{p_x, p_y} \{ F_{p_1}(a_y^x, b_y^x, p_y^x) \wedge F_{p_2}(b_y^x, c_y^x, p_y^x) \wedge F_{p_3}(c_y^x, a_y^x, p_y^x)\} \\
\qquad where \ (p_x, p_y) \ is \ a \ valid \ grid \ point \ inside \\
\qquad the \ triangle \ abc, \ otherwise.
\end{cases}
$$

Proof. First we consider the case when $(max\{a_x, b_x, c_x\} - min\{a_x, b_x, c_x\} = 0) \vee (max\{a_y, b_y, c_y\} - min\{a_y, b_y, c_y\} = 0)$. Then we assign $F_p(a_y^x, b_y^x, c_y^x) = False$ because a grid of at least area 1×1 is necessary to draw a triangle. The next case is $(max\{a_x, b_x, c_x\} - min\{a_x, b_x, c_x\} \geq 1) \wedge (max\{a_y, b_y, c_y\} - min\{a_y, b_y, c_y\} \geq 1)$ when p is a leaf. Then we assign $F_p(a_y^x, b_y^x, c_y^x) = True$ since area 1×1 is sufficient to draw a triangle. In the next case, p is an internal vertex and there is no valid grid point inside the triangle abc. We assign $F_p(a_y^x, b_y^x, c_y^x) = False$ since p cannot be placed inside C_p. The remaining case is $(max\{a_x, b_x, c_x\} - min\{a_x, b_x, c_x\} > 1) \wedge (max\{a_y, b_y, c_y\} - min\{a_y, b_y, c_y\} > 1)$ when p is an internal vertex. Then we define $F_p(a_y^x, b_y^x, c_y^x)$ recursively according to Lemma 6.   □

One can associate tables to store the computed (x, y)-coordinates of the vertices of G to obtain the point-set embedding of G in polynomial time, if it exists.

Theorem 5. *Given a plane 3-tree G with $n \geq 3$ vertices and a point-set of $k > n$ points, Algorithm **Point-Set** decides whether G admits a point-set embedding on S in time $O(nk^8)$.*

6   Conclusion

We give an $O(n^2 \log n)$ time algorithm that decides whether a plane 3-tree G of n-vertices admits a point-set embedding on a set S of n points or not; and computes a point-set embedding of G if such an embedding exists. We observe that it is possible to obtain a point-set embedding of G on S in $O(n^2)$ time if the points are assumed to be in general position. We prove an $\Omega(n \log n)$ lower bound on the time complexity for obtaining point-set embeddings of plane 3-trees. Beside obtaining a point-set embedding of a plane 3-tree, we consider a generalized problem when the given point-set has more than n points and give a polynomial-time algorithm to solve the problem. It is a challenge to find simpler algorithms for obtaining point-set embeddings of plane 3-trees both in the restricted and generalized cases.

Acknowledgment

This work is done in Graph Drawing & Information Visualization Laboratory of the Department of CSE, BUET established under the project "Facility Upgradation for Sustainable Research on Graph Drawing & Information Visualization" supported by the Ministry of Science and Information & Communication Technology, Government of Bangladesh. We thank Tanaeem Muhammad Moosa for fruitful discussions.

References

1. de Berg, M., van Kreveld, M., Overmars, M., Schwarzkopf, O.: Computational Geometry: Algorithms and Applications. Springer, Heidelberg (2000)
2. Biedl, T., Velázquez, L.E.R.: Drawing planar 3-trees with given face-areas. In: Eppstein, D., Gansner, E.R. (eds.) GD 2009. LNCS, vol. 5849, pp. 316–322. Springer, Heidelberg (2010)
3. Bose, P.: On embedding an outer-planar graph in a point set. Computational Geometry - Theory and Applications 23(3), 303–312 (2002)
4. Cabello, S.: Planar embeddability of the vertices of a graph using a fixed point set is NP-hard. Journal of Graph Algorithms and Applications 10(2), 353–363 (2006)
5. Chan, T.M.: Optimal output-sensitive convex hull algorithms in two and three dimensions. Discrete & Computational Geometry 16(4), 361–368 (1996)
6. de Fraysseix, H., Pach, J., Pollack, R.: How to draw a planar graph on a grid. Combinatorica 10, 41–51 (1990)
7. García, A., Hurtado, F., Huemer, C., Tejel, J., Valtr, P.: On embedding triconnected cubic graphs on point sets. Electronic Notes in Discrete Mathematics 29, 531–538 (2007)
8. Ikebe, Y., Perles, M.A., Tamura, A., Tokunaga, S.: The rooted tree embedding problem into points in the plane. Discrete & Computational Geometry 11, 51–63 (1994)
9. Kaufmann, M., Wiese, R.: Embedding vertices at points: Few bends suffice for planar graphs. Journal of Graph Algorithms and Applications 6(1), 115–129 (2002)
10. Nishizeki, T., Rahman, M.S.: Planar Graph Drawing. World Scientific, Singapore (2004)
11. Pach, J., Gritzmann, P., Mohar, B., Pollack, R.: Embedding a planar triangulation with vertices at specified points. American Mathematical Monthly 98, 165–166 (1991)
12. Schnyder, W.: Embedding planar graphs on the grid. In: The first annual ACM-SIAM symposium on Discrete algorithms, pp. 138–148 (1990)

Improving Layered Graph Layouts with Edge Bundling

Sergey Pupyrev[1], Lev Nachmanson[2], and Michael Kaufmann[3]

[1] Ural State University
spupyrev@gmail.com
[2] Microsoft Research
levnach@microsoft.com
[3] Universität Tübingen
mk@informatik.uni-tuebingen.de

Abstract. We show how to improve the Sugiyama scheme by edge bundling. Our method modifies the layout produced by the Sugiyama scheme by bundling some of the edges together. The bundles are created by a new algorithm based on minimizing the total ink needed to draw the graph edges. We give several implementations that vary in quality of the resulting layout and execution time. To diminish the number of edge crossings inside of the bundles we apply a metro-line crossing minimization technique. The method preserves the Sugiyama style of the layout and creates a more readable view of the graph.

1 Introduction

Layered drawings present directed graphs in a way that the nodes are arranged in horizontal layers. Most approaches for drawing layered graphs follow in practice the algorithm proposed by Sugiyama et al. [14]. This method is intended to produce layouts with a small number of edge crossings and smooth edges. However, when the given graph is dense, even an optimal layout can have numerous edge crossings and edges that are too curved. In fact, the edge clutter quickly makes the drawing useless for understanding the relations among the nodes.

One of the popular methods to improve quality of such drawings is edge bundling [13,9,7,3]. This method can significantly reduce clutter and can also help to highlight high-level edge patterns in a graph. A known shortcoming of the method is that it tends to produce drawings with ambiguity; it becomes hard to visually follow a single edge. Still a good edge bundling improves the readability of a drawing. To the best of our knowledge, our method is the first attempt to apply edge bundling for graphs with multiple layers. The algorithm is relatively simple and fast enough to be used in practice. For dense graphs, in our opinion, the method produces drawings that are easier to analyze than the standard layouts.

The input to our method is the output of the standard Sugiyama algorithm. We deviate some of the edge routes and create edge bundles, thus reducing the clutter. We do not change the positions of the original nodes of the graph, and we preserve the edge homotopy classes relative to the original nodes, i.e. the new routing can be obtained from the previous one by a continuous deformation of the edges without overlapping with the nodes. This visual stability can help the user to analyze the graph, for example, by comparing the original and the bundled drawings. Fig. 1 illustrates these ideas and motivates our approach.

U. Brandes and S. Cornelsen (Eds.): GD 2010, LNCS 6502, pp. 329–340, 2011.

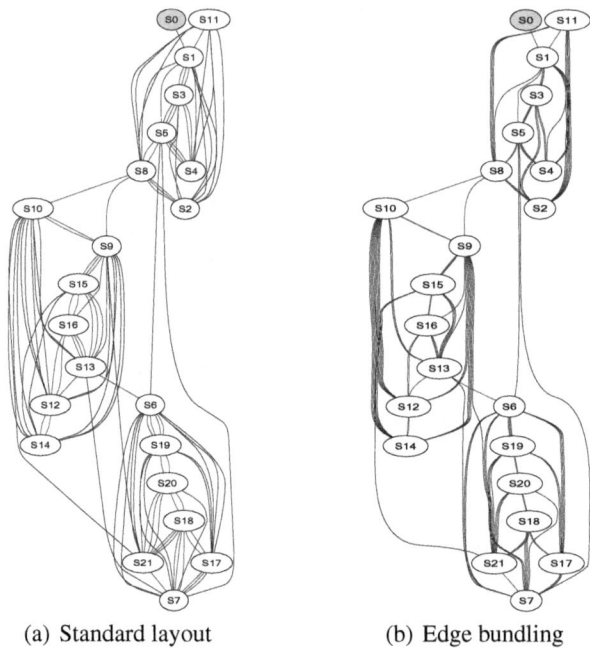

(a) Standard layout   (b) Edge bundling

Fig. 1. The state diagram for the *notepad.exe* model

Recall that before the *Ordering* stage of the Sugiyama scheme the original graph is transformed to a *proper layered graph*. In this graph each original edge is replaced by a connected sequence of *proper* edges passing through added virtual nodes where a proper edge is an edge between *adjacent* layers. The first edge of the sequence starts from the original edge source and the last edge ends at the original target.

The starting point of our algorithm is a layered layout of the graph. We change the positions and the order of virtual nodes to organize edges into bundles. On a high level our method consists of the following three stages: *create edge bundles* of the proper layered graph where each bundle is a subset of edges spanning the same pair of layers; *straighten out the edges* to avoid excessive bends, and finally *reduce the ambiguity in the routing* by sorting the edges and drawing them individually inside the bundles.

In Section 3 we give a detailed explanation of our algorithm. Experimental data and results are presented in Section 4 to demonstrate the effectiveness of the method. Finally, Section 5 concludes the paper and discusses additional aspects and future work. The next section summarizes related work.

2   Related Work

Sugiyama-style layered layouts were proposed in the early 80's [14,4] and improved in many ways during recent years [8,5]. This method is widely used to layout a graph in a monotone fashion where all edges in a directed-acyclic graph follow the same downward direction. The idea of improving the quality of layered layouts with edge bundling is related to edge concentration [13], where a two-layered graph is covered by bicliques reducing the number of drawn edges. Later Eppstein et al. [6] proposed confluent graph drawings, allowing groups of edges to be merged and drawn together.

Recently Holten and van Wijk [9] suggested edge bundling based on an additional tree structure. Unfortunately, not every graph comes with a suitable underlying tree and an artificial one might affect the final layout in an undesirable way. Methods of edge bundling for general graphs were presented by Cui et al. in [3] and by Holten et al. in [10]. We notice that for dense graphs both methods produce ambiguous layouts where it is hard to follow a single edge. In addition, these techniques do not guarantee that the edges are drawn downward which is a requirement of the Sugiyama scheme.

The paper on the metro-line crossing minimization problem by Argyriou et al. [1] inspired us to use the technique of metro-lines crossing minimization to minimize edge crossings inside of the bundles. The paper on circular layouts by Gansner and Koren [7], where bundles were built by minimizing the total ink, also influenced our work.

Unlike the approaches of Holten [10] and Cui [3], we preserve the edge homotopy classes while creating bundles. Moreover, our method takes into consideration the dimensions of the nodes and routes the bundles without overlapping them with the nodes.

3   Edge Bundling

We start with some basic definitions. A directed graph G is a pair (V, E), where $V = \{1, \ldots, n\}$ is the set of nodes and $E \subset V^2$ is the set of edges. As usual for the Sugiyama algorithm, we may assume that G is an acyclic graph. A *layering* L of G is a partition of V into sets of layers $L_1, \ldots, L_h$ such that for every edge $(u, v) \in E$ with $u \in L_i$ and $v \in L_j$ holds $i < j$. From G and L, we build a *proper layered graph* G^p: For each $e = (u, v)$ of G with $u \in L_i$, $v \in L_j$ we add to G^p nodes $u = d_i, d_{i+1}, \ldots, d_j = v$, add d_k to L_k for $k \in [i, j]$ and add to G^p edges (d_k, d_{k+1}) for $i \leq k < j$. Nodes d_k with $i < k < j$ are called *virtual nodes*. They are unique for every edge. We denote by $D(e)$ the sequence $d_i, d_{i+1}, \ldots, d_j$. Nodes of G^p which are also nodes of G are called *original nodes*. Edges of G are called *original edges*.

The Sugiyama algorithm assigns a point in the plane to each node of G^p such that the nodes of the same layer have the same y-coordinate. For the purpose of our discussion we assume that the edges are drawn as polylines; for an edge e of G the polyline is defined by the positions of nodes of $D(e)$.

Suppose S is a subset of edges of G^p connecting a pair of adjacent layers. We define *bundling* of S as a procedure where we horizontally shift the ends of the edges of S to make them coincide (Fig. 2(a)). After bundling the edges of S together we call S a *bundle*, the edges of S are called *bundled* edges, and those nodes that as a result of the shift acquired the same positions are called *bundled* nodes. We define the graph *ink* [15] as the sum of the lengths of line segments needed to draw the graph edges.

The order of the nodes on a layer allows us to define an equivalence relation on the layer nodes where each original node is equivalent only to itself, and any two virtual nodes are equivalent if and only if there is no original node between them. Consider two adjacent layers. We subdivide the edges of G^p between these two layers into mutually disjoint *groups*; two edges belong to the same group Gr if and only if their sources are equivalent in the upper layer and their targets are equivalent in the lower layer, see

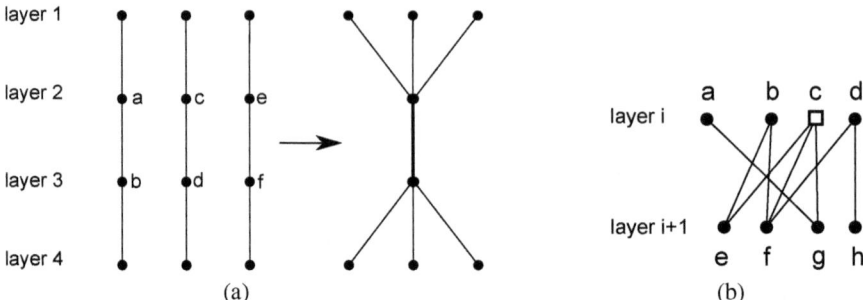

Fig. 2. (a) Bundling of edges (a,b), (c,d), and (e,f). (b) The original node c divides the edges into three groups: $\{(a, g), (b, e), (b, f)\}$, $\{(c, e), (c, f), (c, g)\}$, and $\{(d, f), (d, h)\}$.

Fig. 2(b). We bundle together some of the edges of the same group if routing them on top of each other saves ink. In order to preserve the homotopy classes of the edges we never bundle together edges from different groups. Note that we will not try to solve the ink minimization problem optimally but propose several greedy heuristics. We address the complexity issue in Section 5.

3.1   Identifying Edge Bundles

We use the output of the Sugiyama layout algorithm as the starting point. More precisely, we are given a layering L and the node positions of graph G^P.

We process the graph layers downward, consider each group Gr separately and bundle together some of the edges in Gr. Initially, for each edge, we create a bundle containing only this edge. Iteratively we try to find a pair of bundles whose merge maximally reduces the ink, and, if a pair is found, we create a new bundle from the pair components. The group processing stops when the ink does not decrease anymore or there is only one bundle in the group. The procedure on a high level is sketched below.

Edge Bundling(proper layered graph G^P)
    for $j = 1$ to $h - 1$ **do**
        for every group Gr_i with sources at $Layer_j$ **do**
            $bundles \leftarrow \{\{e_1\}, \ldots, \{e_{m_i}\}\}$;
            do $(b_1, b_2) \leftarrow$ valid pair of bundles whose merge produces the maximal ink gain;
                $bundles \leftarrow bundles - \{b_1, b_2\} + \{$**UniteBundles**$(b_1, b_2)\}$;
            while ink improvement > 0
        end for
    end for

UniteBundles(b_1, b_2) returns the union of b_1 and b_2. It also shifts the ends of the edges of the union to make the edges coincide. The new coordinates of the edge sources (targets) are set to the average of the coordinates of edge sources (targets) of $b_1 \cup b_2$.

Our experiments show that this strategy alone does not always produce good drawings. In a layout with little ink original edges may be highly curved. To address this issue we restrict the bundling with the following *angle constraint*:

- Bundling of two edges is allowed only if it does not introduce sharp bends on the polylines of the original edges participating in the bundling. In our implementation we do not allow for an original edge polyline to turn more than on $\pi/4$ after bundling (Fig. 3(a)).

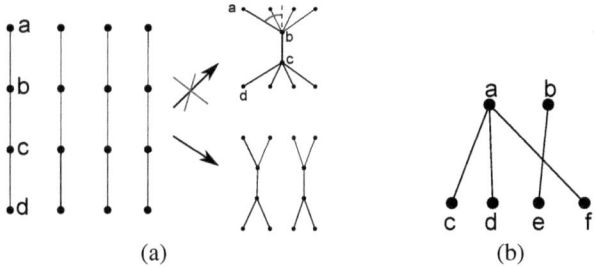

Fig. 3. (a) A polyline turns more than on $\pi/4$ at b; the bundling is forbidden. (b) Edge (a, c) is compatible only with edge (a, d); edge (a, d) is compatible with edges (a, c),(a, f), and (b, e).

The most expensive part of the bundling scheme described above is the lookup for a suitable pair of bundles whose merge gives the maximal ink gain. We consider three different implementation varying in the time complexity and the quality.

A naive lookup. The brute force approach is to try all the pairs of bundles from the same group Gr_i of size m_i and find a pair whose merge gives the maximal ink gain. Let t be the time needed to calculate the ink gain for the merge of two bundles. In the simplest case when bundles contain only one edge $t = O(1)$. However, in the worst case $t = O(m)$, where m is the number of the edges in G^p, because moving a bundle involves moving the edges adjacent to the bundle above and below. A performance analysis of processing Gr_i gives us $O(m_i^3 t)$ since before each merge we consider all $O(m_i^2)$ bundle pairs, we perform $O(m_i)$ merges as the number of bundles is reduced by one at every merge, and each ink gain computation costs $O(t)$ time. That gives us $O(m^3 t)$ for all the groups since $\sum m_i = m$ and therefore $\sum m_i^3 \leq m^3$. Note that this is a worst-case estimation; our experiments show a much better behaviour in practice.

More efficient lookup. The idea here is to avoid calculating ink gain for distant pairs of bundles of the group. For convenience below we denote a bundle by the x-positions of its source and target. For two bundles $e = (x, y)$ and $e'' = (x'', y'')$, we say that $e' = (x', y')$ lies between e and e'' if and only if both endpoints of e' lie between the corresponding endpoints of e and e''; that is $\min(x, x'') \leq x' \leq \max(x, x'')$ and $\min(y, y'') \leq y' \leq \max(y, y'')$. This definition includes the case where e and e'' cross each other (Fig. 3(b)). We call two bundles *compatible* if there is no other bundle lying between them. We noticed that the Sugiyama scheme usually produces such outputs that if we reduce ink by uniting bundles e, e'' and e' lies in between then we would also save ink bundling e, e' and e, e''. The heuristic here is based on this observation.

We enumerate over all the pairs of bundles from group Gr but check the ink gain only for compatible ones. Then we merge a pair of bundles giving the maximal ink gain. In the worst case group Gr_i contains $O(m_i^2)$ compatible pairs. In practice we found that only a small fraction of all possible pairs are compatible. By more involved arguments, it can be seen that the number of compatible pairs is $O(m_i + cr)$, where cr denotes the number of edge crossings within group Gr_i. As a result, we compute the ink gain rarely comparing with the naive implementation, and obtain a significant speedup.

Fast restricted lookup. In the third heuristic we avoid enumerating over all bundle pairs of a group but only consider compatible pairs having a common endpoint.

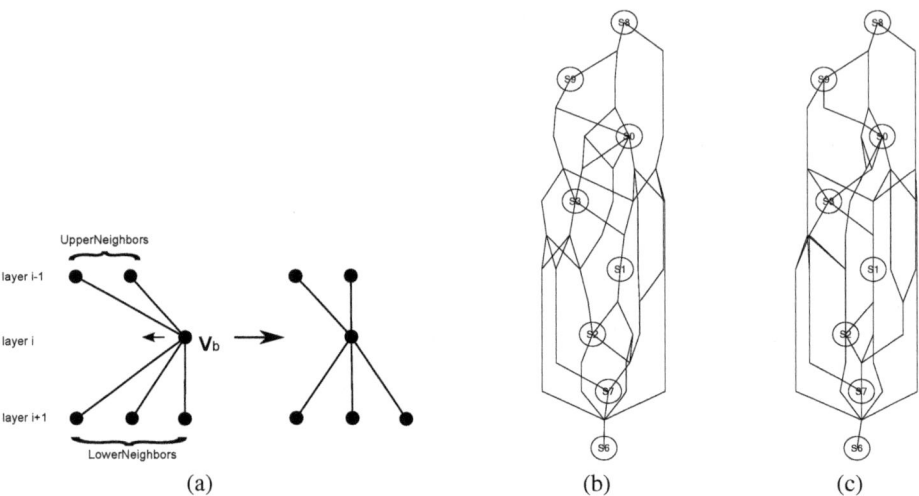

Fig. 4. (a) Shift of node-set V_b results in edges with less turns. (b) Graph layout before straightening phase. (c) Graph layout after straightening phase.

For each node v we order its adjacent edges $(v, u_i), 1 \leq i \leq d_v$ by the x-coordinates of the targets. Here d_v is the out-degree of the node. Then we compute the ink gain of bundling each pair $(v, u_i), (v, u_{i+1}), 1 \leq i < d_v$ and store these values in a balanced tree. A pair giving the maximal ink gain is merged to a bundle. Ink gains of the pairs referencing merged bundles are removed from the tree. Ink gains of merging the new bundle with its left and right neighbors, if they exist, are inserted into the tree. When merging does not decrease the ink, or there is only one bundle left, we stop the procedure and proceed to the next node of the layer. To process the whole graph we sweep top-down layer-by-layer and bundle only pairs with the common source. After the first pass we sweep bottom-top and merge only pairs with a common target.

Processing node v takes $O(td_v \log d_v)$ steps, where t is the time to compute the ink gain. Note that the ink gain calculation might be cheaper for this special case than in general since we move only one end of the bundle. Processing group Gr_i requires $O(tm_i \log m_i)$ steps. Summing over the groups, we achieve a bound of $O(tm \log m)$.

In section 4 we give more details of quality and practical performance of the suggested heuristics.

3.2 Straightening Edges

The polylines generated by the method described above often have too many sharp turns. To resolve this problem we use a modification of the median heuristic [8,5], producing smoother and more vertically aligned edges. We iterate over the graph layers in upward direction and try to shift each set of bundled nodes to align its position with the median position of its neighbors from the layer below (Fig. 4).

The shift follows two rules: (I) the set of bundled nodes is shifted only if the new position is outside of the shapes of the original nodes and (II) the number of turns of the edge polylines passing through the shifted nodes does not increase. After sweeping the

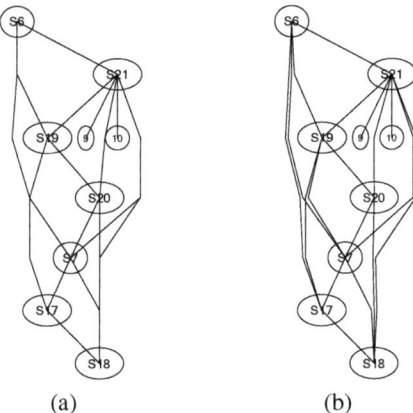

(a)   (b)

Fig. 5. Bundled graph before (a) widening with high level of ambiguity and after (b) with optimal crossing number

graph upward we sweep the layers downward. Our approach is sketched below under the assumption that the top layer has index 1.

Edge Straightening
    do
        for $i = h - 1$ **to** 1 **do** *%% upward iteration*
            for every set of bundled nodes V_b in $Layer_i$ **do**
                $x \leftarrow$ sorted array of x-coordinates of adjacent nodes to V_b on $Layer_{i+1}$;
                ProcessNodes(V_b, x);
            end for
        end for
        for $i = 2$ **to** h **do** *%% downward iteration*
            for every set of bundled nodes V_b in $Layer_i$ **do**
                $x \leftarrow$ sorted array of x-coordinates of adjacent nodes to V_b on $Layer_{i-1}$
                ProcessNodes(V_b, x)
            end for
        end for
        $iteration \leftarrow iteration + 1$
    while $iteration < Max\_iterations$

ProcessNodes(V_b, x)
    $x_{median} \leftarrow (x_{\lfloor (k+1)/2 \rfloor} + x_{\lceil (k+1)/2 \rceil})/2$ *%% median x-position of neighbors V_b*
    if moving nodes of V_b agrees with rules (I) or (II) **then**
        for every v in V_b **do** $v_x \leftarrow x_{median}$

The straightening step requires $Max\_iterations$ iterations, which is 10 in our implementation, over the graph nodes. Processing each node takes $O(d \log d)$ steps, where d is the out-degree of the node. Summing over the nodes gives $O(m \log m)$ time. Here we take into account that the sum of all out-degrees of the nodes is equal to m.

3.3 Metro-Map Widening

After uniting edges we have a bundled graph where the ambiguity is often too high (see Fig.5(a)). In this section, we propose a bundle "widening" technique which helps to reduce the ambiguity and highlight the importance of a single bundle.

We draw each edge within the same bundle individually. This is performed by horizontally moving apart coinciding edge ends to slightly separate the bundled edges.

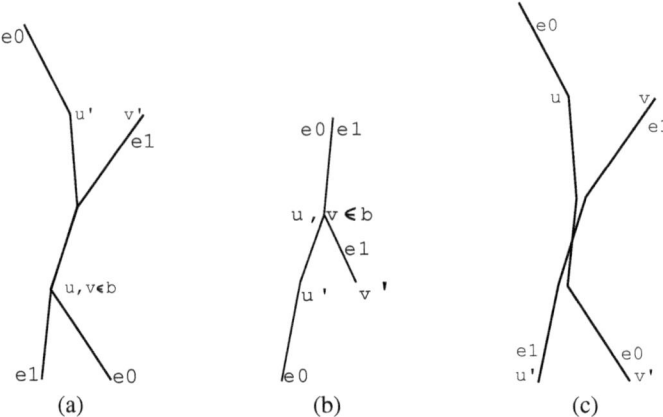

Fig. 6. (a), (b) $e_0 < e_1$; (c) e_0 and e_1 have to cross each other

To avoid introducing unnecessary edge crossings, we order the edges at bundled nodes and spread the nodes according to this order. The problem of finding such orders is similar to the metro-line crossing minimization problem [2,1].

To define the order, consider two original edges e_0 and e_1. Suppose e_0 has node u, and e_1 has node v, which are bundled together to the node set b. To define the order between e_0 and e_1 at b, we first walk up the layers simultaneously over e_0 and e_1 starting from u, v until we find nodes u' and v', which are not bundled together, or reach the source of at least one of the edges. If we find u' and v', as in Fig.6(a), we order e_0 and e_1 at b according to the x-coordinates of u' and v'. If such nodes are not found, that is the edges have the same prefix before b, we walk down the edges, again looking for the first fork. If not bundled nodes u' and v' are found, as in Fig.6(b), we order e_0 and e_1 at b according to the x-coordinates of u' and v'. If no forks are found then two edges belong to the common multi-edge; we order e_0 and e_1 at b arbitrarily but keep the same order for every other common bundled node of e_0 and e_1.

To use the calculated orders, we uniformly, with some gap given in advance, spread by x the nodes of a bundled node set. We keep the same average of the x-positions of the nodes of a set. After widening only unavoidable edge crossings remain. An unavoidable crossing occurs for a pair of original edges e_0, e_1 if there are nodes u, v, u', and v' such that u, v belong to e_0, u', v' belong to e_1, u, v belong to the same layer, and u', v' belong to the same layer, and $(x_u - x_v)(x_{u'} - x_{v'}) < 0$ holds (see Fig. 6(c)). The result of widening is illustrated in Fig. 5(b).

We now discuss the computation cost of this step. The comparison of any two edges can be done in $O(h)$ time, where h is the number of layers. Therefore, the sorting of k edges needs $O(hk \log k)$ time. We need to order the edges at each bundled node set. Let k_i be the number of edges passing through the i-th bundled node set. Then the total complexity of our algorithm is $O((k_1 \log k_1 + \ldots + k_n \log k_n)h)$, where $k_1 + \ldots + k_n = n$ and n is the number of virtual nodes in the proper layered graph. Here we take into account that through every virtual node passes only one original edge. The maximum of $k_1 \log k_1 + \ldots + k_n \log k_n$ is $n \log n$; it is reached when some $k_i = n$. Therefore, the algorithm works in $O(hn \log n)$ steps.

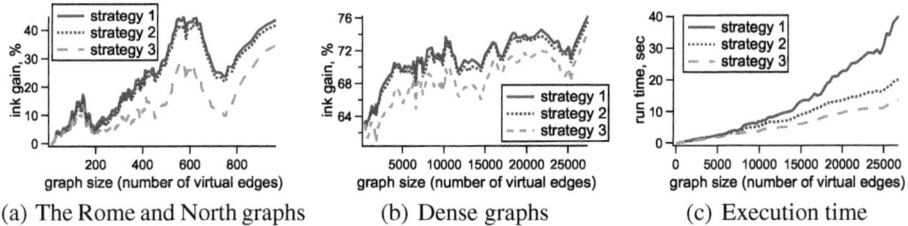

(a) The Rome and North graphs   (b) Dense graphs   (c) Execution time

Fig. 7. (a-b) The quality of the ink minimization strategies; strategy 1 is the naive lookup, strategy 2 is the efficient lookup, and strategy 3 is the fast restricted lookup. Each data point refers to the average ink improvement of all the graphs in the collection with the same number of edges in the proper layered graph G^p. (c) shows the comparison of execution times of strategies.

At the final stage we follow[12]; we insert new points into the polylines to avoid edge-node overlaps, and beautify each polyline by fitting cubic Bezier segments in its corners. We do not have an estimate of the complexity of this step, but in practice it takes only an insignificant fraction of the whole time.

4 Experiments

We implemented our method in tool MSAGL [12], and used the tool in our experiments. We applied the edge bundling to some real-world graphs and three graph collections:

- **Rome** [16] contains 11528 graphs with $10 - 100$ nodes and $9 - 158$ edges. They are obtained from a basic set of 112 real-world graphs. The graphs are originally undirected; we orient the edges according to the node order given in the input files.
- **North** [16] contains 1175 graphs grouped into 9 sets, where set $i = 1, \ldots, 9$ contains graphs with $10i$ to $10i + 9$ edges.
- **Dense** is a collection of power-law graphs as used in the social network analysis. The collection contains 116 graphs with $65 - 95$ nodes and $350 - 850$ edges; it was generated with the benchmark framework described in [11].

The performance and quality analysis of three edge bundling schemes is presented at Fig. 7. It shows the comparison of the ink minimization strategies; the average run-time and the ink gain for the graphs having the same number of virtual edges. The experiments have been run on an Intel Core-2 Duo 1.83GHz with 2GB RAM. The simplest implementation reduces the total ink by four times within 40 seconds for the graphs with 25000 virtual edges. The second strategy works two times faster, while its quality of the ink minimization is almost the same. The implementation with restrictions is the fastest: our largest graphs are processed within $10 - 12$ seconds. Typical layouts obtained with different ink minimization schemes are demonstrated by Fig. 8. We conclude that the second and third strategies are both practical. The former is more aggressive: it bundles larger amount of edges and creates less cluttered images. The latter bundles the edges with common sources or targets only, and thus produces the layouts with less ambiguity.

We compared the time complexity of the edge bundling with other stages of the Sugiyama algorithm. It can be seen that building of bundles is the most time-consuming

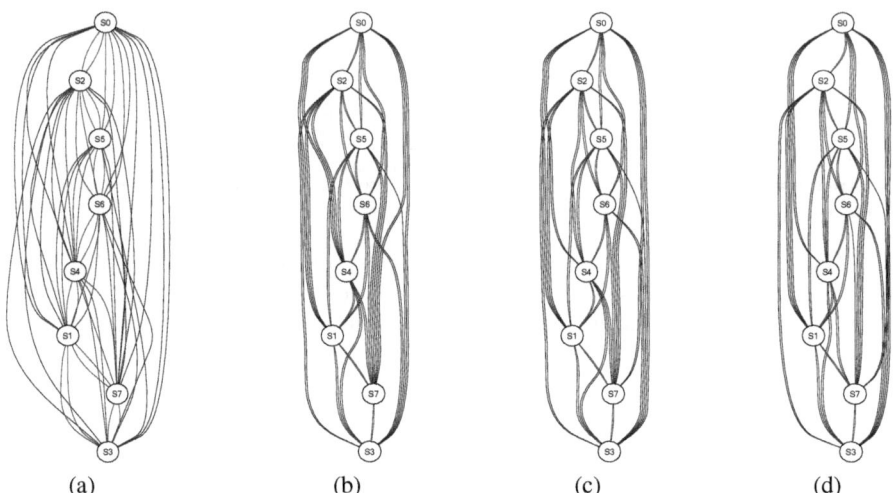

Fig. 8. A comparison of bundle strategies. (a) The standard layout. (b) The naive lookup. (c) The efficient lookup. (d) The fast restricted lookup.

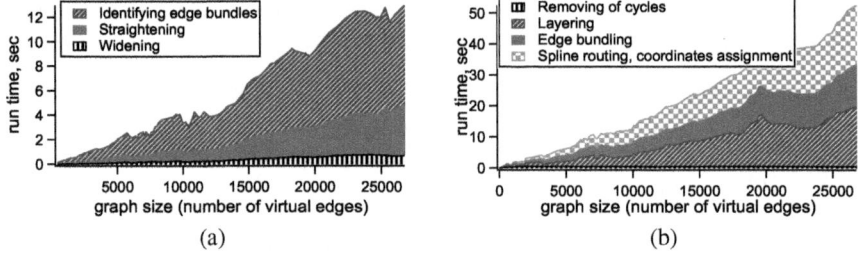

Fig. 9. Execution time of algorithm stages

part of the edge bundling (Fig. 9(a)). This step requires around 65% of the total time. In the whole, the edge bundling time is comparable with the time required by other Sugiyama stages as layering and x-coordinate assignment (Fig. 9(b)). The analysis of performance shows that our algorithm handles well medium-sized graphs containing up to several hundred nodes and a thousand edges.

A real-world example of layered layout is state machine diagram. A tool [17] created a model of the standard Windows application *notepad.exe*. We applied our method to show the state machine of the model. The nodes of the graph depict all possible states of the Notepad model and the edges correspond to the user actions like adding/deleting characters, selecting text, etc. The original and bundled layouts are shown in Fig. 1. Our method reveals the edge patterns and clarifies the structure of the state machine. Another example of edge bundling is given in Fig. 10.

We conclude that our technique is beneficial for dense graphs. For such graphs we utilize the white space better, and the bundles help to visualize the high-level edge

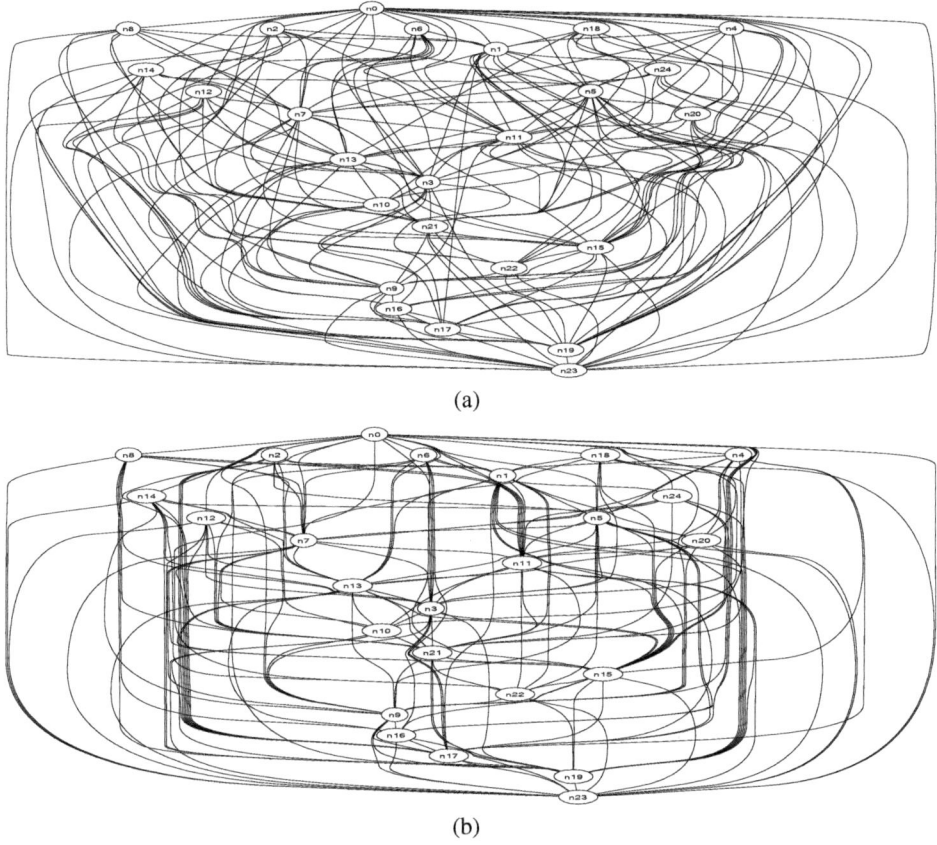

(a)

(b)

Fig. 10. Graph from the **North** collection with 25 nodes and 184 edges

patterns. Our method might introduce some ambiguity as, for example, for edges between $S19$ and $S7$ in Fig. 1(b); interactive edge highlighting can help here.

5   Conclusion, Discussion and Future Work

We have presented an edge bundling scheme for layered graphs that preserves the initial node positions and the homotopy classes of the edges. It ensures that nodes do not overlap with the edge bundles and that the resulting bundles are relatively straight, making them easy to follow visually. The resulting layout highlights the edge routing patterns and shows significant clutter reduction. Our method can improve layouts produced by other layered graph layout algorithms. The experiments show that the method is fast enough to process medium-sized graphs.

We found yet another application of our method to the Sugiyama algorithm, as a post-processing step for the crossing minimization of the *Ordering* phase. The idea is for each layer to *bundle* all virtual nodes which are not separated by an original node. After that we apply the *widening* technique described in Section 3.3. This results in an order with the minimum number of the edge crossings subject to the given order of the original nodes. In some cases we found up to 30% of edge crossings reduction.

There are several directions for future work. One could be a global multi-layer bundling processing the edges by taking into consideration the whole graph. We expect to receive drawings of a better quality by analyzing edges globally. Performance of our method depends on the number of virtual edges which can be large and vary significantly for graphs of the same size. The multi-layer approach might lead to an algorithm which does not depend on the number of virtual nodes so heavily. Finally, we would like to understand whether the problem of ink minimization is solvable in a polynomial time. It would be interesting to know how close the ink of drawings produced by our methods is to the minimal ink.

References

1. Argyriou, E., Bekos, M.A., Kaufmann, M., Symvonis, A.: Two polynomial time algorithms for the metro-line crossing minimization problem. In: Tollis, I.G., Patrignani, M. (eds.) GD 2008. LNCS, vol. 5417, pp. 336–347. Springer, Heidelberg (2009)
2. Bekos, M.A., Kaufmann, M., Potika, K., Symvonis, A.: Line crossing minimization on metro maps. In: Hong, S.-H., Nishizeki, T., Quan, W. (eds.) GD 2007. LNCS, vol. 4875, pp. 231–242. Springer, Heidelberg (2008)
3. Cui, W., Zhou, H., Qu, H., Wong, P.C., Li, X.: Geometry-based edge clustering for graph visualization. IEEE Transactions on Visualization and Computer Graphics 14, 1277–1284 (2008)
4. Eades, P., Sugiyama, K.: How to draw a directed graph. Journal of Information Processing 14(4), 424–437 (1990)
5. Eiglsperger, M., Siebenhaller, M., Kaufmann, M.: An efficient implementation of sugiyamas algorithm for layered graph drawing. Journal of Graph Algorithms and Applications 9(3), 305–325 (2005)
6. Eppstein, D., Goodrich, M.T., Meng, J.Y.: Confluent layered drawings. In: Pach, J. (ed.) GD 2004. LNCS, vol. 3383, pp. 184–194. Springer, Heidelberg (2005)
7. Gansner, E.R., Koren, Y.: Improved circular layouts. In: Kaufmann, M., Wagner, D. (eds.) GD 2006. LNCS, vol. 4372, pp. 386–398. Springer, Heidelberg (2007)
8. Gansner, E.R., Koutsofios, E., North, S.C., Vo, K.-P.: A technique for drawing directed graphs. IEEE Transactions on Software Engineering 19(3), 214–230 (1993)
9. Holten, D.: Hierarchical edge bundles: Visualization of adjacency relations in hierarchical data. IEEE Transactions on Visualization and Computer Graphics 12(5), 741–748 (2006)
10. Holten, D., van Wijk, J.J.: Force-directed edge bundling for graph visualization. In: Computer Graphics Forum, vol. 28, pp. 983–990 (2009)
11. Lancichinetti, A., Fortunato, S., Radicchi, F.: Benchmark graphs for testing community detection algorithms. Physical Review E 80(1) (2008)
12. Nachmanson, L., Robertson, G., Lee, B.: Drawing graphs with GLEE. In: Hong, S.-H., Nishizeki, T., Quan, W. (eds.) GD 2007. LNCS, vol. 4875, pp. 389–394. Springer, Heidelberg (2008)
13. Newbery, F.J.: Edge concentration: A method for clustering directed graphs. In: Proc. 2nd Int. Workshop on Software Configuration Management, pp. 76–85 (1989)
14. Sugiyama, K., Tagawa, S., Toda, M.: Methods for visual understanding of hierarchical system structures. IEEE Trans. on Systems, Man, and Cybernetics 11(2), 109–125 (1981)
15. Tufte, E.R.: The Visual Display of Quantitative Information. Graphics Press, CT (1983)
16. The AT&T graph collection. http://www.graphdrawing.org.
17. http://research.microsoft.com/en-us/projects/specexplorer.

Confluent Drawing Algorithms Using Rectangular Dualization[*]

Gianluca Quercini[1] and Massimo Ancona[2]

[1] Institute for Advanced Computer Studies,
University of Maryland College Park, MD, USA
`quercini@umiacs.umd.edu`
[2] Dipartimento di Informatica e Scienze dell'Informazione,
Università di Genova Genova, Italy
`ancona@disi.unige.it`

Abstract. The need of effective drawings for non-planar dense graphs is motivated by the wealth of applications in which they occur, including social network analysis, security visualization and web clustering engines, just to name a few. One common issue graph drawings are affected by is the visual clutter due to the high number of (possibly intersecting) edges to display. Confluent drawings address this problem by bundling groups of edges sharing the same path, resulting in a representation with less edges and no edge intersections. In this paper we describe how to create a confluent drawing of a graph from its rectangular dual and we show two important advantages of this approach.

Keywords: Confluent Drawing, Rectangular Dualization, Orthogonal Drawing, Clustered Graphs.

1 Introduction

Drawing graphs is a challenging problem, as witnessed by the wealth of approaches (recently surveyed in [6]) that have been proposed over the last two decades. Far from being mere and meaningless collections of nodes and edges, graphs are effective ways to describe relationships between objects (e.g. people, molecules, computers); consequently, visualizing a graph is an important step to clearly reveal all its information. In order to be readable, a drawing must comply with precise aesthetic criteria on the way nodes and edges are visualized; nodes (and edges) should not be too close to one another, symmetries should be highlighted, the total area should be minimized and so on. While optimizing all such constraints is clearly impossible, a good balance is usually the right way to go to create nice drawings. Large and dense graphs are inherently difficult to draw, as the amount of information that can be visualized is limited by the size of the medium (computer screen or paper) used to render the drawing. One

---

[*] This work was supported in part by the National Science Foundation under Grants IIS-10-18475, IIS-09-48548, IIS-08-12377, CCF-08-30618, and IIS-07-13501.

U. Brandes and S. Cornelsen (Eds.): GD 2010, LNCS 6502, pp. 341–352, 2011.

common issue is the visual clutter due to the high number of (possibly intersecting) edges, which prevents the human eye from easily following their paths. Recently a new drawing style called *confluent drawing* has been introduced to draw non-planar graphs with no edge crossings by bundling groups of intersecting edges into *tracks* [4]. Since not all graphs are *confluent* (e.g. can be drawn confluently) and no polynomial exact algorithm is known to decide whether a graph is confluent or not, most of the research in this area focuses on investigating new heuristics to recognize confluent graphs as well as studying new classes of confluent graphs [4, 7, 8, 14, 15]. However, it would be interesting to investigate how edges can be bundled not only to eliminate crossings but also to visualize as few segments/curves as possible, thus reducing the visual clutter in both non-planar and planar dense graphs.

Rather than improving the existing heuristics to recognize confluent graphs, our paper proposes an algorithm that is specifically designed for creating confluent drawings, bundling the edges so as to significantly reduce the amount of drawing elements to display. The algorithm creates a confluent drawing of a graph from its rectangular dual, a graph representation mostly used in VLSI floorplanning [13, 16–18]. We claim two major contributions of this approach:

(a) We create orthogonal-like confluent drawings with large angular resolution ($\geq \pi/2$) of graphs with unbounded maximum degree, such that nodes are visualized as points in the plane and edges as sequences of vertical and horizontal segments connected through *curved bends* (Fig. 4);
(b) We outline how to draw confluently a clustered graph with rectangular clusters, by extending the definition of rectangular dual to clustered graphs [19].

As far as (a) is concerned, in the case of planar graphs we may think of our drawings as a generalization of orthogonal drawings to graphs with maximum degree ≥ 4, even though they are not perfectly orthogonal, due to the curved bends. We are fully aware that approaches exist that create perfectly orthogonal drawings for graphs with unbounded maximum degree. One of the best approaches is the Kandinsky model [11], where the nodes are represented as squares of equal size placed on a grid and edges as sequences of vertical and horizontal segments. However, if the size of the squares is too small the edges incident with a high-degree node may appear too close to one another which we avoid by keeping a large angular resolution. Tree-like confluent graphs [15] and delta-confluent graphs [7] can also be drawn with angular resolution $\geq \pi/2$, but they are limited to chordal bipartite graphs and distance hereditary graphs respectively, while our drawings can represent every confluent graph [1].

The remainder of the paper is organized as follows. The notation and definitions used in the paper and relevant related work are introduced in Section 2. Theory of rectangular dualization is detailed in Section 3. In Section 4 we describe our drawing algorithms and we show preliminary results. In Section 5 we outline our method to create confluent drawings of clustered graphs. Section 6 concludes the presentation.

---

[1] Henceforth when we say "every confluent graph" we mean every graph that can be recognized as confluent by the existing heuristics.

2   Background and Related Work

In a *confluent drawing* edges are merged together into *tracks* which are the union of locally-monotone curves [4]. A *locally-monotone curve* is one that does not self intersect and contains no point with left and right tangents forming an angle ≤ 90 degrees. More precisely, in a confluent drawing D of an undirected graph $G = (V, E)$ each node $v \in V$ is drawn as a point v' in the plane, every edge $(v_i, v_j) \in E$ is represented as a locally-monotone curve connecting v_i' and v_j' and no two locally-monotone curves can cross.

A *clustered graph* C is a graph with a recursive partitioning of its node set $V(G)$ into groups called *clusters*. More formally $C = (G, T)$ consists of an *underlying graph* G and a tree T, describing the inclusion relation between clusters, such that every node ν of T is a cluster $V(\nu)$ of the nodes of G that are leaves of the subtree rooted at ν [9, 10]. *High-level clusters* are those included in no other clusters. Typically each cluster $V(\nu)$ is visualized as a region R in the plane enclosing the drawing of $G(\nu)$ (and the drawing of all subclusters). When the drawing of an edge e crosses the boundary of region R more than once and both endpoints of e are outside R we have a *edge-region crossing*. A drawing with no edge crossings nor edge-region crossings is said *c-planar* and C is *c-planar* if it admits such a drawing. Finally, C is *c-connected* if $G(\nu)$ is connected, for each $\nu \in T$ [9, 10]. Two important algorithms to draw clustered graphs using rectangular regions are described in [1, 5].

3   Rectangular Dualization

Let $G = (V, E)$ be a plane connected graph with n nodes and m edges. A *rectangular dual* $RD(G) = (\Gamma, f)$ of G (Fig. 1) is a *rectangular subdivision system* Γ – a partition of a rectangle into a set of non-overlapping rectangles no four of which meet at the same point – with a one-to-one correspondence $f : V \to \Gamma$ such that two nodes u and v are adjacent in G if and only if their corresponding rectangles $f(u)$ and $f(v)$ share a common boundary [16]. Not every planar graph admits a rectangular dual and, if so, it has many. Different groups of authors independently discovered and proved necessary and sufficient conditions for a graph to have a rectangular dual [16–18]. We prefer the formulation in [16] as we use the resulting algorithm to create rectangular duals.

Theorem 1. *A plane graph G has a rectangular dual with four rectangles on the boundary if and only if (a) every inner face of G is a $3-$cycle and there are exactly four nodes on the outer cycle; (b) G has no separating triangles – $3-$cycles containing at least one node in its interior.*

A graph which complies with Theorem 1 is called a *Proper Triangular Planar* (PTP) graph; we denote the four nodes on the outer cycle of G in counterclockwise order as v_N, v_W, v_S and v_E. By Theorem 1 only a limited subclass of planar graphs admits a rectangular dual; in order to support our statement that confluent drawings can be obtained from a rectangular dual of every confluent graph, we need to explain how we force a graph to comply with the admissibility conditions.

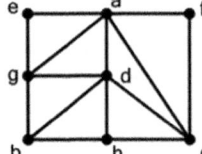 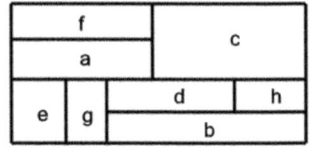

Fig. 1. A graph and its rectangular dual

3.1 Enforcing the Admissibility Conditions

First, Theorem 1 requires G to be planar; if it is not the case, we use the planarization procedure described in [4]. For every clique or biclique C of G, the
edges of C are removed and the nodes of C are connected to a new node v, which
we call a *clique crossover node* if C is a clique and *biclique crossover node* if C is
a biclique; we say that the nodes of C are adjacent *through* the crossover node v.
If the resulting graph is not planar, the algorithm concludes that G is not confluent and terminates. As pointed out in [4], this planarization procedure runs
in $O(n)$ time assuming that G has bounded *arboricity*, which is the minimum
number of forests into which the edges of G can be partitioned.

In order to turn G into a PTP graph G^*, we use a procedure described in [18]
which goes through the following steps:

(a) Edges are added to make G biconnected.
(b) The four nodes v_N, v_W, v_S, v_E are added so that they form a new outer
     cycle of G. New edges are introduced to link the new nodes to those of the
     former outer cycle of G.
(c) Separating triangles are searched [3] and *broken* (see below).
(d) All inner faces are triangulated using the $O(n)$ algorithm described in [2].

As for step (c), the algorithm described in [3] finds all 3-cycles of G, including
not only the separating triangles but also the 3-faces (e.g. faces bounded by a
3-cycle) of G. Since every 3-cycle can be described as a triple of integer numbers,
each integer identifying a node (or an edge), the list L of 3-cycles and the list F
of 3-faces can be lexicographically sorted using radix sort and compared against
each other in $O(n)$ time to remove from L any item of F. To break a separating
triangle $\Delta = (v, w, z)$, one of its edges (say (v, w)) is replaced by two new edges
(v, c) and (c, w), c being a new node called *ST crossover node*, and we say that
v and w are adjacent *through* c. In order to break in $O(n)$ time all separating
triangles of G by adding as few crossover nodes as possible we use the heuristics
described in [19].

Using the algorithm described in [16] the rectangular dual $RD(G^*)$ is computed while removing the rectangles corresponding to v_N, v_W, v_S and v_E. It
stands to reason that a rectangular dual of G^* is not a rectangular dual of G.
Since new edges are added, indeed, two rectangles $f(u)$ and $f(v)$ may be adjacent in $RD(G^*)$ even if the two corresponding nodes u and v are not adjacent in

G; similarly, due to the introduction of a crossover node c, $f(u)$ and $f(v)$ may not be adjacent in $RD(G^*)$ while $(u, v) \in E$. In the first case, since there is no relationship between u and v in G we can simply ignore that $f(u)$ is adjacent to $f(v)$; in the second case, we can think of $f(c)$ as a *gate*, through which one can go from $f(u)$ to $f(v)$, so as they can be considered as adjacent. Therefore, although it is not theoretically sound, we will refer to $RD(G^*)$ as the rectangular dual of G and we will denote it as $RD(G)$. Notice that exactly two nodes are adjacent through a *ST gate* (corresponding to a ST crossover node), while more than two are typically adjacent through *clique* and *biclique* gates.

4   From Rectangular Dual to Confluent Drawing

In this section we describe two methods to create a confluent drawing D of a graph $G = (V, E)$ from a rectangular dual $RD(G) = (\Gamma, f)$ of G. Both use an orthogonal-like drawing style, where nodes are represented as points in the plane and edges as sequences of horizontal and vertical segments, except for the edges connecting nodes adjacent through gates which are bundled into *traffic circles* (Fig. 2) as done in [4]. We remark that, while in an orthogonal drawing a *bend* is a point at which an horizontal and vertical segment meet, in our confluent drawings a bend is represented as a small diagonal segment (Fig. 2) connecting an horizontal to a vertical segment (or the other way round). The reason of this choice is that in a confluent drawing a segment may be shared by two or more edges that may follow different paths, whose direction is indicated by the bends.

Before describing the two approaches, we need some more notation. Every rectangle R in $RD(G)$ is bounded by four lines $x = x_{west}(R)$, $x = x_{east}(R)$, $y = y_{north}(R)$, $y = y_{south}(R)$, such that $x_{west}(R) < x_{east}(R)$ and $y_{south}(R) < y_{north}(R)$. Consequently R has four sides: the *north* and *south* sides lie on lines $y = y_{north}(R)$ and $y = y_{south}(R)$ respectively and the *west* and *east* sides lie on lines $x = x_{west}(R)$ and $x = x_{east}(R)$ respectively. Two rectangles R and Q are adjacent if they share one side or part of it. If R shares (part of) its north side with Q, R is *south of* Q (Q is *north of* R) and we write $south(R, Q)$ ($north(Q, R)$); if R shares (part of) its west side with Q, R is *east of* Q (Q is *west of* R) and we write $east(R, Q)$ ($west(Q, R)$).

In the following, when no ambiguity arises, we will use the same notation to denote a node or an edge and its pictorial representation (a point and a sequence of segments respectively).

4.1   Baseline Approach

The first method we will be describing creates D using no explicit technique to reduce the number of bends nor to bundle as many edges as possible so as to reduce the visual clutter. For this reason we refer to it as the *baseline approach*. To create the nodes, a point u is drawn at the center of every non-gate rectangle $f(u)$. As we will see, this choice heavily impacts on the number of resulting bends. In order to keep the bends at a readable length, if the x-coordinates $x(u)$

and $x(v)$ of two points u and v are such that $|x(u) - x(v)| < \epsilon$, for a small constant ϵ, the position of u and v is modified so that they appear aligned along the same vertical line. The same goes for $y(u)$ and $y(v)$. Next, we draw a traffic circle inside each gate. While traffic circles are the only possible option for clique and biclique gates, as more than two nodes are adjacent through them, they are not so necessary for ST gates, which connect exactly two nodes. However, we found cases where edges running through gates had too many unpleasant bends, which a traffic circle helped to hide. Finally for each point u we create its incident edges as follows (Fig. 3(a)). Without losing generality, we assume that at least one non-gate rectangle $f(v)$ is such that $north(f(v), f(u))$ and $(u, v) \in E$ (recall that since G may have been turned into a PTP graph $f(u)$ and $f(v)$ may be adjacent even if $(u, v) \notin E$). The same reasoning can be replicated if $f(v)$ has a different position with respect to $f(u)$. If u and v are aligned along the same vertical line (e.g. $x(u) = x(v)$), edge (u, v) is simply a segment between them (it has no bends). If u and v are not aligned, we create a new point (called *confluence point*) $c_{north}(u)$ with coordinates $(x(u), y_{north}(f(u)) - \epsilon)$ and we connect u and $c_{north}(u)$ by a segment s. Assuming s oriented from u to $c_{north}(u)$, v may be either left or right of segment s; in the first case a *boundary point* $c^l_{north}(u)$ is created on the north side of $f(u)$ with coordinates $(x(u) - \epsilon, y_{north}(f(u)))$ and connected to $c_{north}(u)$ by a segment; otherwise a boundary point $c^r_{north}(u)$ is created with coordinates $(x(u) + \epsilon, y_{north}(f(u)))$ and connected to $c_{north}(u)$. A segment connecting a confluence point to a boundary point is a bend. Finally, if v has been already processed (meaning that $c^l_{south}(v)$ or $c^r_{south}(v)$ already exist), a segment is drawn between $c^l_{north}(u)$ and $c^l_{south}(v)$ (if v is at the left of segment s) or $c^r_{north}(u)$ and $c^r_{south}(v)$ (if v is at the right of segment s). The resulting edge (u, v) will have exactly two bends. The same procedure is used to create

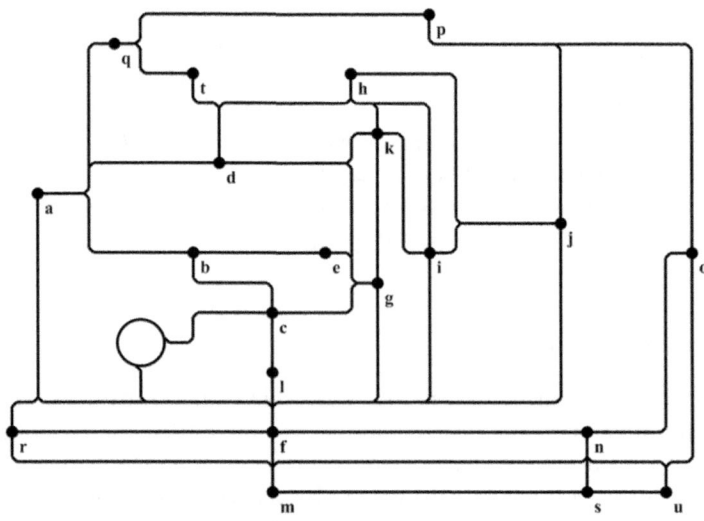

Fig. 2. Drawing produced by the baseline approach

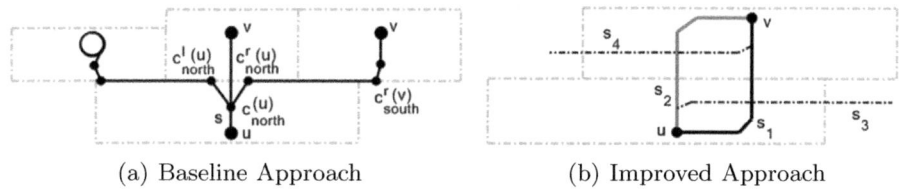

(a) Baseline Approach   (b) Improved Approach

Fig. 3. How the baseline and the improved approaches work

an edge e between u and any traffic circle, with the only difference that e needs
an extra bend to connect to the traffic circle, which is necessary to understand
the direction e has to take inside the circle.

Theorem 2. *The baseline approach creates a confluent drawing D of graph G
in $O(n)$ time.*

Proof. The number of gates is $O(n)$ and to create the edges the adjacency list
of every node is crossed at most once, hence the linearity of the method. As
for confluency, we notice that smooth lines are created only between points
corresponding to nodes that are adjacent in G. In any edge e incident with a
point u, the segment incident with u always run along a line passing through u;
the other edges run along the boundaries of the rectangles and cross a boundary
only to reach an endpoint. Given that, it is impossible that two sequences of
segments cross.

4.2 Improved Approach

As it is clear from Figure 2, the baseline approach creates many unwanted bends
(look for example at edges (n, o) (o, p) and (p, q)) and it does not effectively
reduce the visual clutter. Motivated by these observations, we propose a second
approach, which we refer to as the *improved approach* that goes through the
same steps as the baseline approach (creates the nodes, the traffic circles and,
for each point, the edges incident with it) but optimizes through heuristics the
position of the nodes as well as the layout of the edges to reduce the number of
bends and to bundle as many edges as possible (Fig. 4).

In order to make sure that we do not create edge crossings, we stick with our
principle that the drawing of node u must be inside the corresponding rectangle
$f(u)$ and that the drawing of edge (u, v) must cross the boundary of no rectangle
but $f(u)$ and $f(v)$. Instead of blindly drawing point u at the center of $f(u)$, we
place u within $f(u)$ so as to minimize the number of bends of its incident edges.
Although this does not necessarily imply that we minimize the number of bends
across the whole drawing, we found that eliminates many unnecessary bends,
such as the ones mentioned before. As for the edges, we remark that in the
drawings generated by the baseline approach any edge between two nodes not
adjacent through a gate either has no bends or two. Before creating a two-
bend edge, the improved approach checks whether it is possible to lay it down
with only one. Let us assume again, without loss of generality, that $north(v, u)$.

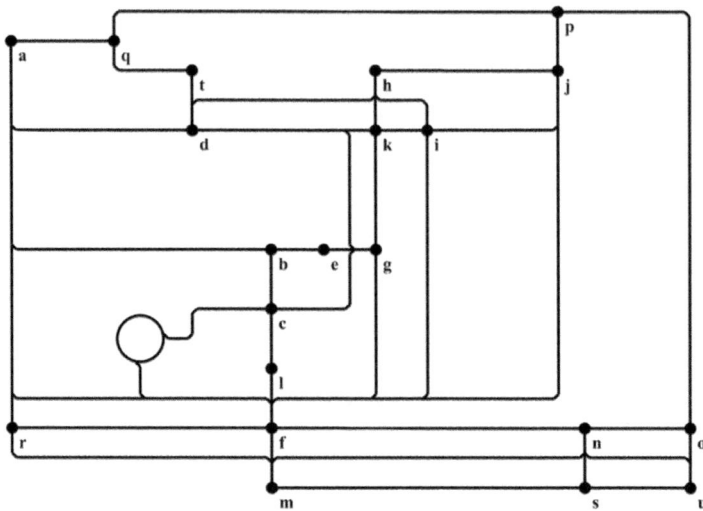

Fig. 4. Drawing produced by the improved approach

There are two ways to draw edge (v, u) with one bend (Fig. 3(b)): using the sequence of black thick segments denoted as s_1 or the sequence of gray thick segments denoted as s_2. s_2 is not a good solution, as it may cross sequences of segments incident with v (such as s_4); on the other hand, s_1 cannot cross any segment incident with v because the segment of s_1 incident with v runs on a line passing through v, as in the baseline approach, where crossings are not possible. However, s_1 may still cross a previously created 1-bend edge s_3 incident with u. Since both s_1 and s_3 represent valid 1-bend edges incident with u, we create the one which maximizes the number of 1−bend edges incident with u; in other words, we create s_1 if the edges following the same path as s_1 are more than the ones following the same path as s_3. Finally, all edges that cannot be created with only one bend are drawn with two-bends as described in the baseline approach. It is immediate to verify that the improved approach creates confluent drawings in linear time.

4.3   Implementation and Results

Both approaches described in the previous section have been implemented in C and included in our software OcORD, which has been originally created to study the properties of rectangular dualization and now has been extended to the graph drawing domain [19]. While introducing this paper, we stated that the goal of our drawing algorithm is to reduce the visual clutter due to the high number of edges in large graphs. To support this statement, we need a measure to assess such a reduction. One way to go here is to compare the number of drawing elements (e.g. segments) used to create the confluent drawing against the number of segments that would have been necessary to create the same drawing without bundling the

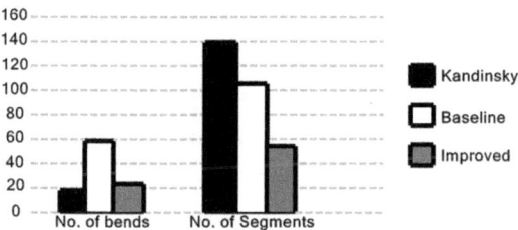

Fig. 5. Experimental results

edges. Therefore we selected 100 graphs from the $AT\&T$ data set[2], whose order ranges from 10 to 100 nodes, and for each of them we created, besides a confluent drawing using both the baseline and the improved approach, a Kandinsky-like drawing using the same drawing conventions for the edges as in the confluent drawings described in the previous section (e.g. edges represented as sequences of vertical and horizontal segments with diagonal bends). For all drawings we then measured the total number of bends and the number of segments used to draw the edges. Figure 5 reports on the average of these values over all data set. As expected, the improved approach outperforms the baseline approach both in terms of reduction of the segments and in terms of total number of bends. On average the drawings produced with the improved approach have up to 30% less segments than the drawings produced without bundling the edges, while in the drawings created by the baseline approach this reduction drops to 20%. Moreover, the improved approach halves the number of bends created with the baseline; however, looking at Fig. 4, we are still able to find unwanted bends, partly due to the traffic circles and partly due to the constraint that edges cannot cross the boundaries of other rectangles than the ones corresponding to their endpoints. As bends are also responsible for visual clutter, we are planning on substituting heuristics with an optimal algorithm which minimizes the bends through a flow based method, like it is done in [12]. We conclude the discussion of the results with two important remarks. First, we found the quality of the drawings is sometimes questionable. Referring to Fig. 4, nodes tend to crowd in some areas, while other regions appear empty; this is due to the different size of the rectangles in the rectangular dual, which is also responsible for the excessive length of some edges (such as edge (f, n)). Finally, a graph has typically many rectangular duals, which obviously lead to different drawings. It would be interesting to investigate how different rectangular duals impact on the quality of the drawing (especially in terms of number of gates and thus bends). Both aspects will be object of future research.

5   Confluent Rectangular Drawing for Clustered Graphs

In [19] the *clustered rectangular dual* (*c-rectangular dual* for short) has been introduced as a generalization of the rectangular dual for c-planar clustered

---

[2] http://www.graphdrawing.org/

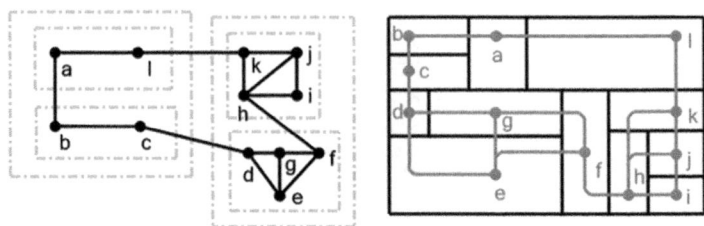

Fig. 6. A clustered graph with its c-rectangular dual and confluent drawing (in gray)

graphs. Intuitively, a c-rectangular dual of $C = (G, T)$ is a rectangular dual of G such that the rectangles corresponding to the nodes in $V(\nu)$ form a rectangular dual of $G(\nu)$ (Fig. 6). More formally:

Definition 1. *A c-rectangular dual $RD(C) = (\Gamma, f)$ of a c-planar clustered graph $C = (G, T)$ is a rectangular dual of G such that, for each node ν of T: (A) The union of the rectangles in $\Gamma(V(\nu))$ is a rectangular dual of $G(\nu)$, where $\Gamma(V(\nu)) = \{R \in \Gamma | f^{-1}(R) \in V(\nu)\}$; (B) If $V(\mu) \subset V(\nu)$, $\Gamma(V(\mu)) \subset \Gamma(V(\nu))$.*

The outline of the admissibility conditions require additional theoretical background which is out of the scope of this paper; the interested reader is referred to [19]. Here we limit ourselves to a short description of the algorithm that creates a c-rectangular dual of every c-planar graph $C = (G, T)$. Without losing generality we assume that the underlying graph G is a PTP graph, whose outer nodes belong to no cluster (they are mere construction nodes). The idea of the algorithm is extremely simple; first we create the rectangular dual $RD(Q)$ of the *quotient graph Q* of C, where each rectangle corresponds to a high-level cluster $V(\nu)$ of C, and next each rectangle is "filled" with the rectangular dual of $G(\nu)$ (if $G(\nu)$ is a plain graph) or the c-rectangular dual of $G(\nu)$ (if $G(\nu)$ is in turn a clustered graph). We recall that Q is the graph having a node v_ν for each high-level cluster $V(\nu)$ of C and an edge between two nodes v_ν, v_μ if and only if there is an edge in C linking a node in $V(\nu)$ to a node in $V(\mu)$. Figure 7 (a) and (b) show the quotient graph of the clustered graph in Fig. 6 and the rectangular dual respectively. It stands to reason that G being a PTP does not imply that Q is a PTP graph; separating triangles and even multiple edges may need to be broken using additional crossover nodes (which may be thought of as empty clusters). Similarly, the rectangular dual of $G(\nu)$, for each $V(\nu)$, must be such that the rectangles on the boundary fulfill the adjacency constraints with the rectangles on the boundary of adjacent clusters. Referring to Fig. 6, the rectangular dual of $G(\{a, b, c, l\})$ can be rotated so that rectangle l appears in the position where b, c are now (and vice versa), and still we would have a valid rectangular dual of $G(\{a, b, c, l\})$; but rectangle l would not be adjacent to rectangle k anymore and, similarly, c would not be adjacent to d. Again, for details on this point we refer the interested reader to [19]. Once a c-rectangular dual has been created, one of the algorithms described in Section 4 is used to create a confluent drawing of the underlying graph. Figure 6 shows that the confluent drawing of $G(\nu)$, for

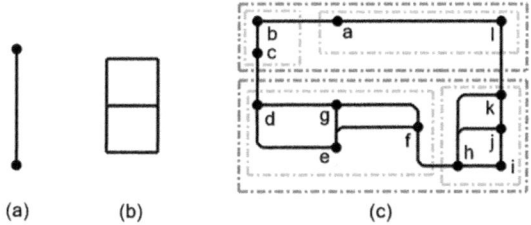

Fig. 7. (a) Quotient graph Q of C. (b) $RD(Q)$. (c) Confluent R-drawing of C.

every cluster, is completely contained in the interior of the rectangular region delimited by the rectangular dual of $G(\nu)$. All we are left to do is to separate the sides of these rectangular regions so as to obtain the *confluent rectangular drawing* (or confluent R-drawing) shown in Figure 7, where each cluster is represented as an independent rectangular region. The advantage of a a confluent R-drawing is that it reduces the visual clutter due to the edges and the visual clutter due to the representation of the clusters; using simple convex regions such as rectangles, in fact, improves the readability of the graph, as pointed out in [5]. We have no experimental results on this part so far, as the algorithm that creates the c-rectangular dual is not perfect yet. In particular, it seems to create too many gates, which, as seen before, have such a negative impact on the number of bends. This will be the object of our future work.

6   Conclusions

In this paper we discussed two algorithms that create a confluent drawing of a graph G from its rectangular dual. Preliminary results indicate that one of the approaches successfully decreases the amount of segments needed to visualize the edges, thus reducing visual clutter. Much work is still left to do to reduce the number of bends as well as applying the described techniques to clustered graphs.

References

1. Angelini, P., Frati, F., Kaufmann, M.: Straight-Line Rectangular Drawings of Clustered Graphs. In: Dehne, F., Gavrilova, M., Sack, J.-R., Tóth, C.D. (eds.) WADS 2009. LNCS, vol. 5664, pp. 25–36. Springer, Heidelberg (2009)
2. Biedl, T.C., Kant, G., Kaufmann, M.: On Triangulating Planar Graphs under the Four-Connectivity Constraint. In: Schmidt, E.M., Skyum, S. (eds.) SWAT 1994. LNCS, vol. 824, pp. 83–94. Springer, Heidelberg (1994)
3. Chiba, N., Nishizeki, T.: Arboricity and Subgraph Listing. SIAM Journal on Computing 14(1), 210–223 (1985)

4. Dickerson, M., Eppstein, D., Goodrich, M.T., Meng, J.Y.: Confluent Drawings: Visualizing Non-planar Diagrams in a Planar Way. In: Liotta, G. (ed.) GD 2003. LNCS, vol. 2912, pp. 1–12. Springer, Heidelberg (2004)
5. Eades, P., Feng, Q.-W., Nagamochi, H.: Drawing Clustered Graphs on an Orthogonal Grid. Journal on Graph Algorithms and Applications 3(4), 3–29 (1999)
6. Eades, P., Gutwenger, C., Hong, S.-H., Mutzel, P.: Graph Drawing Algorithms. In: Algorithms and Theory of Computation Handbook, 2nd edn., vol. 2, CRC Press, Boca Raton (2009)
7. Eppstein, D., Goodrich, M.T., Meng, J.Y.: Delta-Confluent Drawings. In: Healy, P., Nikolov, N.S. (eds.) GD 2005. LNCS, vol. 3843, pp. 165–176. Springer, Heidelberg (2006)
8. Eppstein, D., Goodrich, M.T., Meng, J.Y.: Confluent Layered Drawings. Algorithmica 47(4), 439–452 (2007)
9. Feng, Q.-W., Cohen, R.F., Eades, P.: Planarity for Clustered Graphs. In: Spirakis, P.G. (ed.) ESA 1995. LNCS, vol. 979, pp. 213–226. Springer, Heidelberg (1995)
10. Feng, Q.: Algorithms for Drawing Clustered Graphs. Ph.D. thesis, University of Newcastle, Australia (1997)
11. Fössmeier, U., Kaufmann, M.: Drawing High Degree Graphs with Low Bend Numbers. In: Brandenburg, F.J. (ed.) GD 1995. LNCS, vol. 1027, pp. 254–266. Springer, Heidelberg (1996)
12. Garg, A., Tamassia, R.: A New Minimum Cost Flow Algorithm with Applications to Graph Drawing. In: North, S.C. (ed.) GD 1996. LNCS, vol. 1190, pp. 201–216. Springer, Heidelberg (1997)
13. He, X.: On Floorplans of Planar Graphs. In: Proceedings of the 29th Annual ACM Symposium on Theory of Computing, pp. 426–435. ACM, New York (1997)
14. Hirsch, M., Meijer, H., Rappaport, D.: Biclique Edge Cover Graphs and Confluent Drawings. In: Kaufmann, M., Wagner, D. (eds.) GD 2006. LNCS, vol. 4372, pp. 405–416. Springer, Heidelberg (2007)
15. Hui, P., Pelsmajer, M.J., Schaefer, M., Stefankovic, D.: Train Tracks and Confluent Drawings. Algorithmica 47(4), 465–479 (2007)
16. Kant, G., He, X.: Two Algorithms for Finding Rectangular Duals of Planar Graphs. In: van Leeuwen, J. (ed.) WG 1993. LNCS, vol. 790, pp. 396–410. Springer, Heidelberg (1994)
17. Koźmiński, K., Kinnen, E.: Rectangular Duals of Planar Graphs. Networks 15(2), 145–157 (1985)
18. Lai, Y.-T., Leinwand, S.M.: Algorithms for Floorplan Design via Rectangular Dualization. IEEE Transactions on Computer-Aided Design 7(12), 1278–1289 (1988)
19. Quercini, G.: Optimizing and Visualizing Planar Graphs via Rectangular Dualization. Ph.D. thesis, University of Genoa, Genoa, Italy (2009)

How to Draw a Tait-Colorable Graph

David A. Richter

Western Michigan University, Kalamazoo MI 49008, USA

Abstract. Presented here are necessary and sufficient conditions for a cubic graph equipped with a Tait-coloring to have a drawing in the real projective plane where every edge is represented by a line segment, all of the lines supporting the edges sharing a common color are concurrent, and all of the supporting lines are distinct.

1 Introduction

The complete bipartite graph $K_{3,3}$ has 9 edges and the Pappus configuration has 9 lines. In fact, we may embed $K_{3,3}$ in the Pappus configuration, as in Figure 1.

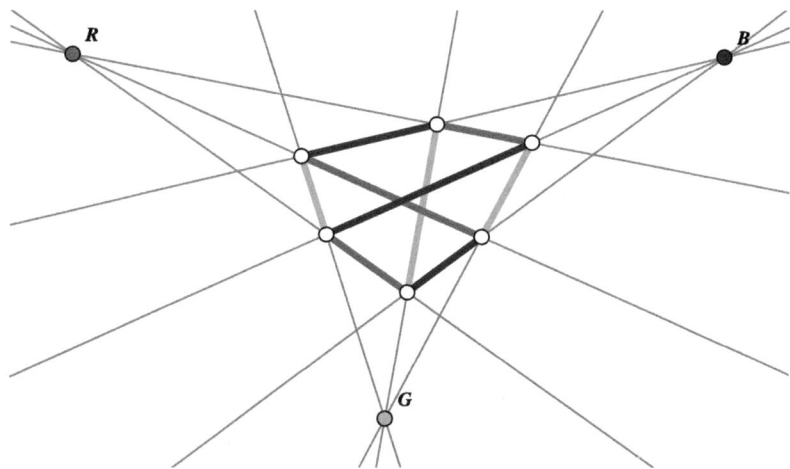

Fig. 1. How to draw $K_{3,3}$

Notice that $K_{3,3}$ has a Tait-coloring, meaning that the graph is cubic and the edges may be colored with three colors in such a way that edges of all three colors meet at every vertex [13]. Given a graph G with a Tait-coloring, we ask in general whether it is possible to represent G by this type of drawing. Specifically, we require that the drawing in the real projective plane have the properties that (a) every edge is represented by a line segment, (b) all of the lines supporting the edges sharing a common color are concurrent, and (c) every supporting line contains exactly 2 vertices of the graph. The main result here

U. Brandes and S. Cornelsen (Eds.): GD 2010, LNCS 6502, pp. 353–364, 2011.

gives combinatorial necessary and sufficient conditions for when such a graph has this type of drawing. Since $K_{3,3}$ appears as a special case, this is a generalization of Pappus's classical hexagon theorem.

This is a particular type of straight-line drawing. Since every bipartite cubic graph is Tait-colorable, the main result yields the fact that every 3-connected bipartite cubic graph has a drawing with slope number less than or equal to 3. Moreover, the main result here singles out a particular class of graphs studied in [9], where it is proved that every graph with degree not exceeding 3 has slope number less than or equal to 5. A significant source for inspiration for this work was the recent study [5], [6] on xyz graphs and xyz polyhedra. According to [5], an xyz graph is a cubic graph which can be represented in $\mathbb{R}^3$ in such a way that every axis-parallel line contains either zero or two points of the graph. Every xyz graph is Tait-colorable, and it is well known that every cubic graph equipped with a Tait-coloring yields a corresponding compact surface, cf. [4]. Hence, an xyz surface is the compact surface associated to an xyz graph. Obviously the definitions of xyz graphs and projective drawings are closely related. Thus, for example, one sees that every xyz graph has a projective drawing, but a graph which has a projective drawing may fail to be an xyz graph. The complete bipartite graph $K_{3,3}$ furnishes an example of this. Thus, our class extends the class of xyz graphs.

The question of drawing graphs in this way arose during a reading of [11], where it is proved that realization spaces of 4-dimensional convex polytopes are universal. In the course of this proof, it is necessary to construct a 4-dimensional polytope with a non-prescribable octagonal face. In particular, imposing certain incidences among the edges and vertices forces slopes of the edges of the octagon to be a harmonic set. Having seen the construction of this octagon, it is natural to inquire how many incidences may be similarly prescribed.

By serendipity, the main result here is a close relative of a theorem which was discovered while studying a phenomenon which may be called "ghost symmetry". Roughly, one says that a subset of a Euclidean space has a ghost symmetry if some projection of that object has an unexpected symmetry group. The main result in [10] gives a necessary and sufficient condition for a cubic graph equipped with a Tait-coloring to correspond to a subset of the plane having ghost symmetries specified by the coloring. Even though the idea behind this "ghost symmetry prescribability theorem" is not nearly as intuitive as that of drawing a graph, it is stated below for the sake of completeness.

The drawings which appear here were produced using Cinderella [12].

2   Theory of Tait-Colored Graphs

Suppose n is a positive integer. Define a *Tait-colored graph* as a pair $G = (S, T)$, where $S = \{1, 2, 3, ..., 2n - 1, 2n\}$, and $T : \{1, 2, 3\} \to H$ is a function into the set $H \subset \mathrm{Perm}(S) \cong S_{2n}$ of all fixed-point-free permutations of S of order 2. It is convenient to designate the image $T(\{1, 2, 3\})$ by $\{b, g, r\}$ and write $T = \{b, g, r\}$.

In graph-theoretic terminology, H coincides with the set of all perfect matchings of the complete graph with vertex set S. Define the Θ-*graph* as the unique Tait-colored graph with 2 vertices.

Notice that we use the term "Tait-colored graph" instead of the more cumbersome phrase "cubic graph equipped with a Tait-coloring". Indeed, every graph which we consider is already equipped with a Tait-coloring, and we do not address the issue of Tait-colorability. The reason we regard a Tait-colored graph as a triple of permutations is due to the connection to ghost symmetries described below.

Define a *component* of a graph $G = (S, T)$ as an orbit in the group generated by T. Call a graph *connected* if it has exactly one component. Define a *monochromatic pair* of a graph $G = (S, T)$ as a pair of transpositions appearing in the cycle decomposition of one of the elements of T. Graph-theoretically, a monochromatic pair is a pair of edges having the same color. Thus, one may designate a monochromatic pair by $\pi = \{(x, \sigma x), (y, \sigma y)\}$, where $\sigma \in T$ and x and y are vertices. A pair of edges of a connected graph G is a *2-edge cut set* if deleting the edges disconnects the graph. If G is Tait-colored and π is a 2-edge cut set, then a 2-edge cut set is necessarily a monochromatic pair.

The language required in the case a graph is not bipartite is streamlined by using the bipartite double cover of a graph, as defined, for example, in [2]. We specialize this to Tait-colored graphs. Thus, suppose $G = (S, T)$ is a Tait-colored graph. Introduce disjoint sets S_1 and S_2 such that there exist bijections $\beta_i : S \to S_i$. The bipartite double cover of G has vertex set $S_1 \cup S_2$ and edges $\{(\beta_1(z), \beta_2(\sigma z)) : z \in S, \sigma \in T\}$. Alternatively, one constructs the bipartite double cover by introducing a monochrome pair $\{(\beta_1(z), \beta_2(\sigma z)), (\beta_2(z), \beta_1(\sigma z))\}$ for every edge $(z, \sigma z)$ of G. Notice that the Tait-coloring of G furnishes a natural Tait-coloring of the bipartite double cover.

In general, a connected graph is bipartite if and only if its bipartite double cover is disconnected; if a graph is bipartite, then the bipartite double cover is a disjoint union of two copies of the original graph. Call a Tait-colored graph $G = (S, T)$ is *strongly non-bipartite* if the deleted graph $G\backslash\pi$ is non-bipartite for every monochromatic pair π. Notice that a graph is strongly non-bipartite if the bipartite double cover of $G\backslash\pi$ is connected for every monochromatic pair π. Finally, note that the bipartite double cover of a connected non-bipartite graph G has a 2-edge cut set if and only if $G\backslash\pi$ has both a bipartite component and a non-bipartite component for some monochromatic pair π.

3   Projective Drawings

The purpose of this section is to state and prove the main results of this article.

Suppose $G = (S, T)$ is a Tait-colored graph and assume $T = \{b, g, r\}$. Define a *parallel drawing* of G as a pair (ι, ϕ) where $\iota : S \to \mathbb{R}^2$ is a function and $\phi = \{\phi_b, \phi_g, \phi_r\}$ is a triple of projections to distinct 1-dimensional subspaces of $\mathbb{R}^2$ such that such that $\phi_\sigma(\iota(z)) = \phi_\sigma(\iota(\sigma z))$ for all $z \in S$ and all $\sigma \in T$. Call a monochromatic pair $\{(x, \sigma x), (y, \sigma y)\}$ *degenerate* for (ι, ϕ) if the lines $\overline{\iota(x), \iota(\sigma x)}$

and $\overline{\iota(y), \iota(\sigma y)}$ coincide. Given a Tait-colored graph $G = (S, T)$, define the *trivial drawing* of G by specifying $\iota(z) = 0$ for all $z \in S$. Clearly every monochromatic pair of a graph is degenerate for its trivial drawing. Call a parallel drawing *faithful* if it has no degenerate pairs. In this terminology, we are interested in general in when one may find a faithful parallel drawing of G.

The space of parallel drawings of a fixed graph $G = (S, T)$ carries an action by the group $GL(2, \mathbb{R})$, induced by the usual action on $\mathbb{R}^2$. Thus, suppose (ι, ϕ) is a drawing of G and $g \in GL(2, \mathbb{R})$. Define a pair (ι^g, ϕ^g) by $\iota^g(x) = g\iota(x)$ and $(\phi^g)_\sigma = (g^{-1})^T \phi_\sigma$ for all $x \in S$ and $\sigma \in T$. Then (ι^g, ϕ^g) is a drawing of T. Notice (ι^g, ϕ^g) is faithful whenever (ι, ϕ) is faithful. The action of $GL(2, \mathbb{R})$ factors through to an action of $PGL(2, \mathbb{R})$ which is triply transitive on $\mathbb{R}P^1$. Thus, one may prescribe $\{\phi_b, \phi_g, \phi_r\}$ to be the projections to any three distinct 1-dimensional subspaces of $\mathbb{R}^2$. For this reason, we assume throughout that these projections are fixed and suppress mention of them unless necessary.

The main results are handled in two cases, depending on whether or not a given graph is bipartite:

Theorem 1. *A connected, bipartite, Tait-colored graph which is not the Θ-graph admits a faithful parallel drawing if and only if it does not have a 2-edge cut set.*

There is a similar characterization when the graph is non-bipartite:

Theorem 2. *A connected, non-bipartite, Tait-colored graph admits a faithful parallel drawing if and only if it is strongly non-bipartite and its bipartite double cover does not have a 2-edge cut set.*

In the proofs of each of these, one must establish two directions. The "combinatorial" direction is to show that if a graph has a certain pathology, then the graph cannot be drawn faithfully. The converse "constructibility" direction is to show that if a graph lacks the pathology, then one may draw it faithfully.

Call a monochromatic pair of a Tait-colored graph G *forced* if it is degenerate in every parallel drawing of G. Forced pairs allow an alternate statement of the main results:

Corollary 3. *Suppose G is a connected, Tait-colored graph with at least 4 vertices and π is a monochromatic pair of G. (a) If G is bipartite, then π is forced if and only if π is a cut set. (b) If G is non-bipartite, then π is forced if and only if $G \backslash \pi$ has a bipartite component.*

In particular, a monochromatic pair π of a connected non-bipartite graph G can be forced in one of two ways. Either $G \backslash \pi$ is connected and bipartite or π is a cut set and exactly one of the two components of $G \backslash \pi$ is bipartite.

Given a Tait-colored graph $G = (S, T)$, define a *projective drawing* of G as a pair (ι, λ) where $\iota : S \to \mathbb{R}P^2$ is a function and $\lambda : T \to \mathbb{R}P^2$ is an injection such that λ_σ lies on $\overline{\iota(z), \iota(\sigma z)}$ for all $z \in S$ and $\sigma \in T$. Call a projective drawing *faithful* if the lines $\overline{\iota(z), \iota(\sigma z)}$ correspond bijectively with the edges $(z, \sigma z)$. A projective drawing becomes a parallel drawing when the three points λ are collinear. Since faithfulness and non-collinearity are open conditions, one may perturb a parallel drawing so that the three "vanishing points" λ are non-collinear while preserving faithfulness. In fact, if a graph admits a faithful parallel drawing, then one may find a projective drawing for any choice of three distinct points λ. The converse of this, however, is not true. For example, the complete quadrangle is a projective drawing of the complete graph K_4 equipped with its unique Tait-coloring, and it is often taken as an axiom in projective geometry that the three diagonal points not lie on a common line.

3.1   The Combinatorial Direction

Proposition 4. *Suppose $G = (S, T)$ is a connected Tait-colored graph and π is a monochromatic pair of G. (a) If G is bipartite and π is a cut set, then π is forced. (b) If G is non-bipartite and $G \backslash \pi$ has a bipartite component, then π is forced.*

Proof. (a) Let $\{S_1, S_2\}$ be the bipartition of S and let V be the vertices of one of the components of $G \backslash \pi$ Without loss of generality, assume that $\pi = \{(x, bx), (y, by)\}$, where $x \in S_1$ and $y \in S_2$. Let ι be any drawing of G. Notice $g(S_1 \cap V) = S_2 \cap V$ and $g(S_2 \cap V) = S_1 \cap V$, so we may write

$$\sum_{z \in S_1 \cap V} \phi_g(\iota(z)) = \sum_{z \in S_1 \cap V} \phi_g(\iota(gz)) = \sum_{z \in S_2 \cap V} \phi_g(\iota(z)) \ .$$

By a similar token, we also have

$$\sum_{z \in S_1 \cap V} \phi_r(\iota(z)) = \sum_{z \in S_1 \cap V} \phi_r(\iota(rz)) = \sum_{z \in S_2 \cap V} \phi_r(\iota(z)) \ .$$

Since ϕ_g and ϕ_r are linearly independent, this yields

$$\sum_{z \in S_1 \cap V} \iota(z) = \sum_{z \in S_2 \cap V} \iota(z) \ .$$

In particular, this implies

$$\sum_{z \in S_1 \cap V} \phi_b(\iota(z)) = \sum_{z \in S_2 \cap V} \phi_b(\iota(z)) \ .$$

Subtracting off all terms corresponding to blue edges of G, this yields $\phi_b(\iota(x)) = \phi_b(\iota(y))$. However, ι is a drawing of G, so $\phi_b(\iota(x)) = \phi_b(\iota(bx))$ and $\phi_b(\iota(y)) = \phi_b(\iota(by))$. Therefore π is a forced pair.

Part (b) follows in an analogous manner, although there are two parts to show. If $G \backslash \pi$ has two components, then one may use the argument above on the bipartite component. If $G \backslash \pi$ is connected, then it is bipartite and there is a uniquely determined bipartition $\{S_1, S_2\}$ of the vertices of $G \backslash \pi$. Again one may show as above that the centroid of $\iota(S_1)$ coincides with the centroid of $\iota(S_2)$. One then projects according to ϕ_b, subtracts terms corresponding to blue edges, and concludes that the vertices of π must be collinear. $\square$

This establishes the combinatorial direction for both theorems.

3.2   The Constructibility Direction

Given a Tait-colored graph $G = (S, T)$, define a *cycle* of G as a sequence $\gamma = (z_1, z_2, ..., z_{2k})$ of vertices such that $z_{i+1} = \sigma_i z_i$ for some $\sigma_i \in T$ for each i, regarding subscripts modulo $2k$. Call a cycle of length $2k$ *simple* if it has $2k$ distinct vertices.

Suppose (ι, ϕ) is a parallel drawing of $G = (S, T)$ and $\gamma = (z_1, z_2, ..., z_{2k})$ is a simple cycle of G. Assuming $T = \{b, g, r\}$, choose non-zero vectors $v_\sigma \in \ker(\phi_\sigma)$ such that $v_b + v_g + v_r = 0$. Define $\tau_i \in T \backslash \{\sigma_{i-1}, \sigma_i\}$, the unique third color at z_i which is not equal to σ_{i-1} or σ_i. Next, assign $s : \{1, 2, 3, ..., 2k\} \to \{-1, 1\}$ as follows. First let $s_1 = 1$. Then, assuming s_i has been defined, let

$$s_{i+1} = \begin{cases} s_i & \text{if } \tau_{i+1} = \tau_i , \\ -s_i & \text{if } \tau_{i+1} \neq \tau_i . \end{cases}$$

For ease of notation, write $v_i = v_\tau$ if $\tau = \tau_i$ for each i. Next, for $t \in \mathbb{R}$, let $\iota_{\gamma, t}(z) = \iota(z)$ for all $z \notin \{z_1, z_2, ..., z_{2k}\}$ and

$$\iota_{\gamma, t}(z_i) = \iota(z_i) + t s_i v_i$$

for all $i \in \{1, 2, 3, ..., 2k\}$. It is routine to verify that $\iota_{\gamma, t}$ is a parallel drawing of G for any choice of cycle γ satisfying the assumptions above. Call $\iota_{\gamma, t}$ a *perturbation* of ι along γ. Notice that a perturbation is defined only for cycles of even length.

One may quickly establish:

Proposition 5. *Suppose $G = (S, T)$ is a Tait-colored graph, ι is a parallel drawing of G, and γ is a simple cycle of G. (a) If z is a vertex of γ and $t \neq 0$, then $\iota_{\gamma, t}(z) \neq \iota(z)$. (b) If z is not a vertex of γ, then $\iota_{\gamma, t}(z) = \iota(z)$ for all t. (c) If $(z, \sigma z)$ is an edge of γ and $t \neq 0$, then $\overline{\iota_{\gamma, t}(z), \iota_{\gamma, t}(\sigma z)} \neq \overline{\iota(z), \iota(\sigma z)}$. (d) If z is a vertex of γ and $(z, \sigma z)$ is not an edge of γ, then $\overline{\iota_{\gamma, t}(z), \iota_{\gamma, t}(\sigma z)} = \overline{\iota(z), \iota(\sigma z)}$ for all t.*

This proposition indicates how the constructibility direction of the proof works. Our aim in each case is to show that one may perturb the trivial drawing along simple cycles and inductively remove every degenerate pair under the given hypotheses. In each case, one chooses the perturbation carefully so as not to introduce any further degeneracies. Note that if $\pi = \{(x, \sigma x), (y, \sigma y)\}$ is a degenerate pair for a drawing ι and there exists a pair (γ, t) such that π is not a degenerate pair for $\iota_{\gamma, t}$, then the general perturbation of ι along γ has strictly fewer degenerate pairs than ι. This follows because the condition for a pair to be degenerate for $\iota_{\gamma, t}$ is a linear equation in t. Since there are only finitely many monochromatic pairs, the number of values of t for which $\iota_{\gamma, t}$ has more or equal numbers of degenerate pairs than ι is finite.

The Bipartite Case. Suppose first that $G = (S, T)$ is a connected, bipartite, Tait-colored graph. Furthermore, assume that G does not have a 2-edge cut set. Let ι be any drawing of G and suppose $\pi = \{(x, \sigma x), (y, \sigma y)\}$ is degenerate for ι. Since π is not a 2-edge cut set, there is a simple cycle γ in G which contains the edge $(x, \sigma x)$, but not the edge $(y, \sigma y)$. Hence, if G has a degenerate pair, one may always choose a non-zero perturbation $\iota_{\gamma, t}$ which has fewer degenerate pairs than ι. This completes the proof of the main result in the case when G is bipartite.

The Non-Bipartite Case. This follows analogously, although it is more cumbersome. Suppose $G = (S, T)$ is a connected, non-bipartite, Tait-colored graph, ι is a drawing of G, and $\pi = \{(x, \sigma x), (y, \sigma y)\}$ is a degenerate pair for ι. Assume moreover that the deleted graph $G \backslash \pi$ is non-bipartite and does not have a bipartite component. Showing the existence of a desirable cycle depends on whether or not π is a cut set.

Suppose first that $G \backslash \pi$ is connected. Since $G \backslash \pi$ is not bipartite, there is a simple cycle γ in G having the property that $(x, \sigma x)$ is an edge of γ and $(y, \sigma y)$ is not an edge of γ. (If such a cycle didn't exist, then $G \backslash \pi$ would be bipartite or disconnected.) Hence, one may perturb along γ to eliminate the degeneracy along π without introducing more degeneracies.

Suppose instead that $G \backslash \pi$ is disconnected, consisting of components G_1 and G_2. Assume x is a vertex of G_1 and σx is a vertex of G_2. Since neither G_1 nor G_2 is bipartite, there are integers j, k and cycles $\gamma_1 = (x = z_1, z_2, ..., z_{2j-1}, z_{2j} = x)$ of length $2j - 1$ in G_1 and $\gamma_2 = (\sigma x = z_{2j+1}, z_{2j+2}, ..., z_{2j+2k} = \sigma x)$ of length $2k - 1$ in G_2. Without loss of generality, assume that γ_1 and γ_2 are simple. Define a cycle γ of length $2j + 2k$ by

$$\gamma = (z_1, z_2, ..., z_{2j-1}, z_{2j}, z_{2j+1}, ..., z_{2j+2k}) .$$

Hence γ is a cycle in G which uses the edge $(x, \sigma x)$ but not $(y, \sigma y)$. We must provide a definition of the perturbation along γ because it has repeated vertices $z_1 = z_{2j} = x$ and $z_{2j+1} = z_{2j+2k} = \sigma x$. In this case, we define $\iota_{\gamma, t}$ as above on the non-repeated vertices, but also specify

$$\iota_{\gamma, t}(z_1) = \iota(z_1) + t(s_1 v_1 + s_{2j} v_{2j})$$

and

$$\iota_{\gamma, t}(z_{2j+1}) = \iota(z_{2j+1}) + t(s_{2j+1} v_{2j+1} + s_{2j+2k} v_{2j+2k}) .$$

As above, it is routine to verify that this defines a 1-parameter family of parallel drawings of G. However, one must verify that this perturbation removes the degeneracy along π without introducing any others. Since the vertices of γ_1 and γ_2 are distinct, we have $\tau_1 \neq \tau_{2j}$ and $\tau_{2j+1} \neq \tau_{2j+2k}$. This in turn implies that $s_1 \neq s_{2j}$ and $s_{2j+1} \neq s_{2j+2k}$. Hence, up to a sign,

$$\iota_{\gamma, t}(z_1) = \iota(z_1) \pm t(v_1 - v_{2j})$$

and

$$\iota_{\gamma, t}(z_{2j+1}) = \iota(z_{2j+1}) \pm t(v_{2j+1} - v_{2j+2k}) .$$

Now, recall that the subspaces $\{v_b^\perp, v_g^\perp, v_r^\perp\}$ are distinct. This implies that the vectors $v_1 - v_{2j}$ and $v_{2j+1} - v_{2j+2k}$ are both non-zero. Hence, a non-zero perturbation of ι along γ may remove the degeneracy along π without introducing any additional degeneracies.

This completes the proof of the theorem in the case G is non-bipartite.

4   Subsequent Results

The purpose of this section is to highlight some general observations concerning projective drawings of Tait-colored graphs. These results are not required in the proofs of the main results above, but they are nevertheless interesting in their own right. Given that the main theorem is a generalization of Pappus's theorem, it is thought that these results may play some role in studying geometric configurations.

4.1   Quasi-Faithful Drawings

A faithful projective drawing of a Tait-colored graph is the nicest possible because all supporting lines are distinct. However, even though a certain graph may not admit a faithful drawing, it may admit one that is nearly faithful. Thus, call a drawing ι of a Tait-colored graph $G = (S, T)$ *quasi-faithful* if ι restricts to a bijection on S. For a quasi-faithful drawing, we do not require that the supporting lines $\overline{\iota(z), \iota(\sigma z)}$ be distinct. Call a Tait-colored graph *quasi-faithful* if it admits at least one quasi-faithful drawing but it does not admit a faithful drawing. For example, the graph in Figure 2 is quasi-faithful because the edges $(5, 7)$ and $(2, 10)$ lie on the same supporting line in every drawing.

Whether or not a graph is quasi-faithful depends only on the combinatorics of its forced pairs. For example, if a graph has only one forced pair, then one may always perturb to find a quasi-faithful drawing. In fact, the main results above show that a graph fails to have a quasi-faithful drawing exactly when it has at least two forced pairs π_1 and π_2 with different colors and at least two common vertices.

The graph of a triangular prism, for example, does not even have a quasi-faithful drawing. For, let $G = (S, T)$, where $S = \{1, 2, 3, 4, 5, 6\}$ and $T = \{b, g, r\}$ with $b = (1, 2)(3, 4)(5, 6)$, $g = (1, 6)(2, 3)(4, 5)$, and $r = (1, 3)(2, 5)(4, 6)$. Then G is non-bipartite and both of the graphs $G \backslash \{(1, 2), (5, 6)\}$ and $G \backslash \{(2, 3), (4, 5)\}$ are bipartite. According to the theorem in the case G is non-bipartite, the pairs $\{(1, 2), (5, 6)\}$ and $\{(2, 3), (4, 5)\}$ are forced. Since the points $\{1, 2, 5, 6\}$ are always collinear and the points $\{2, 3, 4, 5\}$ are always collinear, the points $\{2, 5\}$ common to both always coincide in every drawing of G. By symmetry, the points $\{1, 4\}$ are always coincident as are $\{3, 6\}$.

4.2   Forced Triples

For some projective drawings, one sees extraneous points where three supporting lines always concur. For example, in every projective drawing of the graph appearing in Figure 2, the lines $\overline{1, 2}$, $\overline{4, 9}$ and $\overline{6, 7}$ are concurrent. The purpose of this section is to characterize when this happens. Suppose $G = (S, T)$ is a Tait-colored graph and assume $T = \{b, g, r\}$. Call a triple $\tau = \{(x, bx), (y, gy), (z, rz)\}$ of edges with 6 distinct vertices *forced* if the lines $\overline{\iota(x), \iota(bx)}$, $\overline{\iota(y), \iota(gy)}$, and $\overline{\iota(z), \iota(rz)}$ are concurrent in every drawing ι of G.

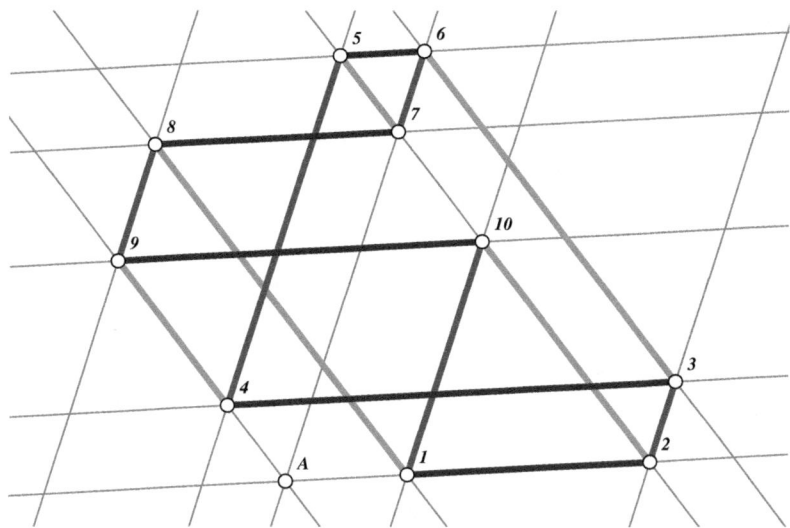

Fig. 2. A quasi-faithful graph with a forced triple

Proposition 6. *Suppose $G = (S, T)$ is a connected Tait-colored graph and τ is a triple of edges with 6 distinct vertices. (a) If G is bipartite, then τ is forced if and only if τ is a cut set. (b) If G is non-bipartite, then τ is forced if and only if $G \backslash \tau$ has a bipartite component.*

Notice in particular that a triple τ in a connected non-bipartite graph G can be forced in one of two ways. Either $G \backslash \tau$ is connected and bipartite or τ is a cut set and exactly one of the two components of $G \backslash \tau$ is bipartite.

The proof of this follows in a manner similar to the main results above. In the combinatorial direction, one uses the style of argument which was used for proposition 4. In the constructibility direction, one shows that if a drawing has a coincident triple of lines which does not arise in one of these ways, then one may perturb the drawing along a carefully chosen cycle to remove the degeneracy.

The presence of a forced triple provides a way to decompose a drawing of a graph into "subdrawings". Suppose for instance that $G = (S, T)$ is connected and bipartite and $\tau = \{(x, bx), (y, gy), (z, rz)\}$ is a forced triple. Then $G \backslash \tau$ has two components. Let S_1 and S_2 be the sets of vertices for each of these components and assume $x, y, z \in S_1$ and $bx, gy, rz \in S_2$. Then one obtains two smaller graphs G_1 and G_2 by introducing vertices, say w_1 and w_2, and edges $\{(x, w_1), (y, w_1), (z, w_1)\}$ in G_1 and $\{(bx, w_2), (gy, w_2), (rz, w_2)\}$ in G_2. The graphs G_1 and G_2 inherit Tait-colorings from G, but they are not subgraphs of G. Nevertheless, due to the proposition above, every projective drawing of G automatically contains projective drawings of G_1 and G_2. There is a similar decomposition when G is non-bipartite.

4.3   Realization Spaces

The purpose here is to discuss the space of all drawings of a given graph. The main point is that such spaces are always topologically trivial.

Given a Tait-colored graph $G = (S, T)$, let $\mathcal{D}(G)$ denote the set of all parallel drawings of G with a fixed triple of projections. Reading the definition, one sees that $\mathcal{D}(G)$ is a vector space. Using perturbations, one may show:

Proposition 7. *Suppose n is a positive integer, $|S| = 2n$, and $G = (S, T)$ is a Tait-colored graph. (a) If G is bipartite, then $\mathcal{D}(G)$ has dimension $n + 1$. (b) If G is non-bipartite, then $\mathcal{D}(G)$ has dimension n.*

Given G, let $\mathcal{F}(G)$ denote the space of all faithful parallel drawings of G. Again, due to the linearity in the definition, one sees that $\mathcal{F}(G)$ is the complement of an arrangement of hyperplanes in $\mathcal{D}(G)$. This yields:

Proposition 8. *Suppose n is a positive integer, $|S| = 2n$, and $G = (S, T)$ is a Tait-colored graph which admits a faithful parallel drawing. (a) If G is bipartite, then $\mathcal{F}(G)$ is a disjoint union of open cells of dimension $n + 1$. (b) If G is non-bipartite, then $\mathcal{F}(G)$ is a disjoint union of open cells of dimension n.*

Currently the number of connected components of $\mathcal{F}(G)$ as a function of G is unknown.

5   Conclusion

5.1   Higher Degree

One may extend the notions above and ask about parallel drawings of d-edge-colored regular graphs. However, even the case of $d = 4$ appears very difficult to handle:

Conjecture 9. The decidability problem of whether or not a 4-edge-colored quartic graph has a faithful parallel drawing is polynomially equivalent to the existential theory of the reals.

See [3] for a definition. The intuition behind this conjecture is based on several facts. Consider that the space of faithful drawings of a Tait-colored graph is always a complement of a hyperplane arrangement, but this property does not appear to be predictable for quartic graphs. For example, there is no simple relationship between the dimension of the realization space and the number of vertices for quartic graphs. By a similar token, the realization space of a Tait-colored graph always contains realizations with integer coordinates, but, by contrast, the realization space of a quartic graph often uses an extension of the rationals. Another source of intuition comes from the theory of graph-encoded manifolds and crystallizations [7]. In this theory, one represents a pseudomanifold of dimension d with a $(d + 1)$-edge-colored graph of degree $d + 1$, and then uses these representations to study the pseudomanifold. Thus, each 4-edge-colored quartic graph represents a certain 3-dimensional pseudomanifold. Even in dimension 3, several decision problems are known to be **NP**-complete [1,8]. Thus, it is thought that decision problems in 3-dimensional manifolds must translate via polynomial-time algorithms to decision problems for 4-edge-colored quartic graphs. Finally, it was shown in [11] that any semialgebraic variety can be approximated by the realization space of a 4-dimensional convex polytope. This is in contrast to the case in dimension 3, where the realization space of every convex polytope is an open cell. Thus, things appear to "go bad" when one increases the dimension. Due to the connection to crystallizations of pseudomanifolds, increasing the degree from 3 to 4 is akin to increasing the dimension.

5.2   Ghost Symmetry in the Plane

This section explains the connection from the main result from [10] to parallel drawings. Suppose $G = (S, T)$ is a Tait-colored graph. Define a *GS realization* of G as a pair (ι, ϕ) where $\iota : S \to \mathbb{R}^2$ is a function and $\phi = \{\phi_b, \phi_g, \phi_r\}$ is a triple of projections to distinct 1-dimensional subspaces of $\mathbb{R}^2$ such that such that $\phi_\sigma(\iota(z)) = -\phi_\sigma(\iota(\sigma z))$ for all $z \in S$ and all $\sigma \in T$. The 1-dimensional subspaces in this definition are the *lines of ghost symmetry* because, while the 2-dimensional point configuration $\iota(S)$ may have a trivial symmetry group, there are three distinct 1-dimensional shadows $\phi_\sigma(\iota(S))$ which each have bilateral symmetry. Call a GS realization *faithful* if each of the projections ϕ_σ restricts to a bijection on $\iota(S)$. Here is the main theorem from [10]:

Theorem 10. *A Tait-colored graph admits a faithful GS realization if and only if it does not have a 2-edge cut set.*

Obviously the statement of this theorem is similar to the main results above on parallel drawings. The main difference is that the theorem does not resort to the notion of bipartiteness. One may prove this analogously. In the combinatorial direction, one shows that a 2-edge cut set makes faithfulness impossible, and in the constructibility direction one shows that one may perturb along cycles (of arbitrary length) to obtain a faithful drawing. Likewise, there are analogous statements for realization spaces and forced triples. Having seen this theorem, it should be clear why we regard a Tait-colored graph as a triple of fixed-point-free involutions: After projecting one of these configurations $\iota(S)$ down to one of the three prescribed subspaces, one obtains a linear configuration whose two-fold symmetry yields the corresponding permutation of those points.

Here is the connection to parallel drawings. Suppose $G = (S, T)$ is a connected, bipartite, Tait-colored graph which does not have a 2-edge cut set. Then G has a faithful GS realization, say ι. Let $\{S_1, S_2\}$ be the bipartition of the vertices. Define a map $\iota' : S \to \mathbb{R}^2$ by

$$\iota'(z) = \begin{cases} \iota(z) & \text{if } z \in S_1, \\ -\iota(z) & \text{if } z \in S_2. \end{cases}$$

Then ι' is a faithful parallel drawing of G.

References

1. Agol, I., Hass, J., Thurston, W.: The computational complexity of knot genus and spanning area. Trans. Amer. Math. Soc. 358(9), 3821–3850 (2006), (electronic), http://dx.doi.org/10.1090/S0002-9947-05-03919-X
2. Biggs, N.: Algebraic graph theory. cambridge Tracts in Mathematics, vol. 67, Cambridge University Press, London (1974)
3. Björner, A., Vergnas, M.L., Sturmfels, B., White, N., Ziegler, G.M.: Oriented Matroids, 2nd edn. Cambridge University Press, Cambridge (2000)
4. Bonnington, C.P., Little, C.H.C.: The Foundations of Topological Graph Theory. Springer, New York (1995)
5. Eppstein, D.: The topology of bendless three-dimensional orthogonal graph drawing. In: Tollis, I.G., Patrignani, M. (eds.) GD 2008. LNCS, vol. 5417, pp. 78–89. Springer, Heidelberg (2009)
6. Eppstein, D., Mumford, E.: Steinitz theorems for orthogonal polyhedra (December 2009) arXiv:0912.0537
7. Ferri, M., Gagliardi, C., Grasselli, L.: A graph-theoretical representation of pl-manifolds—a survey on crystallizations. Aequationes Math. 31(2-3), 121–141 (1986)
8. Ivanov, S.V.: The computational complexity of basic decision problems in 3-dimensional topology. Geom. Dedicata 131, 1–26 (2008), http://dx.doi.org/10.1007/s10711-007-9210-4
9. Keszegh, B., Pach, J., Pálvölgyi, D., Tóth, G.: Drawing cubic graphs with at most five slopes. Comput. Geom. 40(2), 138–147 (2008), http://dx.doi.org/10.1016/j.comgeo.2007.05.003
10. Richter, D.A.: Ghost symmetry and an analogue of Steinitz's theorem. Contrib. Alg. Geom. (to appear, 2010)
11. Richter-Gebert, J.: Realization Spaces of Polytopes. Lecture Notes in Mathematics, vol. 1643. Springer, Heidelberg (1996)
12. Richter-Gebert, J., Kortenkamp, U.H.: The interactive geometry software Cinderella. Springer-Verlag, Heidelberg (1999); with 1 CD-ROM (Windows, MacOS, UNIX and JAVA-1.1 platform)
13. Tutte, W.T.: Graph Theory, Encyclopedia of Mathematics and its Applications, vol. 21. Addison-Wesley Publishing Company, Menlo Park (1984)

Universal Pointsets for 2-Coloured Trees*

Mereke van Garderen[1], Giuseppe Liotta[2], and Henk Meijer[1]

[1] Roosevelt Academy, The Netherlands
{m.vangarderen,h.meijer}@roac.nl
[2] Università di Perugia, Italy
liotta@diei.unipg.it

Abstract. Let R and B be two sets of distinct points such that the points of R are coloured red and the points of B are coloured blue. Let $\mathcal{G}$ be a family of planar graphs such that for each graph in the family $|R|$ vertices are red and $|B|$ vertices are blue. The set $R \cup B$ is a universal pointset for $\mathcal{G}$ if every graph $G \in \mathcal{G}$ has a straight-line planar drawing such that the blue vertices of G are mapped to the points of B and the red vertices of G are mapped to the points of R. In this paper we describe universal pointsets for meaningful classes of 2-coloured trees and show applications of these results to the coloured simultaneous geometric embeddability problem.

1 Introduction

Let G be a planar graph with n vertices whose vertex set is partitioned into subsets $V_0, \ldots, V_{k-1}$ for some positive integer $1 \leq k \leq n$ and let S be a set of n distinct points in the plane partitioned into subsets $S_0, \ldots, S_{k-1}$ with $|S_i| = |V_i|$ ($0 \leq i \leq k-1$). We say that each index i is a *colour*, G is a *k-coloured planar graph*, and S is a *k-coloured set of points compatible with G*. A *k-coloured point-set embedding of G on S* is a polyline drawing of G such that each vertex of V_i is mapped to a distinct point of S_i and no two edges cross.

Let $\mathcal{G}$ be a family of k-coloured graphs such that for each colour all graphs in $\mathcal{G}$ have the same number of vertices of that colour, and let S be a k-coloured set of points compatible with each $G \in \mathcal{G}$. Set S is an *h-bend universal pointset* for the family $\mathcal{G}$ if each $G \in \mathcal{G}$ has a k-coloured point-set embedding on S such that every edge of G is drawn as a polyline having at most h bends. *h-Bend universal pointsets* are the subject of extensive research in the graph drawing and combinatorial geometry literatures.

Pach and Wenger [17] consider the family of n-coloured planar graphs and prove that any pointset in general position is $O(n)$-bend universal for this family. In the same paper, the authors show that for some n-coloured planar graphs and for some n-coloured compatible sets of points $\Omega(n)$ bends per edge may be necessary. Kaufmann and Wiese [16] prove that every set of n distinct points in the plane is 2-bend universal for all (1-coloured) planar graphs. Everett et al. [6] show how to construct a set of n distinct points that is 1-bend universal for all planar graphs. On the negative side, De Fraysseix, Pach, and Pollack [4] show that a 0-bend universal pointset does not

---

* Research supported in part by MIUR under project AlgoDEEP prot. 2008TFBWL4.

exist for the family of planar graphs. Gritzman, Mohar, Pach, and Pollack [10] prove that every set of n distinct points in the plane is 0-bend universal for the outerplanar graphs with n vertices and that this is the largest subfamily of the planar graphs having this property.

Several results are also known for h-bend universal pointsets of k-coloured planar graphs with $1 < k < n$ and $h \geq 1$. Di Giacomo et al. [9] show that every set of n distinct points is 1-bend universal for the properly 2-coloured caterpillars, the properly 2-coloured wreaths, the 2-coloured paths, and the 2-coloured cycles. Di Giacomo et al. [7] also prove that every set of n distinct points is 5-bend universal for the 2-coloured outerplanar graphs, while Badent et al. [1] prove that not all sets of n distinct points can be h-bend universal for 2-coloured non-outerplanar graphs for any fixed constant h.

This paper is devoted to the study of 0-bend universal pointsets for 2-coloured trees. Since we only consider pointsets that support straight-line drawings, we shall simply say *universal pointset* to mean a 0-bend universal pointset. Also, for consistency with existing literature, we name the two colours of our trees as red and blue. Although 2-colored trees may appear as a somewhat restricted subfamily of the k-colored planar graphs, there is a rich literature concerning universal pointsets of these trees. While the interested reader is referred to the survey by Kaneko and Kano [14], we briefly recall here some of the most recent findings that are more closely related to our results.

Abellanas et al. [2] show that any 2-coloured pointset S such that either the convex hull of S consists of all red points and no blue points or S is a linearly separable bipartition (i.e., there exists a line that separates all blue points from the red ones) is a universal set for the properly 2-coloured paths. Brandes et al. [3] extend this result and show that a linearly separable bipartition is universal for the (non necessarily properly) 2-coloured paths. Results by Ikebe et al. [11] and by Kaneko and Kano [12] show that every set of n distinct points in general position is universal for the 2-colored forests of size n and consisting of at most two trees each having exactly one red vertex. Follow-up papers by Kaneko and Kano [13,15] extend this last result to forests with more than two trees under the assumption that the forest consists of either star-trees or trees whose sizes differ from one another by at most one vertex. Di Giacomo et al. [8] also study forests such that every tree has exactly one red vertex and prove that any set of points where the red points are in convex position is universal for these graphs. Finally, Estrella-Balderrama, Fowler, and Kobourov [5] describe universal pointsets for different families of 2- and 3-coloured trees, namely the 2-coloured spiders, the 3-coloured caterpillars, and the 3-coloured radius-two stars.

This paper describes three new, fairly general families of 2-coloured trees for which universal pointsets exist. Our results are as follows.

- There exists a universal pointset for the properly 2-coloured trees whose leaves all have the same colour (Theorem 1, Section 2).
- There exists a universal pointset for the properly 2-coloured full binary trees (Theorem 2, Section 2).
- There exists a universal pointset for the rooted 2-coloured trees where every vertex has at most one child having a colour different from its own (Theorem 3, Section 3).

Finally, an application of the our results to the *k-coloured simultaneous geometric embeddability problem* [3] is given in Section 4 (Corollaries 1, 2, and 3).

2   Properly 2-Coloured Trees

In this section we consider universal pointsets of properly 2-coloured trees, i.e. 2-coloured trees in which no two vertices of the same colour are adjacent. Unless stated otherwise, the trees considered in this section are not rooted. Throughout the paper, the blue vertices will be depicted as black dots and the red vertices as white dots. We start by considering properly 2-coloured trees where all leaves have a same colour.

Theorem 1. *There exists a universal pointset for the properly 2-coloured trees in which all leaves have the same colour.*

Proof. Let $\mathcal{T}$ be the set of properly 2-coloured trees such that each tree in the set has n_r red vertices and n_b blue vertices and such that all leaves have the same colour. Without loss of generality assume that the leaves are red, which implies $n_r \geq n_b$. If $n_r = n_b$ the tree is a path and the statement holds by the result of Abellanas et al. [2]. Hence we prove the theorem for the case that $n_r > n_b$.

We draw a "downwards pointing" circular arc ρ and a horizontal line σ below it so that each point on ρ is visible from each point on σ as illustrated in Figure 1. Let S be a point set defined as follows. Place an alternating sequence of red and blue points on ρ with n_b red and n_b blue points. We place $n_r - n_b$ red points on σ. We now prove that S is universal for $\mathcal{T}$ by presenting an algorithm that, for any tree $T \in \mathcal{T}$, computes a 2-coloured point set embedding of T on S with straight-line edges. For an illustration see Figure 1, where some edges have been purposely drawn as curves to make the picture more readable.

We start with decomposing T in a set of paths. A total of $n_r - n_b$ red leaves will not be included in this decomposition. Since T is properly 2-coloured, any path in T alternates between red and blue vertices. Choose an arbitrary path in T from a leaf v to a vertex w adjacent to a leaf. Add the path from v to w to the decomposition. The path from v to w is called the *first path* of T. Mark the vertices in the first path and repeat the decomposition into paths as follows. For any marked vertex v that has an unmarked neighbour w that is not a leaf, choose a path starting at w and ending at a leaf u. If w is blue, add the path to the decomposition and mark all vertices in this path. If w is red, remove u from the path, add the shortened path to the decomposition and mark all vertices in this path. Notice that all paths start and end at vertices of different colours. We will draw all such paths on arc ρ, and the remaining $n_r - n_b$ red leaves on σ.

Let $v_0, v_1, \ldots, v_{p-1}$ be the first path. Place v_0 on the left or right-most point of ρ, depending on which point has the correct colour. Place the remaining vertices of the first path on neighbouring points of S on ρ. We will now consider T as a rooted tree, rooted at an arbitrary vertex on this path. Let $w_0, w_1, \ldots, w_{k-1}$ be the vertices in T that do not lie on the first path, that are neighbours of vertices in the first path, and that are not leaves. Each w_i is an endpoint of one of the paths in the decomposition. Assume that we have numbered these vertices from one end of the path to the other, i.e. for all $i < j$, if w_i and w_j are neighbours of vertices v_a and v_b in the first path then $a \leq b$. Let n_i be the number of vertices in the subtree rooted at w_i for $0 \leq i < k$. Processing the unused points on ρ starting at the unused point besides v_{p-1} reserve the first n_{k-1} points for the paths in the subtree of w_{k-1}, reserve the next n_{k-2} points for the paths

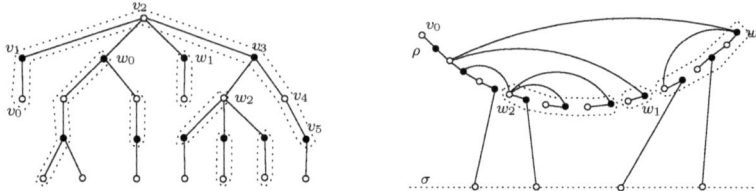

Fig. 1. Illustration for the proof of Theorem 1. The red points are white dots, the blue points are black dots. The first path has length 6 and $k = 3$, $n_0 = 6$, $n_1 = 2$ and $n_2 = 6$. Some edges have been drawn as curves to make the figure more readable.

in the subtree of w_{k-2}, and so on. Then, for all i place w_i on the left or right most point of its reserved group of points, depending on which point has the correct colour. Map the remaining vertices on the path containing w_i to the neighbouring points of S on ρ. We then continue recursively. Finally, we can connect the points of S on σ to their neighbours on ρ, since each point on σ can see all points on ρ. Since the resulting drawing is a 0-bend point-set embedding of T on S, it follows that S is a universal pointset for $\mathcal{T}$.   □

Motivated by Theorem 1, it is natural to ask whether there exists a universal pointset also for properly 2-coloured trees whose leaves can be either red or blue. The next theorem answers this question for the family of properly 2-coloured *full binary* trees, i.e. properly 2-coloured trees whose vertices have degree either three or one. The proof is omitted here because of space limitations.

Theorem 2. *There exists a universal pointset for the properly 2-coloured full binary trees.*

3   Almost Mono-Chromatic Rooted Trees

In this section we study universal pointsets of 2-coloured trees where vertices of the same colour may be adjacent. The *almost monochromatic trees* are those 2-colored rooted trees where each parent has at most one child of a different colour than its own.

Theorem 3. *There exists a universal pointset for the almost monochromatic trees.*

Sketch of Proof: Let $\mathcal{T}$ be the set of almost monochromatic trees having n_r red vertices and n_b blue vertices. Let S be a set of points defined as follows. Draw two circular arcs c_r and c_b so that they form an "hourglas", as illustrated in Figure 2. Place n_r red points on c_r and n_b red points on c_b. Assume without loss of generality that the root v_0 is red. We prove that S is universal for $\mathcal{T}$ by showing how to compute a 0-bend point set embedding on S of any tree $T \in \mathcal{T}$. Call *unfinished* a vertex of T whose children are not all drawn. We compute a 0-bend point set embedding of T on S by maintaining the following invariant: For each unfinished vertex $v \in T$ there exists a convex region that intersects c_r and/or c_b and such that v is either mapped to the lowest red point or to the lowest blue point on c_r or c_b inside this region. The region associated with v

contains the correct number of vertices for the subtree of v. Map v_0 to the bottom red point on c_r. Notice that the invariant holds after the first step. If v_0 has a blue child v_b, assign a convex region to v_b containing the unused bottom-most points from c_r and c_b. The number of red and blue points in this region is equal to the number of red and blue vertices in the subtree rooted at v_b. Place v_b in its region at the bottom unused point of c_b. If v_0 has red children, then for each red child v_r, assign a convex region to v_r containing the next group of bottom-most points from c_r and from c_b. Map the children of v_0 to the bottom most points on c_r in their regions. Since the invariant holds for each child of v_0, each subtree can be recursively drawn inside the corresponding convex regions. Hence S is a universal pointset for $\mathcal{T}$.   □

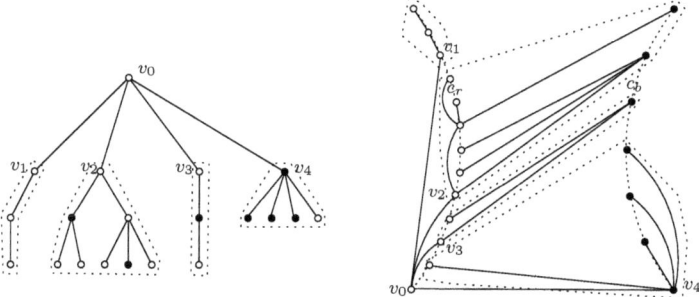

Fig. 2. Illustration for the proof of Theorem 3. The red points are white dots, the blue points are black dots. Some edges have been drawn as curves to make the figure more readable.

4   Coloured Simultaneous Geometric Embeddings

The *k-coloured simultaneous geometric embeddability problem* is defined as follows. The input is a set of k-coloured planar graphs $G_1 = (V, E_1)$, $G_2 = (V, E_2)$, ..., $G_r = (V, E_r)$ on the same vertex set V. The goal is to find planar straight-line drawings D_i of G_i using the same $|V|$ points in the plane for all $i = 1, \ldots, r$, such that vertices of colour i are mapped to points of colour i. The set of drawings D_i $(i = 1, \ldots, r)$ is a *k-coloured simultaneous geometric embedding* of the graphs $G_1 = (V, E_1)$, $G_2 = (V, E_2)$, ..., $G_r = (V, E_r)$. The k-coloured simultaneous geometric embeddability problem was first defined by Brandes et al. [3] and subsequently investigated by Estrella-Balderrama, Fowler, and Kobourov [5]. By using the universal pointsets of Theorems 2, 1, and 3 the following results are immediate.

Corollary 1. *Any number of properly 2-coloured full binary trees admits a 2-coloured simultaneous geometric embedding.*

Corollary 2. *Any number of properly 2-coloured trees whose leaves all have the same colour admits a 2-coloured simultaneous geometric embedding.*

Corollary 3. *Any number of rooted almost-monochromatic 2-coloured trees admits a 2-coloured simultaneous geometric embedding.*

5   Open Problems

It is unknown whether there are universal pointsets for properly 2-coloured (binary) trees, or more generally whether there are universal pointsets for 2-coloured trees.

References

1. Badent, M., Giacomo, E.D., Liotta, G.: Drawing colored graphs on colored points. Theor. Comput. Sci. 408(2-3), 129–142 (2008)
2. Abellanas, M., Garcia-Lopez, J., Hernández-Peñver, G., Noy, M., Ramos, P.A.: Bipartite embeddings of trees in the plane. Discrete Applied Mathematics 93(2-3), 141–148 (1999)
3. Brandes, U., Erten, C., Fowler, J.J., Frati, F., Geyer, M., Gutwenger, C., Hong, S.-H., Kaufmann, M., Kobourov, S.G., Liotta, G., Mutzel, P., Symvonis, A.: Colored simultaneous geometric embeddings. In: Lin, G. (ed.) COCOON 2007. LNCS, vol. 4598, pp. 254–263. Springer, Heidelberg (2007)
4. de Fraysseix, H., Pach, J., Pollack, R.: How to draw a planar graph on a grid. Combinatorica 10(1), 41–51 (1990)
5. Estrella-Balderrama, A., Fowler, J.J., Kobourov, S.G.: Colored simultaneous geometric embeddings and universal pointsets. In: Proceedings of the 21st Canadian Conference on Computational Geometry (CCCG 2009), pp. 17–20 (2009)
6. Everett, H., Lazard, S., Liotta, G., Wismath, S.K.: Universal sets of points for one-bend drawings of planar graphs with vertices. Discrete & Computational Geometry 43(2), 272–288 (2010)
7. Giacomo, E.D., Didimo, W., Liotta, G., Meijer, H., Trotta, F., Wismath, S.K.: k-colored pointset embeddability of outerplanar graphs. J. Graph Algorithms Appl. 12(1), 29–49 (2008)
8. Giacomo, E.D., Didimo, W., Liotta, G., Meijer, H., Wismath, S.K.: Point-set embeddings of trees with given partial drawings. Comput. Geom. 42(6-7), 664–676 (2009)
9. Giacomo, E.D., Liotta, G., Trotta, F.: On embedding a graph on two sets of points. Int. J. Found. Comput. Sci. 17(5), 1071–1094 (2006)
10. Gritzmann, P., Mohar, B., Pach, J., Pollack, R.: Embedding a planar triangulation with vertices at specified points. Amer. Math. Monthly 98(2), 165–166 (1991)
11. Ikebe, Y., Perles, M.A., Tamura, A., Tokunaga, S.: The rooted tree embedding problem into points in the plane. Discrete & Computational Geometry 11, 51–63 (1994)
12. Kaneko, A., Kano, M.: Straight-line embeddings of two rooted trees in the plane. Discrete & Computational Geometry 21(4), 603–613 (1999)
13. Kaneko, A., Kano, M.: Straight line embeddings of rooted star forests in the plane. Discrete Applied Mathematics 101(1-3), 167–175 (2000)
14. Kanenko, A., Kano, M.: Discrete geometry on red and blue points in the plane - a survey. In: Discrete and Computational Geometry. Algorithms and Combinatorics, vol. 25, pp. 551–570. Springer, Heidelberg (2003)
15. Kaneko, A., Kano, M.: Semi-balanced partitions of two sets of points and embeddings of rooted forests. Int. J. Comput. Geometry Appl. 15(3), 229–238 (2005)
16. Kaufmann, M., Wiese, R.: Embedding vertices at points: Few bends suffice for planar graphs. Journal of Graph Algorithms and Applications 6(1), 115–129 (2002)
17. Pach, J., Wenger, R.: Embedding planar graphs at fixed vertex locations. Graph and Combinatorics 17, 717–728 (2001)

The Quality Ratio of RAC Drawings and Planar Drawings of Planar Graphs

Marc van Kreveld

Department of Information and Computing Sciences,
Utrecht University, The Netherlands
`marc@cs.uu.nl`

Abstract. We study how much better a right-angled crossing (RAC) drawing of a planar graph can be than any planar drawing of the same planar graph. We analyze the area requirement, the edge-length ratio, and the angular resolution. For the first two measures, a RAC drawing can be arbitrarily much better, whereas for the third measure a RAC drawing can be 2.75 times as good.

Keywords: Right-angled crossing drawing, planar graphs, quality.

1 Introduction

Right-angled crossing drawings of graphs were introduced recently, motivated by the fact that good drawings may have crossings, as long as the crossing edges have a large crossing angle. In a *right-angled crossing drawing* (RAC drawing), every two edges that cross must do so at a right angle. It was shown that RAC drawings of graphs with n vertices can have up to $4n - 10$ edges, and this bound is tight in the worst case [4].

Although RAC drawings can be drawings of non-planar graphs, one could also use a RAC drawing of planar graph in order to get a better angular resolution (for instance). For example, the K_4 has a RAC drawing whose smallest angle is $\pi/4$ (the square with two diagonals), while any planar drawing has an angle of at most $\pi/6$ (the optimum is realized by the equilateral triangle with the center point). Hence, the angular resolution of a RAC drawing can be 1.5 times as good as the angular resolution of any planar drawing of the same planar graph.

Let Φ be a quality measure of a drawing like area requirement, edge-length ratio, or angular resolution. In this paper we study how much better a RAC drawing of a planar graph can be than any planar drawing of the same graph. In particular, we study the *quality ratio*

$$QR(\Phi) = \sup_{G \text{ planar}} \frac{\Phi_{\mathrm{RAC}}(G)}{\Phi_{\mathrm{planar}}(G)} ,$$

where $\Phi_{\mathrm{RAC}}(G)$ is the quality of the RAC drawing that is optimal for graph G and measure Φ, and $\Phi_{\mathrm{planar}}(G)$ is the quality of the planar drawing that is optimal for graph G and measure Φ. While research on RAC drawings has

U. Brandes and S. Cornelsen (Eds.): GD 2010, LNCS 6502, pp. 371–376, 2011.

considered both the case of straight-line drawings and drawings with bends in the edges [1,4], we consider only straight-line drawings.

Firstly, we consider the area requirement (Section 2). We show that for any n, there is a planar graph with n vertices that admits a RAC drawing of area $O(n)$, but any planar drawing requires area $\Omega(n^2)$. This implies that a RAC drawing can be arbitrarily much better than a planar drawing of the same planar graph. If the area requirement quality $\Phi^A(G)$ is defined as the reciprocal of the minimum area needed in a drawing of G, then $QR(\Phi^A) = \infty$.

Secondly, we consider the edge-length ratio (Section 3). We give a class of graphs that has constant edge-length ratio for RAC drawings but an unbounded edge-length ratio for all planar drawings when $n \to \infty$. We define the edge-length ratio quality $\Phi^E(G)$ as the longest possible length of the shortest edge in any drawing of G, assuming the longest edge in that drawing has length 1. Then we will show that $QR(\Phi^E) = \infty$.

Thirdly, we consider the angular resolution (Section 4). Let $\Phi^\alpha(G)$ be the largest possible smallest angle in a drawing of G. It follows from the results of Malitz and Papakostas [6] that for constant-degree planar graphs, the ratio of angular resolution is constant, because a planar drawing with all angles at least $\Omega(1/7^d)$ exists (where d is the maximum degree; see also [5]), while no drawing—planar or not—can do better than $2\pi/d$. We give a planar graph of degree 8 that shows that $QR(\Phi^\alpha) \geq 2.75$. Whether $QR(\Phi^\alpha)$ is bounded from above by a constant for planar graphs of non-constant degree is open.

2   Ratio of Area Requirement

Consider the class of graphs shown in Figure 1 (left). The n vertices can be placed on a $2 \times n/2$ grid for a RAC drawing, and hence the area requirement is linear in n. For any planar drawing, we show that it contains a linear number of nested triangles, and hence the area requirement of any planar drawing is quadratic [3].

Consider the middle K_4 of the sequence, $abcd$. For a drawing to be planar, one of the vertices must be inside the triangle formed by the other three. Since the situation is symmetric, we can assume that c is inside $\triangle abd$ in a planar drawing. Since the vertices e, f are adjacent to both c and d, they must be inside $\triangle abd$, and so must the whole further part of the graph beyond e and f

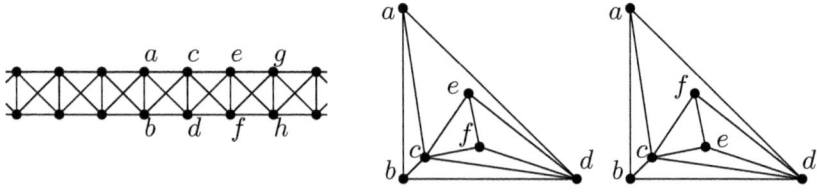

Fig. 1. Area requirement of a RAC drawing and a planar drawing

including g and h. Since the subgraph c, d, e, f must be planar, we either have f inside $\triangle cde$ or e inside $\triangle cdf$, see Figure 1, right. We repeat the argument once more to conclude that h is inside $\triangle efg$ or g is inside $\triangle efh$. In the first case, we have $\triangle abd$ enclosing $\triangle efg$, and the remainder of the graph (right part) is inside $\triangle efg$. The second case is analogous. We conclude that in all cases, any planar drawing has $\Omega(n)$ nested triangles, and hence it has area requirement $\Omega(n^2)$. We conclude that the ratio of area-requirement quality is unbounded.

3   Ratio of Edge-Length Ratio

In this section we show that for a sufficiently large constant R, a graph exists such that a RAC drawing has edge-length ratio at most 3 and a planar drawing has edge-length ratio at least R. The graph consists of nested quadrilaterals, shown as squares in the left part of Figure 2. Nested squares come in pairs; note that the edges between the outer two squares are the same as the edges between the inner two squares. We can extend this construction to $\lfloor n/10 \rfloor$ pairs of squares in a graph with n vertices. The graph has a unique embedding except for the choice of the outer face, and by a standard duplication of the construction we can ensure that any embedding has $\Omega(n)$ pairs of nested quadrilaterals. We will show that the inner quadrilateral of a pair is sufficiently smaller in some sense (area and/or diameter) than the outer quadrilateral of that pair. Since we have $\Omega(n)$ pairs of nested quadrilaterals and n is not bounded by any constant, the innermost one can be made arbitrarily much smaller than the outermost one.

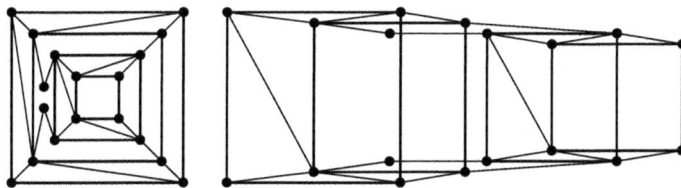

Fig. 2. Planar and RAC drawing of a graph that illustrates that the ratio of edge-length quality is unbounded

Figure 2 (right) shows that a RAC drawing exists where the edge-length ratio is less than 3. The quadrilaterals can be drawn as squares that are not nested, and the drawing can be extended to any number of squares. If the outermost square has edge length 1, then for any $\epsilon > 0$, we can make the innermost square have edge length $1 - \epsilon$, and the longest and shortest edges have lengths approximately $\frac{1}{2}\sqrt{5}$ and $\frac{1}{2}$.

Summarizing, we can prove unbounded ratio of edge-length quality for RAC drawings and planar drawings of the same graph if we can show that in any planar drawing of the graph shown in Figure 2, an edge-length ratio bounded by R implies that the inner quadrilateral of a pair of nested quadrilaterals is smaller by some significant amount than the outer quadrilateral.

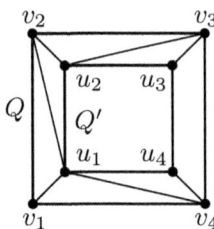 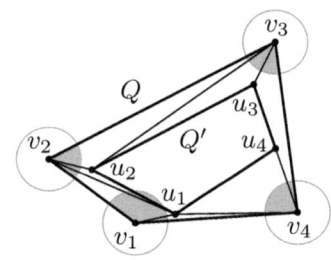

Fig. 3. The graph G and its edges (left). The 1-neighborhood regions of the four vertices of Q.

Lemma 1. *Let G be a planar graph as in Figure 3, using the shown embedding. Then for any embedding-preserving straight-line drawing of G and some large enough constant R, if the edge length ratio of G is at most R, then the diameter ratio of the quadrilaterals $Q = v_1v_2v_3v_4$ and $Q' = u_1u_2u_3u_4$ is greater than $1 + \Omega(1/R^2)$ or the area ratio of Q and Q' is greater than $1 + \Omega(1/R^4)$.*

Proof. Let R be a sufficiently large constant. Consider any embedding-preserving straight-line drawing of G with shortest edge length 1 and longest edge length R. Then the area of Q is in $(0, R^2]$ and the diameter of Q is in $(1, 2R)$. The area and diameter of Q' is always smaller than that of Q, but we will show that at least one of these measures is *significantly* smaller.

We define the 1-neighborhoods of $v_1, \ldots, v_4$ as the intersection of the unit disks centered at $v_1, \ldots, v_4$ with the interior of Q. By the edges of G between Q and Q', u_i cannot be in the 1-neighborhood of v_i for $i = 1, \ldots, 4$, and u_1 cannot be in the 1-neighborhood of v_2 or v_4, and u_2 cannot be in the 1-neighborhood of v_3.

If none of $u_1, \ldots, u_4$ is in the 1-neighborhood of any of $v_1, \ldots, v_4$, then the diameter decreases at least by some constant amount, depending on R. Since R is a constant, the diameter of Q' is smaller by a constant fraction < 1 (some calculation shows that it is at most $1 - \frac{c}{R^2}$ times the diameter of Q, for a constant $c > 0$). So it remains to analyze the cases where at least some $u_1, \ldots, u_4$ are in the 1-neighborhood of some of $v_1, \ldots, v_4$, but with the restrictions on which u_i cannot be close to which v_j.

Consider the six triangles of G in between Q and Q' (we ignore the quadrilateral that is also in between). If any of these triangles has at least a constant area (which may depend on R), then the area of Q' is a constant fraction < 1 less than the area of Q. In particular, this implies that u_1 must be very close to the line through v_1 and v_2 (due to the area of $\triangle u_1v_1v_2$ and the shortest edge length of 1) *and* very close to the line through v_1 and v_4 (due to the area of $\triangle u_1v_1v_4$ and the shortest edge length of 1). In particular, a distance of at least $1/R^2$ implies that the area of a triangle is at least $1/(2R^2)$, which makes the lemma true because the area of Q is at most R^2. So we continue with the case where the distance from u_1 to these lines is less than $1/R^2$. Since u_1 is close to both lines but at least distance 1 from v_1, we see that the angle at v_1 in Q is either very close to 0, or very close to 180, or very close to 360 degrees (within c/R^2 for some constant $c > 0$).

By analyzing these three cases for the angle at v_1 and many subcases, we can always conclude that either the area of Q' or the diameter is significantly smaller than that of Q (or else we get a contradiction with the planarity of G). The rather tedious full proof is in the full paper. □

By applying the lemma above to a graph with k pairs of nested quadrilaterals, the outermost quadrilateral has a diameter that is at least $(1 + \Omega(1/R^2))^{k/2}$ times as large as the innermost one, or the outermost quadrilateral has an area at least $(1+\Omega(1/R^4))^{k/2}$ times as large as the innermost one. By choosing k (and therefore n) large enough, we can achieve a diameter ratio or area ratio larger than any constant, because R is assumed to be constant. Since an upper bound on the diameter or the area (independent of each other) bound the maximum edge length that is possible, we derive a contradiction that follows from the assumption that R is constant. We conclude that the edge-length ratio for any planar drawing of the graph is unbounded, and the quality ratio of edge length is unbounded as well.

4   Ratio of Angular Resolution

That the angular resolution of a RAC drawing can be better than of a planar drawing can easily be seen from the K_4. A RAC drawing can be a square with the two diagonals, giving a smallest angle of 45°, while a planar drawing with optimal angular resolution is an equilateral triangle that has its fourth vertex in the center, giving a smallest angle of 30°. The quality ratio is 1.5. By extending the example we obtain another planar graph that has a RAC drawing with smallest angle 45°, see Figure 4, left. Any planar drawing must have one of a, b, c, d inside the triangle formed by the other three. Since the graph is symmetric, we can assume that d is inside. But then the biconnected component shown below b, d and the biconnected component right of c, d must be inside $\triangle abc$ as well. This implies that $\triangle abc$ has its angle at a partitioned into two angles, its angle

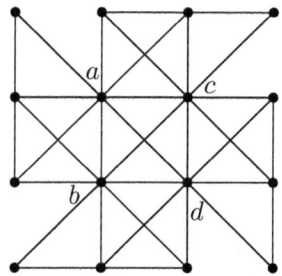
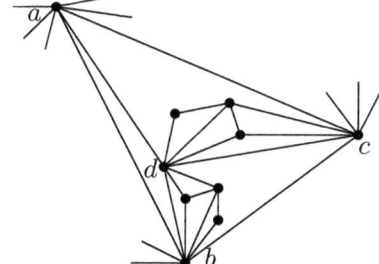

Fig. 4. A planar graph showing that the ratio of angular resolution is at least $11/4$

at b partitioned in five angles, and its angle at c partitioned into four angles by edges from these vertices. In total, the angle of $180°$ for $\triangle abc$ is split into eleven parts, so there will be some angle of at most $\frac{180}{11}°$ in any planar drawing of the graph. Hence, the ratio of angular resolution is at least $11/4 = 2.75$.

5   Conclusions and Further Research

This paper compared quality measures of RAC drawings and planar drawings by defining a ratio, using the best possible RAC drawing and planar drawing of the same graph, maximized over all planar graphs. We showed that the quality ratio for area requirement quality is unbounded, and so is the quality ratio of edge-length. For angular resolution we gave an example that shows that the quality ratio is at least 2.75. It is unknown—and the main open problem arising from this paper—whether this quality ratio is bounded by a constant or not. If it is bounded by a constant, it may still be possible to improve upon the given bound of 2.75.

Several other aesthetic criteria can also be analyzed. A list is given in [2]. Our construction that shows an unbounded quality ratio of edge length also shows that the quality ratio of *total* edge length is unbounded, assuming the shortest edge has unit length. For the criterion of symmetry, a RAC drawing of a K_4 using a square is in a sense "more symmetric" than any planar drawing of a K_4. In a K_4, all vertices are "the same", which is also true in the RAC drawing but not in any planar drawing.

This paper considered only straight-line drawings of planar graphs. One can extend the research to drawings with bends and analyze the ratio of RAC drawings and planar drawings in this case as well. This is left for future research.

References

1. Angelini, P., Cittadini, L., Battista, G.D., Didimo, W., Frati, F., Kaufmann, M., Symvonis, A.: On the perspectives opened by right angle crossing drawings. In: Eppstein, D., Gansner, E.R. (eds.) GD 2009. LNCS, vol. 5849, pp. 21–32. Springer, Heidelberg (2010)
2. Battista, G.D., Eades, P., Tamassia, R., Tollis, I.: Graph Drawing: Algorithms for the Visualization of Graphs. Prentice-Hall, Englewood Cliffs (1999)
3. de Fraysseix, H., Pach, J., Pollack, R.: How to draw a planar graph on a grid. Combinatorica 10, 41–51 (1990)
4. Didimo, W., Eades, P., Liotta, G.: Drawing graphs with right angle crossings. In: Dehne, F., Gavrilova, M., Sack, J.-R., Tóth, C.D. (eds.) WADS 2009. LNCS, vol. 5664, pp. 206–217. Springer, Heidelberg (2009)
5. Garg, A., Tamassia, R.: Planar drawings and angular resolution: Algorithms and bounds (extended abstract). In: van Leeuwen, J. (ed.) ESA 1994. LNCS, vol. 855, pp. 12–23. Springer, Heidelberg (1994)
6. Malitz, S., Papakostas, A.: On the angular resolution of planar graphs. SIAM J. Discrete Math. 7(2), 172–183 (1994)

Convex Polygon Intersection Graphs

Erik Jan van Leeuwen[1] and Jan van Leeuwen[2]

[1] Department of Informatics, University of Bergen
P.O. Box 7803, N-5020 Bergen, Norway
E.J.van.Leeuwen@ii.uib.no
[2] Department of Information and Computing Sciences, Utrecht University
P.O. Box 80.089, NL-3508 TB Utrecht, The Netherlands
j.vanleeuwen@cs.uu.nl

Abstract. Geometric intersection graphs are graphs determined by intersections of geometric objects. We study the complexity of visualizing the arrangements of objects that induce such graphs. We give a general framework for describing geometric intersection graphs, using arbitrary finite base sets of rationally given convex polygons and affine transformations. We prove that for every class of intersection graphs that fits the framework, the graphs in the class have a representation using polynomially many bits. Consequently, the recognition problem of these classes is in NP (and thus NP-complete). We also give an algorithm to find a drawing of the objects in the plane, if a graph class fits the framework.

1 Introduction

A *geometric intersection graph* is the intersection graph of a finite set of geometric objects. There is an edge between two vertices in the graph iff their corresponding objects intersect. The objects form a *representation* of the graph. Classes of geometric intersection graphs are obtained if one only allows objects similar to certain base objects specific for a class.

To visualize a geometric intersection graph, drawing a representation is more informative than just drawing the graph itself. Therefore we study the complexity of visualizing representations. We consider the following problems: do representations in polynomial space exist and if so, how can their drawings be found effectively.

Understanding Geometric Intersection Graphs. Geometric intersection graphs arise naturally in many areas. They are used e.g. in modeling wireless communication networks, where geometric objects model the transmission ranges of the different devices in the network. This has lead to the study of the well-known class of (unit) disk graphs and many other classes [9,20,22,23].

Current geometric intersection graph models use only homothetic copies, thus translations and scalings, of objects. Moreover, normally only a single base object is used. A broader notion of similarity and a larger variety of base objects may be desired in defining a class. Therefore we aim at a more general conceptual framework.

U. Brandes and S. Cornelsen (Eds.): GD 2010, LNCS 6502, pp. 377–388, 2011.
© Springer-Verlag Berlin Heidelberg 2011

Definition 1. *A* signature *is any 2-tuple* $\mathcal{P} = \langle S, T \rangle$ *with:* (a) $S = \{o_1, \cdots, o_m\}$ *is a finite nonempty base set of geometric objects in the plane, with each object in S containing the origin, and* (b) T *maps every object $o \in S$ to a finite set of* similarity templates *that determine how objects similar to o can be obtained.*

Here a similarity template is any family of similarity transforms, e.g. a rotation over some angle, followed by a translation. More generally, a template t will be any parametric family of bi-continuous functions $t(w_1, \ldots, w_k) : \mathbb{R}^2 \to \mathbb{R}^2$ which are shape-preserving in some sense, with the w_i's ranging over e.g. $\mathbb{R}_+$.

Definition 2. *Given a signature $\mathcal{P} = \langle S, T \rangle$, a graph G is called $\mathcal{P}$-intersection graph if it is the intersection graph of a finite set of objects $\mathcal{O}_1, \cdots, \mathcal{O}_n$, where every $\mathcal{O}_i$ ($1 \leq i \leq n$) is similar to an object $o \in S$, i.e. obtained using a transformation conforming to a similarity template in $T(o)$.*

Problem Definitions. In order to visualize $\mathcal{P}$-intersection graphs, we must know the complexity of their representation. In particular, we want to know whether representations exist that require only polynomially many bits. Assume from now on that all objects we consider are fully specified, both for localizing and drawing them, by only finitely many parameters.

Definition 3. *(i) A $\mathcal{P}$-intersection graph with n vertices is said to be* polynomially represented *(using polynomial p), if it is the intersection graph of a finite set of objects $\mathcal{O}_1, \cdots, \mathcal{O}_n$, where every $\mathcal{O}_i$ ($1 \leq i \leq n$) is similar to an object in S according to an allowed template, and has all its specifying parameters equal to rationals $\frac{a}{b}$ with $|a|, |b| \leq 2^{p(n)}$.*
(ii) A class $\mathcal{C}$ of $\mathcal{P}$-intersection graphs is said to be polynomially represented *if there is a polynomial $p = p(n)$ such that every graph in $\mathcal{C}$ is polynomially represented using p.*

Given a graph, we like to determine whether it is a geometric intersection graph of some kind and be able to visualize it by its representation. This leads to the following problems.

$\mathcal{P}$-Intersection Graph Recognition
Given a graph G, decide whether G is a $\mathcal{P}$-intersection graph.

$\mathcal{P}$-Intersection Graph Construction (Visualisation)
Given a graph G that is known to be a $\mathcal{P}$-intersection graph, construct a representation of G by objects in the plane according to signature $\mathcal{P}$.

We consider the complexity of both problems for $\mathcal{P}$-intersection graphs and whether or not such graphs have feasible, i.e. polynomial, representations.

Previous Work. The size of a representation and the complexity of the recognition problem have been studied for many classes of geometric intersection graphs. Some prominent results in this area are shown in Table 1.

Table 1. Some classes of geometric intersection graphs. The first column indicates the graph class, the second column the objects used in representing the class, the third the complexity of the recognition problem, the fourth the size of a representation of the intersection graph (polynomial or exponential). The fifth gives references. Contributions of this paper are marked in italics. We use * to refer to the current paper.

Graph Class	Objects	Recognition	Repr.	Reference
(unit) disk	disks	NP-h, $\in$PSPACE	exp.	[1,14,9,19]
box (rectangle)	rectangles in $\mathbb{R}^2$	NP-c	poly	[13,18]
unit square	unit squares	NP-c	poly	[1,4]
square	squares	NP-h, $\in NP$	*poly*	*
max-tolerance	semi-squares	NP-h, $\in NP$	*poly*	[10], *
polygon intersect.	homoth. conv. polygons	NP-h	-	[16,21]
polygon intersect.	rat. repr. conv. polygons	$\in NP$	*poly*	*
convex intersect.	convex sets $\subset \mathbb{R}^2$	NP-h, $\in$PSPACE	exp.	[12,21]

Recognition can be nontrivial. For example, for disk graphs the problem is algorithmically decidable ([22]), known to be NP-hard [14] and in PSPACE [9,15], but it is open whether the problem actually is in NP. This holds even for the class of unit disk graphs.

The complexity or *size* of a representation poses an equally challenging problem. A first question is whether a class actually has a representation using only rational coordinates. This was shown e.g. for the intersection graphs of all so-called scalable objects [23]. If one only allows objects to touch but not to overlap (*contact graphs*), this is no longer guaranteed [3].

A second question is whether the rationals in these representations can be specified using polynomially many bits. For (unit) disk graphs, this question was answered negatively only recently [19].

Our Results. We apply our framework to define classes that use finite base sets of rationally given convex polygons and templates of rationally constrained affine transformations (see Section 2). We prove that for any such class, the intersection graphs in it have a polynomial representation, even in integers. This contrasts the known fact that intersection graphs of *arbitrary* convex polygons may require exponentially-sized representations in worst case [21].

We also settle, in a general way, the question left open by the recent NP-hardness proof of the recognition problem for intersection graphs of homothetic copies of a single convex polygon [16], namely whether this problem is in NP. Our results immediately imply that this problem is indeed in NP, even for considerably larger classes of intersection graphs. Moreover, we give an algorithm to determine whether a given graph is an intersection graph within the above framework. The algorithm is constructive and returns a visualization of the arrangement of objects representing the given graph, if one exists.

The main result is presented in Section 4. (In Section 3 we show that it is irrelevant whether the objects we consider are open or closed.) By applying the same techniques, one can prove e.g. that max-tolerance graphs and contact graphs of homothetic convex polygons have polynomial representations, and that their recognition problems thus are in NP. Further applications are given in Section 5. More details are given in [25].

2 $\mathcal{P}$-intersection Graphs

$\mathcal{P}$-intersection graphs give a very general framework. We use it to consider $\mathcal{P}$-intersection graphs for signatures $\mathcal{P} = \langle S, T \rangle$, where S is any finite nonempty base set of (closed and) rationally given convex polygons.

We also tune the choice of similarity templates. Similarity templates t were described as parametric families of bi-continuous functions $t(w_1, \ldots, w_k) : \mathbb{R}^2 \to \mathbb{R}^2$ which preserve shapes according to some notion of similarity. Templates should be *smooth*, which means that images of base objects under $t(w_1, \ldots, w_k)$ and $t(z_1, \ldots, z_k)$ must be 'almost equal' if $(w_1, \ldots, w_k)$ and $(z_1, \ldots, z_k)$ are.

Given convex base polygons, we restrict ourselves to linear similarity templates consisting of parameterized affine transformations over $\mathbb{Q}$ only.

Definition 4. *A linear similarity template $t = t_{\alpha,\beta,\gamma,\delta}$ is a family of affine transformations of the form $x \to u + Q(v)x$, where: (i) $\alpha, \beta, \gamma, \delta$ are rationals such that $\alpha\delta - \beta\gamma \neq 0$, (ii) $u = (u_1, u_2)$ is any 2-dimensional vector, and (iii) $Q(v) = v\left(\begin{smallmatrix} \alpha & \gamma \\ \beta & \delta \end{smallmatrix}\right)$ is a 2×2 matrix with v satisfying $v > 0$.*

Linear similarity templates have two parameters: u (the translation vector) and v (the scaling factor of the distortion matrix). Neither of them needs to be rational. The constraint $v > 0$ keeps $Q(v)$ nonsingular and guarantees that all template mappings are topological isomorphisms. Linear similarity templates can be shown to be smooth. From now on, we only consider linear similarity templates.

Even though applying linear similarity transformations $x \to u + Q(v)x$ to objects o amounts to applying regular homothetic transformations to objects $Q(o)$, the framework gives us the conceptual generality we want. Also, the framework allows us to vary templates while keeping the set of base polygons fixed.

An affine transformation $u + Q(v)x \in t$, where t is any template assigned to a polygon in S by T, is called a $\mathcal{P}$-*transformation*. Many familiar transformations (combined with scaling) are $\mathcal{P}$-transformations, aside from shifts and (skewed) scalings: horizontal shears, vertical shears, rotations, and reflexions, where for the latter the coefficients of the appropriate Q must be rounded to keep them rational. T may assign different sets of templates to different objects, without any dependency between them.

Definition 5. *Given a signature $\mathcal{P} = \langle S, T \rangle$, object $\mathcal{O}$ is said to be* similar *to an object $o \in S$ if $\mathcal{O}$ can be obtained from o by applying an allowed $\mathcal{P}$-transformation to it.*

One can show that the notion of similarity is well-founded, i.e. it is decidable, for any signature $\mathcal{P} = \langle S, T \rangle$ and convex polygon $\mathcal{O}$, whether $\mathcal{O}$ is similar to a polygon in S under $\mathcal{P}$-transformation. Consider the (polynomial) representation of any $\mathcal{P}$-intersection graph and an arbitrary object $\mathcal{O}$ occurring in it.

Lemma 1. *Let $\mathcal{P} = \langle S, T \rangle$ be as above, and let $\mathcal{O}$ be similar to $o \in S$.*
(i) If $\mathcal{O}$ is polynomially represented and $\mu > 0$ is a polynomially represented rational, then the scaling of $\mathcal{O}$ by μ is also polynomially represented.
(ii) If $\mathcal{O}$ is polynomially represented and $\rho > 0$ is a polynomially represented rational, then there is an enlargement of $\mathcal{O}$ by an additive margin δ with $0 < \delta < \rho$ that is again polynomially represented. This holds likewise for reductions.

Because the matrices $Q(v)$ in similarity templates are nonsingular, $\mathcal{P}$-transforms map convex polygons 1-1 onto convex polygons. Thus vertices and edges of the latter are images of the vertices and edges of the former, respectively. Recall that a convex polygon can be given by its defining inequalities.

Lemma 2. *(i) Let Q be a nonsingular 2×2 matrix, and o a plane convex polygon containing the origin. When Q transforms o and $\det Q > 0$, then defining inequalities are mapped to defining inequalities with preservation of the inequality sign. If $\det Q < 0$, the inequality signs are reversed.*
(ii) A $\mathcal{P}$-transformation $(u_1, u_2) + Q(v)x$ with $Q(v) = v\left(\begin{smallmatrix} \alpha & \gamma \\ \beta & \delta \end{smallmatrix}\right)$ maps the line $ax + by + c = 0$ onto the line $(a\delta - b\beta)x + (b\alpha - a\gamma)y + (\alpha\delta - \beta\gamma)vc - (a\delta - b\beta)u_1 - (b\alpha - a\gamma)u_2 = 0$.

Lemma 2 enables one to determine exactly how the defining inequalities of a base polygon are transformed under a $\mathcal{P}$-transformation.

3   Open Versus Closed Objects

Let $\mathcal{P} = \langle S, T \rangle$ be a signature. What happens if we let S consist of *open* convex polygons instead of closed ones? For disk graphs, it is known that taking open or closed disks does not change the class of graphs [24]. In [23], this was proved for the intersection graphs of all 'scalable' geometric objects. In the case of (unit) disk graphs, even polynomial representation is preserved.

We consider the case of $\mathcal{P}$-intersection graphs, emphasizing polynomial representation. We show in two steps that for $\mathcal{P}$-intersection graphs the closed and open cases are again equivalent. We use the following facts.

Lemma 3. *Let $\mathcal{O}_1$ and $\mathcal{O}_2$ be two disjoint convex polygons in the plane, both having nonempty interior. The (shortest) distance between $\mathcal{O}_1$ and $\mathcal{O}_2$ is realized as the distance between a vertex of one polygon and an edge of the other.*

Proof. This follows by a simple extension of the proof of Lemma 2.1 in [6].   □

Lemma 4. *Let a, b, c, v_1, v_2 be rationals with their numerator and denominator bounded in absolute value by q for some $q > 0$. If the following fraction is $\neq 0$, then $\frac{|av_1 + bv_2 + c|}{\sqrt{a^2 + b^2}} \geq \frac{1}{2q^5}$.*

We first show one side of the equivalence. Let $\mathcal{P} = \langle S, T \rangle$.

Lemma 5. *Every polynomially represented $\mathcal{P}$-intersection graph using a non-empty base set of closed convex polygons can be obtained as a polynomially represented $\mathcal{P}$-intersection graph using a nonempty base set of open convex polygons.*

Proof. Let $S = \{o_1, \cdots, o_m\}$ be the base set of closed convex polygons. Let G be the intersection graph defined by the objects $\mathcal{O}_1, \cdots, \mathcal{O}_n$, where $\mathcal{O}_i$ ($1 \leq i \leq n$) is similar to $o_{s_i} \in S$ for some $s_i \in \{1, \ldots, m\}$ and obtained by applying an allowed affine transformation of $T(o_{s_i})$ to o_{s_i}. Let $\mathcal{O}_i$ ($1 \leq i \leq n$) only have vertices with rational coordinates $\frac{a_i}{b_i}$ with $|a_i|, |b_i| \leq 2^{p(n)}$.

Let $\mathcal{P}' = \langle S', T \rangle$ be the signature obtained from $\mathcal{P}$ in which every (closed) base polygon o is replaced by its interior o°. G can be viewed as the intersection graph of $\mathcal{O}_1^\circ, \cdots, \mathcal{O}_n^\circ$, provided no intersections of polygons are lost by restricting to the interiors. Intersections are lost precisely when there are (closed) polygons $\mathcal{O}_i$ and $\mathcal{O}_j$ that touch. We show that one can slightly enlarge the polygons $\mathcal{O}_i$ such that this does not occur, while preserving G as the intersection graph.

Suppose one of the closed polygons, say $\mathcal{O}_i$, touches several other polygons $\mathcal{O}_j$. By enlarging $\mathcal{O}_i$ by a small but nonzero margin μ, we can eliminate the touchings and let $\mathcal{O}_i$ overlap nontrivially with each $\mathcal{O}_j$. However, in enlarging it (and enlarging all other polygons for which this step is carried out) we must make sure that no spurious intersections with objects $\mathcal{O}_r$ disjoint from $\mathcal{O}_i$ are created. Suppose $\mathcal{O}_i$ and $\mathcal{O}_r$ are disjoint. By Lemma 3, the distance between them is realized by the distance between a vertex, say $v = (v_1, v_2)$ of one of them and an edge, say $ax + by + c = 0$ of the other. This distance is $\frac{|av_1 + bv_2 + c|}{\sqrt{a^2 + b^2}}$.

The numerators and denominators of v_1, v_2 are $\leq q = 2^{p(n)}$. Also, $ax + by + c = 0$ connects two vertices of a polygon in the set, which have rational coordinates with numerators and denominators $\leq q$. It follows that a, b, c are all rational, with numerators and denominators $\leq 4q^4$. By Lemma 4 the distance is now at least $\frac{1}{dq^{20}}$ for some constant $d > 0$. Hence, if we enlarge $\mathcal{O}_i$ by a nonzero margin of $\mu \leq \frac{1}{3dq^{20}}$, then disjointness with every disjoint $\mathcal{O}_r$ is maintained (taking into account that the latter may also be enlarged by the same factor).

As μ is independent of the specific $\mathcal{O}_i$ chosen, the enlargement can be carried out simultaneously for all polygons. For every $\mathcal{O}_i$, it preserves all the intersections with other polygons, introduces no new ones, and has the effect that every polygon $\mathcal{O}_j$ that it touched, now overlaps nontrivially with it as well (and thus their interiors overlap). By Lemma 1, enlarging every $\mathcal{O}_i$ by a nonzero margin at most μ can be achieved while preserving similarity and polynomial representation. Thus, G is a $\mathcal{P}'$-intersection graph and polynomially represented. $\quad\square$

The converse of the lemma is proved in a similar way. We conclude:

Theorem 1. *Every polynomially represented $\mathcal{P}$-intersection graph with a base set of closed convex polygons can be obtained as a polynomially represented $\mathcal{P}$-intersection graph with a base set of open convex polygons, and vice versa.*

4 Representing $\mathcal{P}$-intersection Graphs

Let G be a $\mathcal{P}$-intersection graph and let some geometric realization of G as $\mathcal{P}$-intersection graph be given. The realization of G can be viewed as a feasible solution of a *model*, namely of a model that defines the exact pattern of intersections and nonintersections between the polygons. We will design a suitable LP model for this such that, if it has a feasible solution (which it has), it also has one that is polynomially represented, thus implying a geometric realization of G with this property. A similar approach was used in [8,24]. We rely on the following fact, which seems folklore (cf. [10,15,25]).

Lemma 6. *Two closed convex polygons in the plane are disjoint iff they can be separated by a line that precisely coincides with an edge of one of them.*

Let G be the $\mathcal{P}$-intersection graph of the convex polygons $\mathcal{O}_1, \cdots, \mathcal{O}_n$, where $\mathcal{O}_i$ $(1 \le i \le n)$ is similar to $o_{s_i} \in S$ (some $s_i \in \{1, \ldots, m\}$). Let $\mathcal{O}_i$ be the result of applying transformation $u_i + Q_i x = \left(\begin{smallmatrix} u_{i,1} \\ u_{i,2} \end{smallmatrix}\right) + v_i \left(\begin{smallmatrix} \alpha_i & \gamma_i \\ \beta_i & \delta_i \end{smallmatrix}\right) \left(\begin{smallmatrix} x \\ y \end{smallmatrix}\right)$ to o_{s_i}, with suitable $u_{i,1}$, $u_{i,2}$, and v_i $(1 \le i \le n)$, all conforming to a template $t = t_i$ applicable to o_{s_i}. Let o_{s_i} $(1 \le i \le n)$ have k_i vertices and (thus) k_i edges. All data related to o_{s_i} (vertices, edges, defining inequalities) will be super-indexed by (i).

4.1 Helpful Inequalities

Consider any two polygons $\mathcal{O}_i, \mathcal{O}_j$ and suppose we want to express that they are *disjoint*. By Lemma 6 there must be a defining inequality of (say) $\mathcal{O}_i$ such that all of $\mathcal{O}_j$ does not satisfy it. Which of $\mathcal{O}_i, \mathcal{O}_j$ to take and which defining inequality, follows from the given geometric realization of G. Say the polygon to take is indeed $\mathcal{O}_i$ and that the defining inequality to take is the one obtained by applying t_i to the defining inequality $a^{(i)}x + b^{(i)}y + c^{(i)} \le / \ge 0$ of o_{s_i}. Lemma 2 implies that this defining inequality of $\mathcal{O}_i$ can then be written as

$$(a^{(i)}\delta_i - b^{(i)}\beta_i)x + (b^{(i)}\alpha_i - a^{(i)}\gamma_i)y + (\alpha_i\delta_i - \beta_i\gamma_i)v_i c^{(i)} - (a^{(i)}\delta_i - b^{(i)}\beta_i)u_{i,1} - (b^{(i)}\alpha_i - a^{(i)}\gamma_i)u_{i,2} \le / \ge 0.$$

Each vertex of $\mathcal{O}_j$ is obtained from a vertex $\left(\begin{smallmatrix} d^{(j)} \\ e^{(j)} \end{smallmatrix}\right)$ of o_{s_j} using $u_j + Q_j x$, and can thus be written as $\left(\begin{smallmatrix} u_{j,1} + \alpha_j d^{(j)} v_j + \gamma_j e^{(j)} v_j \\ u_{j,2} + \beta_j d^{(j)} v_j + \delta_j e^{(j)} v_j \end{smallmatrix}\right)$. To express that $\mathcal{O}_j$ is disjoint of $\mathcal{O}_i$ it now suffices to express that *none* of these k_j vertices of $\mathcal{O}_j$ satisfy the defining inequality. This gives k_j constraints of the form

$$DISJ_{i,j}(u_{i,1}, u_{i,2}, v_i, u_{j,1}, u_{j,2}, v_j) :=$$
$$(a^{(i)}\delta_i - b^{(i)}\beta_i)(u_{j,1} + \alpha_i d^{(j)} v_j + \gamma_i e^{(j)} v_j) + (b^{(i)}\alpha_i - a^{(i)}\gamma_i)(u_{j,2} + \beta_j d^{(j)} v_j + \delta_j e^{(j)} v_j) + (\alpha_i\delta_i - \beta_i\gamma_i)v_i c^{(i)} - (a^{(i)}\delta_i - b^{(i)}\beta_i)u_{i,1} - (b^{(i)}\alpha_i - a^{(i)}\gamma_i)u_{i,2} > / < 0$$

one for each vertex of $\mathcal{O}_j$. We strengthen each inequality to "$\ge$ some positive margin" or "$\le$ some negative margin" respectively (for real nonzero margins), by evaluating the inequalities in the given realization of G. The inequalities are homogeneous in $u_{i,1}, u_{i,2}, v_i, u_{j,1}, u_{j,2}, v_j$. Thus, multiplying all $u_{i,1}, u_{i,2}, v_i, u_{j,1}, u_{j,2}, v_j$ $(1 \le i, j \le n)$ by a factor $\mu \ge 1$ large enough and rescaling the variables, the constraints still express the realization of G, but now with inequalities

$DISJ_{i,j}(u_{i,1}, u_{i,2}, v_i, u_{j,1}, u_{j,2}, v_j) \leq -1/ \geq 1$ for k_j nodes and relevant $1 \leq i, j \leq n$.

(We will in fact choose μ large enough such that a number of further goals w.r.t. the other constraints are achieved as well, as explained below.)

Next, $\mathcal{O}_i, \mathcal{O}_j$ *overlap* iff there is a point $(x_{i,j}, y_{i,j})$ satisfying the defining inequalities of both $\mathcal{O}_i$ and $\mathcal{O}_j$. This leads to $k_i + k_j$ linear constraints

$$IN_i(x_{i,j}, y_{i,j}, u_{i,1}, u_{i,2}, v_i) \leq / \geq 0,$$
$$IN_j(x_{i,j}, y_{i,j}, u_{j,1}, u_{j,2}, v_j) \leq / \geq 0,$$

one for each defining inequality of $\mathcal{O}_i$ and $\mathcal{O}_j$. The inequalities are homogeneous in $x_{i,j}, y_{i,j}, u_{i,1}, u_{i,2}, v_i, u_{j,1}, u_{j,2}, v_j$ and thus scale along with the scaling of the $DISJ$-inequalities.

Observe that the constraints "$v_i > 0$" may be replaced by "$v_i \geq$ some positive margin" in all cases as before, using the data from the given embedding. If we multiply all variables by a $\mu \geq 1$ large enough and rescale the variables accordingly, we can achieve that all constraints continue to express what we want, i.e. the resulting model still realizes G, but now we can also assume w.l.o.g. that $v_i \geq 1$ for $1 \leq i \leq n$.

Finally, note that the arrangement of polygons realizing G can be shifted over any fixed vector we want. Thus, w.l.o.g. we may assume that $u_{i,1}, u_{i,2} \geq 0$ for every $1 \leq i \leq n$.

4.2   Assembling the Model

The model is complete when we define the situation (intersection or not) for every pair $\mathcal{O}_i, \mathcal{O}_j$. First include all constraints of the affine transformations $u_i + Q_i x = \begin{pmatrix} u_{i,1} \\ u_{i,2} \end{pmatrix} + v_i \begin{pmatrix} \alpha_i & \gamma_i \\ \beta_i & \delta_i \end{pmatrix} \begin{pmatrix} x \\ y \end{pmatrix}$, following the templates t_i that are used and taking the scalings into account: $u_{i,1}, u_{i,2} \geq 0$ and $v_i \geq 1$. (The condition $\alpha_i \delta_i - \beta_i \gamma_i \neq 0$ can be assumed for all templates in T.)

Next consider all $\frac{1}{2} n(n-1)$ pairs $\mathcal{O}_i, \mathcal{O}_j$ and express the model inequalities for each pair. For any pair $\mathcal{O}_i, \mathcal{O}_j$ we have k_{ij} or $l_{i,j}$ inequalities respectively of the following form:

if $\mathcal{O}_i, \mathcal{O}_j$ must be disjoint:
$k_{ij} \leq \max\{k_i, k_j\}$ inequalities of type
$\qquad DISJ_{i,j}(u_{i,1}, u_{i,2}, v_i, u_{j,1}, u_{j,2}, v_j) \leq -1/ \geq 1$

if $\mathcal{O}_i, \mathcal{O}_j$ must intersect (m cases):
$l_{ij} = k_i + k_j$ inequalities of type
$\qquad IN_i(x_{i,j}, y_{i,j}, u_{i,1}, u_{i,2}, v_i) \leq / \geq 0$, resp
$\qquad IN_j(x_{i,j}, y_{i,j}, u_{j,1}, u_{j,2}, v_j) \leq / \geq 0$.

Bring the linear system into standard form with nonnegative slack variables $z_i, w_{ij1}, \cdots, w_{ijk_{ij}}, z_{ij1}, \cdots, z_{ijl_{ij}}$, turning inequality into equality constraints:

$\qquad v_i - z_i = 1,$

if $\mathcal{O}_i, \mathcal{O}_j$ must be disjoint:
k_{ij} inequalities $DISJ_{i,j}(u_{i,1}, u_{i,2}, v_i, u_{j,1}, u_{j,2}, v_j) \pm w_{ijr} = \mp 1$ (w_{ijr} used in the r-th inequality),

if $\mathcal{O}_i, \mathcal{O}_j$ must intersect:

l_{ij} inequalities $IN_i(x_{i,j}, y_{i,j}, u_{i,1}, u_{i,2}, v_i) \pm z_{ijr} = 0$ and

$IN_j(x_{i,j}, y_{i,j}, u_{j,1}, u_{j,2}, v_j) \pm z_{ijr} = 0$ (z_{ijr} used in the r-th inequality),

now with the standard constraints: $u_{i,1}, u_{i,2}, v_i, x_{i,j}, y_{i,j}, z_i \geq 0$ $(1 \leq i \leq n)$, $w_{ij1}, \cdots, w_{ijk_{ij}} \geq 0$ $(1 \leq i < j \leq n)$, and $z_{ij1}, \cdots, z_{ijl_{ij}} \geq 0$ $(1 \leq i < j \leq n)$. Note that all linear equations of the model have *rational* coefficients.

4.3   Solving the Model

Because S is finite, there is a constant k such that $k_i \leq k$ for every $1 \leq i \leq n$. As T only has finitely many different templates, there is a constant $q \geq 1$ such that for every $1 \leq i \leq n$, the numerators and denominators of the (rational) coefficients of the defining inequalities of every o_{s_i} and of the rationals $\alpha_i, \beta_i, \gamma_i, \delta_i$ are all $\leq 2^q$. Thus the coefficients in the linear inequalities $DISJ_{i,j}(u_{i,1}, u_{i,2}, v_i, u_{j,1}, u_{j,2}, v_j) \leq -1/ \geq 1$ and $IN_{i,j}(u_{i,1}, u_{i,2}, v_i, u_{j,1}, u_{j,2}, v_j) \leq 0/ \geq 0$ are all rationals with numerators and denominators $\leq 2^{dq}$ for a small integer constant $d \geq 1$. The same bound holds in the standard form.

Let $N = n + \sum_{ij} k_{ij} + \sum_{ij} l_{ij} \leq n + \frac{1}{2}kn(n-1)$. The system of linear equalities can be written as: $\mathbf{Ax} = \mathbf{b}$ with $\mathbf{x} \geq 0$, where

> $\mathbf{A}$ is a N by $N + 3n + 2m$ all-rational matrix, all entries a of $\mathbf{A}$ have numerator and denominator $\leq 2^{dq}$, N columns of $\mathbf{A}$ are unit vectors namely those corresponding to variables $z_i, z_{ij1}, \cdots, z_{ijk_{ij}}, w_{ij1}, \cdots, w_{ijl_{ij}}$ $(1 \leq i \leq n$ and $1 \leq i < j \leq n$ resp.), $\mathbf{x} = (\cdots, u_{i,1}, u_{i,2}, \cdots, v_i, \cdots, x_{i,j}, y_{i,j}, \cdots, z_i, \cdots, w_{ij1}, \cdots, w_{ijk_{ij}}, \cdots, z_{ij1}, \cdots, z_{ijl_{ij}}, \cdots)^T$, and $\mathbf{b} = (\cdots, 1, \cdots, \mp 1, \cdots, 0, \cdots)^T$, with all entries rational, in fact ± 1 or 0.

The term 'unit vector' is used to denote any column that has only one nonzero entry, with this entry being ± 1. Note that $\text{rank}(\mathbf{A}) = N = O(n^2)$.

Theorem 2. *The LP model has an all-rational solution for $u_{i,1}, u_{i,2}, v_i$ with numerators and denominators bounded in absolute value by $2^{\mathcal{O}(n^4)}$.*

Proof (Outline). Because $\mathbf{Ax} = \mathbf{b}$ with $\mathbf{x} \geq 0$ has a feasible solution and $\text{rank}(\mathbf{A}) = N$, it has a *basic* feasible solution with (at least) $3n + 2m$ of the coordinates of $\mathbf{x}$ equal to 0, whereas the N-by-N submatrix $\mathbf{A}'$ consisting of the columns corresponding to the other coordinates is invertible and satisfies $\mathbf{A}'\mathbf{x}' = \mathbf{b}$ (with $\mathbf{x}' \geq 0$), where $\mathbf{x}'$ is the subvector of $\mathbf{x}$ consisting of these other coordinates. Hence, by Cramer's rule, it follows that $(\mathbf{x}')_i = \frac{\det \mathbf{A}'_i}{\det \mathbf{A}'}$, where $\mathbf{A}'_i$ is the matrix formed by replacing the i-th column of $\mathbf{A}'$ by $\mathbf{b}$. Observe that, because $\mathbf{A}'$ and $\mathbf{A}'_i$ are rational matrices, their determinants are rational as well.

The nonzero entries of $\mathbf{A}'$ are all of the form $\frac{f}{h}$ with f, h integer and $|f|, |h| \leq 2^{dq}$. Let H_i be the product of all $|h|$-values that occur as denominators in the i-th column and let $H = \prod_1^N H_i$, thus $H \leq 2^{dq \cdot N \cdot N}$. Then $\det \mathbf{A}' = \frac{1}{H} \det \mathbf{A}''$, where $\mathbf{A}''$ is obtained from $\mathbf{A}'$ by multiplying the elements in the first column by H_1, the elements of the second column by H_2, etc. Using Hadamard's inequality for matrices, $F = |\det \mathbf{A}''| \leq \left(\sqrt{N2^{2dqN}}\right)^N \leq N^{\frac{1}{2}N}2^{dqN^2} \leq 2^{2dqN^2}$. This shows that $\det \mathbf{A}' = \frac{F}{H}$ with F, H integers with $|F|, |H| \leq 2^{gn^4}$ for some constant g. The same bound holds for the other determinants. $\qquad\square$

Theorem 2 gives a sufficient bound and easily leads to the following, main result.

Theorem 3. *Let G be a $\mathcal{P}$-intersection graph. Then G has a polynomial representation, even fully in integers.*

Corollary 1. *The recognition problem for every class of $\mathcal{P}$-intersection graphs is in NP.*

Although the above arguments seem to rely on being given some realization of the $\mathcal{P}$-intersection graph, this is in fact not necessary. It suffices to know for each vertex of G which base polygon of S and which transformations of T to use, and for any nonadjacent pair of vertices which defining inequality of the objects representing these vertices to use to express disjointness. Quantifying over this information suffices to find a realization of $\mathcal{P}$.

Corollary 2. *The construction (i.e. 'drawing') problem of any class of $\mathcal{P}$-intersection graphs can be solved algorithmically, in exponential time.*

5   Applications

The notions of signatures and $\mathcal{P}$-intersection graphs are very useful in modeling classes of intersection graphs, particularly when combined with the generic theorems presented above. We list a few applications.

Square Intersection Graphs: It is known that unit square graphs have polynomial size representations [4]. We can now extend this to *square intersection graphs*. Recall that NP-hardness of the recognition problem of unit square intersection graphs was proved in [1]. For general square graphs, NP-hardness follows from the recent results in [16].

Theorem 4. *Square intersection graphs have polynomial-size integer representations. Their recognition problem is in NP (and thus NP-complete).*

Proof. Define signature $\mathcal{P} = \langle S, T \rangle$ with S consisting of a unit square around the origin, and T consisting of the template $t : u + v\left(\begin{smallmatrix} 1 & 0 \\ 0 & 1 \end{smallmatrix}\right)$. Square intersection graphs are precisely the $\mathcal{P}$-intersection graphs for this signature $\mathcal{P}$. Now apply Theorem 3 and Corollary 1. $\qquad\square$

Polynomial representation for unit square graphs is easily shown directly [4], but also follows from our arguments, by adding equations $v_i = v_j$ to the model.

Max-tolerance Graphs: Kaufmann et al. [10] showed that the max-tolerance graphs are precisely the intersection graphs of so-called semi-squares. (A semi-square is 'a square with one half cut off along the bottom-right to top-left diagonal'.) In the same way as in the previous example, one can show that semi-square intersection graphs, thus max-tolerance graphs, have polynomial-size integer representations and an NP-recognition problem.

As Kaufmann et al. [10] proved the recognition problem of max-tolerance graphs to be NP-hard, it now follows that this problem is in fact NP-complete.

Intersection Graphs of Homothetic Polygons: Kratochvíl and Pergel [16] initiated a general study of the intersection graphs that can be formed using homothetic copies of a single convex polygon P, or P_{hom}-*intersection graphs*. (We assume that P is always finitely given, in rational coordinates.) They show that the recognition problem for P_{hom}-intersection graphs is NP-hard. We can strengthen this as follows.

Theorem 5. P_{hom}-*intersection graphs have polynomial-size integer representations. Their recognition problem is in NP (and thus NP-complete).*

Proof. Define $\mathcal{P} = \langle S, T \rangle$ with $S = \{P\}$ and T assigning the homothetic transformations to P. P_{hom}-intersection graphs are precisely the $\mathcal{P}$-intersection graphs. The result follows. $\square$

In [16], Kratochvíl and Pergel also define P_{hom}-*contact graphs*, where intersections are restricted to being *contacts* only. They pose as an open problem to determine the complexity of recognizing P_{hom}-contact graphs. By modifying the LP model, one can show by the same technique as developed in Section 4 that P_{hom}-contact graphs have polynomial-size integer representations. Thus the recognition problem for P_{hom}-contact graphs is in *NP*. It remains open whether this problem is NP-complete.

Note added in proof: In recent work jointly with Tobias Müller (CWI, Amsterdam), tight upper- and lowerbounds have been obtained on the number of bits needed for representing convex polygon intersection graphs.

References

1. H.: Breu, Algorithmic aspects of constrained unit disk graphs, PhD Thesis, The University of British Columbia, Vancouver (1996)
2. Breu, H., Kirkpatrick, D.G.: Unit disk graph recognition is NP-hard. Computational Geometry 9, 3–24 (1998)
3. Brightwell, G.R., Scheinerman, E.R.: Representations of planar graphs. SIAM Journal of Discrete Mathematics 6(2), 214–229 (1993)
4. Czyzowicz, J., Kranakis, E., Krizanc, D., Urrutia, J.: Discrete realizations of contact and intersection graphs. Int. J. Pure and Applied Mathematics 13(4), 429–442 (2004)
5. Deng, X., Hell, P., Huang, J.: Linear time representation of proper circular arc graphs and proper interval graphs. SIAM Journal of Computing 25, 390–403 (1996)
6. Edelsbrunner, H.: Computing the extreme distances between two convex polygons. J. of Algorithms 6, 213–224 (1985)
7. Golumbic, M.C., Trenk, A.N.: Tolerance graphs. Cambridge University Press, Cambridge (2004)

8. Hayward, R.B., Shamir, R.: A note on tolerance graph recognition. Discrete Applied Mathematics 143, 307–311 (2004)
9. Hliněný, P., Kratochvíl, J.: Representing graphs by disks and balls (A survey of recognition-complexity results). Discrete Mathematics 229, 101–124 (2001)
10. Kaufmann, M., Kratochvíl, J., Lehmann, K.A., Subramanian, A.R.: Max-tolerance graphs as intersection graphs: cliques, cycles, and recognition. In: Proc. 17th Ann. ACM-SIAM Symp. on Discrete Algorithms (SODA 2006), pp. 832–841 (2006)
11. Kozyrev, V.P., Yushmanov, S.V.: Representations of graphs and networks (codings, layouts and embeddings). Journal of Soviet Mathematics 61(3), 2152–2194 (1992)
12. Kratochvíl, J., Matoušek, J.: NP-hardness results for intersection graphs. Commentationes Mathematicae Universitatis Carolinae 30(4), 761–773 (1989)
13. Kratochvíl, J.: A special planar satisfiability problem and a consequence of its NP-completeness. Discrete Applied Mathematics 52(3), 233–252 (1994)
14. Kratochvíl, J.: Intersection graphs of noncrossing arc-connected sets in the plane. In: North, S.C. (ed.) GD 1996. LNCS, vol. 1190, pp. 257–270. Springer, Heidelberg (1997)
15. Kratochvíl, J.: Geometric representations of graphs, Graduate Course, notes, Universitat Politècnica de Catalunya, Barcelona (April 2005),
    http://www.aco.gatech.edu/conference/archive/acokratochvil.ppt
16. Kratochvíl, J., Pergel, M.: Intersection graphs of homothetic polygons. In: Electronic Notes in Discrete Mathematics, vol. 31, pp. 277–280 (2008),
    http://www.canalc2.tv/video.asp?idvideo=7571
17. Lin, M.C., Szwarcfiter, J.L.: Unit circular-arc graph representations and feasible circulations. SIAM J. Discrete Mathematics 22(1), 409–423 (2008)
18. Lingas, A., Wahlen, M.: A note on maximum independent set and related problems on box graphs. Inf. Proc. Letters 93, 169–171 (2005)
19. McDiarmid, C., Müller, T.: The number of bits needed to represent a unit disk graph. In: Thilikos, D.M. (ed.) WG 2010. LNCS, vol. 6410, pp. 315–323. Springer, Heidelberg (2010)
20. McKee, T.A., McMorris, F.R.: Topics in intersection graph theory. SIAM Monographs on Discrete Mathematics and Applications, vol. 2, SIAM, Philadelphia (1999)
21. Pergel, M.: Special graph classes and algorithms on them, PhD Thesis, Dept. of Applied Mathematics, Charles University, Prague (2008)
22. Spinrad, J.R.: Efficient graph representations. In: Field Institute Monographs, vol. 19, American Mathematical Society, Providence (2003)
23. van Leeuwen, E.J.: Optimization and approximation on systems of geometric objects, PhD thesis, University of Amsterdam (2009)
24. van Leeuwen, E.J., van Leeuwen, J.: On the representation of disk graphs, Techn. Report UU-CS-2006-037, Dept. of Information and Computing Sciences, Utrecht University (2006)
25. van Leeuwen, E.J., van Leeuwen, J.: Convex polygon intersection graphs, Techn. Report, Dept. of Information and Computing Sciences, Utrecht University (to appear, 2010)

GraphML-Based Exploration and Evaluation of Efficient Parallelization Alternatives for Automation Firmware

Jürgen Bregenzer

Bosch Rexroth AG, Lohr am Main, Germany
University of Würzburg, Germany
`bregenzer@informatik.uni-wuerzburg.de`

1 Motivation

Graphs are an accepted and popular way of representing and solving problems of various kinds. Thus, applications of graphs as well as related topics such as graph visualization and graph representation are numerous. Our application is located in the domain of industrial automation and driven by the need of parallelizing our controller's firmware for the upcoming multi-core CPUs. As efficiency matters, we thereby aim at gaining maximum performance increases by spending no more implementation effort than necessary. In order to achieve that objective, we at first want to explore, evaluate and visualize those efficient parallelization alternatives by means of a graph-based model of our firmware. Thus, we are currently developing the *EEEPA (Exploration and Evaluation of Efficient Parallelization Alternatives)* tool for this purpose. Thereby, we have chosen and extended *GraphML* [1], a widespread format for graph representation.

2 The EEEPA Tool

At first, the *EEEPA* tool is targeting the static mapping of established tasks and interrupts to CPU cores by modeling the firmware's task system as an extended kind of a task interaction graph (TIG), the *EEEPA.TIG*. Derived from runtime logs, the graph represents each of the firmware's tasks by a node, weighted by the task's fraction of the system's load. An *interaction edge* between two nodes indicates interaction among the corresponding tasks by means of semaphores, events and messages. As *GraphML-Attributes* are restricted to simple types, the *EEEPA-GraphML* scheme redefines the `data-extension.type` for holding the entire interaction profile of an interaction edge as a sequence of complex types. Next, the weights of these edges are derived. This can either happen by taking hardware- and OS-specific benchmark results into account in order to consider the interaction-specific performance impacts of switching to inter-core interaction or by simply adding up the absolute interaction counts. Finally, *effort edges* are parsed from an Excel sheet. These edges indicate the development efforts of implementing proper synchronizations before running the edge-adjacent tasks on different CPU cores. The efforts are commonly estimated by developers.

U. Brandes and S. Cornelsen (Eds.): GD 2010, LNCS 6502, pp. 389–390, 2011.
© Springer-Verlag Berlin Heidelberg 2011

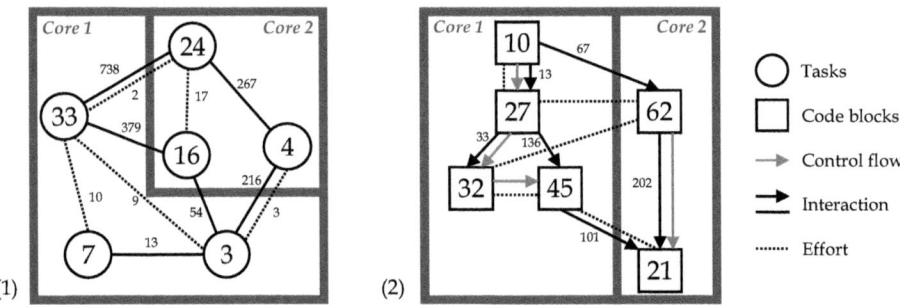

Fig. 1. Visualization of parallelization alternatives: The *EEEPA.TIG* (1) for a mapping of tasks and interrupts and the *EEEPA.CDFG* (2) for the parallel schedule of a task's or interrupt's decomposition.

However, the sole mapping of established tasks and interrupts can cause an imbalanced core utilization due to coarse grain ones, that are consuming a significant fraction of processing time. Thus, decomposition of established tasks and interrupts is also targeted by the EEEPA tool. The employed graph model for this purpose is a special kind of a control/data flow graph (CDFG), the *EEEPA.CDFG*. Again, the graph is constructed on basis of runtime logs: By source code instrumentation, developer-defined code blocks are enclosed by specific logging events. The derived graph comprises a node for each code block, that is weighted by the code block's runtime. Nodes are connected by unweighted *control flow edges* if logging revealed a control flow between them. Weighted *interaction edges* between code blocks indicate interactions in terms of data flows. These edges are derived by means of logging events and source code annotations, that are commonly defined by developers, maybe by assistance of static code analysis. The aforementioned *GraphML* extension again provides means for holding all data flow details between code blocks. Finally, an *effort edge* is added along each control flow edge, as implementation effort is induced by control flow adaptation for running the adjacent code blocks on different cores.

In case of both graph-based firmware models, a multi-objective problem solver is engaged for exploring and evaluating mapping or scheduling alternatives, that are Pareto optimal for a given set of load profiles. For finally selecting an alternative for implementation, not solely their ratings with respect to core balance or schedule length, inter-core interaction and implementation effort are of interest: An appropriate visualization of the alternatives, that highlights and annotates parallelization-specific issues such as core-crossing edges, can be of great advantage. Figure 1 is depicting an according sketch of a visualization.

Reference

1. Brandes, U., Eiglsperger, M., Herman, I., Himsolt, M., Marshall, M.S.: GraphML Progress Report: Structural Layer Proposal. In: Mutzel, P., Jünger, M., Leipert, S. (eds.) GD 2001. LNCS, vol. 2265, pp. 501–512. Springer, Heidelberg (2002)

Automatic Generation of Route Sketches

Andreas Gemsa[1], Martin Nöllenburg[1,2], Thomas Pajor[1], and Ignaz Rutter[1]

[1] Institute of Theoretical Informatics, Karlsruhe Institute of Technology (KIT), Germany
[2] Department of Computer Science, University of California, Irvine, USA

1   Introduction

Generating route sketches is a graph redrawing problem, where we are given an initial drawing of a graph G and want to find a new, *schematized* drawing of G that reduces the drawing complexity while preserving the structural characteristics of the input. The motivation of our work is the visualization of routes in road networks as sketches for driving directions. An important property of a route sketch is that it focuses on road changes and important landmarks rather than exact geography and distances. Typically the start and destination lie in populated areas that are locally reached via a sequence of relatively short road segments. On the other hand, the majority of the route typically consists of long highway segments with no or only few road changes. This property makes it difficult to display driving directions for the whole route in a single traditional map since some areas require much smaller scales than others. The strength of route sketches for this purpose is that they are not drawn to scale but rather use space proportionally to the route complexity.

In summary, we wish to find drawings of routes that have a low visual complexity, yet capture all important parts of a route, such as turning points or road changes. Further, the schematized drawing should in some sense reflect the layout of the original geographic route. We model the problem as a mixed-integer linear program (MIP) that computes schematizations of paths or unions of alternative paths. The solution of the MIP satisfies all mandatory constraints and optimizes the visual quality criteria.

Related Work. Agrawala and Stolte [1] presented a system called *LineDrive* that draws route sketches with heuristic methods based on simulated annealing. Nöllenburg and Wolff [4] used a MIP approach to compute schematic metro maps of public transport graphs. Brandes and Pampel [2] studied a path schematization problem in the presence of orthogonal order constraints for preserving the mental map. They showed that deciding the existence of a rectilinear schematization is NP-hard. Delling et al. [3] gave an efficient algorithm for schematizing monotone paths under orthogonal order constraints.

2   Model

Our aim is to produce a drawing of the input graph that reduces the drawing complexity but at the same time maintains the user's mental map of the route. As a preprocessing step we simplify the input using the standard Douglas-Peucker line-simplification algorithm to reduce the number of edges in the graph while maintaining its overall shape

U. Brandes and S. Cornelsen (Eds.): GD 2010, LNCS 6502, pp. 391–392, 2011.

392 A. Gemsa et al.

as well as all important vertices (e.g., road changes and turning points). For further reducing the drawing complexity, we use a limited set $\mathscr{C}_d = \{z \cdot 90°/d \mid z \in \mathbb{Z}\}$ of admissible edge slopes for $d \geq 1$.

To maintain the mental map, we preserve the *orthogonal order* of the input, i.e., the left-to-right and top-to-bottom order of all vertices. Furthermore, each edge is preferably drawn with the slope in $\mathscr{C}_d$ that is closest to its input slope; an appropriate minimum edge length is used so that each edge is well visible.

We also experimented with two supplementary approaches that help create sketches that are often visually more pleasing. In the original formulation edge lengths in the output do not carry information about the true distances. While it is an important feature of a route sketch that distances are distorted, we may require that the input length order of the edges is preserved in the output. Second, preserving the orthogonal order for vertices that are far apart is of limited importance. So any pair of vertices whose distance in one of the coordinates is at least one third of the total extent of the path in that coordinate does not need to preserve its order in the respective other coordinate.

All these constraints can be modeled as the linear constraints and the linear objective function of a MIP.

3 Evaluation

We have implemented our approach and have tested it on 1000 random routes in the German road network. Two main observations can be made in this study: 1) More than 50% of the instances did not have a valid rectilinear schematization, whereas only 0.7% of the instances were infeasible in the octilinear case $d = 2$ and for any $d \geq 3$ there were at most 0.1% infeasible instances. 2) The average running time of the MIP optimizer increases with increasing d from 107.36 ms for $d = 1$ and 649.49 ms for $d = 3$ to 1347.86 ms for $d = 5$ on a single AMD Opteron 2218 CPU with the MIP optimizer Gurobi 3.0.1.

Additionally, we present examples that display sketches of alternative routes in a single picture.

References

1. Agrawala, M., Stolte, C.: Rendering effective route maps: Improving usability through generalization. In: Proc. 28th Ann. Conf. Computer Graphics and Interactive Techniques (SIGGRAPH 2001), pp. 241–249. ACM, New York (2001)
2. Brandes, U., Pampel, B.: On the hardness of orthogonal-order preserving graph drawing. In: Tollis, I.G., Patrignani, M. (eds.) GD 2008. LNCS, vol. 5417, pp. 266–277. Springer, Heidelberg (2009)
3. Delling, D., Gemsa, A., Nöllenburg, M., Pajor, T.: Path schematization for route sketches. In: Kaplan, H. (ed.) SWAT 2010. LNCS, vol. 6139, pp. 285–296. Springer, Heidelberg (2010)
4. Nöllenburg, M., Wolff, A.: Drawing and labeling high-quality metro maps by mixed-integer programming. IEEE Trans. Visualization and Computer Graphics (2010), doi:10.1109/TVCG.2010.81

Visualizing Differences between Two Large Graphs

Markus Geyer, Michael Kaufmann, and Robert Krug

Wilhelm-Schickard-Institut für Informatik, Universität Tübingen, Germany
{geyer,mk,krug}@informatik.uni-tuebingen.de

1 Introduction

When working with graphs one often faces the problem of comparing two or more graphs. For example in biology, when two protein-protein interaction networks from closely related species have to be compared, the graphs can easily contain several hundred nodes and thousands of edges but very few differences. Our goal was to develop a software-tool that creates an overview of large graphs and maintains the structure but reduces the number of nodes and edges and enables the user to easily find and investigate the areas of interest. This problem has been considered in the past by various groups ([1], [2]) and different heuristics have been proposed. Here, we follow the concept proposed in [3]. For this work, we assume that the input-graphs are relatively large with small local differences and node correspondences are known.

2 Our Method

Condensation of Graphs
We denote the condensation of a graph as the extraction of its structure and the reduction to its important parts. Our first approach identifies important nodes and removes the remaining nodes from the graph. To assess the importance of a node we used different centrality measures from [4]. Neighboring nodes are merged into group nodes as long as their combined centrality value doesn't exceed a pre-defined threshold.

For the second approach a given number of nodes with the highest centrality value is chosen and all other nodes are merged with the closest chosen node. In both cases new edges have to be created to maintain connectivity.

In order to enable the user to investigate interesting areas of the graph, a detailed view of a part of the graph can be created by clicking on a node in the condensed graph. An extra window is then opened which shows the area in the immediate proximity of the chosen node in uncondensed form. By clicking on one of the dashed lines, which represent edges to parts of the graph not yet displayed, the currently shown graph-part can be extended into that direction. Also, by clicking on a normal edge, the subgraph beginning at the target-node of that edge can be removed from the detailed view.

Comparison of Graphs
For the construction of the overview graph the common elements are identified and compacted using one of the previously described methods. Then, for each difference a node is added. For the resulting graph a layout is computed with a force-directed

U. Brandes and S. Cornelsen (Eds.): GD 2010, LNCS 6502, pp. 393–394, 2011.

method. Clicking on one of the difference nodes opens another window that shows the difference and the area around it in detail. It is also possible to extend and contract the part of the graph that is currently shown. For the detailed view, the positions of elements from the overview graph stays the same while new elements are placed with a force-directed method. Once a node has been set to a position, it will always stay there to ease orientation around the graph for the user.

3   Implementation and Results

For our implementation we used Java and the *yFiles* graph library ([5]). Figure 1 depicts the overview graph before and after the condensation as well as a detailed view of a difference node.

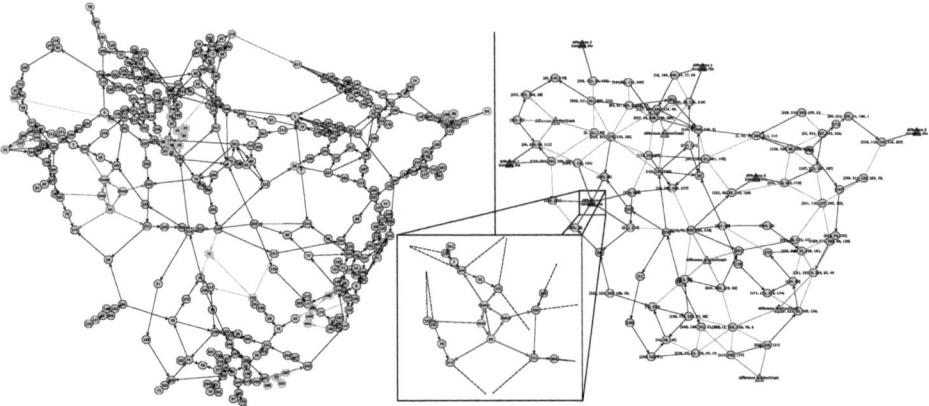

Fig. 1. The overview graph (left) contains 350 nodes while the condensed version (right) contains 75 blue group nodes and 11 red or green difference nodes. In the detailed view in the middle colored elements indicate differences and dashed lines expansion edges.

References

[1] Chimani, M., Jünger, M., Schulz, M.: Crossing minimization meets simultaneous drawing. In: PacificVis, pp. 33–40. IEEE, Los Alamitos (2008)
[2] Tan, K., Shlomi, T., Feizi, H., Ideker, T., Sharan, R.: Transcriptional regulation of protein complexes within and across species. Proceedings of the National Academy of Sciences 104(4), 1283–1288 (2007)
[3] Albrecht, M., Estrella-Balderrama, A., Geyer, M., Gutwenger, C., Klein, K., Kohlbacher, O., Schulz, M.: Visually comparing a set of graphs. In: Borgatti, S.P., Kobourov, S., Kohlbacher, O., Mutzel, P. (eds.) Graph Drawing with Applications to Bioinformatics and Social Sciences, Dagstuhl Seminar Proceedings, Dagstuhl, Germany, vol. 8191 (2008)
[4] Freeman, L.C.: Centrality in social networks: Conceptual clarification. Social Networks 1(3), 215–239 (1979)
[5] yWorks GmbH: yFiles graph library, http://www.yworks.com

Placing Edge Labels by Modifying an Orthogonal Graph Drawing

Konstantinos G. Kakoulis[1] and Ioannis G. Tollis[2]

[1] Dept. of Industrial Design Engineering,
T.E.I. of West Macedonia, Greece
kkakoulis@teikoz.gr
[2] Dept. of Computer Science, University of Crete and
FORTH-ICS, Crete, Greece
tollis@ics.forth.gr

Abstract. In this paper we investigate how one can modify an orthogonal graph drawing to accommodate the placement of overlap-free labels with the minimum cost. We present a polynomial time algorithm that finds the minimum increase of space in one direction, needed to resolve overlaps, while preserving the orthogonal representation of the drawing.

1 Introduction

Automatic labeling is a very difficult problem, and because we rely on heuristics to solve it, there are cases where the best methods available do not always produce an acceptable or legible solution even if one exists. If a solution to the labelng problem is not acceptable one could either redraw the graph taking into account the placement of labels (in [1,2,3] labeling with drawing of orthogonal representations of graphs is combined), or modify the drawing to produce an acceptable label assignment.

In this paper we consider the problem of modifying an existing orthogonal drawing by inserting extra space in order to accommodate the placement of edge labels that are free of overlaps. We will refer to it as the *Opening Space Label Placement* (OSLP) problem. We have chosen orthogonal drawings because they have the most *regular* structure among all layout styles with respect to opening space, which in orthogonal drawings implies inserting rows and/or columns.

2 Solving the OSLP Problem

We want to minimize the extra area needed to resolve label overlaps while preserving the orthogonal representation of the drawing. Because the OSLP problem is NP-Complete [4] we must rely on heuristics to solve it.

Given an orthogonal drawing Γ, first we find an edge label assignment where overlaps are allowed by using existing techniques. Then, we modify Γ by applying a polynomial time algorithm based on minimum flow techniques to find the minimum width needed to eliminate label overlaps in Γ, given a partial order of overlapping objects, while preserving the orthogonal representation of Γ. The case of finding minimum height is analogous.

U. Brandes and S. Cornelsen (Eds.): GD 2010, LNCS 6502, pp. 395–396, 2011.
© Springer-Verlag Berlin Heidelberg 2011

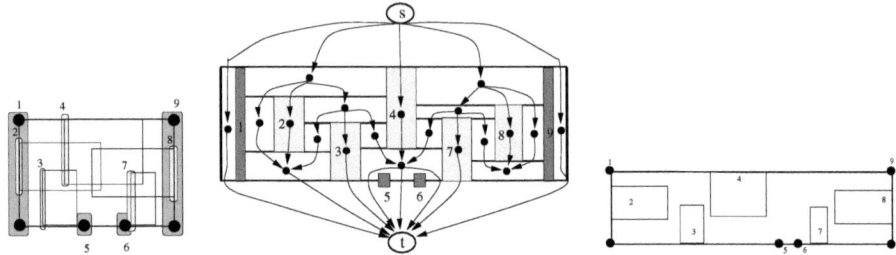

Fig. 1. (a) An input drawing Γ with label overlaps. (b) The flow graph for Γ. (c) Label overlaps have been resolved by running the minimum flow algorithm.

The main idea of our algorithm is the following: We create a directed acyclic graph G_{flow} that transfers flow from the top to the bottom of the drawing in order to insert extra vertical space to resolve horizontal overlaps. Intuitively, if two objects overlap, then we must push between them at least as much flow as the amount of their overlap in the x direction. First we decompose the input drawing Γ into vertical segments (each label, node and vertical edge segment of Γ is a vertical segment). Next, we obtain the partial order of the objects by performing a plane sweep. Then, we create a *separation visibility* graph G_{sv}. For each vertical segment of Γ we insert a node in G_{sv}. Each node in G_{sv} is a rectangle and has the size of its corresponding object in Γ. We insert edges in G_{sv} by expanding (to the left and to the right) the horizontal sides of each node in G_{sv} until they touch another node of G_{sv}. We assign a weight to each edge e of G_{sv} which represents the minimum distance the two objects connected with edge e must be kept apart to avoid overlaps. Next, we create the flow graph G_{flow} from graph G_{sv} by adding into G_{flow}: (*i*) A source node s and a target node t, (*ii*) A node for each face of G_{sv} and (*iii*) An edge for each pair of neighboring faces of G_{sv} that share a horizontal edge segment. For each edge in G_{flow} we assign lower and upper capacity equal to a pair of weights for the only edge in G_{sv} it intersects. We show that the minimum flow of G_{flow} produces the minimum width expansion needed to resolve overlaps.

Theorem 1. *The minimum flow of G_{flow} gives the minimum width of the drawing, in $O(m \log n \ (m + n \log n))$ time, such that all label overlaps are resolved in one direction and the orthogonal representation of the drawing is preserved.*

References

1. Binucci, C., Didimo, W., Liotta, G., Nonato, M.: Orthogonal Drawings of Graphs with Vertex and Edge Labels. CGTA 32(2), 71–114 (2005)
2. Di Battista, G., Didimo, W., Patrignani, M., Pizzonia, M.: Orthogonal and quasi-upward drawings with vertices of arbitrary size. In: Kratochvíl, J. (ed.) GD 1999. LNCS, vol. 1731, pp. 27–37. Springer, Heidelberg (1999)
3. Klau, G.W., Mutzel, P.: Combining Graph Labeling and Compaction. In: Kratochvíl, J. (ed.) GD 1999. LNCS, vol. 1731, pp. 27–37. Springer, Heidelberg (1999)
4. Kakoulis, K.G., Tollis, I.G.: On the Complexity of the Edge Label Placement Problem. Computational Geometry 18(1), 1–17 (2001)

Large Crossing Angles in Circular Layouts

Quan Nguyen[1], Peter Eades[1], Seok-Hee Hong[1], and Weidong Huang[2]

[1] School of Information Technologies, University of Sydney
{qnguyen,peter,shhong}@it.usyd.edu.au
[2] CSIRO ICT Centre, Australia
tony.huang@csiro.au

1 Introduction

Recent empirical research has shown that increasing the angle of crossings reduces the effect of crossings and improves human readability [5]. In this paper, we introduce a post-processing algorithm, namely MAXCIR, that aims to increase crossing angles of circular layouts by using Quadratic Programming. Experimental results indicate that our method significantly increases crossing angles compared to the traditional equal-spacing algorithm, and that the running time is fairly negligible.

2 Algorithm

The post-processing approach MAXCIR in this paper aims to increase the crossing angles after a circular ordering of the vertices is given, such as by a crossing reduction algorithm [3]. With a fixed ordering of the vertices, all of the crossing pairs (e,e') can be pre-determined in linear time. Let Ω denote the set of pairs of crossing edges. We aim to minimise

$$F = \sum_{(e,e')\in\Omega} \left(\alpha_{(e,e')} - \frac{\pi}{2}\right)^2, \qquad (1)$$

where $\alpha_{(e,e')}$ denotes the angle at which edges e and e' cross.

The circular layout is a function $\theta : V \rightarrow [0 .. 2\pi)$ that associates an angle $\theta(u)$ with each vertex u. Suppose that the ordering of the vertices around the circle is $(u_0, u_1, \ldots, u_{n-1})$. We denote $\theta(u_i)$ by θ_i.

For every edge e of G, let e_m and e_M denote the end vertices such that $\theta(e_m) < \theta(e_M)$. For a pair of crossing edges e and e', their crossing angle is given by $\alpha_{(e,e')} = \frac{1}{2}\left(\theta(e_M) - \theta(e'_M) + \theta(e_m) - \theta(e'_m)\right)$ where $\theta(e'_m) < \theta(e_m) < \theta(e'_M) < \theta(e_M)$.

In practice an angular gap g between vertices needs to be preserved, e.g., for avoiding overlapping node labels. The circular ordering and the gap lead to the following constraints:

$$0 \leq \theta_0; \qquad \theta_i + g < \theta_{i+1}, \forall i = 0..n-2; \qquad \theta_{n-1} + g < 2\pi. \qquad (2)$$

Minimizing F in equation (1) subject to the constraints (2) defines a quadratic program.

U. Brandes and S. Cornelsen (Eds.): GD 2010, LNCS 6502, pp. 397–399, 2011.

3   Experiments

We compare our new MAXCIR layout algorithm with EQCIR — a circular layout that places the nodes equally on the circle. We use the Rome data set [2] and use Baur and Brandes's algorithm [3] to generate a vertex ordering. The experiments were conducted on a 3.00GHz Pentium IV CPU and 1GB RAM Solaris/SPARC with cplex solver v10.0.1 [1].

We ran MAXCIR with gap values $g=0$, 0.2γ, 0.4γ, 0.7γ and 0.9γ, where $\gamma = \pi/|V|$. We obtained the *average crossing angle* produced by MAXCIR and EQCIR, over all five gap values. A significant improvement of 61.4% was obtained; the average is slightly reduced when the default gap value is increased. The *average execution time* of MAXCIR varies between 6ms and 13ms, and slightly reduces when the gap value increases.

We measured the angular resolution of MAXCIR layout results. We adapted the measurement used in [4] to circular layouts by defining the optimal angle at a vertex v as $180/\deg(v)$ instead of $360/\deg(v)$. The *average angular resolution* produced by MAXCIR is 49.18 degrees compared to 55.50 degrees produced by EQCIR. When the gap value increases the average angular resolution slightly decreases. A smaller difference implies a better angular resolution as the angles between pairs of coincident edges are closer to optimal.

Figure 1 shows circular layouts of a Rome graph instance produced by MAX-CIR with five gap values. For small gap values, the crossing angles are largely optimised, yet several nodes are placed too close to each other or overlap one another. For larger gap values, the crossing angles and angular resolution are slightly improved, but the nodes are better distinguished.

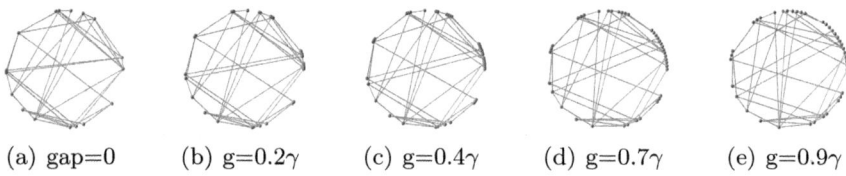

   (a) gap=0   (b) g=0.2γ (c) g=0.4γ (d) g=0.7γ (e) g=0.9γ

Fig. 1. Applying MAXCIR with different gap values on Rome graph 10005_39

4   Conclusion

We have presented a post-processing method based on quadratic programming for circular layouts. The method significantly improves the crossing angles of a circular layout with negligible overheads.

Our future work will incorporate angular resolution into MAXCIR model. Maximising angular resolution is particularly interesting since crossing angle competes with angular resolution when two adjacent edges cross the same edge.

References

1. cplex (2010), http://www.ilog.com/products/cplex/
2. The Rome graphs (2010), http://www.graphdrawing.org/data/index.html
3. Baur, M., Brandes, U.: Crossing reduction in circular layouts. In: Hromkovič, J., Nagl, M., Westfechtel, B. (eds.) WG 2004. LNCS, vol. 3353, pp. 332–343. Springer, Heidelberg (2004)
4. Finkel, B., Tamassia, R.: Curvilinear graph drawing using the force-directed method. In: Pach, J. (ed.) GD 2004. LNCS, vol. 3383, pp. 448–453. Springer, Heidelberg (2005)
5. Huang, W., Hong, S.H., Eades, P.: Effects of crossing angles. In: IEEE Pacific Visualization Symposium, pp. 41–46 (2008)

GVSR: An On-Line Guide for Choosing a Graph Visualization Software

Bruno Pinaud[1] and Pascale Kuntz[2]

[1] Bordeaux I Univ., CNRS UMR 5800 LaBRI, INRIA Bordeaux Sud-Ouest (Gravité), France
bruno.pinaud@labri.fr
[2] CNRS UMR 6241 LINA, site école polytechnique de l'université de Nantes, France
pascale.kuntz@univ-nantes.fr

Abstract. It is easy to find graph visualization applications for all sorts of uses. However, choosing an appropriate application may be difficult. This poster presents a website (http://gvsr.polytech.univ-nantes.fr/) built to help users to choose a program adapted to their problems. So far, this site references eighty programs and aims at helping users both in their choices and in comparing the programs. The site is also designed as a tool repository helping the community to access and compare the available tools, and benchmark new techniques and algorithms.

Keywords: Graph Visualization Software, On-Line Repository.

1 Introduction

The profusion of available graph visualization applications may even confuse an expert in this field. Some programs have been developed in close partnership with the scientific community (Pajek, Cytoscape), others are purely commercial, or some are general graph manipulation and visualization software (Tulip). Generally speaking, the choice of a program well-adapted to both the data and the methodology is difficult. Some books can be used as guides [1, 2], and several websites present lists of programs [3, 4]. However, those websites plainly list the existing software, or make them accessible through snapshots. Consequently much effort is required to compare the various programs before choosing the best one for the problem considered.

Those observations led us to develop GVSR (http://gvsr.polytech.univ-nantes.fr). Its added value is to offer users query about existing software based on commonly used criteria such as scalability, implementation issues or type of uses. Our objectives are to facilitate the users' choices and to compare programs with common criteria. The website also presents the programs with a uniform text-based description. This site keeps evolving and so far contains eighty various software descriptions. In addition, the site allows users to propose new programs by simply completing an enclosed form. The site is also designed as a tool repository helping the user to access and compare the available Graph Visualization tools, and benchmark new techniques and algorithms. The whole community can benefit from the ability to reproduce published results, and from comprehensive comparisons with previous work. Thus, GVSR can be seen as a contribution to improving both the accessibility and quality of graph visualization tools.

U. Brandes and S. Cornelsen (Eds.): GD 2010, LNCS 6502, pp. 400–401, 2011.

2   How to Use the Site?

The site proposes four ways to find a software : 1) a tag cloud with the software names (the more a software page is accessed, the bigger its name is displayed); 2) the "Software List" link gives access to a simple alphabetical order list; 3) the "Advanced Search" link gives access to a search engine on the software database; 4) the "Start Browsing Now" button to start navigating in a taxonomy covering all the criteria used to described software, providing a structured exploration mode of the repository.

Each software description (Fig. 1) page is made of a screenshot, general information (*e.g.* website, ...), specific information on the visualization, technical information (*e.g.* license(s), ...) and references. At the bottom of the page, one can write a comment and score the software. After validation, these information will be added on the page.

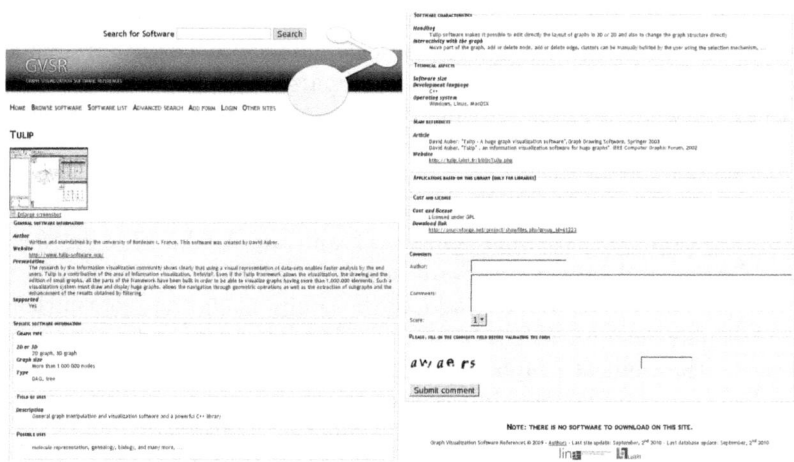

Fig. 1. Example of a software description page

3   Future Works

GVSR keeps evolving by a regular watch on the database, the addition of new programs and functionalities. We are working on an interactive visualization of the taxonomy as a graph. We also plan to directly host samples datasets to benchmark programs.

References

1. Kaufmann, M., Wagner, D. (eds.): Drawing Graphs. LNCS, vol. 2025. Springer, Heidelberg (2001)
2. Mutzel, P., Jünger, M.: Graph Drawing Software. Springer, Heidelberg (2003)
3. http://rw4.cs.uni-sb.de/users/sander/html/gstools.html
4. http://www.manageability.org/blog/stuff/
   open-source-graph-network-visualization-in-java/

IBM ILOG Multi-platform Graph Layout Technology

Georg Sander

IBM ILOG Visualization Group
georg.sander@de.ibm.com

Introduction

Visualization components from IBM provide a comprehensive set of graphics products for creating highly graphical, interactive displays, including diagrams, gantt charts, maps, business dashboards, business charts, telecom displays, SCADA/HMI Screens and many more. User interface developers reduce development risks and implementation time when deploying the IBM visualization technology. The components are available for many different platforms: various C++ platforms, Java Swing applications, Java Eclipse plugins, thin client AJAX applications, for the Microsoft .NET and the Adobe Flex platform.

Graph layout and label layout are key functionalities of many our components. It is part of the algorithmic core of the components that gets enriched by a graphical display layer (classes that actually draw elements on the screen). Graph layout needs the highest graph-theory expertise and is optimized the most. In order to leverage the platform infrastructure in the best way, the graphical display layer was specially developed for the different platforms (Swing, Microsoft .NET, Adobe Flex and various C++ and AJAX platforms). However, for the nonvisual and mathematical part, this is not needed and would be a waste of manpower, since the pure mathematics is always the same, no matter whether you are in C++, Java, C# or ActionScript. On the other hand, IBM ILOG follows a strict *uniform platform strategy* that avoids hybrid technologies (for instance no C++ inside Java). Even the reference documentation must use the native tool (for instance, no Javadoc inside C#). Last but not least, the support engineers must be able to work with the code for debugging or consulting assignments even when they are specialists of only one single platform.

Graph Layout Translation Technology

How to provide graph layout technology in C++, Java, C# and ActionScript without the need for huge development teams that redevelop the same algorithms over and over again? IBM takes an automatic translation approach from a single common source. The algorithms are implemented in a central repository and are translated by a set of tools fully automatically to the different target languages. The advantage: the graph layout team can concentrate on improving the layout

U. Brandes and S. Cornelsen (Eds.): GD 2010, LNCS 6502, pp. 402–403, 2011.

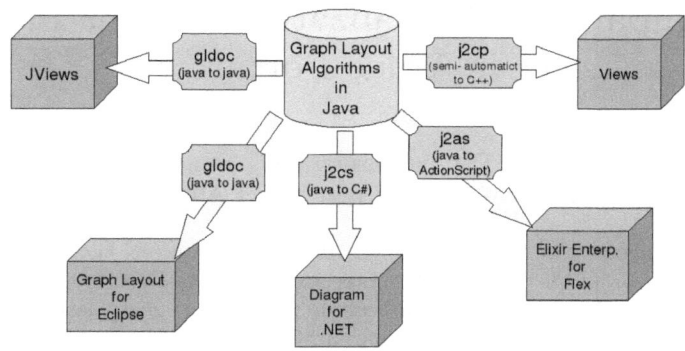

Fig. 1. Graph Layout Translations at IBM ILOG

algorithms instead of spending time on platform coding, and every improvement in the layout algorithms is immediately available on all platforms without effort.

When Microsoft introduced the .NET platform, some simple tools were also introduced to convert Java to C#. Most of these tools were one-shot translations that required manual adaptations of the translation results. As disadvantage, all manual adaptations are lost whenever the common source needs to be retranslated (for instance whenever an incremental improvement was added to a graph layout algorithm). Therefore, IBM ILOG has developed its own tool set that does not require any manual adaptations of the result of the translation. The development is done entirely on the common source that is fed into the translator tools. Some other Microsoft tools converted Java bytecode to C# bytecode, with the disadvantage that the translated code is not human readable and hence unsuitable for the IBM support engineers, and that the translation does not provide any API documentation in the same step. The IBM tool set however translates source to source and converts all the code comments into the canonical format expected by the targeted platform. This allows to produce a reference manual by using the native reference tool on the translated source code.

We develop the graph layout algorithms in a simplified Java based language, a subset of Java enriched by an annotation formalism. Annotations specify which Java methods should become C# properties, or how enumerations are mapped in ActionScript. Since these annotations are comments for Java, it allows for quick turnaround time when testing new algorithms, as the original code can be compiled with Java. However, the production code contains only the result of the source to source translations, even for the Java products (a Java to Java translator converts GUI related code from the common source alternatively to Swing for IBM ILOG JViews Diagrammer, or to SWT/GEF for IBM ILOG JViews Graph Layout for Eclipse). Other tools translate the common source to C# for IBM ILOG Diagram for .NET, to ActionScript for IBM ILOG Elixir Enterprise for Adobe Flex, or to C++[1] for IBM ILOG Views.

---

[1] In the current state, all translations are fully automatic except C++, which requires still small manual adaptations for the memory management.

Comparative Visualization of User Flows in Voice Portals

Björn Zimmer[1], Dennie Ackermann[1], Manfred Schröder[2],
Andreas Kerren[3], and Volker Ahlers[1,*]

[1] University of Applied Sciences and Arts Hannover, Faculty IV,
Dept. of Computer Science, P. O. Box 920251, 30441 Hannover, Germany
[2] HFN Medien GmbH, Ehlbeek 3, 30938 Burgwedel, Germany
[3] Linnæus University, School of Computer Science, Physics and Mathematics,
Vejdes Plats 7, 35195 Växjö, Sweden

1   Introduction

Voice portals are widely used to guide users interactively through an application. Recent portals provide a growing number of functions in one application, thus increasing their complexity. This work presents flow-map-based techniques for the comparative visualization of user flows at different time frames, in order to enable dialog designers to analyze and improve the user interaction with these systems.

Natural Language Systems in Voice Portals: More sophisticated voice portals use natural language systems (NLS), giving users the option to actually "talk" to the system in whole sentences. The system tries to interpret these sentences and interactively asks the user for detailed information, if necessary. Portals using NLS are rather large and complex, making it difficult to analyze their performance. Especially after applying changes to a voice portal or in case of technical problems, it is important to be able to analyze the consequences on user flows in the system.

2   Comparing User Flows

The user flow in a voice portal within a specific time frame corresponds to a weighted graph, where dialogs are represented by nodes and user flows by weighted edges. In order to compare user flows within the same voice portal at different time frames, a joined graph with multiple edges drawn as single, multi-colored arrows is created. The user value passing between two dialogs is visualized by varying either edge width or color saturation. The graph layout is based on a radial tree layout [2], which is manually adjusted. Two different approaches are discussed in the following.

---

* Contact author: `volker.ahlers@fh-hannover.de`
This work is financially supported by the German BMBF (grant no. 17N2809).

U. Brandes and S. Cornelsen (Eds.): GD 2010, LNCS 6502, pp. 404–405, 2011.
© Springer-Verlag Berlin Heidelberg 2011

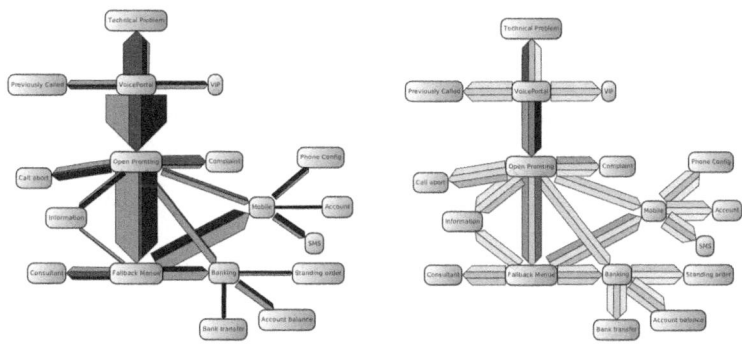

Fig. 1. Comparing user flows via edge width (left) and color saturation (right)

Varying Edge Width: Our first approach uses a flow map-based layout in such a way that the amount of users is represented by edge width, without merging edges that share the same direction [1]. Each color represents a different selectable time frame. The edge width represents the amount of users that have passed between the adjacent dialogs. The result for an example voice portal is shown in Fig.1 (left). The main drawback of this approach is that user flows between central dialogs of the portal lead to very thick edges. Peripheral edges are too thin to actually give a hint of the user flows. Very small flows are nearly invisible. To see the details at peripheral parts, the overall edge width could be increased, with the effect of making the graph look very crowded and unsettled. Additionally, viewing time frames with very different user flows could create the need to adjust node positions in order to route thick edges between them, destroying the mental map of the user [3].

Using Color Saturation: Representing user flows by color saturation avoids varying edge width. The result for the same example voice portal as used above is shown in Fig. 1 (right). Even smaller user flows at peripheral parts are now visible as low-saturated edges. A lower saturation threshold of 0.05 (for the range [0, 1]) is used in order to keep colors distinguishable for small amounts of users. If no users have passed between two dialogs within a certain time frame, the corresponding color is not contained in the edge connecting these dialogs. Changing time frames does not have an impact on the edge width, thus preventing the need to adjust node positions.

References

1. Phan, D., Xiao, L., Yeh, R., Hanrahan, P., Winograd, T.: Flow Map Layout. In: InfoVis 2005, p. 29 (2005)
2. di Battista, G., Eades, P., Tamassia, R., Tollis, I.G.: Graph Drawing. Prentice Hall, Upper Saddle River (1999)
3. Eades, P., Lai, W., Misue, K., Sugiyama, K.: Preserving the Mental Map of a Diagram. Proc. of Compugraphics 91, 24–33 (1991)

Graph Drawing Contest Report

Christian A. Duncan[1], Carsten Gutwenger[2],
Lev Nachmanson[3], and Georg Sander[4]

[1] Louisiana Tech University, Ruston, LA 71272, USA
duncan@latech.edu
[2] University of Dortmund, Germany
carsten.gutwenger@cs.uni-dortmund.de
[3] Microsoft, USA
levnach@microsoft.com
[4] IBM, Germany
georg.sander@de.ibm.com

Abstract. This report describes the 17th Annual Graph Drawing Contest, held in conjunction with the 2010 Graph Drawing Symposium in Konstanz, Germany. The purpose of the contest is to monitor and challenge the current state of graph-drawing technology.

1 Introduction

As in recent years, this year's Graph Drawing Contest was divided into the offline challenge and the online challenge. The offline challenge had three categories: two dealt with edge routing and one was a mystery graph. The data sets for the offline challenge were published months in advance, and contestants could solve and submit their results before the conference started. For the two edge routing categories, the supplied data sets had nodes with fixed positions and nonzero dimensions. The task was to produce an aesthetic routing of the edges using bends or splines. For the mystery graph data set, the task was to determine the meaning of the graph and to produce a suitable drawing.

The online challenge took place during the conference in a format similar to a typical programming contest, where teams were presented with a collection of challenge graphs and had approximately one hour to submit their highest scoring drawings. The topic of the online challenge was to minimize the length of the longest edge in a planar orthogonal grid drawing.

Overall, we received 25 submissions: 9 submissions in the offline challenge and 16 submissions in the online challenge.

2 Edge Routing - Circuit Diagram

The first data set for the edge routing challenge was a circuit diagram of the Apple II+ Video Signal Generator [1]. It consists of 48 nodes and 84 edges. The sizes and positions of the nodes were appoximately the same as in the original

U. Brandes and S. Cornelsen (Eds.): GD 2010, LNCS 6502, pp. 406–411, 2011.

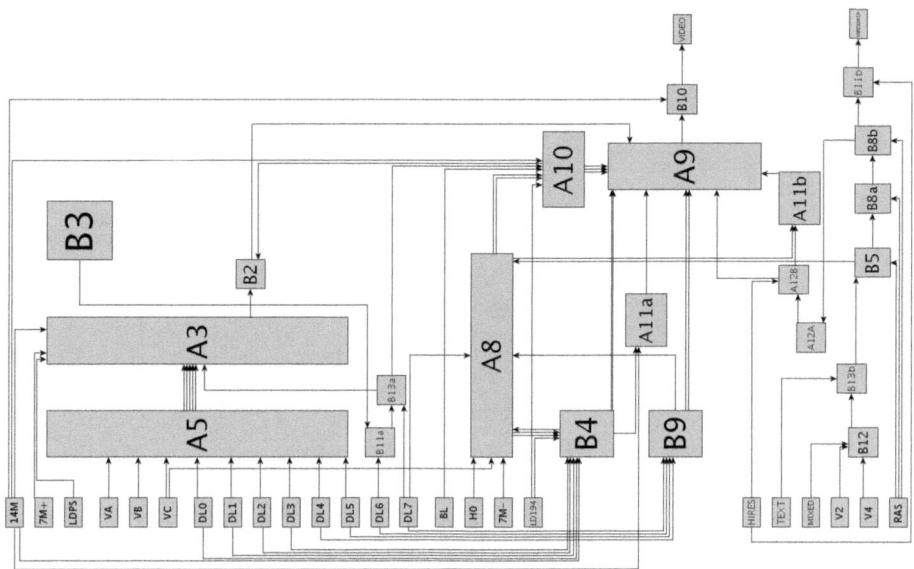

Fig. 1. First place, Circuit Diagram

drawing, except that the original drawing had some node duplicates that were removed from the challenge data set.

The winning submission, from Quan Nguyen from the University of Sydney (Figure 1), used an orthogonal routing style, which is very common for circuit diagrams. The layout was produced using yEd [5] by parametrizing the automated layout with a very large crossing penalty. The layout was then tuned by a few manual adjustments for multiedges.

3   Edge Routing - Author Collaboration Graph

The second data set for the edge routing challenge was the author collaboration graph of the graph drawing community. The 362 nodes represent the authors of select papers published in the graph drawing community. When several people coauthored a paper, edges were created between each pair of authors of that paper, yielding 942 edges. Papers with only a single author do not contribute to the set of edges and were removed from the data. The data was obtained from GDEA [2] for the years 2004 – 2010. The graph contains one big connected component, representing the core graph drawing researchers who have collaborated significantly over the years, and several smaller components from authors from other research communities that probably only occasionally contribute to the graph drawing literature.

The submitted work of Quan Nguyen and Seok-Hee Hong from University Sydney must be honorably mentioned. They analyzed the graph semantically using centrality and k-core techniques to produce confluent drawings. They were

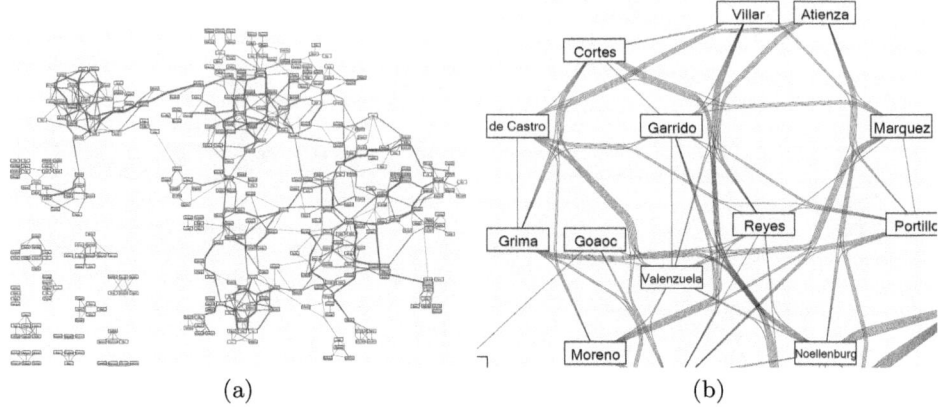

Fig. 2. First place, Author Collaboration Graph, (a) overview and (b) detail

able to detect strongly connected researchers as well as highly connected research groups and could even reverse engineer the related papers from the given data (which contains only information about the authors, not about the papers directly).

However, the winning and most visually pleasing layout (Figure 2) was submitted by Sergey Pupyrev from Ural State University using an edge bundling technique similar to [3]. This layout avoids overlaps between nodes and edges completely and reduces the number of edge crossings.

4   Mystery Graph

The mystery graph was a small bipartite graph with 49 edges. One node set was labeled A to G, and the other node set was labeled 0000 to 1001. Besides delivering an aesthetic drawing, the task was to determine the meaning of the data. Not all submissions found the correct answer that it represents the mapping of the 10 digits, labeled in binary form, to a seven-segment display.

The winning submission came from Michael Baur, Martin Siebenhaller, Roland Wiese and Thomas Wurst from yWorks. Since a straight-forward layout was very unclear due to the relatively large number of edges, they used a dependency analysis to obtain a simpler display. The representation of some digits include other digits. For example, the segments for digit 7 contain all segments of digit 1, and digit 8 requires all seven segments hence contains all other digits. Figure 3a shows these dependencies. To minimize the edge paths by taking advantage of these dependencies, edges were routed to (one of) the nodes of their contained digits whenever possible. The final layout (Figure 3b) was obtained through yFiles [5] by using the orthogonal layout algorithm for the digit nodes while the segment nodes were placed manually, and an orthogonal edge router to combine routes to bundles.

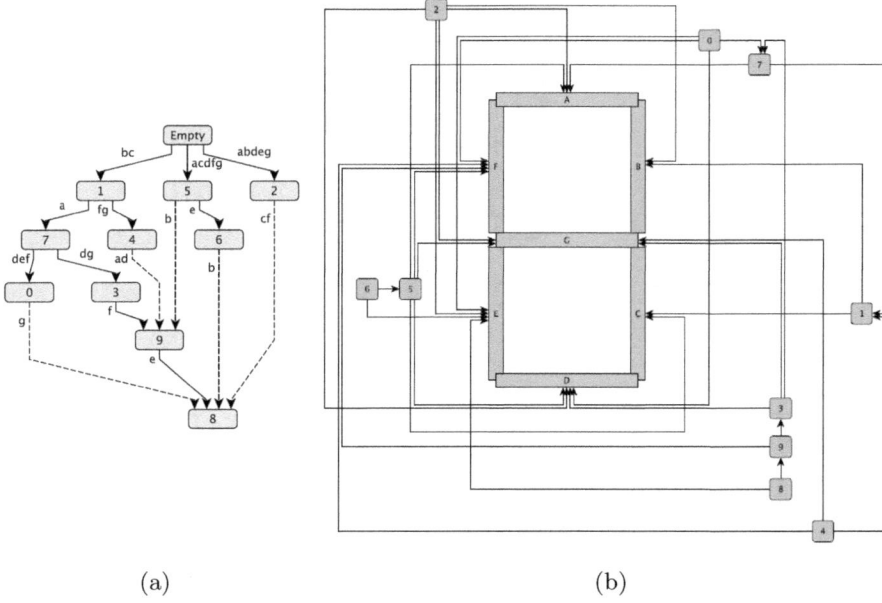

(a)   (b)

Fig. 3. First place, Mystery Graph

5   Online Challenge

The online challenge, which took place during the conference, dealt with minimizing the longest edge in a planar orthogonal drawing. The longest edge can be a bottleneck for many applications, hence minimizing its length is important. The challenge graphs were planar and had at most four incident edges per node. The task was to place nodes and edge bends on integer coordinates so that the edge routing is orthogonal and the layout contains no crossings or overlaps. At the start of the one-hour on-site competition, the contestants were given six graphs with an initial legal planar layout with very long edges. The goal was to rearrange the layout to reduce the length of the longest edge. Only the length of the longest edge was judged; other aesthetic criteria such as the number of edge bends or the area were ignored.

The contestants could participate in one of two categories: *automated* and *manual*. In the automated category, contestants received graphs ranging in size from 69 nodes / 101 edges to 3070 nodes / 4604 edges and were allowed to use their own sophisticated software tools with specialized algorithms. Only one team (Petra Mutzel and Hoi-Ming Wong from TU Dortmund) submitted results in this category and hence was the winner. They submitted only results for the four smallest graphs, which were computed with the tool *Gryphon*, a graph editor based on the OGDF [4] graph drawing library. They applied standard orthogonal graph drawing algorithms for minimizing the number of bends in the drawing, followed by advanced flow-based orthogonal compaction techniques. Notice that the overall optimization goal of this approach is not minimizing the length of the

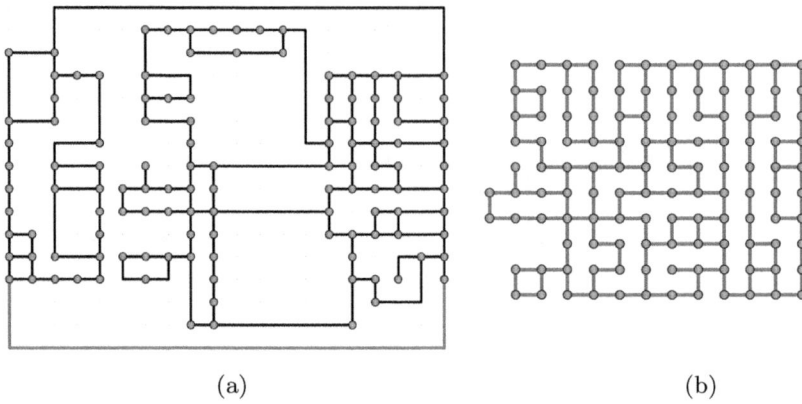

(a)   (b)

Fig. 4. Challenge graph with 120 nodes / 146 edges. (a) The initial layout. (b) The best (and optimal) manually obtained result by team Löffler/Nöllenburg, with longest edge length 1.

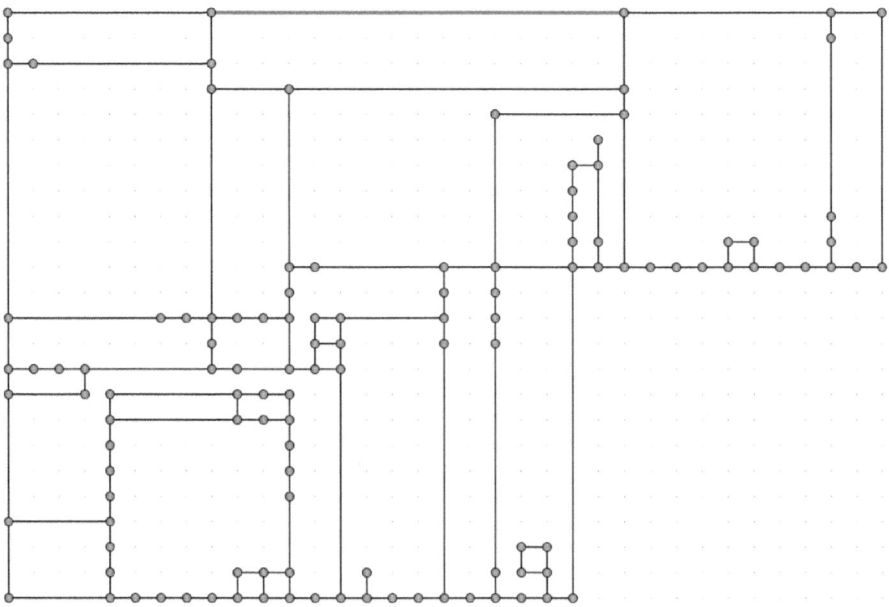

Fig. 5. The best automated result of the same graph as in Fig. 4 by team Mutzel/Wong, with longest edge length 16

longest edge; however, this approach usually leads to short edges. In addition, their bend minimization procedure used a special option which forces two 180 degree angles on nodes of degree two if this does not increase the number of bends. Though this leads to aesthetically pleasing drawings in many cases, it appears to be a bit counter-productive for this contest, since inserting clever "bends" at degree-2 nodes seems to be a basic requirement for achieving short edges (compare manual results in figure 4 with figure 5).

The 15 manual teams solved the problems by hand using IBM's *Simple Graph Editing Tool* provided by the committee. They received graphs ranging in size from 4 nodes / 6 edges to 190 nodes / 284 edges. Two of the larger input graphs were also in the automated category, and the best manual teams scored similar and better than the automated submissions. To determine the winner among the manual teams, the scores of each graph, determined by dividing the longest edge length of the best submission by the longest edge length of the current submission, were summed up. With a score of 4.78, the winner was the team of Maarten Löffler from UC Irvine and Martin Nöllenburg from Karlsruher Institut für Technologie who found the optimal results for four of the six contest graphs.

Figure 4 shows the initial layout and the best manually obtained result of one contest graph with 120 nodes and 146 edges. Figure 5 shows the best automated result of the same graph, since it was used in both the manual and the automated category. For the largest graph in the manual category, we know of a solution with the longest edge having length 13 (Figure 6a), but the best solution found by the manual teams was only 19 (Figure 6b).

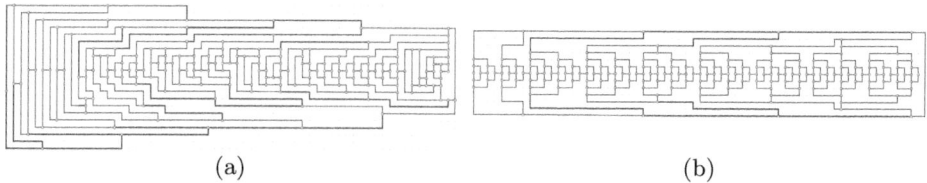

(a)   (b)

Fig. 6. Challenge graph with 190 nodes / 284 edges. (a) A solution with longest edge length 13. (b) The best result, found during the contest, with longest edge length 19.

Acknowledgments. The contest committee would like to thank the generous sponsors of the symposium and all the contestants for their participation.

References

1. Gayler, W.: The Apple II Circuit Description. Sams Technical Publishing (1983)
2. GDEA. Graph drawing e-print archive, http://gdea.informatik.uni-koeln.de/
3. Nachmanson, L., Pupyrev, S., Kaufmann, M.: Improving layered graph layout with edge bundling. In: GD 2010 (this proceeding), Springer, Heidelberg (2010)
4. OGDF. The open graph drawing framework, http://www.ogdf.net
5. yWorks, http://www.yworks.com

Author Index